Harnessing Microbes for Sustainable Food Systems

Food microbiology is the science that comprises the details about various microorganisms associated with food in beneficial and harmful aspects. Spoilage of food by microorganisms has become a problem for many people around the world as foods act as a nutrient source for microorganisms. The past ten years have seen a shift in the field of epidemiology of microbial illness caused by food. This shift has been attributed to the emergence of newly identified foodborne pathogens as well as to changes in lifestyle factors, such as convenience foods, more adventurous eating, and less time spent preparing meals. It is critical to control foodborne illnesses at every stage of the supply chain, from food manufacturing and preparation to customer delivery. Foodborne hazard control can be aided by both established technology and novel strategies like Hazard Analysis Critical Control Point (HACCP) programs and radiation hazard analysis. However, rising microbiological food safety concerns will remain a major public health concern because of microorganisms and the continual evolution of products, processes, and food handling methods.

KEY FEATURES

- To understand the basic information and applied aspects of food microbiology.
- To provide comprehensive knowledge about food spoilage and foodborne pathogens.
- To update readers with modernized food preservation methods and the biotechnology involved, along with quality control standards.
- To describe the latest research ongoing in food fermentation, food safety standards, and emerging issues in the field.

This book is designed for students and research scientists working in the field of food safety and food microbiology, including graduate and postgraduate students, scholars, academicians, scientists, and professionals in food and dairy technology, life sciences, agriculture, pharmaceutical sciences, and biomedical sciences.

Harnessing Microbes for Sustainable Food Systems

Edited by
Priya Sutaoney, Priyambada Singh,
Durgesh Nandini Chauhan,
Nagendra Singh Chauhan, and Kamal Shah

CRC Press
Taylor & Francis Group
Boca Raton London New York

CRC Press is an imprint of the
Taylor & Francis Group, an **informa** business

Designed cover image: iStock by Getty Images [Stock Photo ID:1332011099]

First edition published 2026
by CRC Press
2385 NW Executive Center Drive, Suite 320, Boca Raton FL 33431

and by CRC Press
4 Park Square, Milton Park, Abingdon, Oxon, OX14 4RN

CRC Press is an imprint of Taylor & Francis Group, LLC

ISBN: 978-1-032-88876-7 (hbk)
ISBN: 978-1-032-88878-1 (pbk)
ISBN: 978-1-003-54011-3 (ebk)

DOI: 10.1201/9781003540113

Typeset in Times LT Std
by Apex CoVantage, LLC

Contents

Foreword

Food microbiology is a crucial field that is essential for understanding the microscopic world around the food, as it is vital for public health, food supply spoiling, and quality. Microbiology is more than just science in food production; it is a vital instrument for protecting public health. The perception of microbiology as a topic confined to labs and textbooks is all too common. Food history, fermentation, foodborne illnesses, and microbiological processes that influence our meals are all explored in this engrossing book. Our comprehension of food microbiology's intricacies must advance along with the field. This book brings together cutting-edge research, food microbiome, new diseases and their management, recent developments in food preservation, innovative detection methods, nanomaterials, and contemporary regulatory issues. People have come up with creative techniques to preserve food in every culture in the world.

The book also discusses the microbiome of fresh fruits, vegetables, and dairy products and how microorganisms can cause spoilage. Researching the microbiome of fresh produce improves shelf life, increases food safety, and promotes human health by introducing beneficial microorganisms. Gaining knowledge of these microbial communities promotes improved environmental resilience and food quality, which contribute to sustainable agriculture. The health hazards connected with fungal metabolites in food are discussed in this book, with a focus on the expanding threat of mycotoxins. It highlights global safety standards that are always changing and detecting technology.

Fruits and vegetables are significantly spoiled due to various fungal pathogens, which causes severe economic losses. New control strategies emphasize on utilizing sustainable substitutes for chemical fungicides. This book integrates ecology and food science to reveal how fermentation ecosystems flourish and interact, by examining the varied and changing microbial communities in fermented products. This book will be highly useful for academics, policymakers, and graduate students. I applaud the authors for writing a book that is both current and rigorous for the field of food microbiology.

Dr. Vinay Kumar, ARS
Senior Scientist (Biotechnology)
ICAR – National Institute of Biotic Stress Management,
Raipur (Chhattisgarh)
DARE, Ministry of Agriculture and
Farmers Welfare, Govt. of India

Foreword

Preface

Microorganisms have been used for decades to develop sustainable food systems. In terms of their manufacturing and safety, they are renowned for improving the nutritional value and quality of food. It is necessary to alter current food systems to make them more sustainable. Food is necessary for human survival, and this requires natural resources that have been misused for many years and gradually diminished, damaged, or abolished throughout human history. Our food supply's relationships and interdependencies with the world's ecology and species have gotten less attention, even though they are vital. Incorporating evidence of the importance of plants, the microbiota in humans, animals, and plants, as well as the mutual effects of their connections on food systems is necessary to develop more inclusive solutions for future food systems. The goal of this book is to give readers a thorough understanding of food deterioration and foodborne pathogens while also helping them grasp the fundamentals and practical applications of food microbiology. It also keeps readers informed about quality control standards, biotechnology, and traditional and contemporary food preservation techniques. The most recent studies on food fermentation, food safety regulations, and new developments in the field are also covered.

This book has 19 chapters. In Chapter 1, the historical background of food microorganisms and their function in traditional food processing are discussed, along with the microbial hazards to food safety, contamination, prevention techniques, and their uses in food production and preservation.

In Chapter 2, the significance of fruits and vegetables' microbiomes is covered, along with elements influencing their composition, safety, and disease risk. The various microbes linked to meat and dairy products are discussed in Chapter 3, along with their positive and negative effects and how to control them in the food industry to reduce contamination.

Chapter 4 addresses how phytopathogenic fungi impact the quantity and quality of fruits and vegetables, how to find new bioresources and how they work against phytopathogens, and how to use pre- and post-harvest technologies like edible coatings and biopesticides to increase product stability.

Chapter 5 describes the bacteriological spoiling of food products, how to understand how bacteria cause food to deteriorate, and how to control the presence of pathogens, including research and investment in new detection techniques to address issues brought on by dangerous bacteria.

The widespread problem of fungal contamination and the mechanism of fungal decay in fruits and vegetables are covered in Chapter 6, which also offers a thorough analysis of detection techniques as well as preventative and control measures. Chapter 7 provides an overview of mycotoxin exposure in dairy products, their carcinogenic potential, and the application of various physical, chemical, and biological techniques to prevent fungal growth and cleanse food and feed of toxic compounds.

In order to improve food security and waste, Chapter 8 explains common food preservation techniques, such salting, smoking, chilling, freezing, and sugaring. These techniques also lead to more reasonably priced food preservation alternatives.

The application of current advancements in food preservation methods in the food sector is covered in Chapter 9. This chapter also discusses the benefits and drawbacks of new packaging techniques and preservation methods like microwave, cold plasma, and ultrasound. The use of natural substances in packaging is also included in this.

In Chapter 10, the microbial dynamics that underlie this age-old culinary art are fascinatingly explored. The complex interactions between these microscopic powerhouses and their significant impact on the properties of fermented foods are revealed. The metabolic pathways of various microorganisms leading to fermentation processes are covered in this chapter, along with information on managing and regulating fermentation.

The various extrinsic and intrinsic factors influencing the development of microbial cells are covered in Chapter 11, along with how improving the quality and safety of microbial growth might prolong the shelf life of food.

The characteristics and illnesses of foodborne fungal pathogens, their toxicological effects, the improvements in diagnostic methods related to these pathogens, the mechanism by which they cause disease, their control and prevention measures, previous outbreaks, and future implications are all covered in Chapter 12. The types and epidemiology of foodborne viral infections, together with the developments in diagnosis and management strategies to protect public health, are compiled in Chapter 13.

Chapter 14 covers foodborne parasites, such as helminths, trematodes, and protozoans, as well as their life cycle and dynamics of transmission. It also covers a variety of diagnostic tools, such as immunological and molecular methods, and preventative measures for parasitic illnesses. The emergence of antimicrobial resistance in the chicken microbiome is reviewed in Chapter 15, with particular attention to the effects of antibiotic usage in poultry farming. It employs a variety of methods, including microbial tests, which comprise both culture-dependent and culture-independent methods.

Chapter 16 discusses food testing agencies, acts, regulations, and patents that appear to focus on the regulatory framework, legal standards, and organizations involved in food safety and testing. It likely covers national and international organizations responsible for ensuring food safety; laws governing food production, safety, and testing standards; and innovations in food testing and safety technologies protected by intellectual property rights.

In addition to discussing future views in food safety with strong risk assessment and management approaches, Chapter 17 explores the various functions that microbes play in the fermentation of foods and beverages, highlighting their contribution to improving food quality. To guarantee greater safety standards, foodborne pathogens and pollutants must be detected at sufficient quantities. It also narrates how food safety is significantly impacted by crucial techniques, including mass spectrometry, next-generation sequencing, high-throughput screening, and omics, leading to a more comprehensive understanding of the safety implications of fermentation processes.

Meanwhile, a summary of microalgae, bacteria, yeast, and filamentous fungi that provide food protein for our sustainable society is provided in Chapter 18. It also elucidates the development of technologies in the food sector using alternative protein supplied from bacteria. Microbes can produce food protein on a big scale, affordably, and sustainably. Furtherly, recombinant microbes may manufacture high-quality single-cell proteins from carbon dioxide, methane, methanol, and other substrates. The recombinant microbial technology ensures protein supply and demand and increases sensory food quality on the basis of protein nutrition and function.

An overview of nanomaterial-based intelligent packaging systems for detecting and averting food degradation and microbiological contamination is provided in Chapter 19. Additionally, it summarizes how packaging materials with antioxidant and antibacterial qualities contribute to intelligent packaging, which improves food safety by halting the growth of harmful germs in food. The incorporation of nanomaterials into intelligent food packaging, which results in the creation of more efficient packaging systems, is also covered in this chapter.

The editors express their gratitude to each and every author who has supported us throughout this enormous undertaking and enabled us to finish it. The family members' persistence and tenacity are also appreciated by the editors. Readers are encouraged to submit ideas and critiques. We extend our sincere thanks to the production and publishing staff. This endeavor has been made possible by your assistance.

Editors
Priya Sutaoney, Priyambada Singh, Durgesh Nandini Chauhan, Nagendra Singh Chauhan, and Kamal Shah

About the Editors

Priya Sutaoney, PhD in microbiology, merit list holder in postgraduation, CG SET–qualified, has several years of academic experience in teaching UG and PG students in microbiology at Pt. Ravishankar Shukla University, Raipur, Chhattisgarh, in India. She has taught subjects like food microbiology, medical microbiology and immunology, plant physiology, microbial genetics, and genomics and is a member of the Mycological Society of India and the Society of Biological Chemists. She has published more than ten research articles in reputed journals, and four book chapters. She has attended several workshops and presented research papers at several conferences and symposiums and is also active as a reviewer for scientific journals. Currently, she is working as an assistant professor and guide for students in the Department of Microbiology in Kalinga University, Raipur, Chhattisgarh.

Priyambada Singh, PhD in microbiology, gold medalist in microbiology, CSIR-UGC NET-LS, GATE, and CG SET qualified, has several years of academic experience in microbiology at Chhattisgarh Dental College and Research Institute, Rajnandgaon, India. She has taught subjects like medical microbiology, food microbiology, immunology, and virology and has published four research articles in reputed journals and one book chapter. She has attended several workshops and presented research papers at several national and international conferences and symposiums. Currently, she is working as an assistant professor and guide for students in the Department of Microbiology in Kalinga University, Raipur, Chhattisgarh.

Durgesh Nandini Chauhan, MPharm, has several years of academic (teaching) experience at institutes in India in pharmaceutical sciences. She has taught subjects such as pharmaceutics, pharmacognosy, traditional concepts of medicinal plants, drug delivery, phytochemistry, cosmetic technology, pharmaceutical engineering, pharmaceutical packaging, quality assurance, dosage form designing, and anatomy and physiology. She is a member of the Association of Pharmaceutical Teachers of India, SILAE: SocietàItalo-Latinoamericana di Etnomedicina (The Scientific Network on Ethnomedicine, Italy), and others. Her previous research work included "Penetration Enhancement Studies on Organogel of Oxytetracycline HCL." She has attended several workshops, conferences, and symposiums, including the AICTE-sponsored staff development program on "Effects of Teaching and Learning Skills in Pharmacy: Tool for Improvement of Young Pharmacy Teachers." She has written more than 10 articles published in national and international journals, has contributed 15 book chapters, and is an editor in 10 books. She is also active as a reviewer for several international scientific journals and is an active participant at national and international conferences, including Bhartiya Vigyan Sammelan and the International Convention of Society of

Pharmacognosy. Presently, she has joined Ishita Research Organization, Raipur, India, as a freelance writer and guide for students in pharmacy, Ayurvedic, and science in their research projects. She is interested in data analysis, proposal writing, questionnaire design, designing and developing research questionnaires, and policy. She is also actively engaged with accurately monitoring, investigating, and analyzing all statistical information; professionally presenting information; reports and investigating; and documenting key data and information at the request of senior researchers.

Nagendra Singh Chauhan has professional expertise in natural product isolation and hands-on experience in various techniques, like extraction, isolation, and purification of natural products from plant extracts using various chromatographic techniques for the separation of compounds, including column, LC-MS, chromatography, and prep TLC. He specializes in the elucidation of the structure of natural products with the help of various techniques, like UV, mass spectrometry, IR, and NMR. He is well versed with the design, conduct, and data treatment of various in vivo experiments, like aphrodisiac, antidiabetic, antistress, hair growth, BPH, hepatoprotective, and reproductive parameters. He has written more than 50 articles published in national and international journals, contributed 20 book chapters, and edited four books.

Kamal Shah completed his graduation from Rajiv Gandhi Proudyogiki Vishwavidyalaya, Bhopal (MP), and was also awarded the Chancellor Gold Medal and Smt. Kamla Chaturvedi Memorial Award for securing first position in graduation (BPharm). He qualified for GATE 2003 (IIT, Madras) with an 82.13 percentile and GATE 2005 (IIT, Mumbai) with a 92.64 percentile. In 2006, he completed his masters from Dr. H. S. Gour University–Sagar and worked on method development for the determination of pharmacological active components by UV and HPLC. He has completed his doctorate from Dr. A. P. J. Abdul Kalam Technical University, Lucknow. Currently, he is working as an associate professor in the Institute of Pharmaceutical Research, GLA University, Mathura (UP). He has published his research work in various journals of national and international repute and has a total of 26 published research papers to his credit and more than 300 citations. He is a member of various professional and academic bodies, like the Association of Pharmaceutical Teachers of India (APTI), Institutional Animal Ethical Committee (IAEC), and Board of Studies (BOS). He is a reviewer for various national and international journals, like *Medicinal Chemistry Research*, *Pakistan Journal of Pharmaceutical Sciences*, *Journal of Sexual Medicine*, *Natural Product Research*, *African Journal of Pharmacy and Pharmacology*, *International Journal of Pharmacy and Pharmacology*, *International Research Journal of Pure and Applied Chemistry*, *Pharmaceutical Biology*, and *Asian Journal of Pharmaceutics*. He has also written three chapters in the book (special edition) *Medicinal Plants and Its Therapeutic Uses*. His area of interest is the "synthesis and evaluation of prodrugs of NSAIDs."

Contributors

Reza Abedi-Firoozjah
Student Research Committee
Kermanshah University of Medical Sciences
Kermanshah, Iran

Simge Aktop
Engineering Faculty
Department of Food Engineering
Ankara University
Ankara, Türkiye

Ugur Azizoglu
Department of Crop and Animal Production
Safiye Cikrikcioglu Vocational College
Kayseri University
Kayseri, Turkey
Genome and Stem Cell Research Center (GENKOK)
Erciyes University
Kayseri, Turkey

Adon Babu
Central University of Kerala
Kasaragod, Kerala, India

Soumya Basu
Department of International Health
Bloomberg School of Public Health
Johns Hopkins University
Baltimore, Maryland

Jerusalen Betancourt-Rodríguez
División de Estudios de Posgrado e Investigación
Tecnológico Nacional de México/Instituto Tecnológico de Tepic
Tepic, Nayarit, México

Mayuri Bhagawati
Plant Taxonomy and Biodiversity Laboratory
Department of Botany
Tripura University
Tripura, India

Puja Bhoi
Department of Botany
Guri Ghasidas Vishvavidyalaya
Bilaspur, Chhattisgarh, India

Gourav Biswas
Central University of Kerala
Kasaragod, Kerala, India

Romiluyi Bosede
Department of Microbiology
Faculty of Science
University of Lagos, Nigeria

Kabyashree Buragohain
Microbiology Laboratory, Department of Life Sciences
Dibrugarh University
Dibrugarh, Assam, India

Montserrat Calderón Santoyo
División de Estudios de Posgrado e Investigación
Tecnológico Nacional de México/Instituto Tecnológico de Tepic
Tepic, Nayarit, México

Prem Chandra
School of Agriculture
Indira Gandhi National Open University
New Delhi, India

Durgesh Nandini Chauhan
Columbia College of Pharmacy
Raipur, Chhattisgarh, India

Nagendra Singh Chauhan
Drug Testing Laboratory
Avam Anusandhan Kendra
Raipur, Chhattisgarh, India

Pankaj Chetia
Department of Life Sciences
Dibrugarh University
Dibrugarh, Assam, India

Leonardo Covarrubias-Rivera
División de Estudios de Posgrado e Investigación
Tecnológico Nacional de México/ Instituto Tecnológico de Tepic
Tepic, Nayarit, México

Seyma Dadi
Department of Material Science and Nanotechnology Engineering
Abdullah Gül University
Kayseri 38080, Turkey

Badal Kumar Datta
Department of Botany
Tripura University, Suryamaninagar
Tripura, India

Chayan Dhar
Department of Botany
Tripura University
Suryamaninagar, Tripura, India

Enespa
Department of Plant Pathology
School of Agriculture
MPDC, University of Lucknow
Lucknow, UP, India

Chukwuemeka Amarachi Esther
Nigeria Natural Medicine Development Agency
Victorial Island, Lagos, Nigeria

Indrani Gogoi
Molecular Plant Taxonomy and Bioinformatics Research Laboratory
Department of Life Sciences
Dibrugarh University
Dibrugarh, Assam, India

Tony Grace
Central University of Kerala
Kasaragod, Kerala, India

Shraddha Manish Gupta
Department of Pharmaceutical Sciences
School of Health Sciences and Technology
UPES–Dehradun
Uttarakhand, India

Muhammad Wasim Haider
Department of Horticultural Sciences
Faculty of Agriculture and Environment
The Islamia University of Bahawalpur
Punjab, Pakistan

Ahmed Hamad
Department of Food Hygiene and Control
Benha University, Egypt

Mohammad Hashemi
Medical Toxicology Research Center
Mashhad University of Medical Sciences
Mashhad, Iran

Titirsha Kayal
Department of Microbiology
Indira Gandhi Government Medical College and Hospital (IGGMC&H)
Nagpur, Maharashtra 440018, India

Narendra Kumar
Department of Botany
Guri Ghasidas Vishvavidyalaya
Bilaspur, Chhattisgarh. India

Rafael López-Cruz
División de Estudios de Posgrado e Investigación
Tecnológico Nacional de México/ Instituto Tecnológico de Tepic
Tepic, Nayarit, México

Shilpa Malakar
Department of Microbiology
Kalinga University
Raipur, Chhattisgarh, India

Sucheta Mandal
Department of Botany
Banwarilal Bhalotia College
Asansol, West Bengal, India

Neeraja Mano
Central University of Kerala
Kasaragod, Kerala, India

Rolly Mehrotra
Faculty of Food Science
Department of Home Science
Deen Dayal Upadhyay Gorakhpur University
Gorakhpur, Uttar Pradesh, India

Karpumpuli Mili
Department of Life Sciences
Dibrugarh University
Dibrugarh, Assam, India

Ratul Nath
Department of Life Sciences
Dibrugarh University
Dibrugarh, Assam 786004, India

Mustafa Nisari
Department of Medical Biochemistry
Faculty of Dentistry
University of Nuh Naci Yazgan
Kayseri, Turkey

Seyyed Mohammad Ali Noori
Toxicology Research Center
Medical Basic Sciences Research Institute
Ahvaz Jundishapur University of Medical Sciences
Ahvaz, Iran

Ismail Ocsoy
Department of Analytical Chemistry
Faculty of Pharmacy
Erciyes University
Kayseri, Turkey

Bamidele Tolulope Odumosu
Department of Microbiology
Faculty of Science
University of Lagos
Nigeria

Victor Agbo Olofu
Department of Biological Sciences
Trinity University Yaba
Lagos, Nigeria

Emine Ozdemir
Department of Analytical Chemistry
Faculty of Pharmacy
Erciyes University
Kayseri, Turkey

Ashim Jyoti Pathak
Department of Life Sciences
Dibrugarh University
Dibrugarh, Assam, India

Amit Patnaik
Molecular Biotechnology Laboratory
(HoD) Department of Biotechnology
NIST University
Institute Park
Berhampur, Odisha, India

Ponoop L. Kumar Patro
Department of Botany
Guri Ghasidas Vishvavidyalaya
Bilaspur, Chhattisgarh, India

Minakshi Puzari
Department of Life Sciences
Dibrugarh University
Dibrugarh, Assam, India

Juan Arturo Ragazzo-Sánchez
División de Estudios de Posgrado e Investigación
Tecnológico Nacional de México/ Instituto Tecnológico de Tepic
Tepic, Nayarit, México

Ehsan Sadeghi
Department of Food Science and Technology
School of Nutrition Sciences and Food Technology
Research Center for Environmental Determinants of Health (RCEDH)
Health Institute
Kermanshah University of Medical Sciences
Kermanshah, Iran

Duryodhan Sahu
(HoD) Department of Chemistry
NIST University
Institute Park, Berhampur, Odisha, India

Shyamalima Saikia
Molecular Plant Taxonomy and Bioinformatics Research Laboratory
Department of Life Sciences
Dibrugarh University
Dibrugarh, Assam, India

Pınar Şanlıbaba
Engineering Faculty
Department of Food Engineering
Ankara University
Ankara, Türkiye

Indrani Sarkar
Department of Bioinformatics
Kalinga University
Village-Kotni, Naya Raipur, Raipur, Chhattisgarh, India

Dipali Saxena
Food Sciences
Shri Vaishnav Institute of Home Science
Shri Vaishnav Vidyapeeth Vishwavidyalaya Indore
Madhya Pradesh, India

Kamal Shah
Institute of Pharmaceutical Research
GLA University
Mathura, India

Uttam Sharma
Shri Vaishnav Institute of Home Science
Shri Vaishnav Vidyapeeth Vishwavidyalaya
Indore, Madhya Pradesh, India

Priyambada Singh
Department of Microbiology
Kalinga University
Raipur, Chhattisgarh, India

Sukanya Sonowal
Department of Life Sciences
Dibrugarh University
Dibrugarh, Assam, India

Priya Sutaoney
Department of Microbiology
Kalinga University
Raipur, Chhattisgarh, India

Reza Tahergorabi
Food and Nutritional Sciences Program
North Carolina Agricultural and Technical State University
Greensboro, North Carolina

Milad Tavassoli
Department of Nutrition
Faculty of Health and Nutrition Sciences
Yasuj University of Medical Science
Yasuj, Iran

Nimet Temur
Department of Analytical Chemistry
Faculty of Pharmacy
Erciyes University
Kayseri, Turkey

Shahriyar Valizadeh
Food and Nutritional Sciences Program
North Carolina Agricultural and Technical State University
Greensboro, North Carolina

Jianguo Zhang
School of Health Science and Engineering
University of Shanghai for Science and Technology
Shanghai, China

Xiao Zheng
School of Health Science and Engineering
University of Shanghai for Science and Technology
Shanghai 200093, China

1 Introduction to Food Microbiology

Shilpa Malakar, Priya Sutaoney, Priyambada Singh, Shraddha Manish Gupta, Durgesh Nandini Chauhan, Kamal Shah, and Nagendra Singh Chauhan

1.1 INTRODUCTION

Our food and the rest of mankind are greatly influenced by microorganisms that are invisible to the human eye. They can enhance flavor, texture, and safety and keep food from spoiling (Laranjo et al. 2019). Understanding and using these microbial interactions to improve health, extend product shelf life, and lower the incidence of foodborne illnesses has long been a goal of humankind. Food and microorganisms interact in a complicated way; some are good for brewing and food preservation, while others can lead to spoilage or sickness (Dufossé et al. 2022). These dynamics make food microbiology a crucial area of the biological sciences. Understanding the development, persistence, and interactions of microorganisms in dietary items and habitats is the main goal of microbiology. It has real-world uses in fields including agriculture, pharmacy, and biotechnology (McGovern et al. 2004). One way to use microbial activity to prepare and preserve food is through fermentation (Tamang et al. 2020). Basic foodstuffs, such as yogurt, cheese, bread, and alcoholic beverages, can be transformed by bacteria into edible products, increasing shelf life and enhancing nutritional value and sensory qualities (Vitorino and Bessa 2017). Yeast is necessary for making bread and beer, while lactic acid bacteria in dairy products give yogurt its sour taste and creamy consistency. On the other hand, harmful bacteria, such as *Salmonella*, *Listeria monocytogenes*, and *Escherichia coli*, are dangerous because they can cause severe diseases and incorrect food preparation (Macori and Cotter 2018). For consumer protection, it is essential to comprehend the elements that promote or inhibit these harmful organisms.

Today, stringent quality control and food safety laws govern the worldwide food supply chain. Microbes enhance soil fertility and nutrient cycling and are essential to fermentation, food safety, and ecological sustainability (Chilton et al. 2015). The quality of product and crop yield is influenced by a healthy soil microbiome, which in turn influences the human food chain. Microbial biotechnology developments have produced eco-friendly packaging materials and bio-based preservatives, among other innovations that have improved food security and decreased waste (Chen and Alali 2018). Finding good microbes, getting rid of dangerous ones, and creating new

DOI: 10.1201/9781003540113-1

uses for microbial activity to improve food production and preservation are the main goals of this interdisciplinary area, which combines engineering, chemistry, and biology to address real-world problems. Technologies such as proteomics, genome sequencing, and bioinformatics have transformed the study of microbes and opened up new avenues for understanding the microbes for the good of humanity (Tamang et al. 2017). Due to globalization and the quick expansion of food supply chains, food microbiology has become more important in the regulation and monitoring of food safety. Food is exposed to a variety of microbial populations during foreign travel; therefore, constant attention to detail is necessary to prevent contamination and guarantee safe consumption (Suzzi and Corsetti 2020). Microbiological data has become crucial in today's globalized world for food safety monitoring and control. Growing in popularity are probiotics, which are live bacteria that, when consumed in proper amounts, have health advantages.

Understanding the interactions between these advantageous microorganisms and the human body that result in therapeutic advantages is the main goal of research and innovation. The necessity of studying microbes has been brought to light by climate change and its effects on agriculture. Climate change, rainfall patterns, and rising temperatures are altering microbial ecosystems, which has an effect on food supply and quality (Jeanson and Thierry 2015). Comprehending the impact of environmental stimuli on microbial activity is essential for risk mitigation and resilience. Food deterioration is a significant source of waste because of microbial activity, which is why addressing food waste also depends on microbes. Innovative solutions that depend on knowledge of microbial interactions and their role in the breakdown process include improved packaging technologies, natural preservatives, and cold chain logistics.

1.2 HISTORICAL PERSPECTIVE: FERMENTATION AND FOOD PRESERVATION

Through experimentation, ancient societies learned to use fermentation, a vital food preservation method (McGovern et al. 2004). Microorganisms such as bacteria, yeasts, and molds are responsible for this natural process, which has become an essential part of traditional culinary practices all throughout the world (Vallee 1998). By preserving extra food, it enables civilizations to ensure a supply during famine and keeps perishables from spoiling. Foods that undergo fermentation also have higher nutritional and sensory qualities; sourdough bread, cheese, yogurt, and beer are examples of items that have become mainstays (Steinkraus 2002). While inhibiting the growth of dangerous bacteria, these microbes break down complex chemicals in raw materials to produce acids, alcohols, and gases that improve food's flavor and fragrance. Culturally significant, fermented foods like miso, kimchi, and sauerkraut are used in many different types of cooking. Fermentation offers health advantages beyond flavor and preservation by increasing the bioavailability of minerals and reducing antinutritional components in raw materials (Caplice and Fitzgerald 1999). In regions with limited resources, fermented foods are crucial for nutrition because fermentation is an important way to increase vitamins B12 and K2 levels and absorb minerals. As civilizations advanced, the scientific understanding of fermentation grew, forming the foundation of modern food biotechnology (Hugenholtz 2013). These days, fermentation preserves food while producing new goods that support health and sustainability.

1.2.1 The Function of Microbes in Conventional Food Processing

An essential part of conventional food preparation methods is fermentation, an age-old method of food preservation (Borresen et al. 2012). Yeasts, bacteria, and molds are used to change different meals into stable, palatable forms. For decades, people have utilized fermentation methods to preserve food, dietary customs, cultural identity, and perishable goods (Caricilli and Saad 2013). Examples of these methods include milk, barley, and cabbage. By acting as organic catalysts in fermentation, microbes simplify complicated molecules ànd facilitate food preservation and digestion (Doria et al. 2008). For instance, lactic acid bacteria and wild yeasts convert carbohydrates into lactic acid and ethanol for sourdough bread. Similarly, the distinct smells of soy sauce and miso, staple foods, are caused by mold fungi, like *Aspergillus oryzae*, which break down proteins into sugars and amino acids (Zhu et al. 2024). These examples demonstrate how cultures all across the world have long utilized fermentation to enhance and retain their meals. The sustainability and health implications associated with conventional fermentation are being reexamined (Hossain et al. 2007). Advances in food technology are intended to decrease waste, increase food security, and create new sources of protein. Due to their sustainable and healthful diets, cultured plant-based products like tempeh and kimchi are becoming more and more popular (Abildinova et al. 2024). As plant-based diets often consume fewer natural resources and release fewer greenhouse emissions, fermenting food scraps into new commodities helps minimize food waste and the environmental effect of food supply chains (Hellwig et al. 2024). A traditional technique, fermentation is now essential to cutting waste and advancing sustainable diets. It makes it possible to make plant-based substitutes for foodstuffs made from animals, such as natto, tempeh, and plant-based yogurts. These nutrient-dense, high-protein alternatives help combat global concerns by satisfying customer desire for better, more environmentally responsible options.

As consumer preferences change, fermentation is helping businesses change. Customers who are concerned about their health are becoming more interested in fermented foods since they promote digestive health and provide probiotic advantages (Mannaaet al. 2021). While maintaining established techniques, researchers and food producers are always coming up with new ideas. The development of biopreservatives and natural taste enhancers like lactic acid and ethanol, which lessen the need for synthetic chemicals, is also fueled by fermentation (Walsh et al. 2023). The role of microbes in this novel process is what drives it. Sustainable packaging materials such as bioplastics are currently being produced through fermentation, a technique that blends traditional expertise with contemporary technology (Caffrey et al. 2024). By fusing ancient wisdom with modern technology, this creative strategy advances global sustainability objectives. Because of its versatility, fermentation serves as a catalyst for creative advancements and is an essential part of sustainable food systems.

1.3 MICROBIAL RISKS AND FOOD SAFETY

It was suggested that the presence of dangerous microorganisms in food posed serious threats to public health and safety. Furthermore, these microbes may cause major health issues in the human gastrointestinal system (Widiasih et al. 2024). The factors that promote the growth of dangerous microorganisms need to be thoroughly understood in order to guarantee the safety of food. Among the strategies employed to address microbiological hazards include proper packaging, hygiene, and temperature regulation (Ghimpeţeanuet al. 2022). Advanced methods such as pasteurization, irradiation, and the use of antibacterial or antifungal compounds have significantly enhanced the capacity to avoid or prevent infection. To monitor and control the microbiological risks across the food supply chain, food safety authorities and regulatory bodies should also impose strict restrictions (Pancu et al. 2021). Because foods are subjected to a variety of handling and environmental circumstances, the growth of the global food commerce has made managing microbial threats more difficult (Capozzi et al. 2020). In order to lower microbiological hazards and guarantee that food is safe and healthy for customers everywhere, stakeholders are combining scientific knowledge with contemporary technology.

1.3.1 Pathogens and Their Impact on Human Health

Among the biggest threats to food safety are microbial infections, which can cause everything from minor ailments to fatal diseases. These microbes, which include bacteria like *Salmonella*, *Escherichia coli*, and *Listeria monocytogenes*, can contaminate food at several points during the manufacturing, processing, and distribution processes (Cirne et al. 2019). Bacterial foodborne illnesses can result in fever, vomiting, diarrhea, and stomach pain. They can also cause neurological problems, renal failure, and even death (Cocolin et al. 2016). These illnesses affect millions of individuals globally each year, highlighting the importance of appropriate treatment and prevention actions. These infections also affect society and the economy, costing businesses money, destroying supply networks, and eroding customer confidence (Founou et al. 2016). Exposition to pathogens can have serious effects on vulnerable populations, including children, the elderly, and people with compromised immune systems. Strict monitoring and control procedures are therefore essential throughout the food supply chain.

1.3.2 Strategies for Preventing Contamination

Technical developments, regulatory oversight, and preventative measures must all be used to avoid contamination in food production and distribution. Important components include maintaining cleanliness, managing cold chain logistics, adhering to Hazard Analysis Critical Control Point (HACCP) and good manufacturing practices (GMP) regulations, and conducting routine microbial testing (Hughes et al. 2021). Education and training for food handlers are also essential. The diagnosis and prevention of foodborne illnesses have been revolutionized by technological developments, which allow for prompt corrective action through early detection of microbial risks using fast diagnostic methods like polymerase chain reaction (PCR) and next-generation sequencing (NGS) (Nelson et al. 2019). The safety and shelf life of food items are further improved by the creation of antibacterial coatings, natural preservatives, and sophisticated packaging solutions (Nout 1994). Intelligent packaging systems provide a thorough approach to lowering the danger of microbiological contamination by keeping an eye on environmental conditions and sending out real-time alerts in the event of contamination.

1.3.3 Encouraging a Culture of Safety and Customer Knowledge

For businesses to guarantee adherence to food safety regulations, open communication and openness with regulatory bodies must be their top priorities (Nji et al. 2021). Frequent certifications and audits, like ISO 22000, not only maintain safety standards but also increase customer trust in the items being sold. The preservation of a safe food environment depends on cooperation amongst stakeholders, including legislators, business executives, and scientific specialists. Encouraging customers to take responsibility for reducing foodborne diseases is equally vital. Hand cleaning, cooking food to the right temperature, and properly storing perishables are all easy ways to drastically lower your chance of contracting infections. Education programs and public health campaigns are essential for spreading this information (Ayukekbong et al. 2017). By providing consumers with the knowledge and resources they need to handle food properly, the group's efforts to reduce microbiological dangers are strengthened and more successful. All these steps work together to provide a strong foundation for addressing the problems caused by infections and guaranteeing safer food for all.

1.4 MICROORGANISMS IN AGRICULTURE AND SUSTAINABILITY

In agricultural systems, microorganisms are essential because they support crop yield, soil health, and the sustainability of the larger ecosystem. The soil contains enormous populations of these microbes, which include fungi, bacteria, archaea, and others (Jat et al. 2021). They create a complex network that is responsible for many important biochemical processes. Microbes play a crucial role in guaranteeing robust and successful agricultural methods by increasing nutrient availability, strengthening soil structure, and fostering plant health (Rai et al. 2022).

1. *Contribution to soil health.* Microorganisms have a crucial role in preserving the fertility of the soil, which is the cornerstone of agricultural output. The cycling of nutrients is one of their most important functions. Remaining plant and animal stuff is broken down into smaller molecules by microbes (Kumawat et al. 2022). Through this process, vital nutrients, like potassium, phosphorus, and nitrogen, are returned to the soil and made available to plants. Legumes, for example, have symbiotic partnerships with nitrogen-fixing bacteria, such as *Rhizobium*, which transform atmospheric nitrogen into ammonia that plants can absorb and use. Similarly, by decomposing insoluble phosphates in the soil, phosphate-solubilizing bacteria increase

the availability of phosphorus, a nutrient that would otherwise be scarce. Additionally important are fungi, especially mycorrhizal fungi (Rajguru et al. 2024). By spreading their hyphal networks into the soil, these fungi form mutualistic interactions with plant roots. This allows plants to obtain nutrients from normally inaccessible soil zones and increases the surface area available for nutrient absorption.

2. *Promoting crop productivity*. Microorganisms in the soil and plants interact to greatly affect crop output and growth. Plant development is directly stimulated by beneficial microorganisms, such plant growth-promoting rhizobacteria (PGPR), through a variety of processes (Priya et al. 2021). PGPR generate phytohormones that control plant growth processes, such as root elongation and shoot development, such as auxins, gibberellins, and cytokinins. In addition to producing enzymes and antibacterial substances that inhibit hazardous pathogens, some of these microorganisms also naturally fight sickness. Environmental stressors, including drought, heavy metal toxicity, and salinity, can be lessened for plants by microorganisms. Some fungi can store heavy metals in the soil, lowering the absorption of these metals by plants, whereas salt-tolerant bacteria help plants maintain osmotic balance in saline environments (Behera et al. 2024). Particularly in climate change–affected areas, microorganisms promote sustainable farming methods by enhancing resistance to abiotic stresses.

1.4.1 Innovations for Sustainable Food Systems

The agriculture industry is using microbial advances more and more to solve issues with climate change, environmental degradation, and food security (McElwee et al. 2020). These developments take advantage of the inherent capacities of microorganisms to create effective and environmentally friendly crop production and soil management solutions. The creation of biofertilizers and bio-pesticides is among the most promising fields. Beneficial microorganisms are used in biofertilizers to increase nutrient availability and lessen reliance on synthetic fertilizers, which are frequently expensive and harmful to the environment (Saco et al. 2021).

To enhance soil fertility in a sustainable manner, for example, biofertilizer formulations include bacteria that fix nitrogen and solubilize phosphate. While bio-pesticides do not affect the environment or beneficial creatures, they employ microbial agents like *Bacillus thuringiensis* to target certain pests. In sustainable intensification, more yields are obtained with less ecological effect thanks to these microbial-based inputs. Microbial consortia, which are deliberately crafted mixtures of several microorganisms, are another breakthrough that has been used to enhance agricultural results (Qiu et al. 2019). The efficacy of these consortia is increased by encouraging synergistic interactions, which imitate natural soil ecosystems. In combination, for instance, nitrogen-fixing bacteria, mycorrhizal fungi, and PGPR can increase plant growth, disease resistance, and nutrient availability. Additionally, microbial inoculants are being created to solve agricultural issues unique to a certain location. These are mixtures of beneficial microorganisms that are live and customized for certain crops, soil types, or environmental circumstances (Kour et al. 2023).

Microbial solutions are being incorporated into precision agricultural technology more and more, with sensors and data analytics being used to maximize the impact of these inoculants' application (Beattie et al. 2024). Microbial-based solutions reduce the environmental impact of farming techniques, which goes beyond increasing production and supports the objectives of sustainable agriculture. The usage of synthetic pesticides and fertilizers is reduced, reducing greenhouse gas emissions and preventing toxic chemicals from leaking into waterways. In tackling the dual problems of feeding the world's expanding population and reducing the effects of climate change, this is especially crucial (Jagadesh et al. 2024). Moreover, microbial applications aid in carbon sequestration, which is an essential step in the fight against climate change. Organic carbon is stabilized by soil bacteria, which stop it from escaping into the atmosphere as carbon dioxide. Increased microbial activity promotes increased carbon storage in agricultural soils through techniques including cover crops, less tillage, and the use of microbial inoculants (Clagnan et al. 2024).

1.5 MODERN TECHNOLOGICAL ADVANCEMENTS IN MICROBIAL RESEARCH

Technological advancements have greatly improved the understanding of microbes and how they interact with food systems. The study of microorganisms has been transformed by contemporary instruments and methods like genome sequencing, proteomics, and bioinformatics, which offer historically unheard-of insights into their genetic, structural, and functional characteristics (Ali et al. 2022). With the use of these techniques, scientists can identify microbial species with exceptional precision and examine how they behave in different environments, which help them design focused solutions to improve the safety and quality of food (Thao and Tran 2016). Genomic investigations are supplemented by proteomics, the study of proteins generated by microbes, to gain a better understanding of how these genetic instructions are manifested in real-world scenarios (Trivedi et al. 2021). This approach makes it easy to understand how microorganisms respond to key environmental factors for food processing and storage, such as temperature, pH, or nutrient availability (Kostic et al. 2024). The use of computational tools to understand biological data, or bioinformatics, has emerged as a crucial element in microbial research (Jahn et al. 2023). Advanced technologies such as CRISPR-Cas9 have also made it possible for scientists to accurately modify the genomes of microorganisms, creating new opportunities for genetic engineering. Microorganisms with improved fermentation capabilities, reduced spoiling potential, or better stress tolerance are being created using this method (Jin et al. 2024). There will surely be additional chances to optimize microbial uses in food systems as technology develops, addressing concerns related to sustainability, safety, and innovation (Choi et al. 2024).

1.5.1 Tools Like Genome Sequencing and Bioinformatics

Developments in computer technologies and molecular biology have greatly improved the understanding and control of microbial ecology (**Figure 1.1**). In this field, genome sequencing and bioinformatics are two essential technologies that,

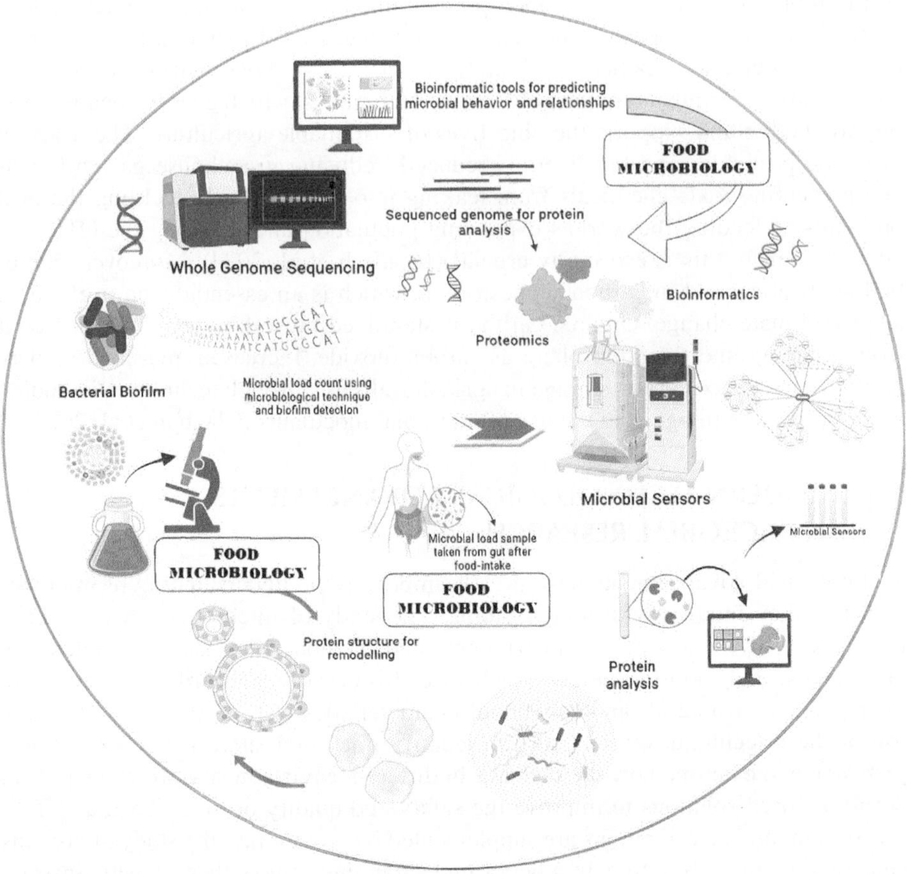

FIGURE 1.1 The diagram illustrates modern technological advancements in microbial research for food microbiology.

Source: Created with the BioRender app.

when combined, offer effective ways to detect, monitor, and control microorganisms in the food sector (Chelliah et al. 2022).

1. *Genome sequencing.* By decoding an organism's whole DNA sequence, this method provides comprehensive insight into its genetic composition. Next-generation sequencing (NGS) and other high-throughput sequencing technologies enable a quick and economical identification of microbes found in food samples (Alkema et al. 2016). Even at low quantities, scientists may identify harmful microbes, like *Salmonella*, *Listeria monocytogenes*, or *Escherichia coli*, by using genetic markers. Additionally, genome sequencing aids in distinguishing across strains of the same species, allowing for accurate source tracking of contamination during epidemics (Van

et al. 2011). For example, whole-genome sequencing (WGS), for instance, is currently often employed in food safety laboratories. It accomplishes this by comparing the genetic fingerprints of bacteria in food with those in a worldwide database, which helps identify the source of contamination (Chelliah et al. 2022). This approach has been extremely effective in reducing outbreaks and improving regulatory reactions.
2. *Bioinformatics.* Analyzing and interpreting the massive volume of data produced by genome sequencing call for advanced computational methods (Taboada et al. 2017). The study of evolutionary relationships, gene identification, and function prediction is aided by bioinformatics through the use of software and algorithms. Molecular evolutionary genetics analysis (MEGA), basic local alignment search tool (BLAST), and other machine learning techniques are used to process microbiological genomes. This is a crucial aspect of bioinformatics work.

These techniques greatly facilitate the identification of genes in pathogens that are responsible for virulence, antibiotic resistance, or rotting potential. Scientists can identify the genes responsible for a disease's virulence, drug resistance, or spoilage potential using these techniques (Wen et al. 2024). Not only can bioinformatics help identify harmful microorganisms, but it also helps identify beneficial bacteria that may be used as biopreservatives or added to probiotics. Examples may be the software applications MetaPhlAn and QIIME, which analyze the microbial communities in fermented foods to ensure that the correct strains are present and that harmful contaminants are absent (Wang et al. 2024).

1.5.2 Applications in Food Production and Preservation

Modern methods for food production and preservation include genome sequencing and bioinformatics. Enhancing food safety, shelf life, and quality is the primary objective of current developments.

1. *Improved fermentation processes.* The activity of advantageous gut bacteria in fermented meals is being increased via the use of CRISPR-Cas9 and other genome editing techniques (Moran 2017). For example, strains of *Lactobacillus* and *Bifidobacterium* have undergone genetic modification to enhance their health benefits, such as enhancing gut health or producing bioactive compounds like vitamins. These advancements ensure the stability and efficacy of probiotics in food products.
2. *Biopreservation.* Chemical preservatives are gradually being replaced by natural microbial treatments. Bacteriocins, antimicrobial peptides produced by certain bacteria, are now used to inhibit the growth of spoilage organisms in packaged foods (Nadon 2017). For instance, **nisin**, a bacteriocin approved for food use, is effective in preventing spoilage in dairy and canned goods.
3. *Predictive modeling and smart packaging.* The convergence of Internet of Things (IoT) technology with bioinformatics has produced breakthroughs,

such as smart packaging (Yousefi et al. 2019). Food packaging sensors that analyze volatile chemicals or microbial growth patterns can identify microbial decomposition in real time. By informing distributors and customers on food quality, these technologies guarantee safety and cut down on waste (Chiu et al. 2024).

1.6 PROBIOTICS AND HEALTH BENEFITS

Probiotics are live microorganisms that, when consumed in adequate amounts, provide health benefits to the host, primarily by supporting the gut microbiome. Often known as "good bacteria," these beneficial microorganisms are naturally present in nutritional supplements and fermented foods, including yogurt, kefir, sauerkraut, and miso. Probiotics' major function is to keep the digestive tract's microbial ecology in balance, which is essential for general health (Maiden et al. 2013). They aid in restoring gut flora that has been disturbed by things like antibiotic usage, poor nutrition, and sickness. The capacity of probiotics to enhance digestive health is among their best-established advantages (Sun et al. 2025). Probiotics improve food absorption and provide defense against dangerous diseases by generating short-chain fatty acids and other metabolites. They have been found effective in managing conditions like irritable bowel syndrome (IBS), diarrhea, and constipation. Research indicates that probiotics are essential for immune system strengthening in addition to aiding with digestion. They provide defense against infections by boosting immune cell activity and antibody production (Plaza-Diaz et al. 2019). Because the gut–brain axis connects gut bacteria to mood regulation and cognitive function, new research also shows that probiotics may be useful in treating mental health conditions. Probiotics may also help manage metabolic diseases, including obesity, and reduce inflammation and improve skin health. Despite the encouraging health advantages, probiotics' effectiveness varies depending on the strain, dose, and individual health problems; therefore, further study is needed to maximize its usage in therapeutic and nutritional applications (Pace 2015).

1.6.1 Role of Beneficial Microbes in Human Well-Being

The substantial influence that probiotics have on human health has garnered a lot of attention. These microorganisms are mostly bacteria, including *Lactobacillus* and *Bifidobacterium*, and some yeasts, like *Saccharomyces boulardii*. From strengthening the immune system and reducing the impacts of harmful infections to promoting gut health and nutrient absorption, they provide a wide range of benefits for human health (Jiménez 2009). These microorganisms are found in the gastrointestinal tract and are a component of the gut microbiota, a complex ecosystem that is essential to general health. Extensive research has been conducted to harness the medicinal potential of beneficial microorganisms and their association with diet. Not only can foods harbor these microorganisms; they also act as a conduit for the health advantages of probiotics (Zhang et al. 2024). Natural probiotics found in fermented foods, like yogurt, kefir, and sauerkraut, are proven to balance the gut microbiota

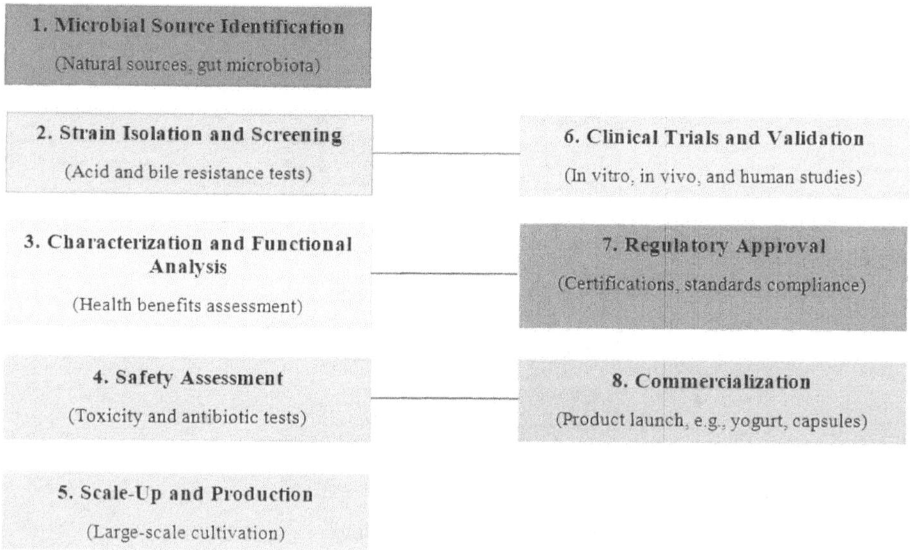

FIGURE 1.2 The flowchart illustrates the stages of probiotic development through microbes, beginning with identifying beneficial strains from natural sources.

Source: Created with Microsoft Word.

and enhance digestive health. Microbes and food are linked, which emphasize the importance of careful study and development to guarantee the organisms' survival and effectiveness when ingested. Beneficial microbe research and use have been transformed by modern technology (**Figure 1.2**). Analyzing biological data, such as probiotic genomic and proteomic data, using computer tools and techniques is known as bioinformatics (Latif et al. 2023).

For instance, bioinformatics enables researchers to identify the mechanisms by which probiotics generate bioactive substances, like vitamins and antimicrobial peptides, or interact with the human immune system. Furthermore, probiotic stability and viability during storage and use are crucially dependent on bioinformatics (Sánchez et al. 2017). Researchers can create novel formulations and delivery methods by examining the effects of environmental variables like temperature, pH, and oxygen exposure on microbial life. Quality control is a critical aspect of probiotic production. Every batch undergoes testing to see whether the microorganisms are viable, comply with safety regulations, and can last in the human digestive system. Researchers spend a lot of money refining the delivery systems for probiotics since they need to continue to work during the digestive process (Huang et al. 2023). Common tactics to guarantee that live bacteria or yeast enter the intestines undamaged include microencapsulation and the application of protective coatings. Because of their adaptability, probiotics are now found in a variety of goods outside of traditional diets (**Figure 1.3**).

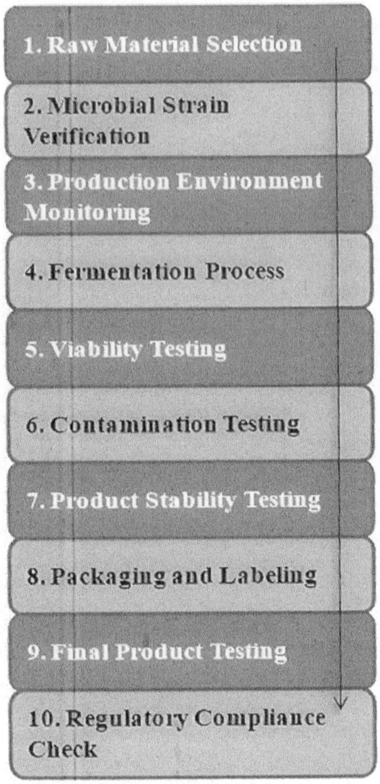

FIGURE 1.3 The flowchart outlines a systematic process for ensuring probiotic quality.

Source: Created with Microsoft Word.

1.7 CLIMATE CHANGE AND ITS EFFECTS ON MICROBIAL ECOSYSTEMS

1.7.1 Influence of Environmental Changes on Microbial Activity

Climate change has emerged as a critical factor influencing microbial ecosystems, directly impacting their growth, distribution, and activity. Due to their great degree of adaptability, microorganisms react quickly to changes in temperature, humidity, and other environmental factors (Xue et al. 2016). For example, rising global temperatures have the potential to speed up microbial metabolic rates, which in turn can increase the activity and reproduction of both helpful and dangerous bacteria. The delicate balance of microbial communities is altered by these changes, which are not uniform; some species may become extinct, while others flourish in warmer areas (Yuan et al.2021). The soil microbiome is one important area where climate change is having an influence. The breakdown of organic matter, cycling of nutrients, and preservation of soil fertility are all attributed to microbial activity in the soil (Zhao et al. 2018).

Climate change indirectly affects weather and crop patterns through agricultural practices. There may be changes in the microbial load in water and soil systems.

Additionally, some of these changes may result in the introduction of hazardous diseases or novel strains of microorganisms into systems used for food production and preservation (Zhou et al. 2010). Foodborne illnesses are more common when crops and water sources are contaminated by dangerous microorganisms due to extreme weather events, like floods (Zhou et al. 2011).

1.7.2 Food Safety and Quality Complications

Variable humidity and high temperatures provide the perfect environment for microbial deterioration, shortening the shelf life of perishable goods, including fruits, vegetables, and dairy (Zhou et al. 2012). The growth of fungi, like *Aspergillus*, for example, which create toxic mycotoxins that infect grains and nuts, can be encouraged by higher temperatures. When consumed, these poisons not only seriously harm one's health but also lower the quality of food. The food trade's globalization makes these problems much more complicated.

In addition to safety concerns, climate change also affects the sensory and nutritional qualities of food through microbial activity (Woodward et al. 2010). Spoilage microbes can alter the taste, texture, and appearance of food, making it unappealing and unacceptable for consumption. Advances in microbial control, enhanced irrigation methods, and crop types resistant to climate change are among the innovations being investigated. Scientists may now take preemptive steps in food safety procedures by using technologies like predictive modeling to better understand how microbial populations react to climatic factors (Wheeler and Von, 2013). In order to overcome these obstacles, multidisciplinary methods are crucial. Developing sustainable solutions requires cooperation between food safety specialists, climate scientists, and microbiologists. Policymakers and industry may create focused interventions by combining microbial research with climate modeling, for example, which can yield important insights into potential hazards.

1.8 FUTURE PERSPECTIVES ON MICROBIAL APPLICATIONS IN FOOD SYSTEMS

With new discoveries being made possible by continuous study, the future of microbial uses in agriculture is bright. Scientists are now able to better comprehend and work with microbial communities because of developments in metagenomics and synthetic biology. The variety and utility of soil microbes are revealed by metagenomic research by Wieder et al. (2015), who concluded that the creation of microorganisms with particular characteristics for agricultural purposes is made possible by synthetic biology. CRISPR-based gene editing and other contemporary methods are used to increase infections' stability and efficacy. Researchers also look at endophytes, which are bacteria that live inside plants and boost growth and resilience. These findings might result in a more robust and ecologically sound food supply.

A notable innovation is the use of natural microbial-based preservatives, which reduce reliance on artificial chemicals. These natural agents, which are produced by helpful microorganisms, prolong the shelf life of perishable foods without compromising their quality or safety by inhibiting the growth of harmful bacteria and fungi

(Tirado et al. 2010). Advances in biotechnology have also enabled new fermentation techniques. These processes not only enhance traditional products like bread, yogurt, and wine but also facilitate the production of alternative protein sources, like fermented plant-based foods and lab-grown meat.

Microbiological solutions for waste management represent yet another cutting-edge area. They are converting organic waste into compost or energy by utilizing specialist bacteria to break it down. In addition to decreasing food waste, these processes provide valuable residues that may be reused back into agricultural systems. Similar to this, microbe-based biodegradable packaging is gaining traction as a way to reduce plastic pollution and improve the overall sustainability of supply networks (Tiedje and Donohue 2008).

Artificial intelligence (AI)–powered systems are superior at predicting microbial activity, enhancing fermentation procedures, and more rapidly and precisely recognizing contamination risks. A safer, more efficient, and more ecologically friendly way to produce food is provided by this combination of biology and technology. Investing in microbial advances can help mankind tackle issues including environmental degradation, public health hazards, and food security, laying the groundwork for a sustainable, resilient food future.

By creating scalable procedures that integrate these microorganisms into manufacturing lines, chemical engineers help guarantee uniformity and quality at the industrial level. Environmental scientists evaluate these advances' ecological effects at the same time, advocating for methods that support sustainable development objectives.

Multidisciplinary efforts are essential in the fight against foodborne diseases. Together, epidemiologists, public health specialists, and microbiologists examine contamination trends to create prediction models that reduce the likelihood of outbreaks. Data science breakthroughs are essential in this case because big data analytics facilitates the processing of enormous volumes of data, providing insights on contamination patterns and microbial activity, revealed by Nychas et al. (2021). Microbiologists, agricultural scientists, and waste management experts all contribute to the fight against food waste. Together, they create systems that recycle organic debris into fertilizer, animal feed, or biofuel by harnessing microbial activity. By maximizing microbial productivity and lowering resource consumption, the combination of AI and machine learning improves these processes even further.

1.9 CONCLUSION

In order to maintain soil health, boost agricultural productivity, and solve environmental problems, microorganisms are essential to agriculture. Microbial technology may be used in agricultural practices to safeguard ecosystems and contribute to global food security. The significance of helpful bacteria in production and preservation is enhanced by technological advancements like as genome sequencing and bioinformatics, which allow for the exact diagnosis of hazardous diseases. Human health, gut health, immunity, and illness prevention all depend on these bacteria. Because of new manufacturing techniques and developments in bioinformatics, probiotics are increasingly being used in therapeutic treatments and functional meals.

It is anticipated that the importance of these organisms in improving human health will only increase as more advantages are discovered via study. By addressing these issues proactively, societies can safeguard public health and ensure the availability of safe, high-quality food for future generations.

REFERENCES

Abildinova, G. Z., Benberin, V. V., Vochshenkova, T. A., Afshar, A., Mussin, N. M., Kaliyev, A. A., Zhussupova, Z., & Tamadon, A. 2024. Global trends and collaborative networks in gut microbiota-insulin resistance research: A comprehensive bibliometric analysis (2000–2024). *Frontiers in Medicine*, 11: 1452227 (https://doi.org/10.3389/fmed.2024.1452227).

Ali, S., Tyagi, A., Mushtaq, M., Al-Mahmoudi, H., & Bae, H. 2022. Harnessing plant microbiome for mitigating arsenic toxicity in sustainable agriculture. *Environmental Pollution (Barking, Essex: 1987)*, 300: 118940 (https://doi.org/10.1016/j.envpol.2022.118940).

Alkema, W., Boekhorst, J., Wels, M., & van Hijum, S. A. 2016. Microbial bioinformatics for food safety and production. *Briefings in Bioinformatics*, 17: 283–292.

Ayukekbong, J. A., Ntemgwa, M., & Atabe, A. N. 2017. The threat of antimicrobial resistance in developing countries: Causes and control strategies. *Antimicrobial Resistance and Infection Control*, 6: 47 (https://doi.org/10.1186/s13756-017-0208-x).

Beattie, G. A., Bayliss, K. L., Jacobson, D. A., Broglie, R., Burkett-Cadena, M., Sessitsch, A., Kankanala, P., Stein, J., Eversole, K., & Lichens-Park, A. 2024. From microbes to microbiomes: Applications for plant health and sustainable agriculture. *Phytopathology*, 114: 1742–1752.

Behera, P. R., Behera, K. K., Sethi, G., Prabina, B. J., Bai, A. T., Sipra, B. S., Adarsh, V., Das, S., Behera, K. C., Singh, L., Mishra, M. K., & Behera, M. 2024. Enhancing agricultural sustainability through rhizomicrobiome: A review. *Journal of Basic Microbiology*, 64: e2400100 (https://doi.org/10.1002/jobm.202400100).

Borresen, E. C., Henderson, A. J., Kumar, A., Weir, T. L., & Ryan, E. P. 2012. Fermented foods: Patented approaches and formulations for nutritional supplementation and health promotion. *Recent Patents on Food, Nutrition & Agriculture*, 4: 134–140.

Caffrey, E. B., Sonnenburg, J. L., & Devkota, S. 2024. Our extended microbiome: The human-relevant metabolites and biology of fermented foods. *Cell Metabolism*, 36: 684–701.

Caplice, E., & Fitzgerald, G. F. 1999. Food fermentations: Role of microorganisms in food production and preservation. *International Journal of Food Microbiology*, 50: 131–149.

Capozzi, V., Fragasso, M., & Russo, P. 2020. Microbiological safety and the management of microbial resources in artisanal foods and beverages: The need for a transdisciplinary assessment to conciliate actual trends and risks avoidance. *Microorganisms*, 8: 306 (https://doi.org/10.3390/microorganisms8020306).

Caricilli, A. M., & Saad, M. J. 2013. The role of gut microbiota on insulin resistance. *Nutrients*, 5: 829–851.

Chelliah, R., Banan-MwineDaliri, E., Khan, I., Wei, S., Elahi, F., Yeon, S. J., Selvakumar, V., Ofosu, F. K., Rubab, M., Ju, H. H., Rallabandi, H. R., Madar, I. H., Sultan, G., & Oh, D. H. 2022. A review on the application of bioinformatics tools in food microbiome studies. *Briefings in Bioinformatics*, 23: bbac007 (https://doi.org/10.1093/bib/bbac007).

Chen, L., & Alali, W. 2018. Editorial: Recent discoveries in human serious foodborne pathogenic bacteria: Resurgence, pathogenesis, and control strategies. *Frontiers in Microbiology*, 9: 2412 (https://doi.org/10.3389/fmicb.2018.02412).

Chilton, S. N., Burton, J. P., & Reid, G. 2015. Inclusion of fermented foods in food guides around the world. *Nutrients*, 7: 390–404.

Chiu, I., Ye, H., Aayush, K., & Yang, T. 2024. Intelligent food packaging for smart sensing of food safety. *Advances in Food and Nutrition Research*, 111: 215–259.

Choi, K. R., Jung, S. Y., & Lee, S. Y. 2024. From sustainable feedstocks to microbial foods. *Nature Microbiology*, 9: 1167–1175.

Cirne, C. T., Tunick, M. H., & Trout, R. E. 2019. The chemical and attitudinal differences between commercial and artisanal products. *NPJ Science of Food*, 3: 19 (https://doi.org/10.1038/s41538-019-0053-9).

Clagnan, E., Costanzo, M., Visca, A., Di Gregorio, L., Tabacchioni, S., Colantoni, E., Sevi, F., Sbarra, F., Bindo, A., Nolfi, L., Magarelli, R. A., Trupo, M., Ambrico, A., & Bevivino, A. 2024. Culturomics- and metagenomics-based insights into the soil microbiome preservation and application for sustainable agriculture. *Frontiers in Microbiology*, 15: 1473666 (https://doi.org/10.3389/fmicb.2024.1473666).

Cocolin, L., Gobbetti, M., Neviani, E., & Daffonchio, D. 2016. Ensuring safety in artisanal food microbiology. *Nature Microbiology*, 1: 16171(https://doi.org/10.1038/nmicrobiol.2016.171).

Doria, A., Patti, M. E., & Kahn, C. R. 2008. The emerging genetic architecture of type 2 diabetes. *Cell Metabolism*, 8: 186–200.

Dufossé, L., & De Medici, D. 2022. Editorial: Insights in food microbiology: 2021. *Frontiers in Microbiology*, 13: 1068066 (https://doi.org/10.3389/fmicb.2022.1068066).

Founou, L. L., Founou, R. C., & Essack, S. Y. 2016. Antibiotic resistance in the food chain: A developing country-perspective. *Frontiers in Microbiology*, 7: 1881 (https://doi.org/10.3389/fmicb.2016.01881).

Ghimpețeanu, O. M., Pogurschi, E. N., Popa, D. C., Dragomir, N., Drăgotoiu, T., Mihai, O. D., & Petcu, C. D. 2022. Antibiotic use in livestock and residues in food-a public health threat: A review. *Foods (Basel, Switzerland)*, 11: 1430 (https://doi.org/10.3390/foods11101430).

Hellwig, M., Diel, P., Eisenbrand, G., Grune, T., Guth, S., Henle, T., Humpf, H. U., Joost, H. G., Marko, D., Raupbach, J., Roth, A., Vieths, S., & Mally, A. 2024. Dietary glycation compounds—implications for human health. *Critical Reviews in Toxicology*, 54: 485–617.

Hossain, P., Kawar, B., & El Nahas, M. 2007. Obesity and diabetes in the developing world—a growing challenge. *The New England Journal of Medicine*, 356: 213–215.

Huang, Y. Y., Liang, Y. T., Wu, J. M., Wu, W. T., Liu, X. T., Ye, T. T., Chen, X. R., Zeng, X. A., Manzoor, M. F., & Wang, L. H. 2023. Advances in the study of probiotics for immunomodulation and intervention in food allergy. *Molecules (Basel, Switzerland)*, 28: 1242 (https://doi.org/10.3390/molecules28031242).

Hugenholtz, J. 2013. Traditional biotechnology for new foods and beverages. *Current Opinion in Biotechnology*, 24: 155–159.

Hughes, A., Roe, E., & Hocknell, S. 2021. Food supply chains and the antimicrobial resistance challenge: On the framing, accomplishments and limitations of corporate responsibility. *Environment and Planning A: Economy and Space*, 53: 1373–1390.

Jagadesh, M., Dash, M., Kumari, A., Singh, S. K., Verma, K. K., Kumar, P., Bhatt, R., & Sharma, S. K. 2024. Revealing the hidden world of soil microbes: Metagenomic insights into plant, bacteria, and fungi interactions for sustainable agriculture and ecosystem restoration. *Microbiological Research*, 285: 127764 (https://doi.org/10.1016/j.micres.2024.127764).

Jahn, L. J., Rekdal, V. M., & Sommer, M. O. A. 2023. Microbial foods for improving human and planetary health. *Cell*, 186: 469–478.

Jat, S., Suby, S., Parihar, C. M., Gambhir, G., Kumar, N., & Rakshit, S. 2021. Microbiome for sustainable agriculture: A review with special reference to the corn production system. *Archives of Microbiology*. 203: 1–23.

Jeanson, S., & Thierry, A. 2015. Editorial: Exploring bacterial colonies in solid foods or model foods using non-destructive techniques. *Frontiers in Microbiology*, 6: 1440 (https://doi.org/10.3389/fmicb.2015.01440).

Jiménez, M. B. 2009. Treatment of irritable bowel syndrome with probiotics. An etiopathogenic approach at last? *Revistaespanola de enfermedadesdigestivas*, 101: 553–564.

Jin, R., Song, J., Liu, C., Lin, R., Liang, D., Aweya, J. J., Weng, W., Zhu, L., Shang, J., & Yang, S. 2024. Synthetic microbial communities: Novel strategies to enhance the quality of traditional fermented foods. *Comprehensive Reviews in Food Science and Food Safety*, 23: e13388 (https://doi.org/10.1111/1541-4337.13388).

Kostic, T., Schloter, M., Arruda, P., Berg, G., Charles, T. C., Cotter, P. D., Kiran, G. S., Lange, L., Maguin, E., Meisner, A., van Overbeek, L., Sanz, Y., Sarand, I., Selvin, J., Tsakalidou, E., Smidt, H., Wagner, M., & Sessitsch, A. 2024. Concepts and criteria defining emerging microbiome applications. *Microbial Biotechnology*, 17: e14550 (https://doi.org/10.1111/1751-7915.14550).

Kour, D., Kour, H., Khan, S. S., Khan, R. T., Bhardwaj, M., Kailoo, S., Kumari, C., Rasool, S., Yadav, A. N., & Sharma, Y. P. 2023. Biodiversity and functional attributes of rhizospheric microbiomes: Potential tools for sustainable agriculture. *Current Microbiology*, 80: 192 (https://doi.org/10.1007/s00284-023-03300-5).

Kumawat, K. C., Razdan, N., & Saharan, K. 2022. Rhizospheric microbiome: Bio-based emerging strategies for sustainable agriculture development and future perspectives. *Microbiological Research*, 254: 126901 (https://doi.org/10.1016/j.micres.2021.126901).

Laranjo, M., Córdoba, M. G., Semedo-Lemsaddek, T., & Potes, M. E. 2019. Food microbiology. *BioMed Research International*, 8039138 (https://doi.org/10.1155/2019/8039138).

Latif, A., Shehzad, A., Niazi, S., Zahid, A., Ashraf, W., Iqbal, M. W., Rehman, A., Riaz, T., Aadil, R. M., Khan, I. M., Özogul, F., Rocha, J. M., Esatbeyoglu, T., & Korma, S. A. 2023. Probiotics: Mechanism of action, health benefits and their application in food industries. *Frontiers in Microbiology*, 14: 1216674 (https://doi.org/10.3389/fmicb.2023.1216674).

Macori, G., & Cotter, P. D. 2018. Novel insights into the microbiology of fermented dairy foods. *Current Opinion in Biotechnology*, 49: 172–178.

Maiden, M. C., Jansen van Rensburg, M. J., Bray, J. E., Earle, S. G., Ford, S. A., Jolley, K. A., & McCarthy, N. D. 2013. MLST revisited: The gene-by-gene approach to bacterial genomics. *Nature Reviews. Microbiology*, 11: 728–736.

Mannaa, M., Han, G., Seo, Y. S., & Park, I. 2021. Evolution of food fermentation processes and the use of multi-omics in deciphering the roles of the microbiota. *Foods (Basel, Switzerland)*, 10: 2861 (https://doi.org/10.3390/foods10112861).

McElwee, P., Calvin, K., Campbell, D., Cherubini, F., Grassi, G., Korotkov, V., Le Hoang, A., Lwasa, S., Nkem, J., Nkonya, E., Saigusa, N., Soussana, J. F., Taboada, M. A., Manning, F., Nampanzira, D., & Smith, P. 2020. The impact of interventions in the global land and agri-food sectors on Nature's Contributions to People and the UN Sustainable Development Goals. *Global Change Biology*, 26: 4691–4721.

McGovern, P. E., Zhang, J., Tang, J., Zhang, Z., Hall, G. R., Moreau, R. A., Nuñez, A., Butrym, E. D., Richards, M. P., Wang, C. S., Cheng, G., Zhao, Z., & Wang, C. 2004. Fermented beverages of pre- and proto-historic China. *Proceedings of the National Academy of Sciences of the United States of America*, 101: 17593–17598.

Moran-Gilad, J. 2017. Whole genome sequencing (WGS) for food-borne pathogen surveillance and control—taking the pulse. *Euro Surveillance: Bulletin Europeen sur les maladies transmissibles = European Communicable Disease Bulletin*, 22: 30547 (https://doi.org/10.2807/1560-7917.ES.2017.22.23.30547).

Nadon, C., Van Walle, I., Gerner-Smidt, P., Campos, J., Chinen, I., Concepcion-Acevedo, J., Gilpin, B., Smith, A. M., Man Kam, K., Perez, E., Trees, E., Kubota, K., Takkinen, J., Nielsen, E. M., Carleton, H., & FWD-NEXT Expert Panel 2017. PulseNet International:

Vision for the implementation of whole genome sequencing (WGS) for global food-borne disease surveillance. *Euro Surveillance: Bulletin Europeen sur les maladies transmissibles = European communicable Disease Bulletin*, 22: 30544 (https://doi.org/10.2807/1560-7917.ES.2017.22.23.30544).

Nelson, D. W., Moore, J. E., & Rao, J. R. 2019. Antimicrobial Resistance (AMR): Significance to food quality and safety. *Food Quality and Safety*, 3: 15–22 (https://doi.org/10.1093/fqsafe/fyz003).

Nji, E., Kazibwe, J., Hambridge, T., Joko, C. A., Larbi, A. A., Damptey, L. A. O., Nkansa-Gyamfi, N. A., Stålsby Lundborg, C., & Lien, T. Q. 2021. High prevalence of antibiotic resistance in commensal Escherichia coli from healthy human sources in community settings. *Scientific Reports*, 11: 3372 (https://doi.org/10.1038/s41598-021-82693-4).

Nout, M. J. R. 1994. Fermented foods and food safety. *Food Research International*, 27: 291–298 (https://doi.org/10.1016/0963-9969(94)90097–3).

Nychas, G. J., Sims, E., Tsakanikas, P., & Mohareb, F. 2021. Data science in the food industry. *Annual Review of Biomedical Data Science*, 4: 341–367.

Pace, F., Pace, M., & Quartarone, G. 2015. Probiotics in digestive diseases: Focus on Lactobacillus GG. *Minerva Gastroenterologica e Dietologica*, 61: 273–292.

Pancu, D. F., Scurtu, A., Macasoi, I. G., Marti, D., Mioc, M., Soica, C., Coricovac, D., Horhat, D., Poenaru, M., & Dehelean, C. 2021. Antibiotics: Conventional therapy and natural compounds with antibacterial activity-A pharmaco-toxicological screening. *Antibiotics (Basel, Switzerland)*, 10: 401 (https://doi.org/10.3390/antibiotics10040401).

Plaza-Diaz, J., Ruiz-Ojeda, F. J., Gil-Campos, M., & Gil, A. 2019. Mechanisms of action of probiotics. *Advances in nutrition (Bethesda, Md.)*, 10(suppl_1): S49–S66 (https://doi.org/10.1093/advances/nmy063).

Priya, P., Aneesh, B., & Harikrishnan, K. 2021. Genomics as a potential tool to unravel the rhizosphere microbiome interactions on plant health. *Journal of Microbiological Methods*, 185: 106215 (https://doi.org/10.1016/j.mimet.2021.106215).

Qiu, Z., Egidi, E., Liu, H., Kaur, S., & Singh, B. K. 2019. New frontiers in agriculture productivity: Optimised microbial inoculants and in situ microbiome engineering. *Biotechnology Advances*, 37: 107371 (https://doi.org/10.1016/j.biotechadv.2019.03.010).

Rai, S., Omar, A. F., Rehan, M., Al-Turki, A., Sagar, A., Ilyas, N., Sayyed, R. Z., & Hasanuzzaman, M. 2022. Crop microbiome: Their role and advances in molecular and omic techniques for the sustenance of agriculture. *Planta*, 257: 27 (https://doi.org/10.1007/s00425-022-04052-5).

Rajguru, B., Shri, M., & Bhatt, V. D. 2024. Exploring microbial diversity in the rhizosphere: A comprehensive review of metagenomic approaches and their applications. *3 Biotech*, 14: 224 (https://doi.org/10.1007/s13205-024-04065-9).

Saco, P. M., McDonough, K. R., Rodriguez, J. F., Rivera-Zayas, J., & Sandi, S. G. 2021. The role of soils in the regulation of hazards and extreme events. *Philosophical Transactions of the Royal Society of London. Series B, Biological Sciences*, 376: 20200178 (https://doi.org/10.1098/rstb.2020.0178).

Sánchez, B., Delgado, S., Blanco-Míguez, A., Lourenço, A., Gueimonde, M., & Margolles, A. 2017. Probiotics, gut microbiota, and their influence on host health and disease. *Molecular Nutrition & Food Research*, 61: 27500859 (https://doi.org/10.1002/mnfr.201600240).

Steinkraus, K. H. 2002. Fermentations in world food processing. *Comprehensive Reviews in Food Science and Food Safety*, 1: 23–32.

Sun, X., Yu, L., Xiao, M., Zhang, C., Zhao, J., Narbad, A., Chen, W., Zhai, Q., & Tian, F. 2025. Exploring Core fermentation microorganisms, flavor compounds, and metabolic pathways in fermented Rice and wheat foods. *Food Chemistry*, 463: 141019 (https://doi.org/10.1016/j.foodchem.2024.141019).

Suzzi, G., & Corsetti, A. 2020. Food microbiology: The past and the new challenges for the next 10 years. *Frontiers in Microbiology*, 11: 237 (https://doi.org/10.3389/fmicb.2020.00237).

Taboada, E. N., Graham, M. R., Carriço, J. A., & Van Domselaar, G. 2017. Food safety in the age of next generation sequencing, bioinformatics, and open data access. *Frontiers in Microbiology*, 8: 909 (https://doi.org/10.3389/fmicb.2017.00909).

Tamang, J. P., Cotter, P. D., Endo, A., Han, N. S., Kort, R., Liu, S. Q., Mayo, B., Westerik, N., & Hutkins, R. 2020. Fermented foods in a global age: East meets West. *Comprehensive Reviews in Food Science and Food Safety*, 19: 184–217.

Tamang, J. P., Holzapfel, W. H., Shin, D. H., & Felis, G. E. 2017. Editorial: Microbiology of ethnic fermented foods and alcoholic beverages of the world. *Frontiers in Microbiology*, 8: 1377 (https://doi.org/10.3389/fmicb.2017.01377).

Thao, N. P., & Tran, L. S. 2016. Enhancement of plant productivity in the post-genomics era. *Current Genomics*, 17: 295–296.

Tiedje, J., & Donohue, T. 2008. Microbes in the energy grid. *Science (New York, N.Y.)*, 320: 985 (https://doi.org/10.1126/science.1159999).

Tirado, M. C., Clarke, R., Jaykus, L. A., McQuatters-Gollop, A., & Frank, J. M. 2010. Climate change and food safety: A review. *FoodResearch International*, 43: 1745–1765 (https://doi.org/10.1016/j.foodres.2010.07.003).

Trivedi, P., Mattupalli, C., Eversole, K., & Leach, J. E. 2021. Enabling sustainable agriculture through understanding and enhancement of microbiomes. *The New Phytologist*, 230: 2129–2147.

Vallee, B. L. 1998. Alcohol in the western world. *Scientific American*, 278: 80–85.

Van HylckamaVlieg, J. E., Veiga, P., Zhang, C., Derrien, M., & Zhao, L. 2011. Impact of microbial transformation of food on health—from fermented foods to fermentation in the gastro-intestinal tract. *Current Opinion in Biotechnology*, 22: 211–219.

Vitorino, L. C., & Bessa, L. A. 2017. Technological microbiology: Development and applications. *Frontiers in Microbiology*, 8: 827 (https://doi.org/10.3389/fmicb.2017.00827).

Walsh, A. M., Leech, J., Huttenhower, C., Delhomme-Nguyen, H., Crispie, F., Chervaux, C., & Cotter, P. D. 2023. Integrated molecular approaches for fermented food microbiome research. *FEMS Microbiology Reviews*, 47: fuad001 (https://doi.org/10.1093/femsre/fuad001).

Wang, B., Shi, Y., Zhang, H., Hu, Y., Chen, H., Liu, Y., Wang, F., & Chen, L. 2024. Influence of microorganisms on flavor substances and functional components of sojae semen praeparatum during fermentation: A study integrating comparative metabolomics and high-throughput sequencing. *Food Research International (Ottawa, Ont.)*, 187: 114405 (https://doi.org/10.1016/j.foodres.2024.114405).

Wen, L., Yang, L., Chen, C., Li, J., Fu, J., Liu, G., Kan, Q., Ho, C. T., Huang, Q., Lan, Y., & Cao, Y. 2024. Applications of multi-omics techniques to unravel the fermentation process and the flavor formation mechanism in fermented foods. *Critical Reviews in Food Science and Nutrition*, 64: 8367–8383.

Wheeler, T., & von Braun, J. 2013. Climate change impacts on global food security. *Science (New York, N.Y.)*, 341: 508–513.

Widiasih, D. A., Pratama, R. P., Drastini, Y., Putri, K., Fatimah, L. N., & Indarjulianto, S. 2024. Rapid testing of antibiotic residues to increase food safety awareness of animal origin. *Veterinary World*, 17: 1177–1183 (https://doi.org/10.14202/vetworld.2024.1177–1183).

Wieder, W. R., Allison, S. D., Davidson, E. A., Georgiou, K., Hararuk, O., He, Y., Hopkins, F., Luo, Y., Smith, M. J., Sulman, B., Todd-Brown, K., Wang, Y-P., Xia, J., & Xu, X. 2015. Explicitly representing soil microbial processes in Earth system models. *Global Biogeochemical Cycles*, 29: 1782–1800 (https://doi.org/10.1002/2015GB005188).

Woodward, G., Perkins, D. M., & Brown, L. E. 2010. Climate change and freshwater ecosystems: Impacts across multiple levels of organization. *Philosophical Transactions of the Royal Society of London. Series B, Biological Sciences*, 365: 2093–2106.

Xue, K., Yuan, M. M., Shi, Z. J., Qin, Y., Deng, Y., Cheng, L., Wu, L., He, Z., Van Nostrand, J. D., Bracho, R., Natali, S., Schuur, E. A. G., Luo, C., Konstantinidis, K. T., Wang, Q.,

Cole, J. R., Tiedje, J. M., Luo, Y., & Zhou, J-Z. 2016. Tundra soil carbon is vulnerable to rapid microbial decomposition under climate warming. *Nature Climate Change*, 6: 595–600 (https://doi.org/10.1038/nclimate2940).

Yousefi, H., Su, H. M., Imani, S. M., Alkhaldi, K., M Filipe, C. D., & Didar, T. F. 2019. Intelligent food packaging: A review of smart sensing technologies for monitoring food quality. *ACS Sensors*, 4: 808–821.

Yuan, M. M., Guo, X., Wu, L., et al. 2021. Climate warming enhances microbial network complexity and stability. *Nature Climate Change*, 11: 343–348 (https://doi.org/10.1038/s41558-021-00989-9).

Zhang, Y., Song, M., Fan, J., Guo, X., & Tao, S. 2024. Impact of probiotics-derived extracellular vesicles on livestock gut barrier function. *Journal of Animal Science and Biotechnology*, 15: 149 (https://doi.org/10.1186/s40104-024-01102-8).

Zhao, L., Zhang, F., Ding, X., Wu, G., Lam, Y. Y., Wang, X., Fu, H., Xue, X., Lu, C., Ma, J., Yu, L., Xu, C., Ren, Z., Xu, Y., Xu, S., Shen, H., Zhu, X., Shi, Y., Shen, Q., Dong, W., & Zhang, C. 2018. Gut bacteria selectively promoted by dietary fibers alleviate type 2 diabetes. *Science (New York, N.Y.)*, 359: 1151–1156.

Zhou, J., Deng, Y., Luo, F., He, Z., Tu, Q., & Zhi, X. 2010. Functional molecular ecological networks. *mBio*, 1: e00169–10 (https://doi.org/10.1128/mBio.00169-10).

Zhou, J., Deng, Y., Luo, F., He, Z., & Yang, Y. 2011. Phylogenetic molecular ecological network of soil microbial communities in response to elevated CO2. *mBio*, 2: e00122–11 (https://doi.org/10.1128/mBio.00122-11).

Zhou, J., Xue, K., Xie, J., Deng, Y., Wu, L., Cheng, X., Fei, S., Deng, S., He, Z., Van Nostrand, J. D., & Luo, Y. 2012. Microbial mediation of carbon-cycle feedbacks to climate warming. *Nature Climate Change*, 2: 106–110 (https://doi.org/10.1038/nclimate1331).

Zhu, Q., Chen, B., Zhang, F., Zhang, B., Guo, Y., Pang, M., Huang, L., & Wang, T. 2024. Toxic and essential metals: Metabolic interactions with the gut microbiota and health implications. *Frontiers in Nutrition*, 11: 1448388 (https://doi.org/10.3389/fnut.2024.1448388).

2 Microbiome of Fresh Fruits and Vegetables

Dipali Saxena, Rolly Mehrotra, Uttam Sharma, and Ahmed Hamad

2.1 INTRODUCTION

In recent years, the microbiome has become a key component of sustainable food production. With rising consumer desire for natural, minimally processed foods, knowledge of the microbial populations associated with fresh produce is critical for improving food safety and quality. From farm methods to post-harvest treatments, every step in the production chain can influence the makeup and function of microbial communities. This complicated ecology has a significant impact on the overall health and quality of food, influencing everything, from rotation rates to nutritional content. While some microbes cause deterioration and foodborne illnesses, others provide protective and symbiotic benefits, including increased shelf life, flavor, and nutritional value.

The microbiomes of fresh fruits and vegetables can affect human health by interacting with the human gut microbiome. Consuming foods rich in beneficial microorganisms can have probiotic effects, improve gut health and immune function, and potentially lower the risk of chronic diseases. In contrast, the consumption of hazardous microorganisms emphasizes the importance of cleaning, handling, and storing fresh produce. The composition and function of human gut microbiota have a long-term impact on human health. Diet is a significant element that influences the composition of gut microbiota (Johnson et al., 2019).

The *microbiome* is a dynamic group of microbes that colonizes fruits and vegetables, which are naturally occurring plant products. This microbiome is comprised of a diverse group of bacteria, fungi, yeasts, and viruses that live on the plant surface (epiphytes), within its tissues (endophytes), and in the rhizosphere. These bacteria are essential for the health and development of fruits and vegetables because they regulate growth, nutritional intake, and resistance to environmental stressors. Furthermore, the microbiome has a significant impact on food quality, post-harvest preservation, and safety, making it an important area of research in food science, agriculture, and microbiology.

2.1.1 Definition of Microbiome

A *microbiome* is a group of microorganisms that exists in a particular habitat, such as the human body, or in a specific region of the body, such as the digestive tract.

DOI: 10.1201/9781003540113-2

Microbes include bacteria, fungi, and viruses, and their genes. The term "microbiome" first appeared in 1952. Microbial communities are typically characterized as a collection of microorganisms that live together. More specifically, microbial communities are multispecies assemblages in which (micro)organisms interact with one another in a shared environment. The term "microbiome," from the words "micro" and "biome," refers to a "characteristic microbial community" in a "reasonably well-defined habitat with distinct physiochemical properties" as their "theatre of activity." This concept significantly advances the definition of a microbial community by defining a microbial community with different traits and functions, as well as its interactions with the environment, which result in the establishment of discrete ecological niches. However, several additional *microbiome* definitions have been published over the last few decades.

2.1.2 Importance of Microbiome in Fruits and Vegetables

The microbial communities found on fruits and vegetables are crucial for various aspects of agriculture, food quality, and human health. These beneficial microorganisms contribute to plant growth by facilitating nutrient absorption, enhancing disease resistance, and boosting overall plant resilience, which reduces the need for chemical fertilizers and pesticides. Additionally, the microbiome influences produce longevity by inhibiting decay, while certain microbial populations can enhance the taste and nutritional content of fresh produce. Furthermore, the microbiome plays a role in food safety by competing with harmful bacteria and reducing contamination risks. Some microorganisms present on fruits and vegetables may offer health benefits to humans, such as improving gut flora and supporting immune system function. Comprehending and regulating the microbiome are essential for sustainable farming practices and improving the quality and safety of agricultural products. Consequently, the microbial ecosystem of agricultural produce is significant in numerous ways and serves multiple important functions.

2.1.2.1 Promotes Plant Growth and Nutrient Uptake

Plant microbiome, particularly the rhizosphere microbiome, plays an important role in nutrient uptake, which is necessary for fruit and vegetable growth. Microbes such as *Pseudomonas* and *Bacillus* species are known to improve nutrient availability by solubilizing phosphorous and fixing nitrogen, which are critical for plant growth (Lugtenberg & Kamilova, 2009).

2.1.2.2 Enhances Disease Resistance

Various microbial populations create protective barriers on the surface of fruits and vegetables, allowing them to fight harmful diseases. Beneficial microbes compete with pathogens for resources and produce antimicrobial chemicals that hinder the growth of harmful species. This can reduce the demand for chemical pesticides and increase crop resilience (Kumar et al., 2024)

2.1.2.3 Improves Post-Harvest Quality and Shelf Life

Certain microorganisms can prolong the freshness of fruits and vegetables by inhibiting decay. Advantageous microbes, including lactic acid bacteria and yeasts, can effectively compete against decay-causing organisms, thus decelerating the breakdown of fresh produce. Well-managed microbial communities can enhance preservation by creating conditions that are unfavorable for decomposing organisms, such as reducing pH levels or generating substances that inhibit their growth (Agriopoulou et al., 2020).

2.1.2.4 Contribution to Food Safety

While beneficial microorganisms play a crucial role in maintaining the health of fruits and vegetables, harmful pathogens like *Salmonella* and *Listeria* pose a significant risk to human well-being. Studies focused on microbiomes contribute to the development of strategies for minimizing contamination in produce, thus enhancing food safety for consumers (Murray et al., 2017).

2.1.2.5 Contributes to Human Health through Probiotic Effects or Vitamin B12

Fruits and vegetables serve as carriers of beneficial bacteria, which may have probiotic benefits in the human gut. The consumption of fresh fruits with a broad microbiome can promote gut health by introducing beneficial bacteria that improve digestion, immunological function, and the overall microbiome diversity.

A recent metagenomic study of fruits and vegetables and human gut microbiomes (Wicaksono et al., 2023) revealed that fruit- and vegetable-derived microbes are a neglected reservoir of environmental microbiota in the human gut and contribute to gut microbiome diversity. In addition to their beneficial nutrients and metabolites, fruits and vegetables carry genes known to positively affect human health, including vitamin B (B12), K, and short-chain fatty acids. The abundance of fruit and vegetable microbes in the human gut microbiome varies between 0.003% and 3.6% in this study. After weaning, children accounted for an average of 2.2% (0.8. 13.6%) of the overall bacterial diversity in the gut. The connection between the gut and plants is mainly established by members of *the Enterobacterales*, *Burkholderiales*, and *Lactobacillales* (Figure 2.1).

2.1.2.6 Supports Sustainable Agriculture

A well-balanced microbiome minimizes the requirement for artificial inputs like fertilizers and pesticides by improving natural nutrient cycles and boosting plant durability. Utilizing the advantageous functions of the microbiome is crucial for sustainable farming practices, as it mitigates the environmental impact of agricultural activities (Suman et al., 2022).

2.2 MICROBIAL COMMUNITIES IN FRESH PRODUCE

The diversity and abundance of microbes on fresh produce vary depending on product type, environmental conditions, and cultivation practices. Different microbial populations can be found on leafy greens, thick-skinned fruits, and root vegetables due to variations in surface area, soil contact, and handling methods. Studies have shown that certain bacterial families, including *Pseudomonas*, *Enterobacteriaceae*, and *Pantoea*, are commonly present on leafy greens. In contrast, fruits like strawberries and bananas exhibit unique microbial profiles, with *Klebsiella* and *Pantoea*

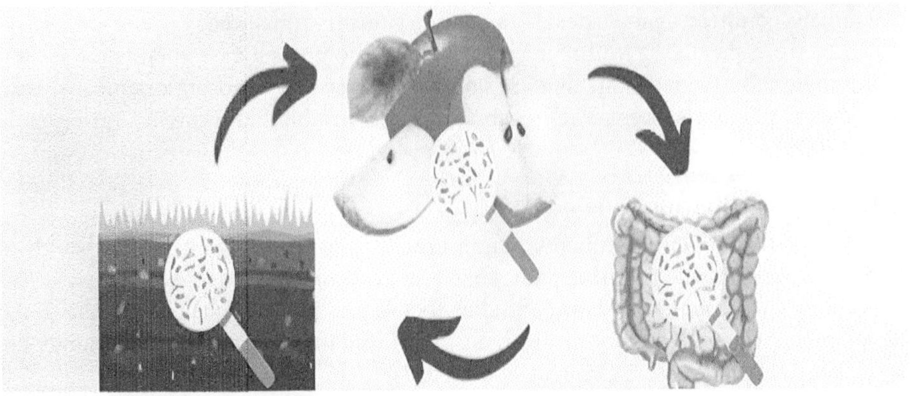

FIGURE 2.1 Microbial connection between the gut and plants.

being more prevalent on berries. The outer leaves of leafy vegetables harbor a greater bacterial diversity compared to inner leaves, increasing contamination risks. Some microorganisms, such as *Salmonella* and *Escherichia*, can be pathogenic.

Fresh produce microbiomes comprise various microbial species, including bacteria, fungi, yeast, and viruses (Martínez & Martínez-Villaluenga, 2021). The most common microbial genera found in fruits and vegetables are:

1. Bacteria: *Pseudomonas*, *Enterobacter*, *Pantoea*, *Bacillus*, *Micrococcus*, and *Lactobacillus* spp. dominate the bacterial microbiome of fresh produce.
2. Fungi: *Penicillium*, *Aspergillus*, *Alternaria*, and *Fusarium* genera are frequently associated with spoilage.
3. Yeasts: *Saccharomyces* and *Candida* can cause fermentation and decay in fruits.

The composition of the microbiome plays a crucial role in both the spoilage and preservation of fresh produce. Spoilage bacteria contribute to food waste, particularly during post-harvest and retail stages. To address spoilage risks, innovative technologies such as smart packaging and microbiome-targeted risk assessments are being explored. These approaches aim to monitor microbial growth, prevent cross-contamination, and extend the shelf life of produce (Karanth et al., 2023).

2.2.1 Categories of Microbiome Fresh Produce

Fresh produce microbiomes can be categorized into two main groups (Fadiji & Babalola, 2020):

- The *phyllosphere microbiome* encompasses bacterial populations on fruits and vegetables. These microorganisms interact with environmental factors, such as air, water, and soil, and are influenced by post-harvest handling and storage practices.

- The *endophytic microbiome* encompasses microbes residing within plant tissues. These endophytes can enhance plant health by facilitating nutrient uptake, secreting growth-promoting hormones, and offering protection against pathogens.

The interplay between phyllosphere and endophytic microbiomes significantly affects the quality and safety of fresh produce. For instance, beneficial endophytes may safeguard plants from diseases, while surface-dwelling spoilage microorganisms can reduce shelf life.

Microorganisms that occur naturally, either as epiphytes on the exterior or endophytes within plant tissues, exhibit significant diversity and encompass a wide range of microbial groups. While some of these microorganisms can be harmful to plants, a larger portion is believed to play a crucial role in the health, quality, and disease resistance of fruits. There remains a gap in understanding regarding the ecology, colonization, survival, and growth of harvested produce.

The surfaces of fruits and vegetables host various microorganisms, including bacteria, yeasts, and filamentous fungi. Certain fresh produce items, such as apples, grapes, lettuce, peaches, peppers, spinach, sprouts, and tomatoes, are known to harbor substantial bacterial populations. The levels of specific bacterial groups can vary depending on the type of produce and cultivar. The composition and abundance of microbial communities found on produce can be influenced by farming practices and storage conditions. Additionally, non-pathogenic microbes present on produce surfaces can interact with and inhibit microbial pathogens.

2.3 FACTORS AFFECTING MICROBIOME COMPOSITION

The microbial composition of fruits and vegetables is constantly changing and is affected by various factors before and after harvesting. Naturally occurring microorganisms contaminate fresh and minimally processed fresh-cut products due to sources such as the agricultural environment, post-harvest handling, and processing activities (Heard, 2002). As a result, fresh-cut produce harbors a wide range of microflora. Harmful pathogens can contaminate produce at multiple stages, including growing, harvesting, processing, and handling. Gram-negative bacteria typically dominate the microbial population in most vegetables, while yeasts and molds primarily constitute the microflora found in raw fruits (Tournas, 2005). The survival or growth of these microorganisms on fresh produce is influenced by factors such as the type of organism, specific crop, and environmental conditions, both in the field and during subsequent storage (figure 2.2). These factors include the following.

2.3.1 Environmental Conditions

As fruits and vegetables are frequently in contact with soil, insects, animals, or humans, they may be contaminated with pathogenic microorganisms at any time from farm to table. Contamination can occur through fecal material, human handling, harvesting equipment, processing, transportation, and distribution (Sothornvit et al., 2009). The protective characteristics of the plant's natural barriers (cell walls

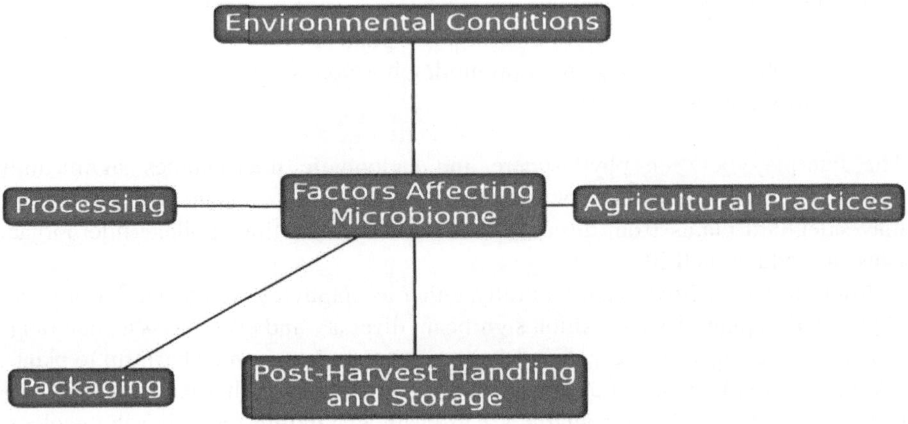

FIGURE 2.2 Different factors affecting microbiome composition.

and outer covering) may prevent the growth of microorganisms on the uninjured outer surfaces of fresh fruits or vegetables. The physical environment surrounding the produce surface in the field is considered inhospitable for the growth and survival of bacteria because of the lack of nutrients and free moisture, temperature and humidity fluctuations, and ultraviolet light (Dickinson, 1986).

However, environmental conditions can significantly influence bacterial populations. The survival and growth of bacterial populations may be promoted by the presence of free moisture on the produce surface, whereas certain conditions, such as sunlight, particularly shorter ultraviolet wavelengths, can damage bacterial cells.

Microbial diversity and composition in crops are influenced by factors such as temperature, humidity, and precipitation patterns. Bacterial and fungal growth tends to thrive in warmer, more humid environments. The microbiome of crops is directly affected by the soil's health and microbial diversity. Compared to conventional farming methods that rely on synthetic inputs, organic farming practices utilizing compost and natural fertilizers often result in a more diverse microbiome. The microbiological contamination of crops is significantly influenced by the origin and quality of irrigation water. Pathogens like *Salmonella* and *Escherichia coli* can be introduced onto fresh produce surfaces through contaminated irrigation water.

2.3.2 Agricultural Practices

Research has shown that the methods used in agriculture substantially influence the microbial composition of fresh produce. Compared to conventional farming, organic practices foster a more varied and resilient microbiome, likely due to decreased pesticide usage and the application of organic fertilizers (Sapkota et al., 2022).

The use of organic fertilizers and the reduced reliance on pesticides in organic farming often lead to greater microbial diversity. In contrast, conventional farming's heavy use of synthetic chemicals may diminish microbial variety and potentially increase the presence of certain harmful organisms. The choice of fertilizer and pesticide can markedly alter fresh produce's microbiome. Organic fertilizers like compost or manure contribute to a more diverse microbial community, while chemical fertilizers tend to suppress this diversity. Broad-spectrum pesticides may unintentionally eliminate beneficial microorganisms while favoring resistant pathogens (Benbrook et al., 2021).

2.3.3 Post-Harvest Management and Storage

Harvest and handling methods. Microbial contamination can occur during harvesting, handling, transport, and storage. Poor hygiene practices during these stages can introduce organisms that cause spoilage and pathogens. Temperature and humidity control during storage is crucial for managing microbial growth. While cold storage can inhibit spoilage organisms, improper temperature regulation can result in microbial proliferation (Gündüz & Gonul, 2017).

2.3.4 Processing

The production of fresh-cut produce involves multiple processing steps, each presenting potential points for microbial contamination. Processing removes the protective epidermal barrier that typically prevents microbial growth on fruit surfaces. Consequently, processing and storage can accelerate decay and quality loss in fresh-cut fruits and vegetables. Processing operations such as cutting, shredding, and slicing damage fruit and vegetable tissues and cellular structures, making fresh-cut produce more vulnerable to microbial attack and causing nutrient and cellular fluid leakage. As a result, minimal processing may increase microbial spoilage of fresh-cut produce by transferring surface microflora to the fruit flesh, which serves as an ideal growth medium (Qadri et al., 2015).

2.3.5 Packaging

The microbial quality of freshly cut products is significantly influenced by packaging. These items are typically stored in modified atmospheric conditions under refrigeration, which can create an environment conducive to the growth of spoilage microorganisms and pathogens, potentially endangering public health. Microbial growth can decrease the economic value of fresh-cut produce by shortening shelf life and increasing the risk of foodborne illnesses (Nguyen-the & Carlin, 2000).

The gaseous composition surrounding fresh-cut produce has a substantial impact on its microbial quality and longevity. Low-oxygen modified atmospheres can inhibit the growth of spoilage microorganisms, thus extending the shelf life of packaged produce. Research on fresh-cut honeydew melon suggested that a gas mixture of 50% O_2 + 50% CO_2 was optimal for prolonging shelf life. Authors investigated how various packaging conditions affected microbial growth and other quality-related factors in fresh-cut pineapples, further emphasizing the importance of packaging in preserving freshness and safety. The respiration rate of the tissue and the internal atmosphere of the package are essential considerations when designing effective packaging for a specific commodity (Conte et al., 2011).

2.4 BENEFICIAL MICROBES AND THEIR ROLES

Beneficial microorganisms in fruits and vegetables are crucial for improving both plant health and human nutrition. These microbes, mainly bacteria and fungi, are part of the natural microbiome and contribute to plant growth by producing phytohormones and aiding nutrient uptake. Microorganisms such as *Pseudomonas* and *Bacillus* spp. also provide biocontrol by producing antimicrobial compounds that shield plants from pathogens, thereby reducing the need for chemical pesticides. Furthermore, certain microbes, like lactic acid bacteria, help prevent spoilage organisms on fruits and vegetables, extend their shelf life, and maintain post-harvest quality. When consumed, the microbial communities on fresh produce can positively affect the human gut microbiome, supporting digestion and immune function and potentially offering protection against diseases. Additionally, these beneficial microbes contribute to sustainable agriculture by decreasing reliance on synthetic fertilizers and pesticides, thus promoting healthier and more sustainable food systems.

2.4.1 Plant Growth Promotion by Microbes

Advantageous microorganisms enhance plant development by improving nutrient accessibility, modulating phytohormones, and boosting plant resilience to stress. Plant growth–promoting microorganisms (PGPM) function as biofertilizers, increasing the availability of macro- and micronutrients. These microbes support plant growth through various mechanisms, including thus:

1. *Cytokinins.* These phytohormones, produced by microorganisms, facilitate cell differentiation, leaf aging, and seed sprouting.
2. *Phosphate-solubilizing microorganisms (PSM).* These microbes transform insoluble phosphorus into a soluble state, thereby supporting plant development.
3. *Mesorhizobium.* These organisms promote plant growth by nitrogen fixation, nutrient provision, and phytohormone synthesis. They also aid in controlling pathogens like fusarium wilt in chickpeas (Khalequzzaman, 2015).
4. *Rhizospheric microorganisms.* These microbes inhabit the rhizosphere, aiding plant growth and development by secreting growth-promoting substances and enhancing nutrient uptake efficiency.
5. *Plant growth–promoting rhizobacteria (PGPRs).* These beneficial microorganisms, found in soil or attached to plant roots, stems, or leaves, help prevent plant diseases, boost crop yields, and foster plant growth.
6. *Soil bacteria.* These microbes assist plants in adapting to poor soil conditions and enhance phytoremediation. They also help reduce Cd translocation from soil to plant shoots and grains.

2.4.2 Resistance to Pathogens and Pests

Microbes are renowned for their roles in pathogen and pest management. Fruit and vegetable microbiomes contribute to pathogen and pest resistance in several ways. Microorganisms such as bacteria, fungi, viruses, and protozoans serve as biocontrol agents against pests. They can eliminate pests, compete with them, or induce diseases. Different microbes target various pests; for instance, some fungi control weeds, while others target specific insects. Rhizosphere microbes can activate the plant's defense mechanisms against pathogens. Microbes residing in floral nectar and insect honeydew generate volatile organic compounds that attract beneficial insects that prey on pests. Modifying DNA methylation can also enhance plant resistance to environmental stress. Small RNA molecules influence plant growth, form, structure, and survival under stressful conditions, like drought, salt stress, insect infestation, and disease (Riseh & Vazvani, 2024).

2.4.3 Role of Microbes in Spoilage

The rapid deterioration of fruits and vegetables is primarily due to microbial spoilage, which significantly limits their shelf life. This process involves the degradation of food caused by the growth and metabolic activities of various microorganisms, including bacteria, yeasts, and molds. These microbes can result in visible decay, unpleasant odors, changes in texture, and a reduction in nutritional content. To develop effective preservation methods, it is essential to comprehend the impact of microbes on fruit and vegetable spoilage. The breakdown of plant cell walls by microorganisms is the main cause of spoilage in fruits and vegetables, leading to softening, discoloration, and the generation of off-flavors (Liao et al., 2020).

Pseudomonas and *Erwinia* species are common spoilage bacteria that degrade pectin, leading to soft rot in vegetables such as carrots, potatoes, and tomatoes. Lactic acid bacteria, such as *Lactobacillus* and *Leuconostoc*, may also contribute to spoilage, especially under anaerobic conditions, producing lactic acid and other organic

TABLE 2.1
Common Spoilage Defects of Fruits and Vegetables

Common Spoilage Defects of Fruits and Vegetables

Defect	Casual Organism	(Dilbaghi & Sharma 2007)
Bacterial soft rot	*Erwinia carotovora*	
Gray mold rot	*Botrytis cinerea*	
Rhizopus soft rot	*Rhizopus nigricans*	
Blue mold rot	*Penicillium* spp.	
Alternaria rot	*Alternaria* spp.	
Pink mold rot	*Trichothecium roseuvnm*	
Green mold rots	*Cladosporium*	
Black rot	*Alternaria*	

acids that cause souring (Tournas, 2023). *Penicillium*, *Aspergillus*, and *Botrytis* species are responsible for post-harvest spoilage in fruits, especially those stored under humid conditions. These fungi produce enzymes that break down the plant cell walls and cause fungal decay. Yeasts, such as *Saccharomyces* and *Candida*, cause fermentation spoilage, leading to the production of alcohol and gas, especially in fruits such as berries and grapes (Zhu et al., 2023).

2.4.4 Biopreservation

Biopreservation uses beneficial microbes to inhibit spoilage and pathogen growth, thereby extending the shelf life of fresh produce. For example, lactic acid bacteria (LAB) produce antimicrobial compounds such as organic acids and bacteriocins, which inhibit spoilage organisms and pathogens (Trias et al., 2008).

Lactic acid bacteria. LAB are commonly used to ferment vegetables such as cabbage (for sauerkraut) and cucumbers (for pickles). Their metabolic products have a lower pH, creating an environment unfavorable for spoilage bacteria and pathogens. Certain yeast species, such as *Candida sake*, are used in post-harvest biocontrol to prevent fungal decay in fruits, such as apples and citrus.

2.5 FOOD SAFETY AND PATHOGEN RISKS

Although fresh fruits and vegetables offer numerous health advantages, they can also serve as carriers for foodborne illnesses. Harmful microbes, including *Salmonella*, *Listeria monocytogenes*, and *E. coli* O157, may contaminate fresh produce during growth, harvesting, and subsequent handling processes. Certain types of produce, such as leafy greens, sprouts, and melons, are particularly susceptible to outbreaks due to their expansive surface areas and close proximity to the ground. Identifying pathogens in fresh produce poses challenges because of the low microbial counts and the intricate nature of microbial communities present. However, advancements in food safety have been achieved through the implementation of current methods, such as rapid detection kits and microbial sampling techniques (Tsakanikas et al., 2018).

2.6 METHODS FOR STUDYING THE MICROBIOME OF FRESH FRUITS AND VEGETABLES

Researchers investigating the microbiome of fresh produce primarily employ DNA sequencing methods, particularly 16S rRNA amplicon sequencing. This technique enables the identification of microbial species through the analysis of a specific gene region. Additionally, scientists utilize DNA extraction procedures, bioinformatics tools, and metagenomic approaches to conduct a comprehensive examination of the microbial community's composition and function on the surface of fruits and vegetables (Shen et al., 2021). The key methods are as follows.

1. *Sampling*. Representative samples were carefully collected from different parts of the fruit or vegetables considering factors such as surface washing and storage conditions.

2. *DNA extraction.* Microbial DNA was isolated from the samples using specialized techniques to ensure high quality and quantity.
3. *PCR amplification using 16S rRNA primers.* Targeting a specific region of the 16S rRNA gene to amplify and sequence microbial DNA allows for species identification.
4. *Next-generation sequencing.* High-throughput sequencing platforms are used to generate large volumes of sequence data from the amplified 16S rRNA fragments.
5. *Bioinformatic analysis.* The sequencing data were processed and analyzed to identify microbial species, calculate relative abundance, and perform phylogenetic analysis.
6. *Harnessing microbiomes for sustainable practices.* Harnessing microbiomes can help with sustainable practices in many ways (Arpan et al., 2022):
   a. *Improving crop productivity.* Microbiome-based biofertilizers and biopesticides can help increase crop yields while reducing the need for chemical inputs.
   b. *Reducing risk and uncertainty.* Crop rotation with a variety of crops can help reduce production risk and uncertainty.
   c. *Improving soil sustainability.* Sustainable agricultural practices such as crop rotation, green manuring, and conservation tillage can help improve soil ecological sustainability.
   d. *Increasing resilience.* Microbiome stewardship can help increase the resilience of wildlife and ecosystems.
   e. *Preventing biodiversity loss.* Microbial stewardship can help reverse the deterioration of biodiversity.
   f. *Treating infections.* Microbiome stewardship can help treat *Clostridium difficile* infections.
   g. *Developing stress-tolerant plants.* Optimizing the plant microbiome through artificial selection can help develop plants that are more tolerant to stress.

2.7 CHALLENGES AND FUTURE PERSPECTIVES

The microbial communities on fruits and vegetables are highly complex and diverse, comprising bacteria, fungi, viruses, and other microorganisms. These communities are influenced by a multitude of factors, including plant type, soil, climate, and agricultural practices. This complexity makes it difficult to identify the key microbial species responsible for beneficial or harmful outcomes. Moreover, many microorganisms are unculturable, making traditional microbial analysis techniques insufficient.

Secondly, there is a lack of standardized protocols for microbiome sampling, DNA extraction, sequencing, and data analyses. This has led to inconsistencies across studies, making it difficult to compare results or draw general conclusions. These inconsistencies are particularly problematic in applied microbiome research, in which reproducibility is crucial for developing scalable solutions.

Microbiomes are highly sensitive to environmental variables, such as temperature, humidity, and exposure to pollutants. Although this sensitivity offers opportunities for manipulating microbial communities, it also poses a challenge for predicting the behavior of microbiomes under varying conditions. Climate change is expected to alter microbial ecosystems, but its exact impact remains poorly understood.

While much progress has been made in cataloging microbial species, less is known about how specific microbes interact with plants. Understanding these interactions is critical for designing microbiome-based solutions, such as biofertilizers and microbial treatments, to improve plant health and yield. Current tools, such as metagenomics, provide only a partial picture, and integrating metabolomics and transcriptomics to better understand these interactions is a significant challenge (Meyer et al., 2023). The use of microbiome-based products, such as microbial treatments or biopesticides, faces regulatory hurdles. In many countries, the regulatory frameworks for these products are still in their infancy, and there is often a lack of clarity regarding safety assessments. Furthermore, consumer acceptance of foods produced using microbial interventions remains a challenge, particularly when genetically engineered microbes are involved (Ricroch et al., 2023).

2.7.1 Future Directions in Microbiome Research

The microbiomes of fruits and vegetables play a crucial role in food quality, safety, and human health. The future directions of microbiome research on fruits and vegetables are highly promising, particularly in the fields of health, sustainability, and disease management. As research in this area advances, several promising directions are emerging, driven by new technologies and a better understanding of microbial interactions in food ecosystems. The key future directions for microbiome research on fruits and vegetables are as follows:

1. *Metagenomics and microbiome analysis.* Advances in sequencing technologies, such as metagenomics and high-throughput sequencing, have improved our understanding of the fruit and vegetable microbiomes. These technologies allow for comprehensive profiling of microbial communities and can be used to track changes in microbial composition throughout the food supply chain. Metagenomic analysis has been used to identify microbial markers for food quality and safety, develop new biocontrol strategies, and understand the interactions between plants and their microbiomes (Gilbert et al., 2012).
2. *Microbiome engineering.* Researchers are exploring microbiome engineering to enhance the beneficial functions of the microbial communities associated with fresh produce production. By introducing or enhancing specific microbial species, it may be possible to improve plant health, extend shelf life, and reduce contamination risks. Engineered microbial consortia can be designed to inhibit pathogens, improve nutrient uptake, and enhance the flavor and nutritional content of fresh produce.
3. *Impact of agricultural practices on microbiomes.* Agricultural practices significantly affect crop microbiomes. There is an increasing interest in understanding how organic farming, no-till agriculture, and reduced pesticide use influence microbial diversity and community composition (Carrion

et al., 2023). Research has shown that organic farming practices promote a richer and more beneficial microbial community, which could lead to more resilient and nutrient-rich crops.

4. *Personalized nutrition and the microbiome.* Personalized nutrition, in which diets are tailored based on an individual's microbiome, is an emerging area of human health research. Future studies will likely focus on how the microbiomes of fruits and vegetables interact with the human gut microbiome and influence health outcomes, such as metabolism, immunity, and disease risk. This could lead to recommendations for specific types of fresh produce to improve health (De Filippis et al., 2022).
5. *Artificial intelligence and microbiome data analysis.* With the increase in microbiome data generated by modern sequencing technologies, artificial intelligence (AI) and machine learning are being used to identify patterns and predict microbial functions (Verma et al., 2023). These tools help manage the vast complexity of microbiome datasets and can assist in developing predictive models for microbial interactions and their effects on fruit and vegetable quality.

2.8 CONCLUSION

Fresh fruits and vegetables harbor microbiomes that are crucial for plant wellness, food quality, safety, and sustainability. Gaining knowledge about these microbial communities' composition and function offers crucial insights into fostering beneficial interactions and reducing pathogenic threats. By harnessing progress in microbiome studies and incorporating microbial management approaches, it becomes feasible to boost crop productivity, decrease chemical usage, and prolong produce shelf life, thus contributing to more robust and sustainable food production systems. Upcoming research should concentrate on investigating intricate interactions between microbiomes and their host environments, creating targeted microbial solutions for agriculture, and applying microbiome-informed practices to safeguard food security and safety. As this field progresses, the opportunity to utilize microbial communities for improving food systems is immense, making it a critical area of interest for sustainable agricultural methods and post-harvest management.

REFERENCES

Agriopoulou, S., Stamatelopoulou, E., Sachadyn-Król, M., & Varzakas, T. 2020. Lactic acid bacteria as antibacterial agents to extend the shelf life of fresh and minimally processed fruits and vegetables: Quality and safety aspects. *Microorganisms*, 8(6), 952. https://doi.org/10.3390/microorganisms8060952.

Benbrook, C., Kegley, S., & Baker, B. 2021. Organic farming lessens reliance on pesticides and promotes public health by lowering dietary risks. *Agronomy*, 11(7), 1266. https://doi.org/10.3390/agronomy11071266.

Carrion, V. J., Perez-Jaramillo, J. E., & Mendes, R. 2023. Sustainable agriculture: How plant-microbiome interactions contribute to ecosystem health. *Annual Review of Microbiology*, 77, 45–63.

Conte, A., Scrocco, C., Brescia, I., Mastromatteo, M., & DelNobile, M. A. 2011. Shelf life of fresh-cut Cime di rapa (Brassica rapa L.) as affected by packaging. *Food Science and Technology*, 44, 1218–1225.

De Filippis, F., Valentino, V., & Ercolini, D. 2022. The interplay between human and food microbiomes in health and disease. *Nature Reviews Microbiology*, 20(10), 712–729.

Dickinson, C. H. 1986. Adaptations of micro-organisms to climatic conditions affecting aerial plant surfaces. In *Microbiology of the phyllosphere*. Cambridge University Press.

Dilbaghi, N., & Sharma, S. 2007. Food spoilage, food infections and intoxications caused by microorganisms and methods for their detection. *Microbiology*. https://doi.org/10.12983/ijsras-2015-p0144-0150

Fadiji, A. E., & Babalola, O. O. 2020. Elucidating mechanisms of endophytes used in plant protection and other bioactivities with multifunctional prospects. *Frontiers in Bioengineering and Biotechnology*, 8.

Gilbert, J. A., et al. 2012. Defining seasonal marine microbial community dynamics. *The ISME Journal*, 6(2), 298–308.

Gündüz, G. T., & Gonul, S. A. 2017. The impact of post-harvest practices on microbial contamination in fresh produce. *Postharvest Biology and Technology*, 133, 48–52.

Heard, G. M. 2002. Microbiology of fresh-cut produce. In A. Lamikanra (Ed.), *Fresh-cut fruits and vegetables: Science, technology and market*. CRC Press, 188–236.

Johnson, A. J., Vangay, P., Al-Ghalith, G. A., Hillmann, B. M., Ward, T. L., Kim, A. D., Shmagel, A. K., Syed, A. N., & Walter, J. 2019. Daily sampling reveals personalized diet-microbiome associations in humans. *Cell Host & Microbe*, 25(6), 789–802.

Karanth, S., Feng, S., Patra, D., & Pradhan, A. 2023, Linking microbial contamination to food spoilage and food waste: The role of smart packaging, spoilage risk assessments, and date labelling. *Frontiers in Microbiology*, 14, 1198124. https://doi.org/10.3389/fmicb.2023.1198124.

Khalequzzaman, K. M. 2015. Seed treatment with rhizobium biofertilizer for controlling foot and root rot of chickpea. *International Journal of Scientific Research in Agricultural Sciences*, 2(6), 144–150.

Kumar, R., Kambhoj, N., & Mitra, D. 2024. Microbial innovations in agriculture: Interdisciplinary approaches to leveraging microbes for food sustainability and security. *Indian Journal of Microbiology Research*, 11(3), 129–139.

Liao, H., et al. 2020. Predictive models of postharvest soft rot caused by *Pectobacterium* species in different storage conditions. *Postharvest Biology and Technology*, 160, 111046.

Lugtenberg, B., & Kamilova, F. 2009. Plant-growth-promoting rhizobacteria. *Annual Review of Microbiology*, 63, 541–556.

Martínez, A., & Martínez-Villaluenga, C. 2021. The microbiome of fresh produce: Examining the role of environmental and handling factors. *Environmental Microbiome*, 6(7), 45–60.

Meyer, F. E., Laing, M. D., & van der Waals, J. E. 2023. The plant microbiome: Future challenges and perspectives for sustainable agriculture. *Applied Microbiology and Biotechnology*, 107(1), 45–61.

Murray, K., Wu, F., Shi, J., Jun Xue, V., & Warriner, K. 2017. Challenges in the microbiological food safety of fresh produce: Limitations of post-harvest washing and the need for alternative interventions, *Food Quality and Safety*, 1(4), 289–301. https://doi.org/10.1093/fqsafe/fyx027.

Nguyenthe, C., & Carlin, F. 2000. Fresh and processed vegetables. *The Microbiological Safety and Quality of Food*, 620–684.

Qadri, O. S., Yousuf, B., Srivastava, A. K., & Yildiz, F. 2015. Fresh-cut fruits and vegetables: Critical factors influencing microbiology and novel approaches to prevent microbial risks—A review. *Cogent Food & Agriculture*, 1(1).

Ricroch, A., Ammann, K., & Kuntz, M. 2023. Regulatory and societal challenges for microbiome-based innovations in agriculture and food production. *Trends in Biotechnology*, 41(6), 743–756.

Riseh, R. S., & Vazvani, M. G. 2024. Review unveiling methods to stimulate plant resistance against pathogens. *Frontiers in Bioscience*, 29(5), 188.

Sapkota, R., et al. 2022. Organic farming improves the plant-associated microbiome by increasing beneficial microbial species. *Frontiers in Microbiology*, 13, 889004.

Shen, Y., Nie, J., Kuang, L., Zhang, J., & Li, H. 2021. DNA sequencing, genomes and genetic markers of microbes on fruits and vegetables. *Microbial Biotechnology*, 14(2), 323–362.

Sothornvit, R., & Kiatchanapaibul, P. 2009. Quality and shelf-life of washed fresh-cut asparagus in modified atmosphere packaging. *Food Science and Technology*, 42, 1484–1490.

Suman, J., Rakshit, A., Devika Ogireddy, A., Singh, S., Gupta, C., & Chandrakala, V. 2022. Microbiome as a key player in sustainable agriculture and human health. *Frontiers in Soil Science. Soil Biogeochemistry & Nutrient Cycling*, 2. https://doi.org/10.3389/fsoil.2022.821589.

Tournas, V. H. 2005. Moulds and yeasts in fresh and minimally processed vegetables, and sprouts. *International Journal of Food Microbiology*, 99, 71–77.

Tournas, V. H. 2023. The role of bacterial spoilage in fresh and minimally processed fruits and vegetables. *Food Microbiology*, 109, 104178.

Trias, R., et al. 2008. Bioprotection of Golden Delicious apples and Iceberg lettuce against foodborne pathogens by lactic acid bacteria. *International Journal of Food Microbiology*, 123(1–2), 50–60.

Tsakanikas, P., Fengou, L. C., Manthou, E., Lianou, A., Panagou, E. Z., & Nychas, G-J. E. 2018. A unified spectra analysis workflow for the assessment of microbial contamination of ready to eat green salads: Comparative study and application of non-invasive sensors. *Computers and Electronics in Agriculture*, 155.

Verma, S., Patel, A., & Sharma, P. 2023. AI-driven insights into plant-microbiome interactions: Harnessing big data for agricultural innovation. *Computational and Structural Biotechnology Journal*, 21, 2109–2119.

Wicaksono, W. A., Cernava, T., Wassermann, B., Abdelfattah, A., Soto-Giron, M. J., Toledo, G. V., Virtanen, S. M., Knip, M., Hyöty, H., & Ber, G. 2023. The edible plant microbiome: Evidence for the occurrence of fruit and vegetable bacteria in the human gut. *Gut Microbes,* 15(2), 2258565

Zhang, J., Cook, J., Nearing, J. T., Zhang, J., Raudonis, R., Glick, B. R., Langille, M. G. I., & Cheng, Z. 2021. Harnessing the plant microbiome to promote the growth of agricultural crops. *Microbiology Research*, 126690.

Zhu, S., Zhang, L., & Yu, D. 2023. Post-harvest fungal spoilage of fresh produce: Mechanisms and control strategies. *Trends in Food Science & Technology*, 134, 125–137.

3 An Overview of Microflora in Dairy and Meat Products

Sucheta Mandal

3.1 INTRODUCTION

Microorganisms, as an essential part of our environment, play a vital role in numerous ecological processes. In the context of food, especially in dairy and meat products, microorganisms are not just inhabitants but are active participants in shaping the characteristics of these foods. Dairy products are actually an assemblage of various bacterial and fungal genera that increase the nutritional profile, aroma, flavor, and shelf life of food. Milk, which is considered a complete food, can cure a variety of diseases for its complex microbial community (Shori, 2012). Most common dairy products are associated with microflora belonging to the genera *Enterococcus*, *Streptococcus*, *Pediococcus*, *Lactobacillus*, *Bifidobacterium*, *Leuconostoc*, *Propionibacterium*, and *Saccharomyces* (Abushelaibi et al., 2017; Mandal & Mandal, 2023). The worth of dairy food is largely affected by several factors, such as the viable microbiota count in fresh milk; the breeding environment, nutritional conditions, breed, and maturity of the animal; the phase of lactation; and milking performance (Khaskheli et al., 2005). The past two decades, the global use of probiotics as functional ingredients has surged across the food, feed, and pharmaceutical industries. The dairy sector, in particular, has emerged as the leading user of probiotics in a wide range of foodstuffs, like fermented milk, curd, cheese, butter, cream, and infant formula. Probiotics are employed in dairy products either as independent starter cultures, alongside traditional starters, or by being added post-fermentation. The majority of probiotic organisms with GRAS (generally recognized as safe) status belong to the lactic acid bacteria (LAB) and *Bifidobacteria* species. Simultaneously, meat and meat-based products, which form a significant portion of human diets for its protein with high digestibility score, considered essential for human growth and development, harbor a diverse range of microbial species. These microbes enhance the nutritional value of meat by secreting bioactive compounds, including essential amino acids, fatty acids, and vitamins, some of which also possess antioxidant properties. To meet the demand of meat, fermented meat products with longer shelf life and increased flavor under specific temperature and humidity came to market. *Staphylococcus carnosus* and *S. xylosus* were reported for increasing the texture and flavor of meat by hydrolyzing fat and protein (Zhou et al., 2018). But the high nutritional value of meat makes it extremely

 DOI: 10.1201/9781003540113-3

susceptible to spoilage and degradation. Meat degradation is a complex process driven by physical, chemical, and biological activities, all of which contribute to the deterioration of the product (Casaburi et al., 2015). Bacteria primarily present in meat slow down the growth of foodborne pathogen through bacterial competition and acid production. Papamanoli et al. (2002) concluded that *Staphylococcus xylosus* inhibits the growth of spoilage bacteria *Listeria monocytogenes* in sausages. Additionally, *Lactobacillus sakei* subsp. *sakei*, starter culture of Sichuan sausage, controls the growth of *E. coli* and Enterobacteriaceae by producing various organic acids. Several studies document that the presence of LAB in meat or dairy products reduces the probability of spoilage by secreting bioactive compounds. The present study encompasses the diversity of microbiome related with different dairy and meat products from raw to storage.

3.2 MICROORGANISMS IN DAIRY PRODUCTS

Microorganisms present in milk vary due to different sources. Bovine and non-bovine milk, in both raw and fermented forms, harbor a diverse range of bacterial and fungal strains. Milk microflora has been extensively studied on cows (Addis et al., 2016; Falentin et al., 2016), buffalo, mice, goats, sheep, camels, reindeer, donkeys, deer (Catozzi et al., 2017; Castro et al., 2020), and humans (Fitzstevens et al., 2017; Corona-Cervantes et al., 2020). The microbial community of different ruminants also differs with the same species, suggesting host microbial adaptation in different environments. A comparative study between bovine and human milk reveals distinct metataxonomic profiles, with both sharing common bacterial genera, such as *Bifidobacterium*, *Corynebacterium*, *Pseudomonas*, *Staphylococcus*, *Streptococcus*, *Propionibacterium*, *Enterococcus*, and *Bacteroides* (Oikonomou et al., 2014).

A culture-independent study revealed that cow milk microbiota is highly complex and diverse, comprising 148 bacterial strains from genera such as *Bifidobacterium*, *Staphylococcus*, *Corynebacterium*, *Enterococcus*, *Streptococcus*, *Propionibacterium*, *Bacteroides*, and *Pseudomonas* (Addis et al. 2016). The shotgun metagenomic approach added some more names to the list of fungal and protozoal strains, like *Malassezia*, *Debaryomyces*, *Candida*, *Cladosporium*, *Rhodotorula*, *Chrysosporium*, *Fusarium*, *Engyodontium*, *Pichia*, *Aspergillus*, *Penicillium*, and *Torrubiella* (Delavenne et al., 2011). The microbial communities and diversity in buffalo milk are characterized by the presence of key taxa, such *as Propionibacterium*, *Staphylococcus*, *Micrococcus*, *Aerococcus*, *Trichococcus*, *Turicibacter*, *Facklamia*, *Clostridium*, *Psychrobacter*, *Acinetobacter*, and *Pseudomonas* (Catozzi et al., 2017). Conversely, the dominant taxa in goat and sheep milk include *Bacillus*, *Leuconostoc (Le)*, *Lactobacillus (Lb)*, *Lactococcus (Lc)*, *Pediococcus*, *Streptococcus*, and *Enterococcus*. *E. faecium*, *E. durans*, *E. hirae*, *E. avium*, *E. casseliflavus*, *L. acidophilus*, *L. lactis* subsp. *lactis*, *L. delbrueckii* subsp. *bulgaricus*, *L. plantarum*, *L. reutei*, *L. fermentum*, *L. casei*, *L. bulgaricus*, *L. brevis*, *L. paracasei*, *L. curvatus*, *L. mesenteroides* subsp. *mesenteroides*, *L. plantarum*, *L. lactis* subsp. *lactis*, *L. raffinolactis*, *P. pentosaceus*, *S. salivarius* subsp. *thermophilus* (Mittu & Girdhar, 2015; Pisano et al., 2019; Patil et al., 2019). Additionally to these genera, camel milk also contains the bacterial genus *Weissella* (Edalati et al., 2019).

The microbial composition of raw milk is influenced by several factors, including the health of the lactating animal, milking practices, and conditions during transportation and storage (Kable et al., 2019). The immediate storage of milk in cold condition helps maintain the native microbial load. Li et al. (2016) observed in his study that *Proteobacteria*, a predominant group of fresh buffalo milk, tends to decrease significantly after 24 hours of cold storage. Similarly, the abundance of *Lactococcus* and *Streptococcus* population increases significantly after refrigeration. To reduce the risk of pathogenic contamination, pasteurization is done, where the genera *Pseudomonas* and *Acinetobacter* are eliminated. The bacterial composition of pasteurized milk remains largely stable over 7 days of storage period, however with an increase in *Lactococcus* population and decrease in *Streptococcus*. At the phylum level, *Firmicutes* and *Proteobacteria* account for over 90% of the microbial community after 7 days of storage (Li et al., 2016). *Paenibacillus*, a dominant taxon in pasteurized milk, begins to decline after 14 days of storage. Apart from raw and pasteurized milk, fermented milk is the richest source of probiotic microorganisms. Fermentation boosts nutrient bioavailability by encouraging the growth of functionally active microbial populations, while also breaking down toxic components, thus improving the safety and extending the biopreservation of the final product (Sun et al. 2020). Different LAB strains associated with fermentation are reported for their antimicrobial and antitoxigenic effect (Mandal et al., 2024; Mandal and Dey Bhowmick et al., 2024b). Gao et al. (2017) identified *Lactococcus* as the most predominant genus and *Lactobacillus* as a subdominant genus in milk samples collected throughout the year. Other genera present in varying proportions include *Leuconostoc*, *Streptococcus*, *Enterococcus*, *Weissella*, *Macrococcus*, *Acetobacter*, *Pseudoclavibacter*, *Dysgonomonas*, and *Corynebacterium*. Additionally, fungal genera such as *Geotrichum*, *Kluyveromyces*, and *Pichia* are consistently found in milk year-round (Gao et al., 2017).

3.3 MICROFLORA OF MEAT AND MEAT PRODUCTS

Meat, which is the muscle tissue of a living animal, is inherently sterile. The microbial load and composition of fresh meat are largely determined by the animal's physiological condition at the time of slaughter, microbial contamination during the slaughter process, and the environmental conditions within the slaughterhouse (Erkmen and Bozoglu, 2016). The primary infection during slaughter occurs when the carcass is cut and the offal is detached. Bacteria present in the gut system, lymph, skin, nodes, handlers, and cutting knives are the potential sources of meat contamination. In their study, Mills et al. (2018) reported that *Carnobacterium* spp. found on lamb skeleton are linked to contamination from the meat processing atmosphere. *Staphylococcus* spp., LAB, *Brocothrix* spp., *Streptococcus* spp., *Psychrobacter* spp., and *Acinetobacter* spp. are dominant microbiota at the slaughterhouse. Proteobacteria, notably *Pseudomonas* spp. and *Enterobacteriaceae*, have emerged as the dominant microflora on carcass surfaces (Stellato et al., 2016). Microflora development on meat leads its spoilage. To raise the shelf life and meet the demand of consumers, fermented meat products came to market. Fermented products like ham and sausage add special flavor in meat. Zhou et al. (2018) reported that enzymes such as esterase, protease, and catalase, secreted by *Staphylococcus carnosus* and *S. xylosus*, alter the proteins and fats in raw meat through biochemical changes. These changes lead to the production of small molecules, such as amino acids, peptides, and short-chain volatile fatty acids, improving the product's texture and flavor. Additionally, mixed starter cultures used in the fermentation of meat help hydrolyze fat and inhibit lipid peroxidation, thus improving the flavor of Harbin dried sausage.

Ham, a non-ready-to-eat fermented meat product, is produced using different bacterial and fungal strains (Table 3.1). Jinhua dry-cured ham is a product of LAB and *Staphylococcus* sp. (Wang et al., 2021a). Dominant microbial strains in Xuanwei ham are *Micrococcus* sp., *Staphylococcus* sp., and *Aspergillus* sp. (Zou et al., 2020). The involvement of different bacterial strains provides separate aroma and flavor to the product.

Sausage is also a non-ready-to-eat meat product from poultry meat following the procedure of mincing, stirring, curing, filling, drying (in shade), and smoking. Sichuan sausages are dominated by bacterial genera *Firmicutes*, *Proteobacteria*, *Bacteroidetes*, and *Actinobacteria* (Table 3.1). Along with bacteria, yeast species like *Debaryomyces* and *Saccharomyces* are also involved during fermentation (Wang et al., 2019). *Staphylococcus epidermidis*, *S. xylosus*, *S. saprophyticus*, *S. cohnii*, and *Bacillus* spp. are responsible for the fermentation of Cantonese sausage (Mao et al., 2024). Similarly, Hubei Enshi sausage contains *Firmicutes*, *Proteobacteria*, *Cyanobacteria*, *Bacteroidetes*, and *Actinobacteria*, with *Firmicutes* and *Proteobacteria* being the dominant phyla. Preserved meat is a non-ready-to-eat product crafted from fresh livestock, primarily relying on raw ingredients. Following some traditional ways to preserve, the meat is quite popularized in China, where *Firmicutes*, Proteobacteria, *Bacteroidetes*, *Actinobacteria*, and *Cyanobacteria* are used. Sichuan traditionally preserved meat contains a diverse microbial community, including fungi from the phyla Ascomycota, Basidiomycota, Glomeromycota, and Rozellomycota, along with various bacteria (Wang et al., 2021a). It is also reported that Hubei Enshi preserved meat contains the presence of *Staphylococcus*, *Micrococcus*, and *Lactobacillus* (Wang et al., 2021b).

In some places of China, pressed duck meat is sold in the market. Jianchang pressed meat primarily includes LAB (*L. plantarum*, *L. acidophilus*, *L. lactis*, *L. casei*, *L. curvatus*, and *L. delbrueckii*), *Micrococcus* (*M. auterisiae*, *M. varians*, *M. epidermidis*, *M. luteus*, *M. candidus*, and *M. roseus*), *Staphylococcus*, yeast, and molds (Lin, 2017).

Apart from meat, fish are also fermented and preserved after scales are removed, followed by drying. *Proteobacteria*, *Firmicutes*, *Bacteroidetes*, and *Actinobacteria* are involved in such preservation (Wang et al., 2021b). Mandal and Mandal (2023) reported about the involvement of LAB strain in preservation.

3.4 MICROFLORA RELATED WITH SPOILAGE OF DAIRY AND MEAT PRODUCTS

The mammary gland of lactating animals is another reservoir of bacteria. Some of them are harmful and responsible for spoilage or sometimes act as foodborne pathogen. The consumption of raw milk or milk products causes health issues as they are contaminated with *Brucella* spp., *Campylobacter jejuni*, *Coxiella burnetii*, *Listeria*

TABLE 3.1
List of Microorganism Related with Meat Products

Type of Meat		Microorganisms Involved	References
Ham	Jinhua ham	*L. pentosus, S. xylosus, P. urinaeequi, P. pentosaceus, S. equus*	He et al., 2007
	Panxian ham	*Psychrobacter, Acinetobacter, Candida, Ochrobactrum, Staphylococcus, M. candidus*	Mu et al., 2019
	Nuodeng ham	*Psychrobacter, Aeromonas, Pseudomonas, Acinetobacter*	Wang et al., 2021a
	Canadian bacon	*Carnobacterium maltaromaticum, Serratia liquefaciens, Leuconostoc mesenteroides,*	Li et al., 2021
	Country ham	*Bacillus, Lactococcus, Micrococcus, Staphylococcus, Streptococcus*	Draughon et al., 1981
Sausage	Bratwurst	*L. curvatus, L. sakei, L. plantarum*	Katsaros and Taoukis, 2021
	Chorizo	*Proteobacteria, Pseudomonas, Streptococcus, Acinetobacter, Brochothrix, Firmicutes*	Juárez-Castelán et al., 2019
	Sichuan-style sausage	*Brochothrix, Lactobacillus, Pediococcus, Staphylococcus, Porphyra, Saccharomyces*	Wang et al., 2021b
	Enshi sausage	*Staphylococcus, Photobacterium, Pseudomonas, Brochothrix, Acinetobacter*	Wang et al., 2021b
Preserved meat	Sichuan	*Staphylococcus, Acinetobacter, Pseudomonas, Brochothrix, Aspergillus, Cupriavidus, Citrobacter, Debaryomyces, Enterobacter, Candida, Penicillium*	Wen et al., 2020
	Hunan	*Staphylococcus, Enterobacter, Leuconostoc, Sphingomonas, Pseudomonas*	Yi et al., 2017
	Jianchang pressed salted duck	*S. simulans, S. xylosus, S. carnosus, L. plantarum, L. acidophilus, L. lactis, L. sakei, L. delbrueckii, L. fermentum, L. bifermentans, L. buchneri, L. helveticus, S. lactis, S. acidilactis, P. acidilactici, P. cerevisiae, S. diacetilactis, M. luteus, P. pentosaceus, M. epidermidis, M. auterisiae, M. varians, M. roseus*	Lin, 2017
	Sun-dried fish	*Fusarium* sp., *Penicillium* sp., *Bacillus* sp., *Klebsiella* sp., *Staphylococcus* sp., *Pseudomonas* sp.	Fitri et al., 2022

Source: Wang et al. (2021b).

monocytogenes, *Salmonella enterica*, *Yersinia enterocolitica*, *Mycobacterium bovis*, *M. paratuberculosis*, and *E. coli* O157:H7. Pasteurization can eliminate these pathogens, but if the contamination occurs after pasteurization, then it will be outburst.

Mutaku et al. (2005) reported about the survivability of *E. coli* strain even at 3°C. *C. jejuni* causes gastroenteritis in the intestinal tract (Facciolà et al., 2017). Similar to dairy products, meat is also spoiled in thew presence of *Flavobacterium* spp., *Aeromonas* spp., *Alcaligenes* spp., *Staphylococcus* spp., *Micrococcus* spp., *Serratia spp.*, *Clostridium* spp., and *Bacillus* spp. (Hassanien, 2004). Not only spoilage, but some of these bacteria, *viz. Bacillus cereus*, *Salmonella* spp., *Campylobacter* spp., *Clostridium perfringens*, *Listeria monocytogenes*, *Yersinia enterocolitica*, *Aeromonas*, *Staphylococcus aureus*, *Enterobacter*, and *Shigella*, also act as foodborne pathogen. Meat contaminated with any species of *Clostridium* is difficult to eliminate. Spores of *Clostridium* after germination produce neurotoxins (Hassanien, 2004). Ready-to-eat products are generally contaminated with *L. monocytogenes*, but the pathogen is heat-sensitive (Hassanien, 2004). Poultry meats are considered a major source of *Salmonella* contamination. Generally, these contaminations occur in the slaughterhouse (Peton & Loir, 2014). The wide range of contamination significantly affects health and welfare, leading to substantial economic losses each year.

3.5 CONCLUSION

Dairy and meat products both harbor a wide range of microorganisms, including bacteria, yeast, molds, parasites, etc. These microfloras show good as well as bad effects. In some cases, they add bioactive compounds, flavor, and aroma to food, but sometimes they act as pathogens. The management of these food sectors has to be made more carefully to minimize contamination. Decontamination strategies have to be implied to reduce the negative effects of microflora.

REFERENCES

Abushelaibi, A., Al-Mahadin, S., El-Tarabily, K., Shah, N. P., & Ayyash, M. (2017). Characterization of potential probiotic lactic acid bacteria isolated from camel milk. *LWT-food Science and Technology*, *79*, 316–325.

Addis, M. F., Tedde, V., Puggioni, G. M. G., Pisanu, S., Casula, A., Locatelli, C., . . . Uzzau, S. (2016). Evaluation of milk cathelicidin for detection of bovine mastitis. *Journal of Dairy Science*, *99*(10), 8250–8258.

Casaburi, A., Piombino, P., Nychas, G. J., Villani, F., & Ercolini, D. (2015). Bacterial populations and the volatilome associated to meat spoilage. *Food Microbiology*, *45*, 83–102.

Castro, I., Alba, C., Fernández, L., García, A. J., & Rodríguez, J. M. (2020). Culture-dependent and metataxonomic analysis of milk from red deer (Cervus elaphus). *International Dairy Journal*, *102*, 104610.

Catozzi, C., Sanchez Bonastre, A., Francino, O., Lecchi, C., De Carlo, E., Vecchio, D., . . . Ceciliani, F. (2017). The microbiota of water buffalo milk during mastitis. *PLoS One*, *12*(9), e0184710.

Corona-Cervantes, K., García-González, I., Villalobos-Flores, L. E., Hernández-Quiroz, F., Piña-Escobedo, A., Hoyo-Vadillo, C., . . . García-Mena, J. (2020). Human milk microbiota associated with early colonization of the neonatal gut in Mexican newborns. *PeerJ*, *8*, e9205.

Delavenne, E., Mounier, J., Asmani, K., Jany, J. L., Barbier, G., & Le Blay, G. (2011). Fungal diversity in cow, goat and ewe milk. *International Journal of Food Microbiology*, *151*(2), 247–251.

Draughon, F. A., Melton, C. C., & Maxedon, D. (1981). Microbial profiles of country-cured hams aged in stockinettes, barrier bags, and paraffin wax. *Applied and Environmental Microbiology*, *41*(4), 1078–1080.

Edalati, E., Saneei, B., Alizadeh, M., Hosseini, S. S., Bialvaei, A. Z., & Taheri, K. (2019). Isolation of probiotic bacteria from raw camel's milk and their antagonistic effects on two bacteria causing food poisoning. *New Microbes and New Infections*, *27*, 64–68.

Erkmen, O., & Bozoglu, T. F. (2016). Spoilage of meat and meat products. In *Food microbiology: Principles into practice*. Chichester: John Wiley & Sons, 279–295.

Facciolà, A., Riso, R., Avventuroso, E., Visalli, G., Delia, S. A., & Lagana, P. (2017). Campylobacter: From microbiology to prevention. *Journal of Preventive Medicine and Hygiene*, *58*(2), E79.

Falentin, H., Rault, L., Nicolas, A., Bouchard, D. S., Lassalas, J., Lamberton, P., . . . Even, S. (2016). Bovine teat microbiome analysis revealed reduced alpha diversity and significant changes in taxonomic profiles in quarters with a history of mastitis. *Frontiers in Microbiology*, *7*, 480.

Fitri, N., Chan, S. X. Y., Che Lah, N. H., Jam, F. A., Misnan, N. M., Kamal, N., . . . Abas, F. (2022). A comprehensive review on the processing of dried fish and the associated chemical and nutritional changes. *Foods*, *11*(19), 2938.

Fitzstevens, J. L., Smith, K. C., Hagadorn, J. I., Caimano, M. J., Matson, A. P., & Brownell, E. A. (2017). Systematic review of the human milk microbiota. *Nutrition in Clinical Practice*, *32*(3), 354–364.

Gao, M. L., Hou, H. M., Teng, X. X., Zhu, Y. L., Hao, H. S., & Zhang, G. L. (2017). Microbial diversity in raw milk and traditional fermented dairy products (Hurood cheese and Jueke) from Inner Mongolia, China. *Genetics and Molecular Research Journal*, *16*(1), 1–13.

Hassanien, F. S. (2004). Bacterial hazards associated with consumption of some meat products. *Benha Veterinary Medical Journal*, *15*(2), 41–54.

He, Z. F., Li, H. Y., Yu, X., Zhou, R. H., & Zhang, M. (2007). Study on the changes in microbial flora of jinhua ham in modern fermentation technology. *Journal of Southwest University*, *29*, 142–150. https://doi.org/10.3969/j.issn.1673-9868.2007.03.030

Juárez-Castelán, C., García-Cano, I., Escobar-Zepeda, A., Azaola-Espinosa, A., Álvarez-Cisneros, Y., & Ponce-Alquicira, E. (2019). Evaluation of the bacterial diversity of Spanish-type chorizo during the ripening process using high-throughput sequencing and physicochemical characterization. *Meat Science*, *150*, 7–13.

Kable, M. E., Srisengfa, Y., Xue, Z., Coates, L. C., & Marco, M. L. (2019). Viable and total bacterial populations undergo equipment-and time-dependent shifts during milk processing. *Applied and Environmental Microbiology*, *85*(13), e00270–19.

Katsaros, G., & Taoukis, P. (2021). Microbial control by high pressure processing for shelf-life extension of packed meat products in the cold chain: Modeling and case studies. *Applied Sciences*, *11*(3), 1317.

Khaskheli, M., Arain, M. A., Chaudhry, S., Soomro, A. H., & Qureshi, T. A. (2005). Physico-chemical quality of camel milk. *Journal of Agriculture and Social Sciences*, *2*, 164–166.

Li, L., Renye Jr, J. A., Feng, L., Zeng, Q., Tang, Y., Huang, L., . . . Yang, P. (2016). Characterization of the indigenous microflora in raw and pasteurized buffalo milk during storage at refrigeration temperature by high-throughput sequencing. *Journal of Dairy Science*, *99*(9), 7016–7024.

Li, X., Xiong, Q., Zhou, H., Xu, B., & Sun, Y. (2021). Analysis of microbial diversity and dynamics during bacon storage inoculated with potential spoilage bacteria by high-throughput sequencing. *Frontiers in Microbiology*, *12*, 713513.

Lin, Q. (2017). Jianchang Duck in the fermentation process of microorganism. *Modern Food, 33*, 66–69. https://doi.org/10.16736/j.cnki.cn41-1434/ts.2017.06.027

Mandal, S., Barman, S., Sen, A., & Mandal, N. C. (2024a). Mitigating aflatoxin contamination in food product using *Limosilactobacillus fermentum* LAB212. *Food Bioscience, 61*, 104750.

Mandal, S., Dey Bhowmik, A., Chattopadhyay, A., & Mandal, N. C. (2024b). Limosilactobacillus fermentum LAB212 effectively ameliorates toxigenicity of aflatoxin. *The Nucleus*, 1–10.

Mandal, S., & Mandal, N. C. (2023). Formulation of food grade *Limosilactobacillus fermentum* for antifungal properties isolated from home-made curd. *Scientific Reports, 13*(1), 20371.

Mao, J., Wang, X., Chen, H., Zhao, Z., Liu, D., Zhang, Y., & Nie, X. (2024). The contribution of microorganisms to the quality and flavor formation of chinese traditional fermented meat and fish products. *Foods, 13*(4), 608.

Mills, J., Horváth, K. M., Reynolds, A. D., & Brightwell, G. (2018). Farm and abattoir sources of *Carnobacterium* species and implications for lamb meat spoilage. *Journal of Applied Microbiology, 125*(1), 142–147.

Mittu, B., & Girdhar, Y. (2015). Role of lactic acid bacteria isolated from goat milk in cancer prevention. *AutoimmunInfec Dis, 1*(2), 2470–1025.

Mu, Y., Su, W., Mu, Y. C., & Jiang, L. (2019). Combined application of high-throughput sequencing and metabolomics reveals metabolically active microorganisms during Panxian Ham processing. *Frontiers in Microbiology, 10*, 3012. https://doi.org/10.3389/fmicb.2019.03012

Mutaku, I., Erku, W., & Ashenafi, M. (2005). Growth and survival of Escherichia coli O157: H7 in fresh tropical fruit juices at ambient and cold temperatures. *International Journal of Food Sciences and Nutrition, 56*(2), 133–139.

Oikonomou, G., Bicalho, M. L., Meira, E., Rossi, R. E., Foditsch, C., Machado, V. S., . . . Bicalho, R. C. (2014). Microbiota of cow's milk; distinguishing healthy, sub-clinically and clinically diseased quarters. *PloS One, 9*(1), e85904.

Papamanoli, E., Kotzekidou, P., Tzanetakis, N., & Litopoulou-Tzanetaki, E. (2002). Characterization of Micrococcaceae isolated from dry fermented sausage. *Food Microbiology, 19*(5), 441–449.

Patil, A., Disouza, J., & Pawar, S. (2019). Shelf life stability of encapsulated lactic acid bacteria isolated from sheep milk thrived in different milk as natural media. *Small Ruminant Research, 170*, 19–25.

Peton, V., & Le Loir, Y. (2014). *Staphylococcus aureus* in veterinary medicine. *Infection, Genetics and Evolution, 21*, 602–615.

Pisano, M. B., Deplano, M., Fadda, M. E., & Cosentino, S. (2019). Microbiota of Sardinian goat's milk and preliminary characterization of prevalent LAB species for starter or adjunct cultures development. *BioMed Research International, 2019*(1), 6131404.

Shori, A. B. (2012). Comparative study of chemical composition, isolation and identification of micro-flora in traditional fermented camel milk products: Gariss, Suusac, and Shubat. *Journal of the Saudi Society of Agricultural Sciences, 11*(2), 79–88.

Stellato, G., La Storia, A., De Filippis, F., Borriello, G., Villani, F., & Ercolini, D. (2016). Overlap of spoilage-associated microbiota between meat and the meat processing environment in small-scale and large-scale retail distributions. *Applied and Environmental Microbiology, 82*(13), 4045–4054.

Sun, Z., Wang, T., Aschalew, N. D., Zhao, W., Chen, X., Zhang, X. F., . . . Qin, G. X. (2020). Effects of yeast cultures with different fermentation times on the growth performance, caecal microbial community and metabolite profile of broilers. *Journal of Animal Physiology and Animal Nutrition, 104*(1), 212–223.

Wang, M., Wichienchot, S., He, X., Fu, X., Huang, Q., & Zhang, B. (2019). In vitro colonic fermentation of dietary fibers: Fermentation rate, short-chain fatty acid production and changes in microbiota. *Trends in Food Science & Technology*, *88*, 1–9.

Wang, Y., Li, F., Chen, J., Sun, Z., Wang, F., Wang, C., & Fu, L. (2021a). High-throughput sequencing-based characterization of the predominant microbial community associated with characteristic flavor formation in Jinhua Ham. *Food Microbiology*, *94*, 103643.

Wang, Z., Wang, Z., Ji, L., Zhang, J., Zhao, Z., Zhang, R., . . . Chen, L. (2021b). A review: Microbial diversity and function of fermented meat products in China. *Frontiers in Microbiology*, *12*, 645435.

Wen, K. Y., Wang, Y., Wen, P. C., Zhu, Y., Yang, M., Zhang, Z. M., & Zhang, W. B. (2020). Study on microbial community structure in Sichuan traditional bacon. *Food and Fermentation Industries*, *46*(3), 36–42.

Yi, L., Su, G., Hu, G., & Peng, Q. (2017). Diversity study of microbial community in bacon using metagenomic analysis. *Journal of Food Safety*, *37*(3), e12334.

Zhou HuiMin, Z. H., Zhang ShunLiang, Z. S., Zhao Bing, Z. B., Li Su, L. S., Pan XiaoQian, P. X., Ren Shuang, R. S., . . . Cheng XiaoYu, C. X. (2018). Effect of starter culture mixture of Staphylococcus xylosus and S. carnosus on the quality of dry-cured meat. *Shipin Kexue/Food Science*, *39*(22), 32–38.

Zou, Y., Liu, S. Y., Wang, G. Y., Pu, Y. H., Ge, C. R., & Liao, G. Z. (2020). Analysis of fungal community structure in Xuanwei ham by PCR-DGGE. *Food and Fermentation Industries*, *46*(6), 269–274.

4 Fungi as Phytopathogens in Fruits and Vegetables
Emerging Control Methods

Rafael López-Cruz, Juan Arturo Ragazzo-Sánchez, Leonardo Covarrubias-Rivera, Jerusalén Betancourt-Rodríguez, and Montserrat Calderón Santoyo

4.1 INTRODUCTION

Fungal diseases produced by phytopathogenic fungi result in serious losses or damage to crops worldwide. Significant losses of fruits and vegetables occur both in the field and during storage due to fungal spoilage (Alamoudi 2023). Fungal diseases related to high humidity, low pH, high nutrients, and inherent resistance to post-harvest decay are responsible for the high loss of fruits and vegetables. Major fungal diseases are caused by the genera *Penicillium*, *Alternaria*, *Colletotrichum*, *Rhizopus*, *Fusarium*, *Botrytis*, *Cladosporium*, *Lasiodiplodia*, *Aspergillus*, *Mucor*, and *Monilinia*, among others (López-Cruz et al. 2023; Oztekin and Karbancioglu-Guler 2021).

Synthetic fungicides have been the predominant control method for diseases caused by phytopathogenic microorganisms. The extensive use of chemical fungicides leads to resistance, high costs for growers, environmental pollution, and toxicity to human health (Pandit et al. 2022). In this sense, it is necessary to study the alternatives to synthetic pesticides that are environmentally safe and economically viable to address this issue. Therefore, researchers are exploring more sustainable, non-toxic, environmentally friendly, and effective antifungal methods (Fenta et al. 2023, Agirman et al. 2023).

There are various emerging control methods used in the management of phytopathogenic fungi. *Biopesticides* are naturally occurring substances (biochemical pesticides), microorganisms and their metabolic products (microbial pesticides), and substances produced by plants containing added genetic material (plant-incorporated protectants) that can control pests (Vero et al. 2023). Biopesticides, such as low-toxicity chemicals (inorganic salts), plant extracts (spices/oils/herbs), animal extracts (peptides/chitosan), and microbial sources (peptides/bacteriocins/enzymes), have been evaluated as alternative emerging control methods to conventional fungicides. In recent years, formulations based on biopesticides have progressively increased in number and diversity and have attracted commercial interest.

DOI: 10.1201/9781003540113-4

Moreover, biological control uses organisms or their metabolites to combat undesirable microorganisms. The use of antagonistic microorganisms is an effective alternative to control a variety of rot pathogens in diverse crops. Biocontrol agents (BCAs) or microbial antagonists prevent infection of the host fruit or vegetable by the pathogen or its establishment. In others, physical methods such as UV-C irradiation and cold plasma also have been proven to be effective against phytopathogenic fungi. These methods have contributed significantly to the outstanding improvements in crop productivity and quality over the past few decades (Thambugala et al. 2020). Further, nanotechnology has emerged as a promising alternative to control fungal infections. This method improves surface area and controlled release, making them ideal candidates for delivering antifungal agents effectively. These tiny particles dissolve better, directly attack plant problems, and remain to protect them for a longer time (Rameez et al. 2024). Finally, edible coatings are one of the applications of active packaging that prevents the surface growth of pathogenic microbes in food by the use of various antimicrobial agents, where a large portion of spoilage and contamination occurs. This system allows a controlled release of antimicrobial agents with a sound barrier and active layers outside the food surface during storage and distribution (Poovazhahi and Thakur, 2022).

This chapter highlights some effective emerging methods to manage fungal diseases and spoilage of fruits and vegetables, like biological control agents and their mechanisms of action, biopesticides, light therapies, nanotechnology, and edible coatings.

4.2 BIOPESTICIDES DERIVED FROM PLANTS, ANIMALS, BACTERIA, AND MINERALS TO CONTROL FUNGAL DISEASES

Biopesticides, derived from natural sources, such as plants, animals, bacteria, and minerals, have established themselves as a viable and sustainable alternative to conventional chemical pesticides to protect fruits and vegetables. Bioactive compounds of plant origin, present in essential oils and extracts, not only inhibit fungal growth but also have low toxicity and high biodegradability. Many plant derivatives act by damaging the fungal cell membrane, inhibiting essential enzymes, inducing oxidative stress, and interfering with sporulation (Matrose et al. 2021). However, the effectiveness of these biopesticides can vary significantly due to factors such as plant species, genotype, plant part used, age, and methods of extraction and handling. These factors highlight the importance of an approach tailored to the specific conditions of the crop and its environment (Goussous et al. 2010). To improve the stability of these biopesticides against adverse environmental conditions, innovative approaches have been developed, such as the application of nanofibers added with jackfruit leaf extract on star fruits, which have shown their efficacy against fungi such as *Cladosporium tenuissimum* and *Aspergillus sydowii* (Ayón-Macías et al. 2023). In addition, the extraction of specific phytochemicals has demonstrated efficacy in the post-harvest stage of fruits. Notable examples include cinnamaldehyde, extracted from *Cinnamomum*, which is effective against *Botrytis cinerea* in peppers (Yang et al. 2023), and citral, a component of the essential oil of *Cymbopogon citratus*, which inhibits *Fusarium pseudocircinatum* in bananas (Iñiguez-Moreno et al. 2022), highlighting the potential of these compounds in fruit and vegetable protection. An additional example of biopesticides of plant origin is the use of peptides from jackfruit leaves obtained by hydrolysis with pepsin. Brion-Espinoza et al. (2021) reported a peptide with six essential amino acids which inhibited mycelial growth and spore germination of *Colletotrichum gloeosporioides* by 99%, one of the main phytopathogenic agents in tropical fruits.

Animal products such as chitosan and antimicrobial peptides have shown great potential as biopesticides for protecting fruits. Chitosan, a biopolymer obtained from the chitin of crustacean shells, is effective in inhibiting fungal growth by altering their cell wall and reinforcing plant resistance. Its use is compatible with integrated pest management programs and has demonstrated efficacy against *Venturia inaequalis* in the pre-harvest stage of apples (DeGenring et al. 2023). Antimicrobial peptides, vital components of the immune system in animals and insects, can alter the cell membrane of pathogens, causing their lysis and eventual death. However, their study has been limited due to high production costs, short shelf life, and sensitivity to environmental factors. Despite these challenges, peptides derived from bovine lactoferricin are effective against *Penicillium digitatum* in mandarins (Muñoz and Marcos 2006). This highlights the need to expand research and develop more efficient extraction techniques to evaluate and enhance their effectiveness in agricultural applications.

On the other hand, bacterial derivatives represent an innovative and promising solution for the control of phytopathogenic fungi. Bacteria that produce antimicrobial metabolites and other bioactive compounds have the potential to inhibit fungal growth, offering a viable option for crop protection. In this context, antimicrobial peptides can bind to the germ tubes and cell membranes of fungi, creating pores that alter their permeability. For example, *Bacillus pumilus* strains PTB180 and *B. velezensis* SH-B74 secrete surfactin, which combat *B. cinerea* and *Magnaporthe oryzae* in tomatoes and rice (Bouchard-Rochette et al. 2022; Sarwar et al. 2018), respectively. *B. subtilis* Z-14 produces iturin A, which destroys *Gaeumannomyces graminis* cells, preventing wheat disease (Xiao et al. 2021). Finally, *B. velezensis* G341 produces bacillomycin-L, effective against *B. cinerea* (gray mold of tomato) and *Colletotrichum coccodes* (anthracnose of red pepper) (Lim et al. 2017).

In recent years, mineral derivatives have emerged as a promising option for the control of phytopathogenic fungi. Depending on the mineral used, they can act by osmotically dehydrating fungal cells, interfering with the integrity of the cell membrane, generating reactive oxygen species, inhibiting key enzymes, or altering cell wall synthesis. Currently, there are commercial products derived from minerals that combine copper (Cu), manganese (Mn), and zinc (Zn), providing a comprehensive solution. For example, the combination of these minerals has been effective for the control of anthracnose in the pre-harvest stage of orange "TaroccoSciré" (Lombardo et al. 2023). Recently, the use of inorganic salts classified as GRAS (generally recognized as safe) has gained relevance. These salts can be combined with biopesticides or coating techniques, which improves their effectiveness by providing a more sustained and specific action in fungal protection. For example, potassium sorbate ($C_6H_7O_2K$), sodium benzoate ($C_7H_5O_2Na$), and potassium silicate (K_2CO_3) salts are effective in controlling anthracnose in oranges and mandarins (Martínez-Blay et al. 2020). Despite significant efforts to develop sustainable alternatives to conventional fungicides, most of these biopesticides are still in the research phase, with the prospect of large-scale production.

4.3 BIOLOGICAL CONTROL: USE OF BENEFICIAL MICROORGANISMS TO SUPPRESS PATHOGENIC FUNGI

Post-harvest fungal diseases have successfully been controlled by various beneficial microorganisms reported as biocontrol agents (BCAs), including bacteria, yeast, and fungi. BCAs are becoming an environmentally friendly and effective method to manage pathogenic fungi as an alternative to chemical fungicides. The biocontrol efficacy of BCAs depends on the antagonistic mechanism of action, dose, method of application, sensitivity of the pathogen, host variability, and environmental conditions (Bonaterra et al. 2022). There are different ways in which biological control is classified: The classic form involves the introduction of foreign organisms into another ecosystem to combat undesirable organisms. *Augmentative* means introducing greater amounts of existing natural antagonists into a habitat to combat pathogens, and *conservative* means manipulating the microenvironment to favor the activity of antagonists (Saldaña-Mendoza et al. 2023).

A wide variety of bacterial genera, including *Bacillus*, *Burkholderia*, *Agrobacterium*, *Alcaligenes*, *Arthrobacter*, *Enterobacter*, *Erwinia*, *Pseudomonas*, *Rhizobium*, *Serratia*, *Stenotrophomonas*, *Streptomyces*, and *Xanthomonas*, have been described to have antagonistic activity against a broad-spectrum fungal pathogen. Disease reduction varies from 20 to 100%, depending on the antagonistic bacteria, target pathogen, host, and dose of application (Bonaterra et al. 2022). These bacteria can use multiple mechanisms implicated in the limitation of plant pathogens development. These mechanisms of action include colonization of infection sites and competitive exclusion of the pathogen; secretion of active antimicrobials, such as antibiotics and cell wall lytic enzymes; and induction of plant resistance (Shi et al. 2021; Lee et al. 2023). Antagonistic capacity *in vivo* against *C. gloeosporioides* on avocado fruit was demonstrated by applying *Bacillus pumilus* AB31 and three strains of *B. thuringiensis* (AB30, AB7, and AB21) (Magallón-Andalón et al. 2024).

Filamentous fungi are also utilized to manage some fungal diseases in fruits, including *Aureobasidium pullulans*, *Clonostachys byssicola*, *Curvularia pallescens*, *Epicoccum dendrobii*, and *Trichoderma* species. Most filamentous fungi were reported to produce antifungal activity via colonization of *Colletotrichum hyphae* and subsequent competition for space (Shi et al. 2021). Also, other mechanisms for biological control against phytopathogens are mycoparasitism, production of secondary metabolites with antibiotic activity, competition for nutrients and space, and induction of the systemic immune response (Saldaña-Mendoza et al. 2023).

Antagonistic yeasts possess advantageous properties concerning bacteria and fungi that make them ideal biocontrol candidates, including strong antagonistic capacity, stress tolerance, rapid colonization of fruit surfaces, genetic stability, applicability, easy large-scale production, and not producing antibiotic substances (Alimadadi et al. 2023). Antagonist yeasts are mostly isolated from the surface of fruits, although they can also be isolated from other sources, such as seawater, roots, and soil. Some genera of yeasts evaluated to manage fungal decay include *Rhodotorula*, *Candida*, *Cryptococcus*, *Metchnikowia*, *Pichia*, *Wickerhamomyces*, *Saccharomyces*, *Cystobasidium*, *Debaryomyces*, *Meyerozyma*, and *Yamadazyma*, among others (Pesce et al. 2018). The mechanisms involved in yeast antagonism include quick colonization to occupy space and nutrients, development of cell biofilm, direct parasitism, secretion of cell wall–degrading enzymes (chitinases, glucanases, and proteases), killer toxins and VOCs production, and induction of host-resistance systems (Ma et al. 2023). In this sense, yeasts provide a combination of the antifungal mechanisms shown by both bacteria and filamentous fungi together (Shi et al. 2021; Herrera-Balandrano et al. 2023).

The successful bioformulation development implicate isolation and screening of new microorganisms, assays of antagonistic efficacy for possible use, mode of action and toxicological profiles, mass production, development of a suitable liquid or dry formulation, packaging, shelf life, and commercial application tests. Furthermore, the next steps include registration, protection of intellectual property rights, and marketing (Thambugala et al. 2020; Teixidó et al. 2022).

Some studies have been developed as an effective strategy by focusing on common fungal diseases in crops to show their potential for commercial viability. A dry formulation created through the electrospraying process of the yeast *Yamadazyma mexicana*, isolated from the surface of avocado fruit, was tested both before and after harvest to manage anthracnose decay in avocado crops. Results indicated a good performance and consistency in two years of application. The disease incidence and severity were reduced by 100% when the product was applied in the field + postharvest and stored under cool conditions (González-Gutiérrez et al. 2024). A new bioformulation based on the yeast *Meyerozyma guilliermondii* LMA-Cp01 formulated by spray-drying was tested under field commercial conditions to manage anthracnose on mango fruit (López-Cruz et al. 2024). Results obtained were successful at two orchards and growing seasons, reducing natural infections of anthracnose incidence (76%) and severity (92%) on mango fruit at the same level of chemical fungicides.

4.4 LIGHT THERAPIES TO REDUCE FUNGI ON THE SURFACE OF FRESH PRODUCE

Fungal proliferation in fresh produce is a major problem for the food industry, threatening food safety and causing economic losses (Sani et al. 2023). Emerging technologies such as cold plasma and ultraviolet C (UV-C) light could extend shelf life and improve quality by reducing fungal contamination on the surface of fresh produce. UV-C light (200 and 280 nm) is known for its germicidal properties. It can penetrate the cells of microorganisms, including fungi, bacteria, and viruses; therefore, it can cause irreparable damage to their genetic material (Abdel-Latif and Zakarya 2023).

UV-C light produces DNA damage through the formation of pyrimidine dimers and abnormal covalent bonds between thymine and cytosine. This process prevents genetic replication and transcription, reducing the viability of microorganisms and causing their death. This technology is effective in reducing the microbial load on the surface of fruits and vegetables. However, its excessive use can affect the quality of food, altering its texture and flavor (Ramos et al. 2013). On the other hand, cold plasma is a state of matter that is generated at room temperature. It is made up of a mixture of electrons, ions, free radicals, and neutral molecules (Gavahian and Cullen 2020). Cold plasma is generated by applying high-frequency electric fields to gases such as argon, oxygen, or air at atmospheric pressure. This process induces partial ionization of the gas without significantly increasing its temperature, making it effective for disinfecting food surfaces due to its ability to penetrate the cell membranes of pathogens. As a result, cold plasma has been adopted in the food industry for surface sterilization, including on fruits. Both UV-C and cold plasma technologies can inactivate phytopathogenic fungi through damage to microbial membranes, induction of oxidative stress by reactive oxygen (ROS) and nitrogen species (RNS), and destruction of cellular proteins, leading to apoptosis and disruption of fungal metabolic and replication functions. These methods are essential to prolong the shelf life of fresh produce and ensure food safety in the industry (Sharma et al. 2018). Several studies have demonstrated the efficacy of UV-C light and cold plasma in inactivating fungi on fresh produce. Misra et al. (2019) found that 254 nm UV-C light was effective against *Colletotrichum* and *Fusarium* spores on strawberries, while *Penicillium expansum* showed resistance but developed few colonies after 15 seconds of irradiation. Abdel-Latif and Zakarya (2023) reported that UV-C irradiation on grapes significantly reduced spore germination in *Aspergillus carbonarius* and *Alternaria alternata* by 25%, and by more than 70% with only 10 seconds of exposure. Valero et al. (2007) indicated that *Alternaria alternata* shows increased resistance to UV radiation, while *Fusarium* sp. showed a reduction in metabolic activity of more than 95% at higher UV doses. Vanhaelewyn et al. (2020) reported that the germination of *Botrytis cinerea* spores in UV-C and cold plasma–treated blueberries was significantly lower than in untreated samples, reducing up to 50.63%. Likewise, Hayashi et al. (2014) highlighted the role of atomic oxygen generated by cold plasma in the inactivation of *Aspergillus oryzae* and *Penicillium digitatum*, leading to unculturable cells and irregular morphology. These investigations highlight the effectiveness of UV light and cold plasma in controlling pathogenic fungi in fresh produce, although their effectiveness varies by species and treatment conditions. A direct comparison between these methods reveals that while UV-C light is effective on exposed surfaces, cold plasma, on the other hand, can penetrate irregular surfaces and reach difficult-to-treat areas. Thus, these data suggest the benefits of these technologies in controlling phytopathogenic fungi in fresh produce. The use of UV-C light and cold plasma offers significant advantages, such as reducing chemical residues and preserving food quality. These clean technologies inactivate microorganisms without compromising the texture or flavor of food, which extends its shelf life. However, the implementation of both technologies faces challenges, including the need for specialized equipment, high initial costs, and regulations. Despite this, some innovations, such as the combination of UV-C light with other preservation methods, are improving their effectiveness and accessibility. Future research should focus on optimizing treatment parameters, understanding long-term impacts, and pinpointing microbial inactivation mechanisms. Overcoming these challenges ensures sustainable and safe practices. As new solutions emerge and new regulations are developed, these technologies could revolutionize fresh produce disinfection globally.

4.5 NANOTECHNOLOGY: EMPLOYING NANOPARTICLES TO PENETRATE AND COMBAT FUNGAL CELLS

In searching for innovative strategies for controlling phytopathogenic fungi in fruits and vegetables, nanotechnology is presented as an effective alternative to conventional fungicides. Nanoparticles (NPs), whether organic or inorganic, have the ability to cross biological barriers, such as cell membranes, and interact directly with critical components of the cell, thanks to their size, which can range from 1 to 100 nm. Some NPs can induce oxidative stress by generating reactive oxygen species, which damages the cell structure and eventually causes the death of the fungus (Kutawa et al. 2021). Carbon-based organic NPs, such as chitosan, besides being biodegradable and having a fungicidal effect, can be applied as nanocarrier NPs for antifungal agents. For example, Saqib et al. (2020) used chitosan-coated iron oxide NPs on peach fruits to inhibit *Rhizopus stolonifer*, showing greater efficacy than a synthetic fungicide (Matalaxyl + mancozeb). Therefore, combining nanotechnologies can favor an integrated management of peach diseases.

On the other hand, inorganic NPs, especially metallic ones, such as copper, silica, and silver, have been the subject of numerous studies due to their high reactivity and durability, which make them effective in agricultural applications. For example, copper NPs have proven effective against *P. digitatum* and *F. solani* in sweet oranges, reducing the incidence of green mold by 88.5% and completely preventing rot caused by *F. solani*. This effect is due to the alteration of the fungus DNA, leading to the death of the pathogen (Khamis et al. 2017). Likewise, silica and chitosan NPs, as well as their combinations, have shown high efficiency in the control of gray mold in table grapes (*Vitis vinifera*), reducing the incidence of the disease. These NPs not only degrade the hyphal structure of the pathogen but also cause DNA alteration (Hashim et al. 2019). Another example of metal NPs is the combination of silver NPs with a plant-derived biofungicide, such as neem extract. When this combination was applied at the post-harvest stage of the banana, a reduction in the disease caused by *C. musae*, the causal agent of anthracnose in the banana, was observed (Jagana et al. 2017).

In this context, these nanoparticles can be used at lower concentrations than conventional fungicides, reducing the amount of chemicals released into the environment. Although the concentrations used are usually minimal, further studies are crucial to assess the persistence and behavior of nanoparticles in the agricultural ecosystem. Another challenge is the high production cost of nanoparticles compared to traditional fungicides, which may limit their use in high-value crops (Ruffo Roberto et al. 2019). Despite these challenges, nanotechnology has the potential to revolutionize the management of diseases caused by phytopathogenic fungi, promoting more sustainable and effective methods for fruit protection.

4.6 EDIBLE COATINGS: APPLYING ANTIFUNGAL FILMS TO EXTEND THE SHELF LIFE OF FRUITS AND VEGETABLES

Fruits and vegetables are susceptible to contamination by microorganisms and are highly perishable due to their high water content. Therefore, the demand to increase their shelf life is increasingly high. Edible coatings (EC) have emerged as an innovative and effective solution (Dhall 2013). EC are thin layers applied directly to fruit surfaces by dipping, spraying, or brushing to create a modified atmosphere. ECs can be consumed as part of the whole product because they are composed of natural polymers, including proteins, polysaccharides, and lipids. Therefore, the composition of EC must comply with the food product regulations in each country (Iñiguez-Moreno et al. 2021). The properties of EC depend mainly on molecular structure and chemical constitution. Among the most important properties considered are the barrier to water vapor, gases, and migration of volatile compounds; sensory properties, such as the appearance of food, shine, and texture; and finally, functional properties, such as physical and mechanical protection capacity. Moreover, the addition of antimicrobial agents, such as essential oils, bioactive compounds, anti-browning, volatile precursors, or nutrients, improves the ability of the coating to inhibit the growth of bacteria and fungi, extending the shelf life of the food.

To prevent contamination of fruits and vegetables, especially by phytopathogenic fungi, and to avoid the use of fungicides, natural edible coatings with antimicrobial agents offer a promising alternative. These coatings help prevent fungal growth and dissemination while improving the quality and safety of the product. Additionally, this property extends the lag phase, suppressing the growth rate of microorganisms and providing the aforementioned benefits. Various biocontrol agents are used, including plant extracts, essential oils, and microorganisms, such as yeasts and/or fungi (Palou et al. 2015). These biocontrol agents are typically combined with natural polymers, such as chitosan (which has antifungal properties against a wide range of phytopathogenic fungi), sodium alginate, and cellulose, providing better support for the biocontrol agents. The mechanisms by which these agents are incorporated follow different paths, such as incorporating volatile and non-volatile antimicrobial agents directly into polymers or using layers of antimicrobial adsorbents within the surfaces of polymers and antimicrobial immobilizers, linked by ionic or covalent bonds. Some of these mechanisms can be divided into two groups: the first involves the migration of active compounds, which partially or fully penetrate the food or surface, and the second involves non-migratory mechanisms, where compounds with antimicrobial activity remain on the surface of the film or coating. These compounds can alter the cell membrane of fungi, inhibiting their ability to proliferate. In addition, it has been reported that microbial activity can destabilize the cell wall of fungi, inhibit fatty acid synthesis, and disrupt carbon metabolism by altering the carbonic anhydrase enzyme, leading to cell death of the pathogen (Pagliarulo et al. 2016). Table 4.1 presents some studies related to the use of edible coatings and the extension of the shelf life of horticultural or minimally processed products.

Application techniques for edible coatings have advanced significantly, combining traditional methods and innovative technologies. In traditional methods, food is dipped into a coating solution, creating a uniform and effective layer to protect fruits and vegetables. This method is simple and inexpensive but can be limited to covering irregular shapes or complex surfaces. On the other hand, more advanced techniques, such as electrospinning, represent important advances in the application of edible coatings (Cruz et al. 2023). Electrospinning uses electric fields to create ultra-fine polymer fibers, forming thin, uniform layers on food, improving coating coverage by ensuring uniform distribution and improved adhesion on complex surfaces.

TABLE 4.1
Practical Cases on the Application of Antifungal Edible Coatings

Edible Coating Materials	Antifungal Agent	Application Techniques	Food Matrix	Reference
Pea starch and potato starch	Potassium sorbate to combat the growth of *Alternaria alternata*	Dipping	Apple, cucumber, tomato	(Mehyar et al. 2014)
Chitosan	Chitosan (1–4%) against selected *Rhizoctonia solani*, *Fusarium oxysporum*, *Collectotrichum acutatum*, and *Phytophthora infestans*	Dipping	Tomato, pepper, eggplant	(Divya et al. 2018)
Pullulan, cashew gum, FucoPol and poly (ethylene oxide) (PEO)	*Meyerozyma caribbica* against six pathogens, including *Colletotrichum gloeosporioides*	Nanofibers by electrospinning process	*In vitro* media and avocado	(Vázquez-González et al. 2024)

While this advanced technology offers precise application and superior food protection, they are challenged by high equipment costs and technical complexity. Future research aims to address this, potentially leading to more sustainable and accessible solutions and further improving the effectiveness of edible coatings in extending shelf life and improving food quality.

4.7 CONCLUSIONS AND FUTURE REMARKS

In recent years, there has been a growing interest in using natural sources to manage fungal decays as an alternative to chemical fungicides. Emerging control methods, such as biopesticides, biocontrol agents, nanotechnology, UVC-light, and edible coatings, have been explored with enormous potential to control fungal diseases in fruits and vegetables. Therefore, the focus has been on searching for new bio-sources, their mechanisms against phytopathogenic fungi, and the effect on disease suppression and maintaining the intrinsic quality of vegetable crops. To ensure the long-term efficacy of emerging control methods against target pathogens, it is important to understand how their effectiveness can be compromised. Furthermore, it is necessary to assess their efficacies under large-scale commercial conditions and to evaluate the cost–benefit ratio of these technologies. Moreover, biocontrol effectivity may be improved by the integration of pre-and post-harvest technologies, such as biopesticides with UV-C light, nanotechnology, and edible coatings, as well as increasing the stability of the product.

REFERENCES

Abdel-Latif, M., and M. Zakarya. 2023. Effect of UV-C light on fungi causing root rot/wilt of sage (*Salvia offecialis* L.). *New Valley Journal of Agricultural Science* 3:573–589.

Agirman, B., E. Carsanba, L. Settanni, and H. Erten. 2023. Exploring yeast-based microbial interactions: The next frontier in postharvest biocontrol. *Yeast* 40:457–475.

Alamoudi, S. A. 2023. Using some microorganisms as biocontrol agents to manage phytopathogenic fungi: A comprehensive review. *Journal of Plant Pathology* 106:3–21.

Alimadadi, N., Z. Pourvali, S. Nasr, and S. A. S. Fazeli. 2023. Screening of antagonistic yeast strains for postharvest control of *Penicillium expansum* causing blue mold decay in table grape. *Fungal Biology* 127:901–908.

Ayón-Macías, K. D., A. J. Castañeda-Andrade, J. A. Ragazzo-Sánchez, and M. Calderón-Santoyo. 2023. Application of nanofibers with jackfruit leaf extract via electrospinning to control phytopathogens in *Averrhoa carambola* L. *Polymer Bulletin* 81:2601–2626.

Bonaterra, A., E. Badosa, N. Daranas, J. Francés, G. Roselló, and E. Montesinos. 2022. Bacteria as biological control agents of plant diseases. *Microorganisms* 10:1759.

Bouchard-Rochette, M., Y. Machrafi, L. Cossus, T. Thuy An Nguyen, H. Antoun, A. Droit, and R. J. Tweddell. 2022. *Bacillus pumilus* PTB180 and *Bacillus subtilis* PTB185: Production of lipopeptides, antifungal activity, and biocontrol ability against *Botrytis cinerea*. *Biological Control* 170:104925.

Brion-Espinoza, I. A., M. Iñiguez-Moreno, J. A. Ragazzo-Sánchez, J. C. Barros-Castillo, C. Calderón-Chiu, and M. Calderón-Santoyo. 2021. Edible pectin film added with peptides from jackfruit leaves obtained by high-hydrostatic pressure and pepsin hydrolysis. *Food Chemistry: X* 12:100170.

Cruz, E. P. da, E. T. Jansen, L. M. Fonseca, H. C. dos S. Hackbart, T. J. Siebeneichler, J. B. Pires, E. A. Gandra, C. V. Rombaldi, E. da R. Zavareze, and A. R. G. Dias. 2023. Red onion skin extract rich in flavonoids encapsulated in ultrafine fibers of sweet potato starch by electrospinning. *Food Chemistry* 406:134954.

DeGenring, L., K. Peter, and A. Poleatewich. 2023. Integration of Chitosan and Biopesticides to Suppress Pre-Harvest Diseases of Apple. *Horticulturae* 9:707.

Dhall, R. K. 2013. Advances in edible coatings for fresh fruits and vegetables: A review. *Critical Reviews in Food Science and Nutrition* 53:435–450.

Divya, K., V. Smitha, and M. S. Jisha. 2018. Antifungal, antioxidant and cytotoxic activities of chitosan nanoparticles and its use as an edible coating on vegetables. *International Journal of Biological Macromolecules* 114: 572–577.

Fenta, L., H. Mekonnen, and N. Kabtimer. 2023. The exploitation of microbial antagonists against postharvest plant pathogens. *Microorganisms* 11:1044.

Gavahian, M., and P. J. Cullen. 2020. Cold plasma as an emerging technique for mycotoxin-free food: Efficacy, mechanisms, and trends. *Food Reviews International* 36:193–214.

González-Gutiérrez, K. N., J. A. Ragazzo-Sánchez, and M. Calderón-Santoyo. 2024. Field and postharvest application of microencapsulated *Yamadazyma mexicana* LPa14: Anthracnose control and effect on postharvest quality in avocado (*Persea americana* Mill. cv. Hass). *Pest Management Science* 80:3459–3469.

Goussous, S. J., F. M. Abu el-Samen, and R. A. Tahhan. 2010. Antifungal activity of several medicinal plants extracts against the early blight pathogen (*Alternaria solani*). *Archives of Phytopathology and Plant Protection* 43:1745–1757.

Hashim, A. F., K. Youssef, and K. A. Abd-Elsalam. 2019. Ecofriendly nanomaterials for controlling gray mold of table grapes and maintaining postharvest quality. *European Journal of Plant Pathology* 154:377–388.

Hayashi, N., Y. Yagyu, A. Yonesu, and M. Shiratani. 2014. Sterilization characteristics of the surfaces of agricultural products using active oxygen species generated by atmospheric plasma and UV light. *Japanese Journal of Applied Physics* 53:05FR03.

Herrera-Balandrano, D. D., S.-Y. Wang, C.-X. Wang, X.-C. Shi, F.-Q. Liu, and P. Laborda. 2023. Antagonistic mechanisms of yeasts *Meyerozyma guilliermondii* and *M. caribbica* for the control of plant pathogens: A review. *Biological Control* 186:105333.

Iñiguez-Moreno, M., M. Calderón-Santoyo, J. C. Barros-Castillo, D. M. Miss-Zacarías, J. A. Díaz, and J. A. Ragazzo-Sánchez. 2022. Nanofibers added with citral: Characterization and their application to postharvest control of *Fusarium pseudocircinatum* in bananas. *Journal of Food Processing and Preservation* 46:e17188.

Iñiguez-Moreno, M., J. A. Ragazzo-Sánchez, J. C. Barros-Castillo, J. R. Solís-Pacheco, and M. Calderón-Santoyo. 2021. Characterization of sodium alginate coatings with *Meyerozyma caribbica* and impact on quality properties of avocado fruit. *LWT* 152:112346.

Jagana, D., Y. R. Hegde, and R. Lella. 2017. Green Nanoparticles—A novel approach for the management of banana anthracnose caused by *Colletotrichum musae. International Journal of Current Microbiology and Applied Sciences* 6:1749–1756.

Khamis, Y., A. F. Hashim, R. Margarita, M. A. Alghuthaymi, and K. A. Abd-Elsalam. 2017. Fungicidal efficacy of chemically-produced copper nanoparticles against *Penicillium digitatum* and *Fusarium solani* on citrus fruit. *Philippine Agricultural Scientist (Philippines) 100:69–78.*

Kutawa, A. B., K. Ahmad, A. Ali, M. Z. Hussein, M. A. Abdul Wahab, A. Adamu, A. A. Ismaila, M. T. Gunasena, M. Z. Rahman, and M. I. Hossain. 2021. Trends in nanotechnology and its potentialities to control plant pathogenic fungi: A review. *Biology* 10:881.

Lee, J., S. Kim, H. Jung, B. Kyoung, K. Jeong, A. H. Ho, and S. Lee. 2023. Exploiting bacterial genera as biocontrol agents: Mechanisms, interactions and applications in sustainable agriculture. *Journal of Plant Biology* 66:485–498.

Lim, S. M., M.-Y. Yoon, G. J. Choi, Y. H. Choi, K. S. Jang, T. S. Shin, H. W. Park, N. H. Yu, Y. H. Kim, and J.-C. Kim. 2017. Diffusible and volatile antifungal compounds produced by an antagonistic *Bacillus velezensis* G341 against various phytopathogenic fungi. *The Plant Pathology Journal* 33:488–498.

Lombardo, M. F., S. Panebianco, A. Azzaro, V. Catara, and G. Cirvilleri. 2023. Assessing copper-alternative products for the control of pre- and postharvest citrus anthracnose. *Plants* 12:904.

López-Cruz, R., J. A. Ragazzo-Sánchez, and M. Calderón-Santoyo. 2024. Field testing of a newly formulated product based on *Meyerozyma guilliermondii* LMA-Cp01 to manage anthracnose in mango fruit. *Journal of Plant Diseases and Protection* 1–12.

López-Cruz, R., G. Segarra, R. Torres, N. Teixidó, J. A. Ragazzo-Sanchez, and M. Calderon-Santoyo. 2023. Biocontrol efficacy of *Meyerozyma guilliermondii* LMA-Cp01 against post-harvest pathogens of fruits. *Archives of Phytopathology and Plant Protection* 56:1003–1020.

Ma, Y., M. Wu, X. Qin, Q. Dong, and Z. Li. 2023. Antimicrobial function of yeast against pathogenic and spoilage microorganisms via either antagonism or encapsulation: A review. *Food Microbiology* 112:104242.

Magallón-Andalón, C. G., M. Calderón-Santoyo, R. Balois-Morales, V. A. Ochoa-Jiménez, P. P. Casas-Junco, G. G. López-Guzmán, I. F. Pérez-Ramírez, Y. A. Palomino-Hermosillo, and P. U. Bautista-Rosales. 2024. Antagonistic potential of *Bacillus* spp. for the control of *Colletotrichum gloeosporioides* on avocado (*Persea americana* Mill.) "Hass". *Journal of Plant Diseases and Protection* 1–12.

Martínez-Blay, V., M. B. Pérez-Gago, B. de la Fuente, R. Carbó, and L. Palou. 2020. Edible coatings formulated with antifungal GRAS salts to control citrus anthracnose caused by *Colletotrichum gloeosporioides* and preserve postharvest fruit quality. *Coatings* 10:730.

Matrose, N. A., K. Obikeze, Z. A. Belay, and O. J. Caleb. 2021. Plant extracts and other natural compounds as alternatives for post-harvest management of fruit fungal pathogens: A review. *Food Bioscience* 41:100840.

Mehyar, G. F., H. M. Al-Qadiri, and B. G. Swanson. 2014. Edible coatings and retention of potassium sorbate on apples, tomatoes and cucumbers to improve antifungal activity during refrigerated storage. *Journal of Food Processing and Preservation* 38:175–182.

Misra, N. N., B. Yadav, M. S. Roopesh, and C. Jo. 2019. Cold plasma for effective fungal and mycotoxin control in foods: Mechanisms, inactivation effects, and applications. *Comprehensive Reviews in Food Science and Food Safety* 18:106–120.

Muñoz, A., and J. F. Marcos. 2006. Activity and mode of action against fungal phytopathogens of bovine lactoferricin-derived peptides. *Journal of Applied Microbiology* 101:1199–1207.

Oztekin, S., and F. Karbancioglu-Guler. 2021. Bioprospection of *Metschnikowia* sp. isolates as biocontrol agents against postharvest fungal decays on lemons with their potential modes of action. *Postharvest Biology and Technology* 181:111634.

Pagliarulo, C., F. Sansone, S. Moccia, G. L. Russo, R. P. Aquino, P. Salvatore, M. Di Stasio, and M. G. Volpe. 2016. Preservation of strawberries with an antifungal edible coating using peony extracts in chitosan. *Food and Bioprocess Technology* 9:1951–1960.

Palou, L., S. A. Valencia-Chamorro, and M. B. Pérez-Gago. 2015. Antifungal edible coatings for fresh citrus fruit: A review. *Coatings* 5:962–986.

Pandit, M. A., J. Kumar, S. Gulati, N. Bhandari, P. Mehta, R. Katyal, C. D. Rawat, V. Mishra, and J. Kaur. 2022. Major biological control strategies for plant pathogens. *Pathogens* 11:273.

Pesce, V. M., M. C. Nally, G. P. Carrizo, C. Rojo, B. A. Pérez, M. E. Toro, L. I. Castellanos de Figueroa, and F. Vazquez. 2018. Antifungal activity of native yeasts from different microenvironments against *Colletotrichum gloeosporioides* on ripe olive fruits. *Biological Control* 120:43–51.

Poovazhahi, A., and M. Thakur. 2022. Antimicrobial edible packaging: Applications, innovations, and sustainability. Pages 513–526 *Edible Food Packaging*. Springer Nature Singapore, Singapore.

Rameez, M., N. Khan, S. Ahmad, and M. M. Ahmad. 2024. Bionanocomposites: A new approach for fungal disease management. *Biocatalysis and Agricultural Biotechnology* 57:103115.

Ramos, B., F. A. Miller, T. R. S. Brandão, P. Teixeira, and C. L. M. Silva. 2013. Fresh fruits and vegetables—An overview on applied methodologies to improve its quality and safety. *Innovative Food Science & Emerging Technologies* 20:1–15.

Ruffo Roberto, S., K. Youssef, A. F. Hashim, and A. Ippolito. 2019. Nanomaterials as alternative control means against postharvest diseases in fruit crops. *Nanomaterials* 9:1752.

Saldaña-Mendoza, S. A., S. Pacios-Michelena, A. S. Palacios-Ponce, M. L. Chávez-González, and C. N. Aguilar. 2023. *Trichoderma* as a biological control agent: Mechanisms of action, benefits for crops and development of formulations. *World Journal of Microbiology and Biotechnology* 39:269.

Sani, I. K., L. Aminoleslami, S. S. Mirtalebi, M. A. Sani, E. Mansouri, H. Eghbaljoo, A. T. Jalil, R. D. Thanoon, S. M. Khodaei, F. Mohammadi, and B. Kazemzadeh. 2023. Cold plasma technology: Applications in improving edible films and food packaging. *Food Packaging and Shelf Life* 37:101087.

Saqib, S., W. Zaman, A. Ayaz, S. Habib, S. Bahadur, S. Hussain, S. Muhammad, and F. Ullah. 2020. Postharvest disease inhibition in fruit by synthesis and characterization of chitosan iron oxide nanoparticles. *Biocatalysis and Agricultural Biotechnology* 28:101729.

Sarwar, A., M. N. Hassan, M. Imran, M. Iqbal, S. Majeed, G. Brader, A. Sessitsch, and F. Y. Hafeez. 2018. Biocontrol activity of surfactin a purified from *Bacillus* NH-100 and NH-217 against rice bakanae disease. *Microbiological Research* 209:1–13.

Sharma, R. R., S. V. R. Reddy, and S. Sethi. 2018. Cold plasma technology for surface disinfection of fruits and vegetables. Pages 197–209 *Postharvest Disinfection of Fruits and Vegetables*. Elsevier.

Shi, X.-C., S.-Y. Wang, X.-C. Duan, Y.-Z. Wang, F.-Q. Liu, and P. Laborda. 2021. Biocontrol strategies for the management of *Colletotrichum* species in postharvest fruits. *Crop Protection* 141:105454.

Teixidó, N., J. Usall, and R. Torres. 2022. Insight into a successful development of biocontrol agents: Production, formulation, packaging, and shelf life as key aspects. *Horticulturae* 8:305.

Thambugala, K. M., D. A. Daranagama, A. J. L. Phillips, S. D. Kannangara, and I. Promputtha. 2020. Fungi vs. fungi in biocontrol: An overview of fungal antagonists applied against fungal plant pathogens. *Frontiers in Cellular and Infection Microbiology* 10:1–19.

Valero, A., M. Begum, S. L. Leong, A. D. Hocking, A. J. Ramos, V. Sanchis, and S. Marín. 2007. Effect of germicidal UVC light on fungi isolated from grapes and raisins. *Letters in Applied Microbiology* 45:238–243.

Vanhaelewyn, L., D. Van Der Straeten, B. De Coninck, and F. Vandenbussche. 2020. Ultraviolet radiation from a plant perspective: The plant-microorganism context. *Frontiers in Plant Science* 11:597642.

Vázquez-González, Y., C. Prieto, M. Calderón-Santoyo, J. A. Ragazzo-Sánchez, and J. M. Lagarón. 2024. Development of antifungal electrospun nanofiber mats containing *Meyerozyma caribbica*. *Food Hydrocolloids* 147:109343.

Vero, S., G. Garmendia, E. Allori, J. M. Sanz, M. Gonda, T. Alconada, I. Cavello, J. R. Dib, M. A. Diaz, C. Nally, R. S. Pimenta, J. F. M. da Silva, M. Vargas, F. Zaccari, and M. Wisniewski. 2023. Microbial biopesticides: Diversity, scope, and mechanisms involved in plant disease control. *Diversity* 15:457.

Xiao, J., X. Guo, X. Qiao, X. Zhang, X. Chen, and D. Zhang. 2021. Activity of fengycin and iturin aisolated from *Bacillus subtilis* Z-14 on *Gaeumannomyces graminis* Var. tritici and soil microbial diversity. *Frontiers in Microbiology* 12:682437.

Yang, L., X. Liu, H. Lu, C. Zhang, J. Chen, and Z. Shi. 2023. Cinnamaldehyde inhibits postharvest gray mold on pepper fruits via inhibiting fungal growth and triggering fruit defense. *Foods* 12:3458.

5 Bacteriological Spoilage and Detection in Food Products

Simge Aktop and Pınar Şanlıbaba

5.1 INTRODUCTION

Food spoilage refers to the chemical, physical, and microbiological processes that contribute to decreased food quality and safety. Thus, food spoilage is associated with alterations in the flavor, smell, color, texture, and nutritional values of foods. Spoiled food can harm human health. Microbial spoilage happens due to microorganisms like bacteria, yeast, and mold (Odeyemi et al., 2020). Bacterial spoilage is a common cause of foodborne illness and a major risk to food safety and public health, as pathogens can grow in food, leading to infections or poisoning. Eating food with microorganisms like *Escherichia coli*, *Listeria monocytogenes*, *Staphylococcus aureus*, *Vibrio cholerae*, and *Yersinia enterocolitica* can lead to issues such as fever, dizziness, nausea, throwing up, diarrhea, and more serious health problems. Some microorganisms are responsible for the production of harmful toxins (Moi et al., 2022). *Clostridium botulinum*, *C. perfringens*, and enterotoxigenic *E. coli* (ETEC) are examples of these. Bacteria that cause food spoilage can negatively affect the sensory properties of the food. In particular, negative changes in smell, taste, and appearance are not welcomed by the consumer. In addition, foods may experience nutritional value declines due to bacterial spoilage. All these negative changes cause food waste and economic losses. To prevent this situation, spoilage bacteria must be identified. It is very important to manage the process according to the bacteria found later. Many different methods, from cultural methods to molecular methods, are being applied and developed in detecting bacteria in foods. Among these, methods such as biosensors and polymerase chain reaction (PCR) have attracted the attention of researchers in recent years. These methods are frequently used because they provide fast and sensitive results in finding bacteria (Jayan et al., 2020; Odeyemi et al., 2020).

5.2 BACTERIA CAUSING FOODBORNE MICROBIAL DISEASES AND FOOD SPOILAGE

Foodborne pathogens both cause food safety hazards and negatively affect food quality. Therefore, they are biological threats that need to be taken into consideration. Some bacteria produce acid and gas. This situation causes negative changes in the

DOI: 10.1201/9781003540113-5

properties of the food, such as taste and smell. While *L. monocytogenes* is responsible for the flavor change in milk and meat, bacteria such as *E. coli* and *S. aureus* are responsible for acid and gas production in foods. *B. cereus* creates toxins, especially in starchy foods, and thus causes taste and texture changes. These not only affect the taste and smell of foods but also change the color and its texture. This may result in the food being disliked and not consumed. In addition, these bacteria can cause many foodborne diseases. The ingestion of food contaminated with pathogenic microorganisms or toxins can result in foodborne illnesses. This chapter examines foodborne diseases in the following three main categories: foodborne infections, toxicoinfections, and intoxications (Comi, 2017; Majumdar et al., 2018).

5.2.1 Foodborne Infections

5.2.1.1 *Salmonella*

Salmonella species, recognized as one of the most significant foodborne pathogens globally, are characterized as Gram-negative, motile, mesophilic, facultatively anaerobic, and non-spore-forming bacilli (Stella and Ifeanyi, 2018). More than 2,500 serovars of *Salmonella* have been identified and are considered pathogenic. Most *Salmonella* serovars can grow under the following optimal conditions: temperature 7–48°C, pH 6.5–7.5, water activity (a_w) values 0.94–0.99 (Kemal, 2014; Moura-Alves et al., 2020).

Salmonella can be transmitted to humans from animal sources, such as meat, milk, and eggs, as well as from foods contaminated with *Salmonella*, including green vegetables, condiments, and drinking water. Salmonellosis, a major global cause of mortality, results from the consumption of foods contaminated with *Salmonella*. The annual numbers of salmonellosis cases and salmonellosis-related deaths are approximately 93.8 million and 155,000, respectively (Mkangara, 2023). The two primary serotypes associated with salmonellosis are *Salmonella enterica* and *Salmonella bongori* (Kemal, 2014). Symptoms of salmonellosis encompass headache, sudden-onset fever, nausea, abdominal pain, diarrhea, vomiting, enteritis, and septicemia (Pal et al., 2015; Sykes and McDonough, 2021).

5.2.1.2 Pathogenic *Escherichia coli*

Escherichia coli, a member of the *Enterobacteriaceae* family, is characterized as a Gram-negative, facultatively anaerobic, non-spore-forming, primarily motile, capsulated, short rod-shaped, mesophilic, catalase-positive, and oxidase-negative bacterium. Some *E. coli* strains are non-motile. The optimal growth conditions for *E. coli* are as follows: temperature 35–40°C, a_w values 0.95, pH close to neutral (Sharma et al., 2016; Kamal et al., 2018; Thakur et al., 2024).

Non-pathogenic strains of *E. coli* are constituents of the intestinal microbiota in humans and warm-blooded animals. *E. coli* is considered an indicator of fecal contamination when it is detected outside of its natural environment. *E. coli* has been identified in various foods, including raw or undercooked beef, unpasteurized milk and dairy products, as well as fruits and vegetables irrigated with contaminated water. The use of contaminated water and fertilizer and unhygienic practices result in *E. coli* contamination of these foods. *E. coli* contamination can cause food to spoil and lead to health problems. When food gets contaminated by *E. coli*, it changes the smell, flavor, color, and texture of the food, negatively impacting the quality and safety of the products (Abdel-Aziz et al., 2016; Galli et al., 2016; Hassan et al., 2021).

Intestinal pathogenic *E. coli* is classified into six groups based on their pathotypes: enterohemorrhagic *E. coli* (EHEC), enteropathogenic *E. coli* (EPEC), enterotoxigenic *E. coli* (ETEC), enteroinvasive *E. coli* (EIEC), diffusely adherent *E. coli* (DAEC), and enteroaggregative *E. coli* (EAEC). EPEC and ETEC cause foodborne toxicoinfections, while EHEC, EIEC, EAEC, and DAEC are associated with foodborne infections (Kim et al., 2020; Hrustemović et al., 2021).

EHEC can lead to hemolytic uremic syndrome (HUS) and hemorrhagic colitis (Amin et al., 2022). Symptoms of EHEC infections include fever, intense stomach pain, diarrhea (which can be bloody or not), and vomiting. EHEC infections may negatively impact the central nervous system. The primary EHEC type that leads to these issues is O157:H7. This group produces Shiga toxin (Stx), which is responsible for HUS. Varieties in this group are known as Stx toxigenic *E. coli* (STEC) or verotoxigenic *E. coli* (VTEC) due to their harmful effects on Vero cells. The toxins enter the bloodstream and can harm organs like the intestines, brain, kidneys, and liver (Shinde et al., 2020; Hassan et al., 2021; Shen et al., 2022).

EIEC group types go into the epithelium and grow, causing inflammation in the mucosa. The ingestion of these strains results in symptoms such as shigellosis-like dysentery, high fever, diarrhea, abdominal cramps, and headache. In certain instances, mucus and blood might be detected in the stool. The most common serotype of EIEC is O:124 (Bhattacharya et al., 2020b; Govindarajan et al., 2020; Saad et al., 2023).

EAEC infection is frequently associated with long-term diarrhea in newborns and children in developing countries. The EAEC strains produce heat-resistant cytotoxins. Infections with EAEC strains are frequently associated with afebrile watery diarrhea. Meanwhile, infection with DAEC, which can invade intestinal epithelial cells, can cause diarrhea, especially in young children. The primary symptom of a DAEC infection is watery diarrhea; however, other symptoms, including fever, abdominal cramps, and vomiting, may also occur (Cabrera-Sosa and Ochoa, 2020; Hassan et al., 2021).

5.2.1.3 *Listeria monocytogenes*

Listeria monocytogenes is a Gram-positive, facultatively anaerobic, psychrotrophic, non-encapsulated, catalase-positive, oxidase-negative, halotolerant, and non-spore-forming bacterium known to cause foodborne illness and contribute to food spoilage (Şentürk et al., 2022). The optimal growth conditions for *L. monocytogenes* are as follows: temperature 0–45°C, pH 6.0–8.0, but can grow in the pH range of 4.1–9.6. *L. monocytogenes* can produce acid from carbohydrates but do not produce gas. Based on their O and H antigens, *L. monocytogenes* can be divided into several serotypes (groups 1/2a, 1/2b, 1/2c, 3a, 3b, 3c, 4a, 4ab, 4b, 4c, 4d, 4e, and 7). Serotypes 1/2a, 1/2b, and 4b are the most important pathogenic strains (Şanlıbaba and Uymaz, 2015).

Ready-to-eat and long-term refrigerated foods, including raw or pasteurized milk, red meat, fish, shellfish, soft cheese, ice cream, fermented meat products, vegetables, fruits, and poultry, are frequently contaminated with *L. monocytogenes*.

This bacterium has also been detected in various environments, such as soil, water, sewage, decaying vegetables, and silage. Additionally, *L. monocytogenes* has been isolated from the feces of infected domestic and wild animals. Notably, *L. monocytogenes* can thrive at refrigerator temperatures and in humid, saline, and dry conditions (Bakış and Derleme, 2021; Taylor and Zhu, 2021).

The mortality rate of listeriosis, which is caused by *L. monocytogenes*, is approximately 20–30% worldwide (Osek et al., 2022). Symptoms of listeriosis include fever, headache, vomiting, and flu-like manifestations, which may occur with or without diarrhea. *L. monocytogenes* can lead to meningoencephalitis and septicemia in newborns. Furthermore, this bacterium has the ability to cross the feto-placental barrier, potentially resulting in spontaneous abortions or stillbirths during pregnancy. Individuals at higher risk for developing listeriosis include newborns, pregnant women, fetuses, the elderly, and patients with compromised immune systems, such as those with cancer (Wang et al., 2021; Valenti et al., 2021).

5.2.1.4 *Campylobacter*

Campylobacter is a Gram-negative, non-spore-forming, and encapsulated bacterium characterized by polar flagella at one or both ends (Liu, 2021). Most species of *Campylobacter* are microaerophilic, necessitating 5–10% carbon dioxide for optimal growth. These bacteria exhibit a respiratory-type metabolism that utilizes amino acids or intermediates of the tricarboxylic acid cycle in place of carbohydrates. Although the optimal growth temperatures for *Campylobacter* species can vary, they typically range from 37 to 42°C (Ammar et al., 2021; Huang and Garcia, 2022).

The risk factors for *Campylobacter* infections are drinking raw milk and contaminated water, the consumption of animal products, and contact with other animal species. In industrialized countries, approximately 50 to 70% of *Campylobacter* infections are attributed to the consumption of poultry meat and poultry products. This high prevalence is due to the presence of *Campylobacter* in the gastrointestinal tracts of birds, particularly chickens and turkeys (Facciolà et al., 2017).

Campylobacteriosis, caused by *Campylobacter*, is one of the foremost widespread infectious diseases in the last century. The most prevalent symptoms of campylobacteriosis include fever, nausea, diarrhea, abdominal pain, and cramps. In infants and travelers, *Campylobacter* infections can cause diseases such as meningitis, HUS, endocarditis, colitis, pancreatitis, and cholecystitis. The infectious dose for campylobacteriosis is as low as 500 *Campylobacter* cells (Facciolà et al., 2017; Zhang and Sahin, 2020). *Campylobacter jejuni* is the cause of many human enteritis cases. In rich nations, *C. jejuni* infection often leads to white blood cells and other signs, like fever, blood in the stool, and diarrhea with inflammation (Man, 2011; Mandal et al., 2018).

5.2.1.5 *Shigella*

Shigella is a bacterium that is part of the *Enterobacteriaceae* family. It is Gram-negative, facultatively anaerobic, and does not form spores. *Shigella* grows best at a temperature of 37°C and a pH level of 5 to 9 (Khan et al., 2022). The main ways *Shigella* infections spread are by eating or drinking food and water with fecal material and through direct contact with infected individuals. *Shigella* can often be found in raw milk, chicken, salads, seafood, vegetables, and uncooked foods. Usually, food gets contaminated due to lack of hygiene when food is being prepared or served (Odo et al., 2021).

Shigella infections are estimated to cause around 1.1 million deaths annually worldwide (Lampel, 2009). The types *S. dysenteriae*, *S. boydii*, *S. flexneri*, and *S. sonnei* cause shigellosis, which has symptoms like belly pain, vomiting, fever, watery diarrhea, and blood in the stool (Kotloff et al., 2018; Uddin et al., 2023). Other diseases caused by *Shigella* may also occur. The main ones are ulcer and hemolytic uremic syndrome (HUS). Shigellosis is also an important cause of diarrhea. Children and babies under the age of 5 are especially affected by this condition. Shigellosis cases seen in elderly individuals are associated with malnutrition (Bhunia, 2018a).

5.2.1.6 *Vibrio*

V. parahaemolyticus and *V. vulnificus* frequently cause foodborne illnesses. Of these, *V. parahaemolyticus* has Gram-negative characteristics. They do not have the ability to form spores. Its optimal growth occurs at temperatures from 5°C to 43°C and pH levels at or below 5.0. It does not grow well in salt concentrations over 10%. Conversely, halophilic bacteria grow rapidly in 3% NaCl. Eating raw or undercooked fish and fish products contaminated with *V. parahaemolyticus* can lead to various illnesses. Moreover, *V. parahaemolyticus* can contaminate cooked foods from raw ones, which can cause gastroenteritis. Typical symptoms of *V. parahaemolyticus* infections are headache, nausea, vomiting, fever, stomach cramps, and severe watery diarrhea. Rarely, it can breach the intestinal barrier and enter the bloodstream, potentially causing septicemia (Kalburge et al., 2014; de Souza Santos et al., 2015; Yaashikaa et al., 2016).

V. vulnificus is a halophilic bacterium that does not produce spores. It grows optimally between 8°C and 43°C and requires *aw* values near 0.98. Infections from *V. vulnificus* are uncommon and generally mild, mostly linked to raw oysters (Baker, 2016; da Silva et al., 2021). However, severe cases can be fatal. The key problems from *V. vulnificus* are wound infections, which can arise from the gastrointestinal system or injured skin, and septicemia. Severe diarrhea, sometimes with blood, is another common issue. Individuals with weakened immune systems are at a higher risk of developing septicemia from *V. vulnificus* (Coerdt and Khachemoune, 2021).

5.2.1.7 *Yersinia enterocolitica*

Y. enterocolitica is a Gram-negative, non-capsulated, non-spore-forming bacterium in the *Enterobacteriaceae* family. The optimum growth temperature for this bacterium is between 30°C and 37°C. The pH range it needs to grow is 4.2–9.6 (King, 2011; Ozdemir et al., 2020). Cases of yersiniosis caused by *Y. enterocolitica* are mostly caused by pork and dairy products. Apart from these foods, the yersiniosis agent is also found in chicken meat, seafood, vegetables, and fruits (Chlebicz and Śliżewska, 2018). Symptoms of yersiniosis are generally the same as other foodborne illnesses. Fever, abdominal pain, diarrhea, and sometimes vomiting are among these symptoms. Serious health problems may occur in people with weakened immune systems infected with *Y. enterocolitica*, which is also associated with gastrointestinal diseases. Septicemia and myocarditis are some of these problems. In healthy individuals, the clinical picture is generally mild (Platt-Samoraj, 2022).

5.2.2 Foodborne Toxicoinfections

5.2.2.1 Enterotoxigenic *Escherichia coli* (ETEC) and Enteropathogenic *Escherichia coli* (EPEC)

ETEC, which results in traveler's diarrhea, exerts adverse effects on health by producing toxins. The symptoms of ETEC infections are headache, shivering, nausea, vomiting, and bloating. The pathogen adheres to and colonizes the intestinal epithelial cells via pili or the colonization factor antigen. The subsequent production of heat-stable enterotoxins (ST) and heat-labile enterotoxins (LT) promotes the development of disease (Rašková, 2020). The main risk groups for ETEC infections are infants and young children. ETEC infections frequently lead to deaths in infants and children due to malnutrition and dehydration. Humans serve as the reservoir for ETEC. However, the consumption of food or water in contact with an infected individual leads to the development of symptoms. Foods such as crab meat, mayonnaise, deli meats, Brie cheese, and salads are frequently contaminated with ETEC (Smith and Fratamico, 2017).

EPEC, which adheres to the mucosal surface and induces symptoms including fever, abdominal pain, vomiting, and diarrhea, is the main cause of watery diarrhea in infants and young children. The strains of this group are especially prevalent in areas with poor sanitation measures. Several foodborne EPEC infection outbreaks are associated with serotypes such as O111:H12 and O55:H6. Similar to ETEC, EPEC grows in meat, mayonnaise, salads, and deli meats, causing disease in humans (Navarro and Estrada-Garcia, 2010; Mare et al., 2021).

5.2.2.2 *Clostridium perfringens*

C. perfringens is a Gram-positive, anaerobic, and spore-forming bacterium that can lead to enteric infections in mammals and poultry. The optimum growth temperature range for *C. perfringens* is 43–47°C. Salt is a growth-limiting factor for *C. perfringens*, which cannot grow in the presence of approximately 6% NaCl (Juneja et al., 2021; Hwang et al., 2024). *C. perfringens* spores are resistant to environmental stress factors. Sporulation is important for the expression of toxins. The following six toxins are produced by *C. perfringens*: α-toxin (CPA), β-toxin (CPB), ε-toxin (ETX), iota-toxin (ITX), enterotoxin (CPE), and necrotic B-like toxin (NetB). These toxins are important in the initiation and progression of infections (Navarro et al., 2018).

C. perfringens infection is a common foodborne illness. In addition, cases of food poisoning caused by CPE are also common. Spores of *C. perfringens* have a durable structure. It can withstand conditions such as high heat. Undercooked meat or poultry often plays a role in outbreaks. Cells can regrow as a result of germination of *C. perfringens* spores. Temperatures between 10°C and 54°C are ideal for this (Bhattacharya et al., 2020a; Tian et al., 2022). Symptoms caused by *C. perfringens* may vary depending on the type of toxin. They might consist of fever, stomachache, throwing up, and diarrhea. Additionally, *C. perfringens* can cause harm to the liver and kidneys, gas gangrene, and gut illnesses (Fu et al., 2022). ETX produced by *C.*

perfringens type B and D strains is one of the most potent clostridial toxins, while CPB produced by type C strains causes pigbel disease (necrotizing enteritis). The disease outbreak is usually associated with pork feasts where pork is undercooked (Bokori-Brown et al., 2011; Gohari and Prescott, 2022).

5.2.2.3 *Bacillus cereus*

B. cereus is a Gram-positive, facultative anaerobic, and endospore-forming bacterium. The optimal growth conditions for *B. cereus* are as follows: temperature 4–50°C, a_w values ≥0.92, pH 4.5–9.3, NaCl concentrations <10% (Tewari and Abdullah, 2015). The spread of *B. cereus* can be attributed to the resistance of its spores to environmental stress factors. *B. cereus* is frequently isolated from various food products, including rice, pasta, meat and meat products, milk and dairy products, pasteurized liquid eggs, spices, ready-to-eat vegetables, dried foods, and sauces (Easa, 2011; Jovanovic et al., 2021). Most foodborne outbreaks of *B. cereus* occur at levels of $\geq 10^5$ CFU/g (EFSA, 2016).

Diarrhea and vomiting can occur following the consumption of food contaminated with *Bacillus cereus*. In the diarrheal syndrome associated with *B. cereus*, the ingestion of spores or viable cells leads to toxin production and bacterial colonization of the small intestine. This syndrome connects with three enterotoxins (Gharib et al., 2020; Jovanovic et al., 2021). Hemolysin BL (Hbl) is an enterotoxin complex made up of L1, L2, and B proteins which has hemolytic activity. These three components combine independently and form a transmembrane pore, promoting osmotic lysis (Jessberger et al., 2020). The production of enterotoxins happens in a temperature range of 10 to 43°C, with the best temperature being 32°C. The toxins are sensitive to heat and can break down quickly at a pH of 3. Also, these toxins can be hydrolyzed by proteolytic enzymes in the digestive tract. As a result, the diarrheal syndrome arises from eating cells instead of pre-existing toxins. However, in young children, the toxins in food are not destroyed due to their lower stomach acid levels (Griffiths and Schraft, 2017; Rusnan et al., 2020; Mohammadi et al., 2022).

In the emetic syndrome, the binding of preformed toxins to 5-HT3 receptors induces vomiting. These receptors stimulate the vagus afferent nerve, activating the vomiting center of the brain and triggering the vomiting reflex (Haque et al., 2021). Thus, the emetic syndrome resembles *Staphylococcus*-like food poisoning, which can be challenging for differential diagnosis. The toxin is produced during the early stationary phase of optimal growth (El-Arabi and Griffiths, 2021). The optimal temperature range for the production of the emetic *B. cereus* toxin is 12–37°C. When inadequate heat treatment is applied, toxin production may occur. As a result, cases of food poisoning may occur (Finlay et al., 2000).

5.2.2.4 *Vibrio cholerae*

V. cholerae is a Gram-negative, non-spore-forming, aerobic bacterium from the *Vibrionaceae* family. It can create gas, acid, and hydrogen sulfide from glucose and is affected by high temperatures, acid, and dryness. This bacterium can survive in cold conditions, especially in food at temperatures of 5 to 10°C, and it can grow at temperatures ranging from 10 to 43°C. It thrives in foods with a pH between 5 and 9.6 (Yaashikaa et al., 2016; Ebob, 2020).

V. cholerae is responsible for cholera and is divided into four serogroups: O1, O139, non-O1, and non-O139. The O1 and O139 serogroups can lead to severe dehydration due to their production of large amounts of watery diarrhea (Weil and LaRocque, 2020). In contrast, non-O1 and non-O139 serogroups are typically linked to diarrhea that resembles cholera. Consuming seafood, rice, vegetables, and ice contaminated with *V. cholerae* can result in cholera. Furthermore, cholera can also spread between people through infected feces (Arteaga et al., 2020).

5.2.3 Foodborne Intoxications

5.2.3.1 *Clostridium botulinum*

C. botulinum is a Gram-positive bacterium that does not need oxygen and can form spores. It generates a strong toxin called botulinum neurotoxin (BoNT), which leads to botulism, a serious foodborne illness that can result in high death rates (Meurens et al., 2023). There are different types of neurotoxins, labeled A to G, produced by *C. botulinum*. This bacterium is grouped into different categories based on the types of BoNTs they make: Group I produces A, B, and F types; group II has B, E, and F types; group III includes C and D types; and group IV produces type G BoNT (Bowe et al., 2022). Group I strains of *C. botulinum* can break down proteins and use amino acids for energy. Their spores are also heat-resistant. In contrast, group II strains do not break down proteins and are known as saccharolytic bacteria. The spores of group II are less heat-resistant compared to those of group I (Brunt et al., 2020; Poortmans et al., 2022; Popoff and Brüggemann, 2022). Group III strains typically affect birds and other animals. Group IV is now recognized as a new species, called *C. argentinense* (Rawson et al., 2023).

The toxins produced by *C. botulinum* cause diseases spanning from mild, self-limiting diarrhea to severe and life-threatening diseases, such as botulism, enterotoxemia, gangrene, and pseudomembranous colitis. BoNTs inhibit the release of acetylcholine from motor nerve endings, leading to impairment of the peripheral nervous system (Bhunia, 2018b; Poulain et al., 2020). Early signs of the disease include weakness, dizziness, and dry mouth, which can be followed by symptoms such as blurred vision, difficulty swallowing and speaking, constipation, mydriasis, and ptosis. Neurological symptoms, such as respiratory paralysis, may develop with disease progression (Baljić et al., 2020; Liu, 2024). The four clinical forms of botulism poisoning in humans are foodborne botulism, infant botulism, wound botulism, and adult intestinal colonization. The most important cause of foodborne botulism is the improper canning of foods at home. Heat treatments applied to commercially prepared low-acid canned foods have minimized *C. botulinum* infection outbreaks (Rasetti-Escargueil et al., 2020; Semenko et al., 2021).

5.2.3.2 *Staphylococcus aureus*

S. aureus is a Gram-positive, facultatively anaerobic, and non-spore-forming pathogen. The optimal growth conditions for *S. aureus* are as follows: temperature 6–48°C, pH 4–10, a_w value 0.83. Additionally, salt-tolerant *S. aureus* strains can grow in salt concentrations of up to 20%. To produce enterotoxin, the a_w value must be at least 0.86. Staphylococcal enterotoxins are protease-resistant protein toxins.

The protease-resistant property of enterotoxins is critical for their emetic effect as they pass from the stomach to the intestinal tract after ingestion (Pinchuk et al., 2010; Grispoldi et al., 2021).

Staphylococcal food poisoning (SFP), a common serious foodborne illness worldwide, results from the consumption of enterotoxin produced by *S. aureus* in contaminated food stored at appropriate temperatures. The most frequently implicated foods in SFP outbreaks include meat and meat products, milk and dairy products, egg products, salads, and cream-filled pastries (Kadariya et al., 2014; Argaw and Addis, 2015). Symptoms of SFP include severe nausea, abdominal cramps, vomiting, and fatigue, often accompanied by diarrhea or acute gastroenteritis. Although the mortality rate of SFP is low, it is a risk factor for infants, the elderly, and immunocompromised individuals (Abdallah et al., 2024).

5.3 BACTERIAL SPOILAGES ADVERSELY AFFECTING FOOD QUALITY

Some bacteria, including *Pseudomonas*, *Proteus*, *Lactobacillus*, *Leuconostoc*, *Moraxella*, *Acinetobacter*, *Alcaligenes*, *Flavobacterium*, and *Serratia* species, adversely affect the physical, chemical, sensory, and textural properties of foods rather than causing diseases. These bacteria can grow in foods such as meat, milk, eggs, fermented vegetables and fruits, and cereal products, causing spoilage-related alterations, such as stickiness; undesirable odor, color, and taste formation; softening; gas formation; swelling; and roping.

5.3.1 Meat and Meat Products

Meat and meat products, which are a good source of nutrients for bacteria, are one of the most common food groups that undergo bacterial spoilage. Bacteria can contaminate meat and meat products at every stage, from slaughter to consumption. After the animal is killed, the meat can get microorganisms from outside of the animal and from bacteria in its gut (Sallam et al., 2020). Also, microorganisms can come from clothes, knives, and tools used in making the meat. Meat spoiled by different bacteria shows changes in its physical, chemical, and sensory properties, such as changes in color and surface, sticky texture, bad smell and taste, fat oxidation, and putrid deterioration (Morshdy et al., 2023; Madhumala and Soraganvi, 2024).

Color change is a clear sign of meat going bad. The red of the meat can change to brown or green because of chemicals like hydrogen sulfide and peroxide that bacteria make, including *Lactobacillus* and *Leuconostoc*, when meat is contaminated. Stickiness on the surface also shows spoilage and comes from bacteria like *Acinetobacter*, *Alcaligenes*, *Bacillus*, *Leuconostoc*, *Micrococcus*, *Moraxella*, and *Streptococcus* (Ramamoorthi et al., 2010; Odeyemi et al., 2020; Madhumala and Soraganvi, 2024). Bacteria on meat can create putrid taste and smell. For meat to smell bad and feel sticky, the bacteria levels need to be at 10^6 CFU/cm^2. Bacteria like *Pseudomonas* and *Achromobacter* lead to fat oxidation (Pellissery et al., 2020). This fat oxidation produces acids and aldehydes, which create unpleasant smells in meat. Additionally, certain pigmented bacteria and *Photobacterium* species change the color on the surface of the meat. Bacteria such as *Photobacterium* species produce phosphorescent substances in meat. The growth of microorganisms such as some *Clostridium* species and the coliform bacteria in meat promotes the production of butyric, formic, and propionic acids, which confer a sour taste and odor to the meat (Comi, 2017). *Clostridium* species are also involved in putrefactive spoilage. During putrefactive spoilage, compounds such as ammonia, indole, amine, and hydrogen sulfide are released due to the breakdown of proteins by bacteria, resulting in the development of undesirable odor. Some species of *Alcaligenes*, *Clostridium*, and *Pseudomonas* are involved in putrefactive spoilage (Ajayeoba et al., 2020; Shao et al., 2021).

5.3.2 Milk and Dairy Products

Milk and dairy products are a good source of nutrition for spoilage bacteria. Bacteria such as *Pseudomonas*, *Lactobacillus*, *Proteus*, *Lactococcus*, *Enterococcus*, *Bacillus*, *Clostridium*, *Micrococcus*, and *Alcaligenes* species can cause spoilage in milk and dairy products under suitable conditions. The contamination of milk and dairy products with these bacteria leads to gas formation; roping; proteolysis; undesirable odor, taste, and color formation; and changes in milk fat and alkali formation (Dasriya et al., 2024; Kumar and Patyal, 2024).

Milk that has been milked under aseptic conditions contains few bacteria. However, milk can be contaminated with spoilage bacteria at any stage after milking. The principal carbohydrate of milk is lactose. The number of bacteria containing lactose-hydrolyzing enzymes in milk is low. Lactic acid bacteria (LAB), coliform bacteria, and some Gram-negative bacteria can hydrolyze lactose, promoting milk spoilage. Lactic acid, which is formed by the hydrolysis of lactose in the environment, causes souring and coagulation in milk. LAB, such as *Lactobacillus lactis*, are the most common bacteria involved in milk souring. After milking, milk can be quickly cooled and stored for a few days. During this cold storage period, *Alcaligenes*, *Flavobacterium*, and *Pseudomonas* species, and some coliform bacteria, can grow in milk and alter the taste (Muhović et al., 2023; Kumar and Patyal, 2024). These bacteria, which can metabolize milk proteins, confer bitter, fruity, and undesirable aromas to the milk. Bacteria such as *Bacillus*, *Clostridium*, *Enterococcus*, *Proteus*, and *Pseudomonas* species can grow in milk that is not cooled rapidly after milking (Angelidis, 2015; Clark, 2017). If *Lactobacillus* and *Lactococcus* species grow along with these bacteria, spoilage-related alterations, such as gas formation, lipolysis, and proteolysis, occur in milk. Some *Alcaligenes* species and coliform bacteria cause roping in milk. The production of viscous extracellular polysaccharides by the bacteria contributes to roping (Ledenbach and Marshall, 2009; Kumar and Patyal, 2024).

In addition to raw milk, bacterial spoilage can also be observed in pasteurized and ultra-high-temperature milk. Pasteurization generally inactivates various acid-forming bacteria in milk (Kumar and Patyal, 2024). However, pasteurization cannot prevent the growth of thermoduric bacteria, including *Enterococcus*, *Streptococcus*, *Micrococcus*, and *Lactobacillus* species, which promote lactic acid fermentation in milk. Coliform bacteria cause spoilage by producing compounds such as lactic acid, CO_2, H_2, acetic acid, and alcohol in milk. Similarly, *Clostridium* species cause spoilage by producing compounds, such as butyric acid, CO_2, and H_2 in milk (Ledenbach and Marshall, 2009; Júnior et al., 2018).

Heat treatment during the production of dairy products, such as concentrated milk, thickened milk, and milk powder, inactivates the vegetative cells of most bacteria, but not thermophilic bacteria and their spores. The germination of thermophilic bacterial spores under suitable conditions promotes spoilage in thickened and dried dairy products. Meanwhile, the growth of some *Bacillus* species results in sedimentation-type spoilage in milk. *Clostridium* species cause package explosions by producing gas. Additionally, both *Bacillus* and *Clostridium* species are responsible for the formation of bitter taste in these products (Lopez-Brea et al., 2017; Deeth, 2020; Bilska et al., 2024).

5.3.3 Fish and Seafood

Fish meat, which is a good source of energy for bacteria owing to its rich nutritional content, is more sensitive to bacterial contamination than red meat. Bacterial spoilage in fish begins after the development of rigor mortis. To extend the shelf life of fish, the rigor mortis process can be extended using processes, such as cooling and lowering the pH, delaying bacterial spoilage. Fish contains high levels of protein, glycogen, non-protein nitrogenous compounds, and peptides. Additionally, the pH values of fish are high. These factors establish a suitable environment for the growth of bacteria (Comi, 2017; Yu et al., 2020).

Facultative anaerobic bacteria and Gram-negative aerobic rods cause spoilage in fish. As fish are generally stored at low temperatures, *Pseudomonas* species can grow in fish meat and cause spoilage. *Moraxella*, *Acinetobacter*, and *Flavobacterium* species can cause spoilage in products stored at cold temperatures (Salama and Chennaoui, 2024). Bacteria such as *Bacillus*, *Clostridium*, *Escherichia*, *Micrococcus*, *Proteus*, and *Serratia* species are responsible for spoilage in fish that are inadequately cooled or maintained at room temperature. *Pseudomonas fluorescens*, *Sarcina*, and *Bacillus* species and some micrococci are responsible for discoloration in fish. The growth of these bacteria confers the fish meat with colors, such as yellow, red, pink, and green. Also, changes like stickiness, rancidity, and loss of muscle can happen in fish because of bacteria spoilage (Erkmen and Bozoglu, 2016; Odeyemi et al., 2020; Veljanovski and Blazhekovikj-Dimovska, 2023).

Seafood can get spoiled by bacteria. Shellfish like crab, lobster, and shrimp have lots of non-protein nitrogenous compounds that bacteria like to consume. The main bacteria found in shellfish are Gram-negative rods and *Pseudomonas* species (Zhuang et al., 2021; Kios et al., 2023). However, lobsters and crabs rarely spoil because they are frozen right after processing or sold alive. Bacteria such as *Pseudomonas*, *Bacillus*, *Alcaligenes*, *Moraxella*, *Acinetobacter*, *Flavobacterium*, *Vibrio*, and *Micrococcus* can grow on these types of shellfish and cause spoilage. Shrimp are more prone to spoilage as they die soon after being caught. Bacterial spoilage in shrimp is often caused by *Vibrio*, *Moraxella*, *Pseudomonas*, and *Acinetobacter*. When bacteria act on shellfish, it leads to issues like stickiness, bad taste, color loss, and softening of the shellfish meat. Mollusks (such as oysters and scallops) can also be spoiled by Gram-negative rods, *Pseudomonas*, *Moraxella*, *Acinetobacter*, *Micrococcus*, *Vibrio*, and *Flavobacterium* species (Erkmen and Bozoglu, 2016; Parlapani et al., 2024; Salama and Chennaoui, 2024).

5.3.4 Eggs and Egg Products

In good breeding situations, the inside of the egg is thought to be clean. But if the eggshell is broken, microorganisms can get into the egg through tiny holes on the surface. This can cause the egg to go bad if environmental factors do not stop the microorganisms from growing. The usual microorganisms that spoil eggs are Gram-negative bacteria. Spoilage from Gram-positive bacteria is uncommon because lysozyme in egg whites can stop them. Besides lysozyme, eggs have other substances that inhibit bacteria, like conalbumin and antivitamin proteins. As these inhibitory substances occur in egg whites, bacterial growth is inhibited in egg whites. In contrast, egg yolk is a good growth medium for bacteria owing to its rich nutritional content and high pH. Therefore, egg yolk is often associated with bacterial spoilage (Baron et al., 2023). Bacteria that cause egg spoilage include *Proteus*, *Enterobacter*, *Pseudomonas*, *Moraxella*, *Acinetobacter*, *Alcaligenes*, *Aeromonas*, and *Serratia* species. Bacterial activity in the egg promotes spoilage-related changes, such as undesirable color, taste, and odor formation. These bacteria transform the color of the egg yolk to green, pink, black, blue, and red and confer a rotten odor (Bist and Bist, 2021).

Egg products are widely used in various foods, such as cakes, pasta, biscuits, and sauces, due to their various functional properties. Additionally, eggs can be separated from the shell and used as liquid or dried in powder form. As dried eggs have low a_w values, bacterial growth is inhibited under physiological conditions. In contrast, liquid eggs can undergo bacterial spoilage. Liquid eggs are subjected to pasteurization, which inactivates most bacteria. Although these products are maintained in cold storage after pasteurization, Gram-negative bacteria can cause spoilage. Bacterial spoilage leads to undesirable taste and odor formation in liquid eggs (Jan et al., 2018; Baron et al., 2023).

5.3.5 Fermented Products

Fermented products are obtained after the microbial fermentation of certain raw materials. The microflora of these fermented products comprise a high number of microorganisms. The metabolites produced by these microorganisms enhance the shelf life of the fermented product. This is because metabolites such as alcohol and organic acids establish an acidic environment, which prevents microbial growth. However, fermented products can be spoiled by some bacteria (Gänzle, 2015; Sahu and Bala, 2017).

Bacteria such as *Lactobacillus*, *Enterococcus*, *Weissella,* and *Leuconostoc* species cause spoilage in fermented meat products. Mesophilic bacteria, such as *Bacillus* and *Clostridium*, can also cause spoilage in these products. In particular, the slow growth of homofermentative LAB during fermentation is conducive to this type of spoilage (Fessard and Remize, 2017; Kaveh et al., 2023). The spoilage is characterized by slippery layer formation, color change, and sour taste. Bacteria that cause this spoilage include *Brochothrix thermosphacta* and *Lactobacillus*, *Enterococcus*, and *Weissella* species (Ibrahim et al., 2021; Woraprayote et al., 2023). *B. thermosphacta* and *Lactobacillus sakei* are linked to sausage spoilage (Dias et al., 2013). Also, certain *Lactobacillus* and *Enterococcus* species that grow beneath the product coating lead to a sour taste and change the color of foods. Some LAB, like *Enterococcus*, *Lactobacillus*, and *Weissella* spp., can turn the color of fermented meat products green (Danilović and Savić, 2017).

Fermented dairy products, such as cheese, yogurt, and butter, can undergo bacterial spoilage. Cheese spoilage is associated with gas formation; undesirable color, taste, and odor formation; and textural deterioration. Some *Bacillus* and *Clostridium* species cause gas formation in cheese. Gas formation can cause undesirable changes in cheese aroma and curd spoilage. Acid-tolerant proteolytic bacteria promote the formation of a bitter taste in cheese (Ledenbach and Marshall, 2009; Sreeja and Gawai, 2019; Kaya et al., 2023). However, species such as *Lactobacillus cremoris* and *Lactococcus lactis* cause bitter taste formation in semi-hard cheese (Benson, 2024). Porous structure formation is frequently observed during the ripening of cheese, which can be attributed to the growth of some *Clostridium* species, heterofermentative LAB, and coliform bacteria. Cheese spoilage is also associated with color alterations. Some *Lactobacillus* and *Propionibacterium* species that grow in various cheeses contribute to the formation of yellow, pink, and brown colors (Ledenbach and Marshall, 2009; del Olmo et al., 2018). In cheese with low acidity, some *Clostridium* species are responsible for putrefaction (Carminati et al., 2024).

In yogurt, a fermented milk product, most bacteria cannot grow due to high acidity. However, some LAB can cause yogurt spoilage. *Lactobacillus delbrueckii* subsp. *bulgaricus*, which grows in yogurt, can cause bitter aroma formation and stickiness. Acetaldehyde, which is formed with the growth of *Lactobacillus* and *Streptococcus* species, can adversely affect yogurt aroma. Flavors such as sour and buttery aromas can develop in yogurt (Purwandari, 2009; Delavenne et al., 2015). Most bacterial spoilage in butter originates from the cream from which the butter is prepared. Several bacteria, especially LAB, grow in butter if the cream is maintained under suitable conditions for a long time. Bacterial spoilage seen in milk can also occur in cream. However, coliform bacteria grow in unsalted butter and negatively affect the aroma (Wan Ismail et al., 2022; Sadri-Saeen et al., 2024).

Bacterial spoilage is frequently observed in fermented vegetables. The spoilage of fermented products such as pickles and olive brine is mediated by halophilic bacteria because these products contain high levels of salt. In products with low salt concentrations, coliform bacteria and heterofermentative LAB can grow, produce CO_2, and cause swelling in the product (Chen, 2021; Safkan et al., 2024). *Enterobacter*, *Aeromonas*, and *Bacillus* species are involved in pickle spoilage and can alter the flavor, texture (such as softening), and color. *Bacillus nigrificans* is frequently associated with the formation of black color in pickles (Sultana et al., 2014; Liu et al., 2022; Oyedeji et al., 2023). Spoilage-related alterations such as softening, color transformation, swelling, and undesirable odor and taste formation are frequently observed in olives, which are frequently fermented fruits. Olive spoilage is mediated by coliform bacteria and heterofermentative LAB. Moreover, the cause of yeast or white spots seen in olives is usually *Lactobacillus plantarum* (Todorov et al., 2010; Arroyo-López et al., 2012).

Bacterial spoilage happens in drinks like wine, beer, and vinegar. In vinegar, stickiness is from acetic acid bacteria. These bacteria can change acetic acid in vinegar into water and CO_2 (Giudici et al., 2017; Wang et al., 2023). Wine spoilage often happens because of aerobic and facultative anaerobic bacteria. Acetic acid bacteria break down alcohol in wine, resulting in compounds like acetic acid, acetaldehyde, and CO_2, which create a vinegar smell (Osborne and Edwards, 2005; Bartowsky and Henschke, 2008; Cruess, 2009). Also, *Lactobacillus* species in wine can cause changes such as cloudiness, poor color, and unwanted smells (Rubio-Bretón et al., 2017).

Beer spoilage is usually caused by LAB. *Lactobacillus* and *Pediococcus* species often spoil beer (Garofalo et al., 2015). *Lactobacillus* species in beer can cause problems like roping, cloudiness, and off-flavors, while *Pediococcus* species lead to higher acidity and turbidity. *Acetobacter* and *Gluconobacter* species also cause spoilage issues like cloudiness, sour taste, and stickiness. *Clostridium* species might occur during beer production and create butyric acid fermentation, resulting in an unpleasant odor (Raspor and Goranovič, 2008; Brožová et al., 2018; de Lima et al., 2023).

5.3.6 Cereals and Cereal Products

The low moisture contents and a_w values of cereals limit the growth of most bacteria. Increased moisture content promotes the growth of bacteria, leading to spoilage. While cereal products, such as bread, pasta, and dough, undergo bacterial spoilage under high-moisture-content conditions, foods with low moisture content, such as cereal grains and cereal flours, do not undergo microbial spoilage as long as the moisture content does not exceed 12%. However, bacteria can grow in these products when their moisture content exceeds 17%. Coliform bacteria cause spoilage in cereal grains and cereal flours by mediating acid fermentation, which leads to the development of undesirable aromas. Bacterial growth is inhibited in other cereal products, such as pasta, cake, pastry, and similar products, due to the low a_w value and the applied heat treatment. However, LAB and coliforms can grow in pasta produced under inappropriate production conditions before the drying process and cause acid fermentation (Alegbeleye et al., 2022; Panza et al., 2023).

Bread is frequently associated with bacterial spoilage. Most bacteria are inactivated during the heat treatment applied during bread production. Under suitable temperature and humid climate conditions, bacteria can grow in bread and cause spoilage-related alterations, such as roping (Pacher et al., 2022). Roping promotes stickiness, changes the color of bread from yellow to brown, and induces a soft texture. Some *Bacillus* species are mainly involved in bread spoilage. The spores of these *Bacillus* species are resistant to the heat treatment applied during the baking of the bread. Therefore, these *Bacillus* species are not inactivated and subsequently grow under suitable conditions and cause spoilage. Bacterial spoilage of bread is also associated with the formation of a red color, which is usually caused by *Serratia marcescens* (Pereira et al., 2020; Madhumala and Soraganvi, 2024).

5.3.7 Fruits and Vegetables

Fruits and vegetables can frequently undergo bacterial spoilage owing to their high water and nutrient contents. Although the inner tissues of fruits and vegetables do not comprise microorganisms, specific microflora are detected on their outer surfaces. Bacteria can pass from these regions to the inner tissues of fruits and vegetables, which are damaged for various reasons. Some spoilage-related alterations, such as undesirable color, aroma, and texture formation, can occur in these foods. As fruits have high acidity, acid-tolerant bacteria such as LAB along with *Acetobacter* and *Gluconobacter* species are involved in fruit spoilage. The acidity of vegetables is lower than that of fruits; therefore, vegetables frequently undergo bacterial spoilage. Important bacteria that cause vegetable spoilage include *Erwinia*, *Pseudomonas*, *Xanthomonas*, *Corynebacterium*, and *Pectobacterium* species. In vegetables, bacterial species promote spoilage-related alterations, such as soft rot, black discoloration, and undesirable odor formation (Jay et al., 2005; Barth et al., 2009; Alegbeleye et al., 2022).

Fruit products, such as jam, fruit juice, and fruit concentrates, generally have low pH values (2.5–4.0) and high sugar contents (Askar and Treptow, 2013), which limit the growth of bacteria in these products. However, acid-resistant bacteria, such as *Acetobacter*, *Lactobacillus*, and *Leuconostoc* species, can grow in these products and cause spoilage. Bacteria that grow in fruit juices stored at room temperature cause alcoholic or acidic fermentation, adversely affecting the color, taste, and aroma of the juice (Alegbeleye et al., 2022).

5.4 METHODS FOR DETECTION OF BACTERIA

Several bacteria cause food spoilage and foodborne illnesses. Microorganisms in food can provide information about faulty processes in production. It can also be used to determine final product quality. Therefore, it is critical to determine the bacterial load in foods quickly and precisely. In this section, information is given about traditional and rapid methods.

5.4.1 Traditional Methods

Traditional methods are frequently used to identify bacteria. The identification of bacteria using the culture method and a microscope is within the scope of traditional methods. These methods have many advantages, some of which are that they are cheap and can be used in many laboratories.

5.4.1.1 Culturing and Microscopy

Identification of bacteria using the culture method and a microscope is a traditional method that has been used for many years. In the culturing method, bacteria are grown in special environments for a specific type of bacteria. These bacteria can then be observed as colonies. MacConkey agar and xylose lysine deoxycholate (XLD) agar are examples of these special media (Váradi et al., 2017; Pandey et al., 2022). In these special environments, bacteria can be identified according to certain characteristics.

When identifying bacteria using a microscope, information can be obtained about the properties of the bacterial cell. Techniques such as Gram staining can be used to obtain information about cell walls and other properties. In this method, bacterial shapes and sizes can be observed. Therefore, this method is very effective in detecting bacteria. Traditional methods can be used separately or together. Thus, more comprehensive and reliable results can be obtained. In addition, both numerical and morphological data of bacteria in foods can be collected (Roane and Pepper, 2015).

The standard plate count (SPC) method is an established technique for measuring live bacteria in food. It involves diluting a bacterial suspension several times, then spreading or pouring a measured amount onto a medium and incubating it at a specific temperature for a designated time. The bacteria are counted by observing how many colonies develop. While effective for detecting bacteria in food, the SPC method can be labor-intensive, requires various materials, and consumes significant time (Basak and Shetty, 2021).

The most probable number (MPN) method is a traditional statistical technique used primarily to estimate low bacteria counts (Alexander and MacLean, 2020). In this method, a sample is combined with a fixed amount of media at different dilution levels. Bacterial growth is observed at each dilution, using statistical models to predict the total bacteria count based on growth data. This approach is commonly used for checking coliform bacteria in drinking water. The MPN method offers good sensitivity and reliability for detecting low bacteria levels and can analyze multiple samples simultaneously (Tambi et al., 2023).

The methylene blue dye reduction test (MBRT) assesses bacterial activity and quantity. It measures how quickly bacteria can reduce redox dyes, such as methylene blue, present in a medium. MBRT is useful for assessing milk quality. Methylene blue remains blue in oxygen but changes color during bacterial growth, indicating activity. The advantages of this method include rapid bacterial detection, ease of use, and low cost, but it fails to identify some specific bacteria types (Nandy and Venkatesh, 2010; Mayra et al., 2020).

Direct microscopic counting (DMC) is a technique for quickly determining bacterial counts using a microscope. DMC employs specialized counting tools, such as the hemocytometer and the Petroff–Hausser counting chamber (Basak and Shetty, 2021; Rahman and Butzin, 2024). The hemocytometer has a defined area that allows for the counting of bacterial cells when viewed under a microscope. Thus, the number of bacterial cells in a specific volume can be determined (Murugalatha et al., 2022). The Petroff–Hausser counting chamber allows for a quick and precise counting of bacterial cells in the chamber area observed under the microscope. Additionally, fluorescence microscopy uses specific dyes or fluorescent markers, enhancing the accuracy of bacterial counting (Jenkins and Maddocks, 2019; Basak and Shetty, 2021).

5.4.2 Rapid Methods

Traditional methods to find bacteria use much time, often taking days for results. Faster methods can provide results more quickly than traditional methods. This is beneficial for rapid actions and decisions in urgent situations, such as food poisoning cases. Additionally, rapid methods are usually more sensitive and precise. Rapid bacteria detection methods include molecular techniques, immunological techniques, bioluminescence techniques, spectroscopic techniques, and biosensors (Jayan et al., 2020; Vasavada et al., 2020).

5.4.2.1 Molecular Methods

Molecular identification methods are modern technologies that focus on the genetic material to identify bacteria. These methods, which analyze bacterial DNA or RNA, deliver quick, accurate, and effective results. In recent times, various molecular techniques, such as PCR and DNA microarray, have enabled faster bacterial detection (Law et al., 2015). PCR and quantitative real-time PCR (qRT-PCR) can identify bacteria by amplifying specific genetic sections. DNA microarray technology permits the simultaneous examination of multiple bacterial species. Molecular methods yield quicker results compared to traditional culture techniques and can detect low amounts of bacterial material. Thus, molecular methods are crucial for food safety (Rasooly and Herold, 2008; Zhang et al., 2021).

PCR is the most common DNA-based technique. It involves separating DNA strands at high temperatures, then allowing primers to attach and synthesize new strands with the DNA polymerase enzyme at various temperatures (Shahzad et al., 2020). In the initial denaturation step, DNA samples are heated to around 95°C to separate the double strands. The temperature is then lowered to 50–60°C for primers to attach to the DNA. Lastly, the temperature is increased to 72°C for DNA polymerase to synthesize new strands. This process happens over and over at temperatures between 30°C and 40°C to produce numerous copies of the DNA segment. The DNA generated can be utilized for various techniques, like DNA sequencing and gel electrophoresis (Korkmaz and Merey, 2023).

Multiplex PCR allows the simultaneous amplification of multiple target DNA regions. In this method, multiple primer sets are used to amplify various DNA sequences in a single reaction mixture. This aids in the simultaneous detection of several genetic targets. It saves both time and money in genetic tests. Also, multiplex PCR helps look at many pathogens or genetic variations in complex samples at the same time. This technique is popular in food safety because it quickly finds the pathogens that cause illnesses. However, multiplex PCR has some downsides, like cross-reactions or not amplifying targets efficiently. Hence, primers and reaction conditions need careful adjustment (van Pelt-Verkuil et al., 2008; Tao et al., 2020; Chen et al., 2021).

qRT-PCR is an advanced PCR method that can monitor DNA amplification in real time. In contrast to traditional PCR, qRT-PCR measures the fluorescent signal at each cycle during DNA amplification. The intensity of the signal directly indicates the amount of target DNA. In qRT-PCR, fluorescent dyes or markers, such as SYBR Green, emit a fluorescent signal upon binding to DNA (Adams, 2020; Fu et al., 2020). Additionally, specific probe systems can be used to obtain signals specific to the target DNA. In this method, while the primer is being extended, a fluorescent dye–labeled probe is allowed to hybridize with the target DNA. The endonuclease activity of Taq DNA polymerase cleaves the probe. Thus, the reporter dye is released, and the amount of amplified PCR products is analyzed using a laser detector (Mackay and Landt, 2007; Overbergh et al., 2010).

Hybridization methods are nucleic acid–based methods that provide high sensitivity and specificity by using specific pairing of genetic material in the detection of bacteria. Hybridization occurs in the following three stages: the blocking of the completed membrane to minimize non-specific binding, the binding of target DNA to the probe, and the visualization of signals after the washing of unhybridized probes. Probes have to be exact for the target bacteria's gene or gene area and must match completely with the target DNA or RNA during hybridization. Techniques like southern blotting, northern blotting, dot blotting, and in situ hybridization (ISH) are commonly used in molecular biology labs. These techniques have advanced the study of genetics and helped in identifying bacterial pathogens (Bhat et al., 2020; Nuovo, 2020; Young et al., 2020).

DNA microarray is used for collecting large amounts of data in genetic studies. It involves placing many DNA or RNA probes on one surface simultaneously. The probes typically work to cover the complete genetic sequence of a bacterial cell. The sample DNA or RNA bonds with these probes. Specific binding locations are shown with fluorescent tags. This approach can examine many genetic traits at once, such as gene expression and genetic differences, and it can also detect pathogens. Thus, large volumes of data can be analyzed quickly and efficiently (Kumar et al., 2020; Pastinen, 2023).

5.4.2.2 Immunological Methods

Immunological techniques are useful for identifying pathogens. Diagnosis involves using antibodies or antigens linked to bacteria. Enzyme-linked immunosorbent assay (ELISA) employs specific antibodies for particular bacteria to detect infections by measuring the levels of produced antibodies or antigens (Alhajj et al., 2023). Additional methods, such as immunofluorescence, radioimmunoassay, immunodiffusion, agglutination tests, and immunomagnetic separation, can quickly and sensitively detect bacteria (Betts Carpenter, 2011).

Enzyme immunoassay (EIA) is a biochemical technique for detecting specific antigens or antibodies. In this technique, antibodies or antigens specifically attach to target molecules and are marked by binding with an enzyme. A color change occurs from the enzyme–substrate reaction, indicating the presence and quantity of the target molecule. EIA is widely used in clinical diagnosis and research due to its high sensitivity and specificity. This technique is fundamental to ELISA (Dita, 2021; Alhajj et al., 2023).

ELISA functions through antigen–antibody interactions, measuring either antigens or antibodies. There are two ELISA types: direct and indirect. In indirect ELISA, the sample's antigens bond to a microtiter plate surface and are then mixed with unlabeled primary antibodies. Then enzyme-labeled antibodies are added, and the mixture is incubated for a set period. After removing unbound labeled antibodies, a chromophore is introduced to the medium. The resulting color change shows how much antigen is present. For indirect ELISA, sample antigens stick to the plate and are mixed with unlabeled primary antibodies for a known time. After unbound primary antibodies are washed away, samples are incubated with enzyme-labeled secondary antibodies for a specific duration. After washing again to remove unbound secondary antibodies, they are mixed with the substrate, and the presence of antibodies is assessed via color change (Kohl and Ascoli, 2017). A specific pre-enrichment medium may be required to identify pathogenic bacteria using ELISA. This technique can detect species such as *Salmonella* and *Shigella*, botulinum toxin, and staphylococcal enterotoxins (Law et al., 2015; Lin, 2015; Kohl and Ascoli, 2017).

Immunofluorescence is an important method in immunology for identifying bacteria. It detects target molecules through specific antigen–antibody interactions. Antigens or antibodies are placed on surfaces. After they bind to the target bacteria, a signal arises from a labeled enzyme or fluorescence probe (Quintero-Ronderos et al., 2013; Zhu et al., 2014). Bacteria attached to labeled antibodies emit fluorescent signals observable under a fluorescence microscope. The signal intensity indicates the presence and amount of the target molecule (Fang et al., 2022). There are two key types of immunofluorescence analysis: direct and indirect. In the direct method, the target antigen coats the test surface and binds to sample antibodies, creating a complex that produces a direct signal. In the indirect method, the target antigen is initially applied to the surface, where it binds to sample antibodies. Following this, samples are treated with an enzyme-labeled secondary antibody that recognizes the primary antibodies. The secondary antibody's binding to the primary antibody generates a signal (Hsieh and Rao, 2017; Dellavance and Andrade, 2019; Ahmadi et al., 2024).

The radioimmunoassay technique is effective for detecting antigens or antibodies. This approach requires target items to attach to antibodies or antigens that feature radioisotopes. The sample goes through a specific interaction with these labeled antibodies. Antigens and antibodies are used along with radioactively labeled substances on the test surface (Ahirwar et al., 2022). The common radioisotope for RIA is radioactive iodine (^{131}I). After binding through incubation is allowed, samples are washed to eliminate unbound radioactive substances. The signals from the bound radioactive compounds are detected using a special counting device. Radioimmunoassay can identify and quantify target bacteria (Taylor, 2022; Balzer and Whitehurst, 2023).

Reverse passive latex agglutination (RPLA), immunomagnetic separation (IMS), and immunodiffusion are methods to detect bacteria. In the RPLA test, antigens interact with antibodies that are attached to latex particles. Agglutination occurs when the target antigens are present (Rajkovic et al., 2020). The IMS method employs antibodies that are linked to magnetic particles to isolate target cells or molecules. A magnetic field is used to separate these target substances from other materials (Nekounam et al., 2022). The immunodiffusion method relies on the interaction between antigens and antibodies via diffusion in an agar gel. This method reveals a precipitate band when either antibodies or antigens in the gel interact, indicating the presence of specific bacterial antigens or antibodies (Basha, 2020; Singh, 2022).

5.4.2.3 Spectroscopic Methods

Spectroscopic methods find bacteria by examining how food reacts with light. These techniques provide fast and accurate results and work in different areas. Compared to traditional culture methods, spectroscopic techniques can quickly and sensitively identify bacteria. Still, these methods are not commonly used in food analysis due to certain drawbacks, such as the need for special equipment and prior preparations to accurately identify the components (Korkmaz and Merey, 2023).

Popular methods for detecting bacteria include Fourier transform infrared (FT-IR) spectroscopy, ultraviolet-visible (UV-Vis) spectroscopy, and Raman spectroscopy. FT-IR spectroscopy studies how the chemical components of samples vibrate. Conversely, UV-Vis spectroscopy measures how samples absorb light in the UV-Vis range. These methods analyze the movement and light absorption of various bacterial components to determine bacterial types. Raman spectroscopy looks at the light that reflects off molecules in samples when they are exposed to intense laser light. Thus, Raman spectroscopy can identify bacteria if their molecules respond to Raman light (Santi et al., 2021; Korkmaz and Merey, 2023).

5.4.2.4 Bioluminescence Methods

The bioluminescence method is a fast and sensitive way to detect bacteria and measure cellular adenosine triphosphate (ATP) levels. This method is based on the principle that ATP in a sample emits light during a luminescent reaction (Efremenko et al., 2022). The technique utilizes biological light-emitting molecules, such as luciferin and luciferase from fireflies and green fluorescent protein from jellyfish. Additionally, lab-modified luminescent proteins can also be used as these biological molecules. These molecules produce light when they react with specific target molecules, creating a measurable signal. A luminometer or detector captures this light signal, allowing for the calculation of ATP levels. Bioluminescence method application spans various fields, including bacterial detection, food safety, and quality control (Ghabru et al., 2023).

5.4.2.5 Biosensors

In the last few years, using biosensors to find bacteria in food has grown. Biosensors have parts that are biological and parts that change signals. The bioreceptors in these devices are made for specific molecules. These biological molecules can be aptamers, antibodies, or enzymes. When a bioreceptor meets the target molecule, it produces a signal (Olejnik et al., 2021; Chadha et al., 2022). Transducers can then turn these signals into measurable data. The performance of the biosensor is influenced by how well the bioreceptor works with the transducer. Consequently, biosensors can provide fast, accurate, and sensitive measurements (Sharma et al., 2021; Kulkarni et al., 2022). Thanks to these features, biosensors can also be used to detect pathogens. There are many types of biosensors, especially electrochemical and immune sensors. Optical, nano-, magnetic, and DNA sensors are among the other types of biosensor (Xu et al., 2021).

Electrochemical biosensors are widely used types of biosensors for the rapid detection of bacteria. The working principle of these biosensors is that the interaction between the bioreceptor and the analyte is converted into measurable signals by a transducer. The interaction mentioned here is essentially perceived as electrical changes by electrochemical biosensors (Kaya et al., 2021). Signal measurements show the presence of bacteria in the environment. In addition, information about the amount of bacteria can be obtained with these biosensors. Providing fast and sensitive results is an advantage of these biosensors. However, they can be affected by environmental conditions. Another disadvantage is that the efficiency of the bioreceptor decreases over time (Castle et al., 2021).

In immunosensors, bacteria are detected using specific antibodies or aptamers. These antibodies are located on the biosensor surface and interact with specific bacteria. A signal is generated as a result of this interaction. Similarly, aptamers bind to bacteria and interact. The interaction is measured as a signal, and the presence of bacteria is detected. The principle in immunofluorescent sensors is based on the measurement of light released from fluorescent markers attached to bacteria-specific antibodies. In immuno-electrochemical sensors, antibody–bacteria interaction is measured similarly to other biosensors. The interaction mentioned here includes changes on the electrode surface. Immunosensors are frequently preferred because they provide fast and accurate bacterial detection (Subjakova et al., 2021; Beeman, 2021).

Optical biosensors are also types of biosensors used in the detection of bacteria. For this purpose, elements such as optical fibers and surface plasmon resonance are used. The bioreceptor, which consists of a surface and a section that processes optical signals, interacts with bacteria. Changes in light happen a lot. Differences in fluorescence and shifts in plasmonic wavelengths cause these changes. Optical biosensors can detect and measure these changes. They are often used to find pathogens because they analyze quickly and have high resolution (Sharma and Sharma, 2023).

The main idea of nanobiosensors is to detect biomolecules with nanomaterials. Because they are very sensitive and small, nanosensors are effective at detecting bacteria. The nanomaterials in these sensors may be carbon nanotubes or may be metal or magnetic nanoparticles (Pourmadadi et al., 2021). In these nanosensors, the interaction between the bioreceptor and the analyte can be seen as a signal. These biosensors have the benefit of making sensitive and specific measurements (Santi et al., 2021).

The principle of magnetic biosensors is to find biomolecules with a magnetic field. These biosensors use magnetic particles and sensors. They also have biological elements that work as a bioreceptor to find the target substance. The magnetic particle is linked to a base that holds the biological element. Sensors are good at tracking how magnetic particles move. These biosensors are often used to identify biomolecules (Chen et al., 2017).

DNA biosensors can be used especially for the purpose of finding bacterial DNA. These biosensors help determine sensitive sequences in DNA by detecting some specific areas. Special probe sets are used for this purpose. DNA biosensors are generally used with different sensors. These include electronic, optical, or magnetic sensors (Sassolas et al., 2008).

5.5 CONCLUSION

Bacteriological spoilage is the deterioration of food when bacteria grow and infect the food. As a result, some properties of the food, especially its taste and smell, may change. If food is contaminated with pathogens, this can pose health risks. These bacteria can cause many health problems, from general symptoms, such as abdominal pain, vomiting, and diarrhea, to serious problems, such as organ failure. In worsened cases, it can even result in death. Correct methods must be used to know how and by which bacteria food spoils and to take precautions to prevent this from happening. This is also important for food safety and product life. Controlling the presence of pathogens in food is a factor that ensures public health and food safety. Although traditional methods used for these purposes still work, new techniques have the ability to find bacteria more quickly and accurately. Further research and investment in new detection methods are crucial to solving the problems posed by harmful bacteria in the food supply around the world.

REFERENCES

Abdallah, E. M., Sulieman, A. M. E. and Saleh, Z. A. 2024. New discoveries in toxins from gram-positive Bacteria *Staphylococcus aureus*. In *Microbial Toxins in Food Systems: Causes, Mechanisms, Complications, and Metabolism*, ed. Sulieman, A. M. E. and Alshammari, N. I., 235–252. Switzerland: Springer Nature Switzerland.

Abdel-Aziz, S. M., Asker, M. M., Keera, A. A. and Mahmoud, M. G. 2016. Microbial food spoilage: Control strategies for shelf life extension. *Microbes in Food and Health*, 239–264.

Adams, G. 2020. A beginner's guide to RT-PCR, qPCR and RT-qPCR. *The Biochemist*, 42(3): 48–53.

Ahirwar, R., Bhattacharya, A. and Kumar, S. 2022. Unveiling the underpinnings of various non-conventional ELISA variants: A review article. *Expert Review of Molecular Diagnostics*, 22(7): 761–774.

Ahmadi, Y., Ahmadi, H. and Aghebati-Maleki, L. 2024. Increase in the specificity of immunoassay methods by direct targeting different epitopes; sequential chain reactions. *Human Antibodies*, (Preprint): 1–6.

Ajayeoba, T. A., Kaka, O. M. and Ajibade, O. A. 2020. 7 Microbial food spoilage of selected food and food products. *Food Science and Technology: Trends and Future Prospects*, 173.

Alegbeleye, O., Odeyemi, O. A., Strateva, M. and Stratev, D. 2022. Microbial spoilage of vegetables, fruits and cereals. *Applied Food Research*, 2(1): 100122.

Alexander, H. K. and MacLean, R. C. 2020. Stochastic bacterial population dynamics restrict the establishment of antibiotic resistance from single cells. *Proceedings of the National Academy of Sciences*, 117(32): 19455–19464.

Alhajj, M., Zubair, M. and Farhana, A. 2023. Enzyme linked immunosorbent assay. *StatPearls*. https://www.statpearls.com/point-of-care/21178 (accessed September 6, 2024).

Amin, M. A., Hashem, H. R., El-Mahallawy, H. S., Abdelrahman, A. A., Zaki, H. M. and Azab, M. M. 2022. Characterization of enterohemorrhagic *Escherichia coli* from diarrhoeic patients with particular reference to production of Shiga-like toxin. *Microbial Pathogenesis*, 166: 105538.

Ammar, A. M., El-Naenaeey, E. S. Y., Abd El-Hamid, M. I., El-Gedawy, A. A. and Elmalt, R. M. 2021. Campylobacter as a major foodborne pathogen: A review of its characteristics, pathogenesis, antimicrobial resistance and control. *Journal of Microbiology, Biotechnology and Food Sciences*, 10(4): 609–619.

Angelidis, A. S. 2015. The microbiology of raw milk. *A Practical Approach*, 22–69.

Argaw, S. and Addis, M. 2015. A review on staphylococcal food poisoning. *Food Science and Quality Management*, 40(2015): 59–72.

Arroyo-López, F. N., Romero-Gil, V., Bautista-Gallego, J. et al. 2012. Yeasts in table olive processing: Desirable or spoilage microorganisms? *International Journal of Food Microbiology*, 160(1): 42–49.

Arteaga, M., Velasco, J., Rodriguez, S. et al. 2020. Genomic characterization of the non-O1/non-O139 *Vibrio cholerae* strain that caused a gastroenteritis outbreak in Santiago, Chile, 2018. *Microbial Genomics*, 6(3): e000340.

Askar, A. and Treptow, H. 2013. *Quality Assurance in Tropical Fruit Processing*. Berlin: Springer-Verlag.

Baker, G. L. 2016. Food safety impacts from post-harvest processing procedures of molluscan shellfish. *Foods*, 5(2): 29.

Bakış, G. and Derleme, G. 2021. Overview of *Listeria monocytogenes* as a foodborne pathogen: Traditional review. *Turkiye Klinikleri Journal of Veterinary Sciences*, 12(1): 37–48.

Baljić, R., Gazibera, B., Gojak, R., Lukovac, E. and Mulabdić, V. 2020. A family outbreak of foodborne botulism following consumption of smoked meat in Sarajevo, Bosnia and Herzegovina. *Medical Journal/Medicinski Žurnal*, 26.

Balzer, A. H. and Whitehurst, C. B. 2023. An analysis of the biotin–(strept) avidin system in immunoassays: Interference and mitigation strategies. *Current Issues in Molecular Biology*, 45(11): 8733–8754.

Baron, F., Jan, S. and Techer, C. 2023. Microbiology of egg and egg products. In *Handbook of Egg Science and Technology*, ed. Mine, Y., Guyonnet, V., Hatta, H., Nau, F. and Qiu, N., 355–396. Boca Raton: CRC Press.

Barth, M., Hankinson, T. R., Zhuang, H. and Breidt, F. 2009. Microbiological spoilage of fruits and vegetables. *Compendium of the Microbiological Spoilage of Foods and Beverages*, 135–183.

Basak, S. and Shetty, P. H. 2021. Conventional microbial counting and identification techniques. *Techniques to Measure Food Safety and Quality: Microbial, Chemical, and Sensory*, 69–89.

Basha, M. 2020. Immuno-techniques. *Analytical Techniques in Biochemistry*, 77–102.

Bartowsky, E. J. and Henschke, P. A. 2008. Acetic acid bacteria spoilage of bottled red wine-A review. *International Journal of Food Microbiology*, 125(1): 60–70.

Beeman, M. G. 2021. Electrochemical sensors for the detection of whole cell pathogenic bacteria and ethanol. Doctoral dissertation, The University of Utah.

Benson, P. 2024. Understanding the role of *Lactococcus lactis* and *Lactococcus cremoris* on bitter peptide formation in aged cheeses. Masters thesis, Oregon State University.

Betts Carpenter, A. 2011. Immunoassays for the diagnosis of infectious diseases. *Manual of Clinical Microbiology*, 60–72.

Bhat, A. I., Rao, G. P., Bhat, A. I. and Rao, G. P. 2020. Dot-blot hybridization technique. In *Characterization of Plant Viruses*, ed. Bhat, A. I. and Rao, G. P., 303–321. New York: Springer.

Bhattacharya, A., Shantikumar, S., Beaufoy, D. et al. 2020a. Outbreak of *Clostridium perfringens* food poisoning linked to leeks in cheese sauce: An unusual source. *Epidemiology and Infection*, 148: e43.

Bhattacharya, R., Hwang, J. H. and Ko, C. 2020b. Infections of the Gastrointestinal Tract. *Yamada's Handbook of Gastroenterology*, 431.

Bhunia, A. K. 2018a. *Shigella* species. In *Foodborne Microbial Pathogens: Mechanisms and Pathogenesis*, ed. Bhunia, A. K., 331–341. New York: Springer.

Bhunia, A. K. 2018b. *Clostridium botulinum, Clostridium perfringens, Clostridium difficile*. In *Foodborne Microbial Pathogens: Mechanisms and Pathogenesis*, ed. Bhunia, A. K., 209–228. New York: Springer.

Bilska, A., Wochna, K., Habiera, M. and Serwańska-Leja, K. 2024. Health hazard associated with the presence of *Clostridium* bacteria in food products. *Foods*, 13(16): 2578.

Bist, N. S. and Bist, P. 2021. Role of microorganisms in post-harvest loss of agricultural products: A review. *Sustainability in Food and Agriculture*, 2(1): 1–4.

Bokori-Brown, M., Savva, C. G., Fernandes da Costa, S. P., Naylor, C. E., Basak, A. K. and Titball, R. W. 2011. Molecular basis of toxicity of *Clostridium perfringens* epsilon toxin. *The FEBS Journal*, 278(23): 4589–4601.

Bowe, B. K., Wentz, T. G., Gregg, B. M. et al. 2022. Genomic diversity, competition, and toxin production by Group I and II *Clostridium botulinum* strains used in food challenge studies. *Microorganisms*, 10(10): 1895.

Brožová, M., Kubizniakova, P. and Matoulková, D. 2018. Brewing microbiology–bacteria of the genus *Clostridium*. *Kvasny Prumysl*, 64(5): 242–247.

Brunt, J., van Vliet, A. H., Stringer, S. C., Carter, A. T., Lindström, M. and Peck, M. W. 2020. Pan-genomic analysis of *Clostridium botulinum* group II (non-proteolytic *C. botulinum*) associated with foodborne botulism and isolated from the environment. *Toxins*, 12(5): 306.

Cabrera-Sosa, L. and Ochoa, T. J. 2020. *Escherichia coli* diarrhea. In *Hunter's Tropical Medicine and Emerging Infectious Diseases*, ed. Ryan, E. T., Hill, D. R., Solomon, T., Aronson, N. E. and Endy, T. P., 481–485. Elsevier.

Carminati, D., Bonvini, B., Rossetti, L., Mariut, M., Zago, M. and Giraffa, G. 2024. Identification and characterization of the microbial agent responsible of an alteration in spoiled, Grana Padano cheese during ripening. *Food Control*, 155: 110050.

Castle, L. M., Schuh, D. A., Reynolds, E. E. and Furst, A. L. 2021. Electrochemical sensors to detect bacterial foodborne pathogens. *ACS Sensors*, 6(5): 1717–1730.

Chadha, U., Bhardwaj, P., Agarwal, R. et al. 2022. Recent progress and growth in biosensors technology: A critical review. *Journal of Industrial and Engineering Chemistry*, 109: 21–51.

Chen, P. 2021. Lactic acid bacteria in fermented food. In *Advances in Probiotics*, ed. Dhanasekaran, D. and Sankaranarayanan, A., 397–416. Academic Press.

Chen, Y., Wang, Z., Shi, Q. et al. 2021. Multiplex PCR method for simultaneous detection of five pathogenic bacteria closely related to foodborne diseases. *3 Biotech*, 11: 1–8.

Chen, Y. T., Kolhatkar, A. G., Zenasni, O., Xu, S. and Lee, T. R. 2017. Biosensing using magnetic particle detection techniques. *Sensors*, 17(10): 2300.

Chlebicz, A. and Śliżewska, K. 2018. Campylobacteriosis, salmonellosis, yersiniosis, and listeriosis as zoonotic foodborne diseases: A review. *International Journal of Environmental Research and Public Health*, 15(5): 863.

Clark, S. 2017. Sensory evaluation of cow's milk. In *Achieving sustainable production of milk*, ed. van Belzen, N., 159–181. London: Burleigh Dodds Science Publishing.

Coerdt, K. M. and Khachemoune, A. 2021. *Vibrio vulnificus*: Review of mild to life-threatening skin infections. *Cutis*, 107(2): E12–E17.

Comi, G. 2017. Spoilage of meat and fish. In *The Microbiological Quality of Food*, ed. Bevilacqua, A., Corbo, M. R. and Sinigaglia, M., 179–210. Woodhead Publishing.

Cruess, W. V. 2009. The role of microorganisms and enzymes in wine making. *Advances in Enzymology and Related Areas of Molecular Biology*, 3: 349.

Danilović, B. and Savić, D. 2017. Microbial ecology of fermented sausages and dry-cured meats. *Fermented Meat Products: Health Aspects*, 127–66.

da Silva, L. V., Ossai, S., Chigbu, P. and Parveen, S. 2021. Antimicrobial and genetic profiles of *Vibrio vulnificus* and *Vibrio parahaemolyticus* isolated from the Maryland Coastal Bays, United States. *Frontiers in Microbiology*, 12: 676249.

Dasriya, V. L., Ranveer, S. A., Dhillon, H. S. et al. 2024. 4 food contamination in milk. *Food Safety: Contaminants and Risk Assessment,* 54.

Deeth, H. C. 2020. The effect of UHT processing and storage on milk proteins. In *Milk Proteins*, ed. Boland, M. and Singh, H., 385–421. Academic Press.

Delavenne, E., Cliquet, S., Trunet, C., Barbier, G., Mounier, J. and Le Blay, G. 2015. Characterization of the antifungal activity of *Lactobacillus harbinensis* K. V9. 3.1 Np and *Lactobacillus rhamnosus* K. C8. 3.1 I in yogurt. *Food Microbiology*, 45: 10–17.

de Lima, A. C., Brandao, L. R., Botelho, B. G. et al. 2023. Multivariate analysis of the influence of microfiltration and pasteurisation on the quality of beer during its shelf life. *Foods*, 13(1): 122.

Dellavance, A. and Andrade, L. E. C. 2019. Detection of autoantibodies by indirect immunofluorescence cytochemistry on HEp-2 cells. *Autoantibodies: Methods and Protocols*, 19–46.

del Olmo, A., Calzada, J. and Nuñez, M. 2018. The blue discoloration of fresh cheeses: A worldwide defect associated to specific contamination by *Pseudomonas fluorescens*. *Food Control*, 86: 359–366.

de Souza Santos, M., Salomon, D., Li, P., Krachler, A. M. and Orth, K. 2015. *Vibrio parahaemolyticus* virulence determinants. *The Comprehensive Sourcebook of Bacterial Protein Toxins*, 4.

Dias, F. S., Ramos, C. L. and Schwan, R. F. 2013. Characterization of spoilage bacteria in pork sausage by PCR-DGGE analysis. *Food Science and Technology*, 33: 468–474.

Dita, M. C. 2021. Enzyme-Linked Immunosorbent Assay (ELISA): A narrative literature review. *Natural Sciences Engineering and Technology Journal*, 1(2): 24–31.

Easa, S. M. H. 2011. Microorganisms found in fast and traditional fast foods. *Egyptian Journal of Agricultural Sciences*, 62(2): 226–244.

Ebob, T. 2020. A review on diagnostic methods for the identification of *Vibrio cholerae*. *Journal of Advances in Medicine and Medical Research*, 32(8): 136–164.

Efremenko, E., Senko, O., Stepanov, N., Maslova, O., Lomakina, G. Y. and Ugarova, N. 2022. Luminescent analysis of ATP: Modern objects and processes for sensing. *Chemosensors*, 10(11): 493.

EFSA Panel on Biological Hazards (BIOHAZ). 2016. Risks for public health related to the presence of *Bacillus cereus* and other *Bacillus* spp. including *Bacillus thuringiensis* in foodstuffs. *EFSA Journal*, 14(7): e04524.

El-Arabi, T. F. and Griffiths, M. W. 2021. *Bacillus cereus*. In *Foodborne Infections and Intoxications*, ed. Morris, J. G. and Vugia, D. J., 431–437. Academic Press.

Erkmen, O. and Bozoglu, T. F. 2016. Spoilage of fish and other seafoods. In *Food Microbiology Principles into Practice*, ed. Erkmen, O. and Bozoglu, T. F., 301–306. John Wiley and Sons.

Facciolà, A., Riso, R., Avventuroso, E., Visalli, G., Delia, S. A. and Lagana, P. 2017. *Campylobacter*: From microbiology to prevention. *Journal of Preventive Medicine and Hygiene*, 58(2): E79.

Fang, S., Yang, H., Liu, C. et al. 2022. Bacterial coloration immunofluorescence strip for ultrasensitive rapid detection of bacterial antibodies and targeted antibody-secreting hybridomas. *Journal of Immunological Methods*, 501: 113208.

Fessard, A. and Remize, F. 2017. Why are *Weissella* spp. not used as commercial starter cultures for food fermentation? *Fermentation*, 3(3): 38.

Finlay, W. J. J., Logan, N. A. and Sutherland, A. D. 2000. *Bacillus cereus* produces most emetic toxin at lower temperatures. *Letters in Applied Microbiology*, 31(5): 385–389.

Fu, L., Qian, Y., Zhou, J., Zheng, L. and Wang, Y. 2020. Fluorescence-based quantitative platform for ultrasensitive food allergen detection: From immunoassays to DNA sensors. *Comprehensive Reviews in Food Science and Food Safety*, 19(6): 3343–3364.

Fu, Y., Alenezi, T. and Sun, X. 2022. *Clostridium perfringens*-induced necrotic diseases: An overview. *Immuno*, 2(2): 387–407.

Galli, L., Brusa, V., Rodríguez, R., Signorini, M., Oteiza, J. M. and Leotta, G. A. 2016. *Escherichia coli* in food products. *Escherichia coli in the Americas*, 173–203.

Garofalo, C., Osimani, A., Milanović, V., Taccari, M., Aquilanti, L. and Clementi, F. 2015. The occurrence of beer spoilage lactic acid bacteria in craft beer production. *Journal of Food Science*, 80(12): M2845–M2852.

Gänzle, M. G. 2015. Lactic metabolism revisited: Metabolism of lactic acid bacteria in food fermentations and food spoilage. *Current Opinion in Food Science*, 2: 106–117.

Ghabru, A., Rana, N. and Verma, G. 2023. Bioluminesence: "The Living Light". *Molecular Biology Plant Physiology*, 33: 203.

Gharib, A. A., El-Hamid, M. I. A., El-Aziz, N. K. A., Yonan, E. Y. and Allam, M. O. 2020. *Bacillus cereus*: Pathogenicity, viability and adaptation. *Advances in Animal and Veterinary Sciences*, 8(s1): 34–40.

Giudici, P., Bonciani, T., Lemmetti, F. and Corradini, G. 2017. Vinegars: Process, Aging, and Spoilage. In *Post-Fermentation and-Distillation Technology*, ed. Bordiga, M., 207–234. Boca Raton: CRC Press.

Gohari, I. M. and Prescott, J. F. 2022. *Clostridium. Veterinary Microbiology*, 309–334.

Govindarajan, D. K., Viswalingam, N., Meganathan, Y. and Kandaswamy, K. 2020. Adherence patterns of *Escherichia coli* in the intestine and its role in pathogenesis. *Medicine in Microecology*, 5: 100025.

Griffiths, M. W. and Schraft, H. 2017. *Bacillus cereus* food poisoning. In *Foodborne Diseases*, ed. Dodd, C. E. R., Aldsworth, T., Stein, R. A., Cliver, D. O. and Riemann, H. P., 395–405. Academic Press.

Grispoldi, L., Karama, M., Armani, A., Hadjicharalambous, C. and Cenci-Goga, B. T. 2021. *Staphylococcus aureus* enterotoxin in food of animal origin and staphylococcal food poisoning risk assessment from farm to table. *Italian Journal of Animal Science*, 20(1): 677–690.

Haque, M. A., Khan, A. and He, C. 2021. Feed-borne *Bacillus cereus*: An emerging threat to food chain related hazard, safety and pathogenic potentiality. *Veterinary Pathobiology and Public Health*, 1: 251–269.

Hassan, A. O., Ojo, B. O. and Abdulrahman, A. O. 2021. *Escherichia coli* as a global pathogen. *Achievers Journal of Scientific Research*, 3(1): 239–260.

Hrustemović, E., Čaklovica, F. and Đeđibegović, J. 2021. Review of *Escherichia coli* pathogens and food-borne antibiotic resistance to prevention. *III International Agricultural, Biological and Life Science Conference (AGBIOL 2021)*, 1–3 September, Edirne, Turkey, 12–37. Edirne: Trakya University.

Hsieh, Y. P. and Rao, Q. 2017. Immunoassays. *Food Analysis*, 487–502.

Huang, H. and Garcia, M. M. 2022. Isolation and identification of *Campylobacter* species from food and food-related environment. *Campylobacter*, 43.

Hwang, C. A., Huang, L. and Sheen, S. 2024. Modeling the growth probability of *Clostridium perfringens* in cooked cured meat as affected by sodium chloride and sodium tripolyphosphate. *Microbial Risk Analysis*, 26: 100296.

Ibrahim, S. A., Ayivi, R. D., Zimmerman, T. et al. 2021. Lactic acid bacteria as antimicrobial agents: Food safety and microbial food spoilage prevention. *Foods*, 10(12): 3131.

Jan, S., Baron, F., Coat, R. and Gonçalves, O. 2018. Spoilage of egg products. In *Alteration of Ovoproducts*, ed. Gonçalves, O. and Legrand, J., 51–156. Elsevier.

Jay, J. M., Loessner, M. J. and Golden, D. A. 2005. Vegetable and fruit products. *Modern Food Microbiology*, 125–147.

Jayan, H., Pu, H. and Sun, D. W. 2020. Recent development in rapid detection techniques for microorganism activities in food matrices using bio-recognition: A review. *Trends in Food Science and Technology*, 95: 233–246.

Jenkins, R. and Maddocks, S. 2019. *Bacteriology Methods for the Study of Infectious Diseases*. Academic Press.

Jessberger, N., Dietrich, R., Schauer, K., Schwemmer, S., Märtlbauer, E. and Benz, R. 2020. Characteristics of the protein complexes and pores formed by *Bacillus cereus* hemolysin BL. *Toxins*, 12(11): 672.

Jovanovic, J., Ornelis, V. F., Madder, A. and Rajkovic, A. 2021. *Bacillus cereus* food intoxication and toxicoinfection. *Comprehensive Reviews in Food Science and Food Safety*, 20(4): 3719–3761.

Juneja, V. K., Osoria, M., Purohit, A. S. et al. 2021. Predictive model for growth of *Clostridium perfringens* during cooling of cooked pork supplemented with sodium chloride and sodium pyrophosphate. *Meat Science*, 180: 108557.

Júnior, J. R., Tamanini, R., De Oliveira, A. L. M., Alfieri, A. A. and Beloti, V. 2018. Genetic diversity of thermoduric spoilage microorganisms of milk from Brazilian dairy farms. *Journal of Dairy Science*, 101(8): 6927–6936.

Kadariya, J., Smith, T. C. and Thapaliya, D. 2014. *Staphylococcus aureus* and staphylococcal food-borne disease: An ongoing challenge in public health. *BioMed Research International*, 2014(1): 827965.

Kalburge, S. S., Whitaker, W. B. and Boyd, E. F. 2014. High-salt preadaptation of *Vibrio parahaemolyticus* enhances survival in response to lethal environmental stresses. *Journal of Food Protection*, 77(2): 246–253.

Kamal, M. R., Fakhruzzaman, M., Akter, M. R. and Haque, M. A. 2018. Isolation, identification and antibiogram profiles of enterovirulent *Escherichia coli* from diarrhoeic goat in some selected areas of Rangpur district of Bangladesh. *Asian Journal of Medical and Biological Research*, 4(1): 36–43.

Kaveh, S., Hashemi, S. M. B., Abedi, E., Amiri, M. J. and Conte, F. L. 2023. Bio-preservation of meat and fermented meat products by lactic acid bacteria strains and their antibacterial metabolites. *Sustainability*, 15(13): 10154.

Kaya, H. I., Simsek, O. and Akgunoglu, O. 2023. Diversity of *Clostridium* spp. causing late blowing in Kaşar cheese and their behaviour against various antimicrobials. *International Dairy Journal*, 139: 105560.

Kaya, H. O., Cetin, A. E., Azimzadeh, M. and Topkaya, S. N. 2021. Pathogen detection with electrochemical biosensors: Advantages, challenges and future perspectives. *Journal of Electroanalytical Chemistry*, 882: 114989.

Kemal, J. 2014. A review on the public health importance of bovine salmonellosis. *Journal of Veterinary Science and Technology*, 5(1).

Khan, S., Hubab, M., Rehman, A., Hayat, A., Rehman, M. U. and Umar, T. 2022. Isolation, identification and evaluation of antibiotic sensitivity and resistance of *Shigella dysenteriae* isolated from food, water and soil. *Journal of Microbiological Sciences*, 1(01): 05–10.

Kim, J. S., Lee, M. S. and Kim, J. H. 2020. Recent updates on outbreaks of Shiga toxin-producing *Escherichia coli* and its potential reservoirs. *Frontiers in Cellular and Infection Microbiology*, 10: 273.

King, A. M. 2011. Effects of lactic acid and commercial chilling processes on survival of *Salmonella* spp., *Yersinia enterocolitica*, and *Campylobacter coli* in pork variety meats. Doctoral dissertation, Texas A & M University.

Kios, K., Kakasis, S., Syropoulou, F. and Boziaris, I. S. 2023. Seafood and shellfish. In *Functional foods and their implications for health promotion*, ed. Zabetakis, I., Lordan, R., Tsoupras, A. and Ramji, D., 281–302. Academic Press.

Kohl, T. O. and Ascoli, C. A. 2017. Direct and indirect cell-based enzyme-linked immunosorbent assay. *Cold Spring Harbor Protocols*, 2017(5): pdb-prot093732.

Korkmaz, R. and Merey, G. 2023. Gıda Güvenliğinin Önemi ve Bakterilerin Tespiti İçin Yöntemler. *Journal of Health Sciences and Management*, 3(2): 47–53.

Kotloff, K. L., Riddle, M. S., Platts-Mills, J. A., Pavlinac, P. and Zaidi, A. K. 2018. Shigellosis. *The Lancet*, 391(10122): 801–812.

Kulkarni, M. B., Ayachit, N. H. and Aminabhavi, T. M. 2022. Biosensors and microfluidic biosensors: From fabrication to application. *Biosensors*, 12(7): 543.

Kumar, A., Pandey, S. C. and Samant, M. 2020. DNA-based microarray studies in visceral leishmaniasis: Identification of biomarkers for diagnostic, prognostic and drug target for treatment. *Acta Tropica*, 208: 105512.

Kumar, A. and Patyal, A. 2024. An insight on microbial flora of milk and milk products. In *The Microbiology, Pathogenesis and Zoonosis of Milk Borne Diseases* ed. Rana, T., 69–94. Academic Press.

Lampel, K. A. 2009. Shigella. *Pathogens and Toxins in Foods: Challenges and Interventions*, 131–145.

Law, J. W. F., Ab Mutalib, N. S., Chan, K. G. and Lee, L. H. 2015. Rapid methods for the detection of foodborne bacterial pathogens: Principles, applications, advantages and limitations. *Frontiers in Microbiology*, 5: 770.

Ledenbach, L. H. and Marshall, R. T. 2009. Microbiological spoilage of dairy products. *Compendium of the Microbiological Spoilage of Foods and Beverages*, 41–67.

Lin, A. V. 2015. Direct Elisa. *ELISA: Methods and Protocols*, 61–67.

Liu, D. 2021. Molecular identification of *Campylobacter* spp. In *Molecular Food Microbiology*, ed. Liu, D., 213–227. Boca Raton: CRC Press.

Liu, D. 2024. *Clostridium botulinum* and associated neurotoxins. In *Molecular Medical Microbiology*, ed. Tang, Y. W., Hindiyeh, M. Y., Liu, D., Sails, A., Spearman, P. and Zhang, J. R. 933–944. Academic Press.

Liu, L., Tao, Y., Li, Y. et al. 2022. Isolation and characterization of bacteria that produce quorum sensing molecules during the fermentation and deterioration of pickles. *International Journal of Food Microbiology*, 379: 109869.

Lopez-Brea, S. G., Gómez-Torres, N. and Arribas, M. Á. 2017. Spore-forming bacteria in dairy products. *Microbiology in Dairy Processing: Challenges and Opportunities*, 11–36.

Mackay, J. and Landt, O. 2007. Real-time PCR fluorescent chemistries. *Protocols for Nucleic Acid Analysis by Nonradioactive Probes*, 237–261.

Madhumala, Y. and Soraganvi, V. S. 2024. Microbiology of food spoilage. In *Frontiers in Food Biotechnology*, ed. Yaradoddi, J. S., Meti, B. S., Mudgulkar, S. B. and Agsar, D., 67–80. Singapore: Springer Nature Singapore.

Majumdar, A., Pradhan, N., Sadasivan, J. et al. 2018. Food degradation and foodborne diseases: A microbial approach. In *Microbial Contamination and Food Degradation*, ed. Holban, A. M. and Grumezescu, A. M., 109–148. Academic Press.

Man, S. M. 2011. The clinical importance of emerging *Campylobacter* species. *Nature Reviews Gastroenterology and Hepatology*, 8(12): 669–685.

Mandal, B. K., De Mol, P. and Butzler, J. P. 2018. Clinical aspects of *Campylobacter* infections in humans. In *Campylobacter Infection in Man and Animals*, ed. Butzler, J. P., 21–32. Boca Raton: CRC Press.

Mare, A. D., Ciurea, C. N., Man, A. et al. 2021. Enteropathogenic *Escherichia coli*—a summary of the literature. *Gastroenterology Insights*, 12(1): 28–40.

Mayra, P. L., Lucía, Y. V., Holger, P. F., Jimmy, N. P., Rosario, E. V. and Edmundo, R. P. 2020. The extended methylene blue reduction test and milk quality. *Foods and Raw Materials*, 8(1): 140–148.

Meurens, F., Carlin, F., Federighi, M. et al. 2023. *Clostridium botulinum* type C, D, C/D, and D/C: An update. *Frontiers in Microbiology*, 13: 1099184.

Mkangara, M. 2023. Prevention and control of human *Salmonella enterica* infections: An implication in food safety. *International Journal of Food Science*, 2023(1): 8899596.

Mohammadi, B., Gorkina, N. and Smith, S. A. 2022. Pathogenicity, toxin production, control and detection of *Bacillus cereus*. In *Foodborne Pathogens-Recent Advances in Control and Detection*, ed. Lamas, A., Franco, C. M. and Regal, P. IntechOpen.

Moi, I. M., Ibrahim, Z., Abubakar, B. M. et al. 2022. Properties of foodborne pathogens and their diseases. In *Foodborne Pathogens-Recent Advances in Control and Detection*, ed. Lamas, A., Franco, C. M. and Regal, P. IntechOpen.

Morshdy, A. E. M., Mehrez, S. M., Tharwat, A. E., Abdallah, K. M. and Darwish, W. 2023. A review on the microbial surface contaminants of the animal carcasses. *Journal of Advanced Veterinary Research*, 13(6): 1248–1251.

Moura-Alves, M., Machado, C., Saraiva, C. and Silva, J. A. 2020. Determination of D and z values for *Salmonella* Typhimurium inoculated in an egg-based pastry. *Brazilian Journal of Food Technology*, 23: e2019122.

Muhović, B., Islam, N., M Tambuwala, M. and Hromić-Jahjefendić, A. 2023. Characterizing lactic acid bacteria and thermotolerant coliforms in spoiled dairy products from Bosnia and Herzegovina. *Journal of Medical Microbiology and Infectious Diseases*, 11(4): 200–212.

Murugalatha, N., Growther, L., Hena, J. V. et al. 2022. *Microbiological Techniques*. MJP Publisher.

Nandy, S. K. and Venkatesh, K. V. 2010. Application of methylene blue dye reduction test (MBRT) to determine growth and death rates of microorganisms. *African Journal of Microbiology Research*, 4(1): 061–070.

Navarro, A. and Estrada-Garcia, T. 2010. Epidemiology of diarrheagenic *Escherichia coli* pathotypes in Mexico, past and present. *Pathogenic Escherichia coli in Latin America*, 1: 191–208.

Navarro, M. A., McClane, B. A. and Uzal, F. A. 2018. Mechanisms of action and cell death associated with *Clostridium perfringens* toxins. *Toxins*, 10(5): 212.

Nekounam, H., Babaei, M., Kisomi, M. F. et al. 2022. Application of functional magnetic nanoparticles for separation of target materials: A review. *Current Nanoscience*, 18(5): 554–570.

Nuovo, G. J. 2020. *In Situ Molecular Pathology and Co-expression Analyses*. Academic Press.

Odeyemi, O. A., Alegbeleye, O. O., Strateva, M. and Stratev, D. 2020. Understanding spoilage microbial community and spoilage mechanisms in foods of animal origin. *Comprehensive reviews in Food Science and Food Safety*, 19(2): 311–331.

Odo, S. E., Uchechukwu, C. F. and Ezemadu, U. R. 2021. Foodborne diseases and intoxication in Nigeria: Prevalence of *Escherichia coli* 0157: H7, *Salmonella*, *Shigella* and *Staphylococcus aureus*. *Journal of Advances in Microbiology*, 20(12): 84–94.

Olejnik, B., Kozioł, A., Brzozowska, E. and Ferens-Sieczkowska, M. 2021. Application of selected biosensor techniques in clinical diagnostics. *Expert Review of Molecular Diagnostics*, 21(9): 925–937.

Osborne, J. P. and Edwards, C. G. 2005. Bacteria important during winemaking. *Advances in Food and Nutrition Research*, 50: 139–177.

Osek, J., Lachtara, B. and Wieczorek, K. 2022. *Listeria monocytogenes*–how this pathogen survives in food-production environments? *Frontiers in Microbiology*, 13: 866462.

Overbergh, L., Giulietti, A. P., Valckx, D. and Mathieu, C. 2010. Real-time polymerase chain reaction. In *Molecular Diagnostics*, ed. Patrinos, G. P. and Ansorge, W. J., 87–105. Academic Press.

Oyedeji, A. B., Green, E., Jeff-Agboola, Y. A. et al. 2023. Presence of pathogenic microorganisms in fermented foods. In *Indigenous Fermented Foods for the Tropics*, ed. Adebo, O. A., Chinma, C. E., Obadina, A. O., Soares, A. G., Panda, S. K. and Gan, R. Y., 519–537. Academic Press.

Ozdemır, F., Arslan, S. and Erturk, H. G. 2020. Expression of blaA and blaB and susceptibility to Penicillins and cephalosporins in *Yersinia enterocolitica* from different foods. *European Journal of Biology*, 79(2): 83–88.

Pacher, N., Burtscher, J., Johler, S. et al. 2022. Ropiness in bread—a re-emerging spoilage phenomenon. *Foods*, 11(19): 3021.

Pal, M., Merera, O., Abera, F., Rahman, M. T. and Hazarika, R. A. 2015. Salmonellosis: A major foodborne disease of global significance. *Beverage Food World*, 42(12): 21–24.

Pandey, J., Pandey, I. K. and Bhardwaj, N. 2022. Identification and molecular characterization of bacteria near source of potable water. *Biochemical and Cellular Archives*, 22(2).

Panza, O., Lacivita, V., Tarantino, F., Manzi, A., Conte, A. and Del Nobile, M. A. 2023. Fruit and vegetable by-products as source of bioactive compounds to preserve handmade fresh pasta. *LWT*, 190: 115584.

Parlapani, F. F., Boziaris, I. S. and Drosinos, E. H. 2024. Detection of fish spoilage. In *Handbook of Seafood and Seafood Products Analysis*, ed. Toldrá, F. and Nollet, L., 560–585. Boca Raton: CRC Press.

Pastinen, T. 2023. Single nucleotide polymorphism genotyping using microarrays. *Microarrays and Microplates*, 89–113.

Pellissery, A. J., Vinayamohan, P. G., Amalaradjou, M. A. R. and Venkitanarayanan, K. 2020. Spoilage bacteria and meat quality. In *Meat Quality Analysis*, ed. Biswas, A. K. and Mandal, P. K., 307–334. Academic Press.

Pereira, A. P. M., Stradiotto, G. C., Freire, L. et al. 2020. Occurrence and enumeration of rope-producing spore forming bacteria in flour and their spoilage potential in different bread formulations. *LWT*, 133: 110108.

Pinchuk, I. V., Beswick, E. J. and Reyes, V. E. 2010. Staphylococcal enterotoxins. *Toxins*, 2(8): 2177–2197.

Platt-Samoraj, A. 2022. Toxigenic properties of *Yersinia enterocolitica* biotype 1A. Toxins, 14(2): 118.

Poortmans, M., Vanoirbeek, K., Dorner, M. B. and Michiels, C. W. 2022. Selection and development of nontoxic nonproteolytic *Clostridium botulinum* surrogate strains for food challenge testing. *Foods*, 11(11): 1577.

Popoff, M. R. and Brüggemann, H. 2022. Regulatory networks controlling neurotoxin synthesis in *Clostridium botulinum* and *Clostridium tetani*. *Toxins*, 14(6): 364.

Poulain, B., Lemichez, E. and Popoff, M. R. 2020. Neuronal selectivity of botulinum neurotoxins. *Toxicon*, 178: 20–32.

Pourmadadi, M., Yazdian, F., Hojjati, S. and Khosravi-Darani, K. 2021. Detection of microorganisms using graphene-based nanobiosensors. *Food Technology and Biotechnology*, 59(4): 496–506.

Purwandari, U. 2009. Physical properties of functional fermented milk produced with exopolysaccharide-producing strains of *Streptococcus thermophilus*. Doctoral dissertation, Victoria University.

Quintero-Ronderos, P., Arango, M. T., Castiblanco, J., Correa, N. E. and Montoya-Ortíz, G. 2013. Analysis of proteins and antibodies. In *Autoimmunity: From Bench to Bedside*, ed. Anaya, J. M., Shoenfeld, Y., Rojas-Villarraga, A. et al. Bogota (Colombia): El Rosario University Press. https://www.ncbi.nlm.nih.gov/books/NBK459447/ (accessed September 6, 2024).

Rahman, K. T. and Butzin, N. C. 2024. Counter-on-chip for bacterial cell quantification, growth, and live-dead estimations. *Scientific Reports*, 14(1): 782.

Rajkovic, A., Jovanovic, J., Monteiro, S. et al. 2020. Detection of toxins involved in foodborne diseases caused by Gram-positive bacteria. *Comprehensive Reviews in Food Science and Food Safety*, 19(4): 1605–1657.

Ramamoorthi, L., Toshkov, S., Stetzer, A., Tucker, E. V. E. and Brewer, M. S. 2010. Effects of antioxidants on irradiated beef color. *Journal of Food Quality*, 33(3): 283–298.

Rasetti-Escargueil, C., Lemichez, E. and Popoff, M. R. 2020. Toxemia in human naturally acquired botulism. *Toxins*, 12(11): 716.

Rasooly, A. and Herold, K. E. 2008. Food microbial pathogen detection and analysis using DNA microarray technologies. *Foodborne Pathogens and Disease*, 5(4): 531–550.

Raspor, P. and Goranovič, D. 2008. Biotechnological applications of acetic acid bacteria. *Critical Reviews in Biotechnology*, 28(2): 101–124.

Rašková, H. 2020. Enterotoxins. In *Handbook of Toxinology*, ed. Shier, W. T., 309–336. Boca Raton: CRC Press.

Rawson, A. M., Dempster, A. W., Humphreys, C. M. and Minton, N. P. 2023. Pathogenicity and virulence of *Clostridium botulinum*. *Virulence*, 14(1): 2205251.

Roane, T. M. and Pepper, I. L. 2015. Microscopic techniques. In *Environmental Microbiology*, ed. Pepper, I. L., Gerba, C. P. and Gentry, T. J., 177–193. Academic Press.

Rubio-Bretón, P., Garde-Cerdán, T., Martínez, J., Gonzalo-Diago, A., Pérez-Álvarez, E. P. and Bordiga, M. 2017. Wine aging and spoilage. In *Post-Fermentation and-Distillation Technology*, ed. Bordiga, M., 113–158. Boca Raton: CRC Press.

Rusnan, A. N., Nordin, N., Radu, S. and Abdul-Mutalib, N. A. 2020. Pathogenic *Bacillus cereus*, an Overlooked Food Contaminants in Southeast Asia. *Pertanika Journal of Tropical Agricultural Science*, 43(1).

Saad, R. E., Sabike, I. I., Abo EL-Roos, N. S. and Elsabagh, R. 2023. The potential contribution of food contact surfaces to the incidence of *Escherichia coli* in Egyptian meat outlets, as well as their biofilm and pathogenicity characteristics. *Benha Veterinary Medical Journal*, 45(1): 167–171.

Sadri-Saeen, Z., Khani, M. and Fadaei, V. 2024. Improving the microbial quality and sensory properties of pasteurized sweet cream butter during refrigerated storage using chia seed ethanolic extract. *Food Science and Nutrition*, 12(1): 564–573.

Safkan, D., Agirman, B., Safkan, B., Soyer, Y. and Erten, H. 2024. Black-carrot drink şalgam as a habitat for spoilage yeasts. *Fungal Biology*, 128(8): 2471–2478.

Sahu, M. and Bala, S. 2017. Food processing, food spoilage and their prevention: An overview. *International Journal of Life-Sciences Scientific Research*, 3(1): 753–759.

Salama, Y. and Chennaoui, M. 2024. Microbial spoilage organisms in seafood products: Pathogens and quality control. *European Journal of Microbiology and Infectious Diseases*, 1(2): 66–89.

Sallam, K. I., Abd-Elghany, S. M., Hussein, M. A. et al. 2020. Microbial decontamination of beef carcass surfaces by lactic acid, acetic acid, and trisodium phosphate sprays. *BioMed Research International*, 2020(1): 2324358.

Santi, S., Wahab, A. W., Raya, I., Ahmad, A. and Maming, M. 2021. Synthesis, spectroscopic (FT-IR, UV–visible) study, and HOMO-LUMO analysis of adenosine triphosphate (ATP) doped trivalent terbium. *Journal of Molecular Structure*, 1237: 130398.

Sassolas, A., Leca-Bouvier, B. D. and Blum, L. J. 2008. DNA biosensors and microarrays. *Chemical Reviews*, 108(1): 109–139.

Semenko, N., Mokhort, H., Sokolovska, O., Kolesnikova, I., Kuzin, I. and Saylors, K. 2021. Foodborne botulism in Ukraine from 1955 to 2018. *Foodborne Pathogens and Disease*, 18(4): 260–266.

Shahzad, S., Afzal, M., Sikandar, S. and Afzal, I. 2020. Polymerase chain reaction. In *Genetic Engineering-A Glimpse of Techniques and Applications*, ed. Jamal, F., 13–32. IntechOpen.

Shao, L., Chen, S., Wang, H., Zhang, J., Xu, X. and Wang, H. 2021. Advances in understanding the predominance, phenotypes, and mechanisms of bacteria related to meat spoilage. *Trends in Food Science and Technology*, 118: 822–832.

Sharma, G., Sharma, S., Sharma, P. et al. 2016. *Escherichia coli* biofilm: Development and therapeutic strategies. *Journal of Applied Microbiology*, 121(2): 309–319.

Sharma, K. and Sharma, M. 2023. Optical biosensors for environmental monitoring: Recent advances and future perspectives in bacterial detection. *Environmental Research*, 116826.

Sharma, S., Shukla, S. and Rastogi, M. 2021. A review on biosensors and recent development. *Academicia: An International Multidisciplinary Research Journal*, 11(10): 613–620.

Shen, J., Zhi, S., Guo, D. et al. 2022. Prevalence, antimicrobial resistance, and whole genome sequencing analysis of Shiga toxin-producing *Escherichia coli* (STEC) and enteropathogenic *Escherichia coli* (EPEC) from imported foods in China during 2015–2021. *Toxins*, 14(2): 68.

Shinde, D. B., Singhvi, S., Koratkar, S. S. and Saroj, S. D. 2020. Isolation and characterization of *Escherichia coli* serotype O157: H7 and other verotoxin-producing *E. coli* in healthy Indian cattle. *Veterinary World*, *13*(10): 2269.

Singh, J. 2022. Antigen–antibody interactions. In *An Interplay of Cellular and Molecular Components of Immunology*, ed. Singh, I. K. and Sharma, P., 133–149. Boca Raton: CRC Press.

Smith, J. L. and Fratamico, P. M. 2017. *Escherichia coli* as a pathogen. In *Foodborne Diseases*, ed. Dodd, C. E. R., Aldsworth, T., Stein, R. A., Cliver, D. O. and Riemann, H. P., 189–208. Academic Press.

Sreeja, V. and Gawai, K. M. 2019. Shelf-life extension of fermented milk products. In *Engineering Practices for Milk Products*, ed. Goyal, M. R. and Hati, S., 277–310. New York: Apple Academic Press.

Stella, E. I. and Ifeanyi, O. E. 2018. A review on salmonella species and indicator organisms in drinking water. *International Journal of Comprehensive Research in Biological Sciences*, 5(2): 5–23.

Subjakova, V., Oravczova, V., Tatarko, M. and Hianik, T. 2021. Advances in electrochemical aptasensors and immunosensors for detection of bacterial pathogens in food. *Electrochimica Acta*, 389: 138724.

Sultana, S., Iqbal, A. and Islam, M. N. 2014. Preservation of carrot, green chilli and brinjal by fermentation and pickling. *International Food Research Journal*, 21(6): 2405–2412.

Sykes, J. E. and McDonough, P. L. 2021. Salmonellosis. In *Greene's Infectious Diseases of the Dog and Cat*, ed. Sykes, J. E., 750–758. Elsevier.

Şanlıbaba, P. and Uymaz, B. 2015. Gıdalarda *Listeria monocytogenes*'in Biyokontrolünde Faj Uygulaması. *Akademik Gıda*, 13(1): 81–88.

Şentürk, E., Buzrul, S. and Şanlıbaba, P. 2022. Prevalence of *Listeria monocytogenes* in ready-to-eat foods, and growth boundary modeling of the selected strains in broth as a function of temperature, salt and nisin. *International Journal of Food Properties*, 25(1): 2237–2253.

Tambi, A., Brighu, U. and Gupta, A. B. 2023. Methods for detection and enumeration of coliforms in drinking water: A review. *Water Supply*, 23(10): 4047–4058.

Tao, J., Liu, W., Ding, W. et al. 2020. A multiplex PCR assay with a common primer for the detection of eleven foodborne pathogens. *Journal of Food Science*, 85(3): 744–754.

Taylor, D. W. 2022. The inducible defense system: Antibody molecules and antigen-antibody reactions. In *Infection, Resistance, and Immunity*, ed. Kreier, J., 105–129. New York: Routledge.

Taylor, M. H. and Zhu, M. J. 2021. Control of *Listeria monocytogenes* in low-moisture foods. *Trends in Food Science and Technology*, 116: 802–814.

Tewari, A. and Abdullah, S. 2015. *Bacillus cereus* food poisoning: International and Indian perspective. *Journal of Food Science and Technology*, 52: 2500–2511.

Thakur, A., Maroju, P. A., Ganesan, R. and Ray Dutta, J. 2024. Contrasting responses of motile and non-motile *Escherichia coli* strains in resuscitation against stable ultrafine gold nanosystems. *Micro and Nano Systems Letters*, 12(1): 1–13.

Tian, Y., Wu, L., Lu, R. et al. 2022. Virulent phage vB_CpeP_HN02 inhibits *Clostridium perfringens* on the surface of the chicken meat. *International Journal of Food Microbiology*, 363: 109514.

Todorov, S. D. and Franco, B. D. G. D. M. 2010. *Lactobacillus plantarum*: Characterization of the species and application in food production. *Food Reviews International*, 26(3): 205–229.

Uddin, M. N., Khan, W., Qayum, N. et al. 2023. Shigellosis; a clinical perspective. *Zoonosis, Unique Scientific Publishers, Faisalabad, Pakistan*, 4: 544–556.

Valenti, M., Ranganathan, N., Moore, L. S. and Hughes, S. 2021. *Listeria monocytogenes* infections: Presentation, diagnosis and treatment. *British Journal of Hospital Medicine*, 82(10): 1–6.

van Pelt-Verkuil, E., van Belkum, A. and Hays, J. P. 2008. PCR primers. *Principles and Technical Aspects of PCR Amplification*, 63–90.

Vasavada, P. C., Lee, A. and Betts, R. 2020. Conventional and novel rapid methods for detection and enumeration of microorganisms. *Food Safety Engineering*, 85–128.

Váradi, L., Luo, J. L., Hibbs, D. E. et al. 2017. Methods for the detection and identification of pathogenic bacteria: Past, present, and future. *Chemical Society Reviews*, 46(16): 4818–4832.

Veljanovski, K. and Blazhekovikj-Dimovska, D. 2023. The influence of halophilic bacteria on the fish quality and safety. *Journal of Agriculture and Sustainable Rural Development*, 1(1–2): 132–138.

Wang, X., Hu, K., Liu, F. et al. 2023. Isolation and characterization of a gas-producing and acid-resistant bacterium from spoiled vinegar. *International Journal of Food Microbiology*, 394: 110167.

Wang, Z., Tao, X., Liu, S., Zhao, Y. and Yang, X. 2021. An update review on *Listeria* infection in pregnancy. *Infection and Drug Resistance*, 1967–1978.

Wan Ismail, W. M. A. H., Rudi, M. A. and Mohamed, E. 2022. The use of microorganisms in dairy products (butter). *Chemical Process and Sustainability in Industrial Biotechnology*. Penerbit UTHM, 51–66.

Weil, A. A. and LaRocque, R. C. 2020. Cholera and other vibrios. In *Hunter's Tropical Medicine and Emerging Infectious Diseases*, ed. ed. Ryan, E. T., Hill, D. R., Solomon, T., Aronson, N. E. and Endy, T. P., 486–491. Elsevier.

Woraprayote, W., Janyaphisan, T., Adunphatcharaphon, S. et al. 2023. Bacteriocinogenic lactic acid bacteria from Thai fermented foods: Potential food applications. *Food Bioscience*, 52: 102385.

Xu, L., Bai, X. and Bhunia, A. K. 2021. Current state of development of biosensors and their application in foodborne pathogen detection. *Journal of Food Protection*, 84(7): 1213–1227.

Yaashikaa, P. R., Saravanan, A. and Kumar, P. S. 2016. Isolation and identification of *Vibrio cholerae* and *Vibrio parahaemolyticus* from prawn (*Penaeus monodon*) seafood: Preservation strategies. *Microbial Pathogenesis*, 99: 5–13.

Young, A. P., Jackson, D. J. and Wyeth, R. C. 2020. A technical review and guide to RNA fluorescence in situ hybridization. *PeerJ*, 8: e8806.

Yu, D., Wu, L., Regenstein, J. M. et al. 2020. Recent advances in quality retention of non-frozen fish and fishery products: A review. *Critical Reviews in Food Science and Nutrition*, 60(10): 1747–1759.

Zhang, Q. and Sahin, O. 2020. Campylobacteriosis. *Diseases of Poultry*, 754–769.

Zhang, S., Li, X., Wu, J. et al. 2021. Molecular methods for pathogenic bacteria detection and recent advances in wastewater analysis. *Water*, 13(24): 3551.

Zhuang, S., Hong, H., Zhang, L. and Luo, Y. 2021. Spoilage-related microbiota in fish and crustaceans during storage: Research progress and future trends. *Comprehensive Reviews in Food Science and Food Safety*, 20(1): 252–288.

Zhu, K., Dietrich, R., Didier, A., Doyscher, D. and Märtlbauer, E. 2014. Recent developments in antibody-based assays for the detection of bacterial toxins. *Toxins*, 6(4): 1325–1348.

6 Fungal Spoilage and Detection in Food Products

Shahriyar Valizadeh and Reza Tahergorabi

6.1 INTRODUCTION

Fungi were long considered part of the plant kingdom due to similarities like growth patterns and having both underground and aerial structures. However, fungi differ fundamentally from plants as they lack chlorophyll and cannot perform photosynthesis. Fungi sustain themselves by decomposing dead organic matter or extracting nutrients from living organisms. From the advent of agriculture, when early humans started cultivating crops and storing their yields, spoilage fungi have continuously posed a significant challenge. These fungi, manifesting as fuzzes, powders, and slimes in a variety of colors—from white and black to green, orange, red, and brown—have silently infiltrated food supplies. Through processes of acidification, fermentation, discoloration, and disintegration, they have rendered once-nutritious commodities unpalatable and, at times, unsafe. Despite this long history of interaction, fungi were historically regarded as causing merely superficial spoilage. This view remained unchanged despite the association of *Claviceps purpurea* with human illness over two centuries ago, and the recognition of the acute toxicity of certain macrofungi dates back even further (Steinberg and Gurr 2020).

This underestimation of fungi's impact belies their true nature as some of the most resilient spoilage microorganisms in the food industry. These organisms possess an extraordinary ability to circumvent even the most rigorous control measures. Their propagules, which are efficiently dispersed through both water and air, allow them to survive in extreme environments and sustain their growth over time (Snyder and Worobo 2018a). Certain fungal species are exceptionally resilient, flourishing in extreme physicochemical environments and withstanding the rigorous thermal processes applied in commercial food production. Yeasts and filamentous fungi, in particular, exhibit a notable ability to spread and cause cross-contamination throughout food processing facilities. The diverse structural adaptations and survival mechanisms of these fungi enable them to colonize specific ecological niches, making them potent contaminants that can spoil even meticulously processed foods (Snyder and Worobo 2018a).

The impact of fungal spoilage on the food industry is significant, as reflected in various indicators. While fungi in processed foods rarely pose an immediate food safety risk, they play a major role in waste and loss across the supply chain. Fungal

DOI: 10.1201/9781003540113-6

contamination triggers product recalls, accelerates perishability, and diminishes consumer trust in food manufacturers and retailers. Approximately 25% of the global food supply is lost or wasted as a result of post-harvest microbial spoilage. However, this figure varies based on factors like product formulation, processing techniques, packaging, and storage conditions. For instance, a survey conducted among juice manufacturers in the United States and Europe showed that 92% had experienced mold or yeast spoilage in their final products, while 89% reported spoilage at the ingredient stage. The pervasive nature of fungal spoilage underscores its critical impact on the food industry and highlights the need for more effective control strategies (Snyder and Worobo 2018b; Garnier, Valence, and Mounier 2017).

6.2 FUNGAL SPOILAGE IN FOOD AND FOOD PRODUCTS

Fungal spoilage poses a major concern for a broad array of food products, such as fruits, vegetables, dairy items, and cereals. The presence of fungi not only leads to significant economic losses due to reduced shelf life and quality but also poses serious health risks through the production of mycotoxins. Each type of food is susceptible to specific fungal pathogens that thrive under different environmental conditions, leading to distinct patterns of spoilage. Understanding the mechanisms of fungal infection and growth in various food matrices is essential for developing effective control strategies. This section explores the key fungal species responsible for spoilage in different food categories, highlighting their impact on food safety and quality and discussing current methods for detection and prevention.

6.2.1 Fruit

The defense mechanisms in fruits are generally very effective against most fungi, as only a limited number of genera and species can penetrate and cause significant damage. Some of these fungi are highly specialized, targeting only specific types of fruit, while others possess a broader capacity to invade various fruit tissues. As fruits ripen, their vulnerability to fungal infections increases for several reasons: The tissue pH elevates, skin layers soften, soluble carbohydrates build up, and their defense mechanisms weaken. Post-harvest fruit storage is a complex field that involves managing the balance between optimal maturity and storage longevity, ripening versus overripening, and maintaining desirable qualities while preventing fungal contamination and spoilage (Atif et al. 2020; Perez-Nadales et al. 2014).

6.2.1.1 Citrus Fruit

Citrus fruit decay is mainly caused by *Penicillium* rots from *P. italicum* (blue rot) and *P. digitatum* (green rot), which infect fruit post-harvest, particularly when the skin is damaged. The decay spreads via contact between fruits and flourishes from 20 to 25°C, but its growth slows at lower than 5°C or higher than 30°C. Effective control includes careful handling and post-harvest treatments, like heated washes with detergents, alkalis, or fungicides. Additionally, GRAS compounds and biological control with bacteria and yeasts are being considered as alternatives, especially for export markets where certain fungicides are limited (Moraes Bazioli et al. 2019; Kanashiro et al. 2020).

6.2.1.2 Pome Fruits

The primary and most destructive fungal issue affecting pome fruits, like apples and pears, is driven by *Penicillium*, notably *P. expansum*, which causes blue mold. The infection typically starts as a small, soft, discolored spot that swiftly spreads across the fruit's surface and penetrates deep into the tissue. As the fungus proliferates, bluish-green spore-producing structures form on the skin. Although *P. expansum* thrives even at cooler temperatures, refrigeration can only slow down the spoilage, not entirely stop it. Fruits that are either bruised or overly ripened are especially prone to this infection. In advanced cases, diagnosis can be made by visual inspection or by culturing samples on dichloran rose bengal chloramphenicol (DRBC), followed by czapek yeast autolysate (CYA) and czapek sucrose nitrate (CSN) media. If an infected fruit is left at room temperature for several days, it will produce the typical blue-green spore structures characteristic of *Penicillium expansum* (Palou and Smilanick 2019; Díaz-Urbano et al. 2023). Oyom et al. (2022a) provided a comprehensive review of recent advances in post-harvest technology for controlling fungal diseases in Asian pears.

6.2.1.3 Stone Fruits

Stone fruits like peaches, plums, apricots, nectarines, and cherries are particularly prone to brown rot, primarily caused by *Monilia fructicola* and, in certain regions, closely related species, such as *M. fructigena* (Pers.) Pers. and *M. laxa* (Ehrenb.) Sacc. and Voglino. These fungi also go by their teleomorph names: *Monilinia fructicola* (G. Winter) Honey [formerly *Sclerotinia fructicola* (G. Winter) Honey], *Monilinia fructigena* (Aderh. and Rühle) Honey, and *Monilinia laxa* (Aderh. and Rühle) Honey. However, only the *Monilia* form is typically observed on commercial fruit or in laboratory cultures. Brown rot is the most significant post-harvest disease impacting apricots, peaches, and nectarines. Early symptoms include water-soaked spots on the fruit, which turn brown within 24 hours, spreading rapidly and penetrating deeper into the tissue. Eventually, a layer of pale brown spores forms, and the entire fruit can decay in 3 to 4 days. The infection originates in the orchard, making rigorous pre-harvest applications of benomyl or related benzimidazole fungicides crucial for control. Additionally, post-harvest treatments with dips in dichloran, iprodione, or triforine can be beneficial. Storing fruit at temperatures below 5°C is recommended for further control (Martínez-Blay et al. 2021).

6.2.2 Grapes

Botrytis cinerea, often referred to as "noble rot" in certain wine grape varieties, plays a prized role in winemaking; however, it is also a leading cause of spoilage in table grapes. Initially, the fungus infects stems and the interior of the grape berry, eventually breaking through the surface to produce gray conidia. The fungus quickly spreads through densely packed grape clusters in high humidity, leading to large patches of rot. Under a microscope, the distinctive conidiophores of *Botrytis* are easily identifiable. To control this pathogen, pre-harvest treatments with benomyl and rapid post-harvest cooling are recommended. Post-harvest treatments with sulfur dioxide or benomyl are also effective in preventing spoilage (König, Unden, and Fröhlich 2009).

6.2.2.1 Berries

Berries, due to their low-growing nature and direct contact with soil, are highly vulnerable to contamination by soilborne pathogens and fungal spores. Additionally, their delicate structure makes them susceptible to damage during harvesting and post-harvest handling, further increasing the risk of fungal infections. These factors contribute significantly to the challenges of maintaining berry quality and safety throughout the supply chain. Most berries share similar vulnerabilities to fungal diseases and can be grouped together in this regard. The primary fungi responsible for decay in most berry crops include *Botrytis cinerea*, *Rhizopus stolonifer*, and *Mucor piriformis*. *Botrytis* is particularly damaging, causing soft rot in cane berries, like raspberries and loganberries, and a firm, dry rot in strawberries. Infected fruit is typically enveloped in gray mold. Strawberry crops are especially vulnerable, with the fungus spreading rapidly through contact, leading to clusters of rotting fruit. Effective control strategies focus on pre-harvest fungicidal spraying and maintaining cold storage conditions, as post-harvest antifungal treatments tend to have limited efficacy (Kłapeć et al. 2022).

6.2.3 Vegetables

The almost-neutral pH of vegetables makes them more prone to bacterial infections, which places bacterial pathogens on nearly equal footing with fungal ones. Nonetheless, vegetables are vulnerable to a wide array of fungal infections. Bacterial rots are generally identified by their watery or slimy texture, lack of visible mycelium, and foul odor, setting them apart from fungal infections. To aid in diagnosis, wet mounts stained with lactofuchsin can be used as an effective tool for detecting and confirming the presence of bacterial pathogens. This method highlights the bacterial structures and helps differentiate bacterial rot from other types of spoilage (Udoh et al. 2015).

6.2.3.1 Onions

As a vegetable that grows underground, onions are exposed to fungi throughout their growth and maturation. Onions exhibit a robust resistance to fungal invasion, with many fungal diseases only emerging after harvest. While few *Aspergillus* species cause plant diseases, *A. niger* is a notable pathogen in onions, producing black conidia that accumulate between the outer layers. Other fungi, such as *Fusarium* species (*F. solani* and *F. oxysporum*) and *Botrytis* species (*B. allii* and *B. cinerea*), can infect onions during their time in the field and continue to thrive throughout storage, leading to further spoilage and quality degradation. Additionally, various *Penicillium* species are known to cause blue rot in onions worldwide, with *P. glabrum* notably affecting Australian onions and reported in other regions. The effective control of onion diseases includes prompt drying and curing, as well as maintaining lower temperatures and humidity levels (Ji et al. 2018).

6.2.3.2 Potatoes

Potatoes are predominantly affected by bacterial rot, but fungal diseases also pose significant threats. *Fusarium* species, particularly *Fusarium solani* var. *coeruleum* and *Fusarium sulphureum Schltdl*, are the main culprits behind dry rot. Potatoes are also susceptible to silver scurf, caused by *Helminthosporium solani* Durieu and Mont., and skin spot, caused by *Polyscytalum pustulans* (M. N. Owen and Wakef.) M. B. Ellis. In addition to these, other pathogens, such as *Phytophthora*, *Pythium*, *Phoma*, and various other genera, can contribute to the development of potato diseases, complicating control measures. Infection typically occurs through wounds, with brown lesions forming and the tissues shrinking and wrinkling as the decay advances. Effective control measures for managing post-harvest pathogens in crops include careful handling and sorting to minimize physical damage, surface drying to reduce moisture, and refrigerated storage to slow pathogen growth. A hot water treatment at 60°C has also been shown to be highly effective in controlling post-harvest pathogens such as *Fusarium oxysporum*, *Botryodiplodia theobromae*, and *Rhizopus oryzae*, reducing the risk of spoilage and extending shelf life. These combined strategies help ensure better preservation of produce quality during storage and distribution (Tiwari et al. 2020).

6.2.4 Cereals, Nuts

Fungi that colonize crops intended for drying have traditionally been classified as "field" or "storage" fungi. While this distinction has blurred in recent years—especially with findings that some species, such as *Aspergillus flavus*, can thrive in both environments—it still serves as a useful framework. Field fungi primarily act as plant pathogens, infecting seeds or nuts during the growing season prior to harvest. This categorization aids in understanding the timing and conditions under which these fungi impact crop quality and safety. Spoilage or deterioration of a crop often occurs due to a specific fungal invasion, which is facilitated by certain climatic conditions, plant varieties, or agricultural practices that create a conducive environment for large-scale infestation by that fungus. If control measures are available, they typically involve improvements in agricultural practices. Field fungi generally do not play a significant role in further crop deterioration after harvest (Rao et al. 2021).

6.2.5 Dairy Products

Fresh milk, with its neutral pH, is predominantly susceptible to bacterial spoilage, which generally restricts fungal growth. However, when milk is processed into products like cream, cottage cheese, or butter, the proliferation of lactic acid bacteria lowers the pH, creating an environment conducive to the growth of spoilage yeasts. These yeasts can result in undesirable effects, including gas production and off-flavors in cream and cottage cheese, as well as rancidity or other flavor defects in butter. Additionally, *Geotrichum candidum* can cause spoilage in cream, often due to inadequate sanitation practices in machinery used on farms (Lu and Wang 2017).

In UHT dairy products, spoilage due to heat-resistant fungi is uncommon, though contamination from soil is still a possibility. Conversely, yeasts are commonly found in yogurt and can occasionally lead to spoilage, especially in fruit-containing varieties. Species such as *Candida parapsilosis*, *C. famata* (the anamorph of *Debaryomyces hansenii*), *C. diffluens*, *Rhodotorula mucilaginosa*, *Kluyveromyces marxianus*, and *Yarrowia lipolytica* are frequently implicated in this spoilage process. These yeasts can contribute to off-flavors, gas production, and changes in texture, ultimately affecting the quality and consumer acceptance of yogurt products. Moreover, fruit-flavored yogurts can be adversely affected by fungi introduced during the fruit preparation process. For instance, *Mucor circinelloides* can lead to fermentative spoilage, even at refrigeration temperatures, and can thrive in conditions with extremely low oxygen levels. This spoilage not only compromises the quality of the yogurt but can also result in off-flavors and undesirable textures, ultimately impacting consumer satisfaction and product safety (Souza et al. 2022; Shi and Maktabdar 2022).

6.3 FACTORS CONTRIBUTING TO FUNGAL SPOILAGE

Fungal spoilage in food products presents a critical challenge to both food safety and quality management. As shown in **Figure 6.1**, various factors contribute to fungal contamination, manifesting at different phases of food production, storage, and distribution. These contributing elements can be generally classified into three main categories: environmental conditions, food-related attributes, and specific fungal characteristics (Davies et al. 2021; Shi and Maktabdar 2022; Snyder and Worobo 2018a). Temperature is the second most important factor influencing mold growth, following water activity (*aw*). The interplay between temperature and aw significantly affects fungal development. For example, molds like *Aspergillus flavus* and *Aspergillus niger* can grow within a temperature range of 8 to 45°C, with an optimal temperature around 30°C. This makes them particularly concerning for spoilage in shelf-stable foods, especially during hot weather. Conversely, molds that affect chilled foods tend to be more psychrophilic; certain species of *Penicillium*, *Cladosporium*, and *Fusarium* can thrive at temperatures as low as 0°C, posing risks for spoilage in refrigerated products. At the handling and processing levels, air and water within a facility can effectively facilitate the spread of fungi. To minimize contamination, air handling systems may employ positive pressure or high-efficiency particulate air (HEPA) filters. Airborne molds present challenges during the filling process, prompting many facilities to isolate this critical step. Inadequate control of air pressure has been associated with increased fungal spoilage. Additionally, the use of compressed air for cleaning packaging or surfaces, as well as high-velocity air from blast freezers, can contribute to product contamination. It is essential to select sanitizers based on the composition of the food products and the resistance of spoilage organisms to ensure effective microbial control. For instance, *Zygosaccharomyces bailii*, which resists many common sanitizers, requires specific chemicals effective against it, especially for processors like juice manufacturers concerned about yeast spoilage (Dagnas and Membré, 2013).

6.4 MECHANISMS OF FUNGAL SPOILAGE

Many fungal species possess the remarkable ability to produce a variety of spores within a single colony, demonstrating the complexity and adaptability of their reproductive strategies. This capacity to generate different asexual spore types, such

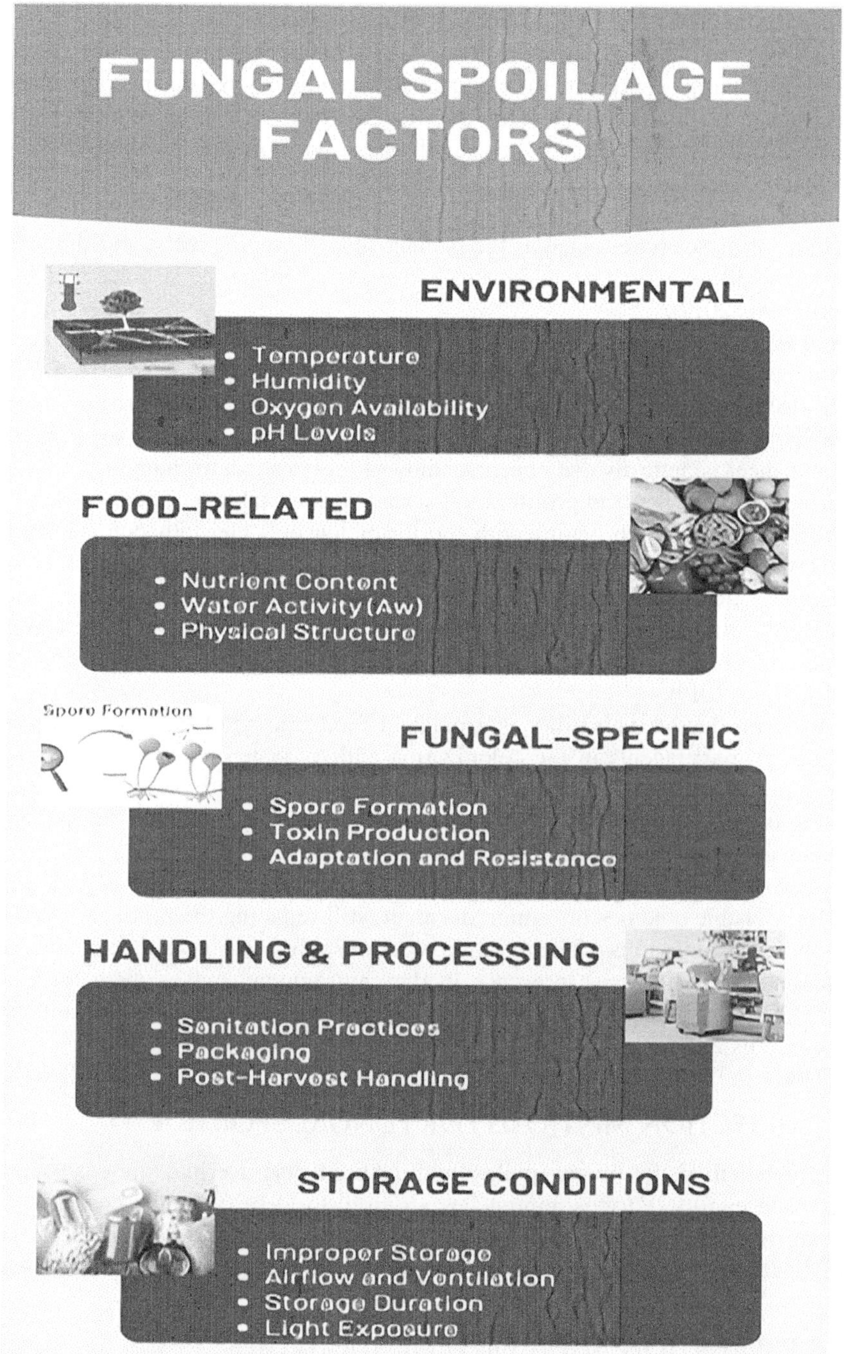

FIGURE 6.1 Factors that influence fungal spoilage of foods.

as aerial conidia and chlamydospores, enhances their ecological versatility. For instance, *Botrytis cinerea*, a prevalent spoilage agent in crops like strawberries, produces both chlamydospores within hyphae and conidial spores on specialized structures known as conidiophores. This multifaceted approach to reproduction allows fungi to efficiently colonize and thrive in diverse environments. Similarly, *Fusarium oxysporum*, which affects cereals, produces both microconidia and macroconidia for dispersal, along with chlamydospores for long-term survival. The diverse shapes and functions of these spores, such as the banana-shaped macroconidia and the smaller microconidia, further illustrate their role in ensuring fungal persistence and spread (Dijksterhuis 2019).

In addition to asexual spores, many fungi are capable of producing sexual spores within the same colony. For example, *Aspergillus nidulans* generates asexual conidia while also having the capacity to form sexual ascospores in response to environmental factors such as light and nutrient availability. This dual reproductive strategy not only enhances genetic diversity but also allows fungi to adapt to changing conditions, ensuring their survival and proliferation across various habitats. The process of sexual reproduction, which begins with the formation of a cleistothecium, is equally complex and involves meiosis and genetic recombination, ensuring the creation of new genetic combinations that promote evolutionary adaptability. The simultaneous production of both sexual and asexual structures within the same colony is a hallmark of fungal flexibility, allowing them to thrive under diverse conditions (Dijksterhuis 2019).

This reproductive versatility is crucial in food microbiology, as fungal spores serve as primary agents in the colonization and spoilage of food products. These spores are ubiquitously present in both outdoor and indoor air, with concentrations varying from hundreds to thousands per cubic meter. Due to their airborne nature, they can easily settle on food surfaces during exposure to the environment, increasing the risk of contamination and spoilage. Once on the food, they may germinate and form visible colonies or remain dormant, still reducing the product's shelf life. The ability of fungi to produce multiple types of spores within a single colony, combined with their pervasive presence in the environment, makes them formidable agents of food spoilage, complicating efforts to maintain food quality and safety (Dijksterhuis 2019).

6.5 DETECTION METHODS FOR FUNGAL SPOILAGE

Detection methods for fungal spoilage in food products are critical for ensuring food safety and quality. Various approaches are used to identify fungal contamination, ranging from traditional culture-based methods to more advanced molecular and analytical techniques. An overview of key detection methods is shown in **Table 6.1**.

6.6 PREVENTION AND CONTROL STRATEGIES

Effectively managing fungal pathogens at various stages of food production is vital for ensuring food security and minimizing economic losses. Implementing a range of prevention and control strategies, from pre-harvest practices to post-harvest

TABLE 6.1
Detection Methods for Fungal Spoilage

Methods	Advantage	Limitation	References
Visual inspection	• Simple • Cost-effective • Immediate results	• Only effective when fungal growth is visible; cannot detect fungal presence in early stages or at the microscopic level	(Dagnas and Membré 2013)
Microscopic examination	• Early detection • Higher accuracy • Low cost	• Limited sensitivity • Requires expertise • Non-specific	(Mendonca et al. 2022)
Culture-based methods	• Allows identification and isolation of specific fungal species	• Time-consuming; typically requiring 3–7 days for colonies to form	(Erfani et al. 2024)
Biochemical methods	• High sensitivity and specificity; able to detect fungal toxins even at low concentrations	• Requires prior knowledge of the fungal species to develop appropriate antibodies	(Franco-Duarte et al. 2019)
Molecular methods	• Highly sensitive and specific; able to detect fungi at very low concentrations	• Requires technical expertise and specialized equipment	(Rodriguez et al. 2015)
Metagenomic sequencing	• Comprehensive and can detect multiple fungal species, even in complex food matrices	• High cost and complexity	(Muñoz-Martinez et al. 2025)
Mycotoxin detection	• Extremely sensitive and precise, allowing quantification of specific mycotoxins	• Expensive and requires skilled personnel	(Maphaisa et al. 2025)
Electronic nose (e-nose)	• Non-invasive, rapid, and able to detect early stages of fungal growth	• Limited species differentiation and requires product-specific calibration	(Jia et al. 2019)
Spectroscopic techniques	• Non-destructive, rapid, and requires minimal sample preparation	• Requires specialized instrumentation and calibration	(Argyri, Panagou, and Nychas 2014)
Biosensors	• Highly sensitive, portable, and can provide real-time results	• Limited to detecting specific fungi or their by-products	(Kumar et al. 2024)

preservation measures, can significantly mitigate the impact of these pathogens. Emphasizing integrated approaches that combine genetic, chemical, biological, and physical methods is essential for sustainable food production practices.

6.6.1 Pre-Harvest Strategies

6.6.1.1 Genetic Resistance Breeding

The development of crop varieties with inherent resistance to fungal pathogens is crucial for pre-harvest management. Genetic resistance can significantly reduce the reliance on chemical fungicides. Specific examples include the incorporation of resistance genes, such as the *mlo* gene in barley, which provides resistance against powdery mildew, and others in bananas targeting *Fusarium oxysporum* (Kusch and Panstruga 2017).

6.6.1.2 Application of Fungicide Synergies

Employing combinations of fungicides with varying modes of action significantly improves the overall effectiveness of fungal control strategies while reducing the risk of resistance development. This integrated approach not only enhances the efficacy of the treatment but also helps ensure long-term management of fungal pathogens, promoting sustainable practices in agricultural and food safety applications. This multifaceted approach not only improves management outcomes but also reduces the overall chemical load applied to crops, supporting both agricultural sustainability and environmental health (Davies et al. 2021).

6.6.1.3 Implementation of Biocontrol Agents

Employing natural antagonistic organisms, particularly those from the *Trichoderma* genus, offers a viable alternative to chemical fungicides. These biocontrol agents work by competing with pathogenic fungi for resources and space, thereby inhibiting their growth. The potential of such biological strategies represents an increasingly popular method within integrated pest management systems (Yao et al. 2023).

6.6.2 Control of Fungal Contamination in Fresh Produce

6.6.2.1 Application of Chemical Sanitizers

Employing chemical sanitizers, such as various chlorine solutions, ozone, or hydrogen peroxide, during the washing of fresh produce serves to significantly lower fungal contamination levels. Caution is necessary to ensure these treatments do not negatively impact produce quality (Davies et al. 2021).

6.6.2.2 Utilization of Electrolyzed Water

Electrolyzed water, produced through the electrolysis of saltwater, serves as an effective sanitizing solution for washing fresh produce. Its application has been shown to significantly decrease fungal loads without leaving harmful chemical residues, making it an ideal choice for food industry practices (de Almeida Lopes et al. 2021).

6.6.2.3 Application of Protective Coatings

The employment of natural or synthetic edible coatings on fresh fruits and vegetables can enhance shelf life and reduce fungal spoilage. These coatings can provide a physical barrier to pathogens and can be embedded with antifungal agents that are slowly released over time, providing ongoing protection (Díaz-Montes and Roberto 2021). Recently, Oyom et al. (2022b) demonstrated that a starch-based antifungal coating effectively mitigated black spot rot induced by *Alternaria alternata* in Asian pears over a 21-day incubation period, thanks to the antimicrobial agents incorporated into the coating formulation.

6.6.3 Preservation Strategies for Stored Foods

6.6.3.1 Biopreservation Techniques

Utilizing natural products, such as essential oils and fermented foods, has garnered attention as a sustainable strategy to inhibit fungal spoilage. These biopreservation methods leverage the antimicrobial properties of natural compounds, thus providing an effective means of maintaining food safety and quality (Valizadeh et al. 2020; Issa et al. 2017).

6.6.3.2 Combating Spoilage with Preservative Combinations

The strategic formulation of preservatives that employ different modes of action enhances their effectiveness against resistant fungal strains. By utilizing various preservatives in conjunction, it is feasible to reduce the required dosages of each, minimizing potential negative impacts on food quality and safety (Taylor and Cunniffe 2023).

6.6.3.3 Risk Assessment in Supply Chains

Implementing quantitative microbial risk assessments (QMRAs) within the food supply chain enables producers to anticipate and mitigate fungal contamination risks effectively. By identifying critical control points, stakeholders can optimize the application of preservatives and other interventions to safeguard food products from fungal spoilage throughout the supply chain (Zambon et al. 2022).

6.7 FUTURE TRENDS AND RESEARCH DIRECTIONS AND CONCLUSION

Future trends in research on fungal spoilage in food products are set to focus on the development and application of emerging detection and control technologies, addressing the critical concerns posed by spoilage fungi and their associated mycotoxins. Advanced tools like biosensors and nucleic acid–based methods, such as real-time PCR and vPCR, are becoming increasingly prioritized for their superior ability to identify pathogens early, surpassing the capabilities of conventional immunoassays. These technological advancements, combined with effective training programs for food surveillance personnel, are essential for optimizing the use of these tools, ensuring their maximum impact in detecting and controlling fungal contaminants.

Furthermore, innovative research directions, such as the use of bacteriophages to specifically target foodborne pathogens, represent a promising area of study. As detection methods evolve, the integration of holistic approaches that blend advanced technologies with natural antimicrobial interventions will be key to achieving robust control over fungal spoilage (Wang et al. 2024; Aladhadh 2023).

Fungal spoilage remains a significant challenge for food safety and quality, necessitating a comprehensive and multifaceted response. The complexity of fungal contamination involves diverse spoilage fungi, infection mechanisms, and numerous contributing factors, which underscores the importance of a deep understanding of these elements. Progress in molecular detection and biosensor technologies offers encouraging prospects for more precise identification and control of fungal outbreaks. Additionally, the adoption of integrated food safety strategies, which merge genetic, biochemical, and biocontrol approaches, is critical for strengthening defenses against spoilage fungi. Continuous research into preservation techniques, including the exploration of natural antimicrobials, is expected to further enhance the capacity to combat fungal spoilage, thereby improving food quality and safety standards (Wang et al. 2024).

In conclusion, effectively tackling fungal spoilage necessitates a proactive and interdisciplinary strategy that encourages collaboration among researchers, industry stakeholders, and regulatory agencies. This approach is essential for developing standardized practices and promoting the exchange of knowledge in the field. Prioritizing the education and training of food safety professionals is crucial for the successful implementation of advanced detection technologies and control strategies. While fungal spoilage remains a persistent threat, the ongoing development of innovative tools and integrated approaches holds the potential to minimize its impact, ensuring the safety and quality of food products for consumers worldwide.

REFERENCES

Aladhadh, M. 2023. A review of modern methods for the detection of foodborne pathogens. *Microorganisms*, *11*(5), 1111. https://doi.org/10.3390/microorganisms11051111.

Argyri, A. A., Panagou, E. Z., and G.-J. Nychas. 2014. Monitoring microbial spoilage of foods by vibrational spectroscopy (FT-IR & Raman). *Novel Food Preservation and Microbial Assessment Techniques*, *386*.

Atif, M., Ilavenil, S., Devanesan, S., AlSalhi, M. S., Choi, K. C., Vijayaraghavan, P., . . . N. F. Alanazi. 2020. Essential oils of two medicinal plants and protective properties of jack fruits against the spoilage bacteria and fungi. *Industrial Crops and Products*, *147*, 112239. https://doi.org/10.1016/j.indcrop.2020.112239.

Dagnas, S., and J. M. Membré. 2013. Predicting and preventing mold spoilage of food products. *Journal of Food Protection*, *76*(3), 538–551. https://doi.org/10.4315/0362-028X.JFP-12-349.

Davies, C. R., Wohlgemuth, F., Young, T., Violet, J., Dickinson, M., Sanders, J. W., . . . S. V. Avery. 2021. Evolving challenges and strategies for fungal control in the food supply chain. *Fungal Biology Reviews*, *36*, 15–26. https://doi.org/10.1016/j.fbr.2021.01.003.

de Almeida Lopes, M. M., de Lucena, H. H., da Silveira, M. R. S., dos Santos Garruti, D., Machado, T. F., de Aragão, F. A. S., and S. E. de Oliveira. 2021. The use of electrolyzed water as a disinfectant for fresh cut mango. *Scientia Horticulturae*, *287*, 110227. https://doi.org/10.1016/j.scienta.2021.110227.

Díaz-Montes, E., and C-M. Roberto. 2021. Edible films and coatings as food-quality preservers: An overview. *Foods*, *10*(2), 249. https://doi.org/10.3390/foods10020249.

Díaz-Urbano, M. Pablo, V., Víctor, M. R., and P. Jorge. 2023. Endophytic fungi in postharvest disease management in fresh produce. In *Postharvest Management of Fresh Produce*, ed. P. Bhim S. A. Singh, S. Garima, and G. Vijai Kumar, 81–112. Academic Press. https://doi.org/10.1016/B978-0-323-91132-0.00004-6.

Dijksterhuis, J. 2019. Fungal Spores: Highly variable and stress-resistant vehicles for distribution and spoilage. *Food Microbiology*, *81*(August), 2–11. https://doi.org/10.1016/j.fm.2018.11.006.

Erfani, A., Shakeri, G., Moghimani, M., and A. Afshari. 2024. Specific species of probiotic bacteria as bio-preservative cultures for control of fungal contamination and spoilage in dairy products. *International Dairy Journal*, *151*, 105863. https://doi.org/10.1016/j.idairyj.2024.105863.

Franco-Duarte, R., Černáková, L., Kadam, S., Kaushik, K. S., Salehi, B., Bevilacqua, A., and M. R. Corbo. 2019. Advances in chemical and biological methods to identify microorganisms—from past to present. *Microorganisms*, *7*(5), 130. https://doi.org/10.3390/microorganisms7050130.

Garnier, L., Valence, F., and J. Mounier. 2017. Diversity and control of spoilage fungi in dairy products: An update. *Microorganisms*, *5*(3), 42. https://doi.org/10.3390/microorganisms5030042.

Issa, A., Ibrahim, S. A., and R. Tahergorabi. 2017. Impact of sweet potato starch-based nanocomposite films activated with thyme essential oil on the shelf-life of baby spinach leaves. *Foods*, *6*(6), 43.

Ji, S. H., Kim, T. K., Keum, Y. S., and S. C. Chun. 2018. The major postharvest disease of onion and its control with thymol fumigation during low-temperature storage. *Mycobiology*, *46*(3), 242–253. https://doi.org/10.1080/12298093.2018.1505245.

Jia, W., Liang, G., Tian, H., Sun, J., and C. Wan. 2019. Electronic nose-based technique for rapid detection and recognition of moldy apples. *Sensors*, *19*(7), 1526. https://doi.org/10.3390/s19071526.

Kanashiro, A. M., Akiyama, D. Y., Kupper, K. C., and T. P. Fill. 2020. Penicillium italicum: An underexplored postharvest pathogen. *Frontiers in Microbiology*, *11*, 606852. https://doi.org/10.3389/fmicb.2020.606852.

Klapec, T., Wójcik-Fatla, A., Farian, E., Kowalczyk, K., Cholewa, G., Cholewa, A., and J. Dutkiewicz. 2022. Mycobiota of berry fruits-levels of filamentous fungi and mycotoxins, composition of fungi, and analysis of potential health risk for consumers. *Annals of Agricultural and Environmental Medicine*, *29*(1). https://doi.org/10.26444/aaem/147297.

König, H., Unden, G., and J. Fröhlich (Eds.). 2009. *Biology of Microorganisms on Grapes, in Must and in Wine* (pp. 3–29). Heidelberg: Springer. https://doi.org/10.1007/978-3-319-60021-5.

Kumar, A., Kulshreshtha, S., Shrivastava, A., and A. Saini. 2024. Biosensors for food spoilage detection: A comprehensive review of current advances. *Journal of Food Chemistry and Nanotechnology*, *10*(Suppl 1), S73–S82. https://doi.org/10.17756/jfcn.2024-s1-009.

Kusch, S., and R. Panstruga. 2017. mlo-based resistance: An apparently universal "weapon" to defeat powdery mildew disease. *Molecular Plant-Microbe Interactions*, *30*(3), 179–189. https://doi.org/10.1094/MPMI-12-16-0255-CR.

Lu, M., and N. S. Wang. 2017. Spoilage of milk and dairy products. In *The Microbiological Quality of Food* (pp. 151–178). Woodhead Publishing. https://doi.org/10.1016/B978-0-08-100502-6.00010-8.

Maphaisa, T. C., Akinmoladun, O. F., Adelusi, O. A., Mwanza, M., Fon, F., Tangni, E., and P. B. Njobeh. 2025. Advances in mycotoxin detection techniques and the crucial role of reference material in ensuring food safety: A review. *Food and Chemical Toxicology*, 115387. https://doi.org/10.1016/j.fct.2025.115387.

Martínez-Blay, V., Taberner, V., Pérez-Gago, M. B., and L. Palou. 2021. Postharvest treatments with sulfur-containing food additives to control major fungal pathogens of stone fruits. *Foods*, *10*(9), 2115. https://doi.org/10.3390/foods10092115.

Mendonca, A., Santos, H., Franco-Duarte, R., and P. Sampaio. 2022. Fungal infections diagnosis–past, present and future. *Research in Microbiology*, *173*(3), 103915. https://doi.org/10.1016/j.resmic.2022.103915.

Moraes Bazioli, J., Belinato, J. R., Costa, J. H., Akiyama, D. Y., Pontes, J. G. D. M., Kupper, K. C., . . . T. P. Fill. 2019. Biological control of citrus postharvest phytopathogens. *Toxins*, *11*(8), 460. https://doi.org/10.3390/toxins11080460.

Muñoz-Martinez, T. I., Rodríguez-Hernández, B., Rodríguez-Montaño, M., Alfau, J., Reyes, C., Fernandez, Y., Ramos, R. T., Franco De Los Santos, E. F., and L. O. Maroto-Martín. 2025. Unlocking the hidden microbiome of food: The role of metagenomics in analyzing fresh produce, poultry, and meat. *Applied Microbiology*, *5*(1), 26. https://doi.org/10.3390/applmicrobiol5010003.

Oyom, W., Xu, H., Liu, Z., Long, H., Li, Y., Zhang, Z., . . . D. Prusky. 2022b. Effects of modified sweet potato starch edible coating incorporated with cumin essential oil on storage quality of 'early crisp'. *LWT*, *153*, 112475. https://doi.org/10.1016/j.lwt.2021.112475.

Oyom, W., Zhang, Z., Bi, Y., and R. Tahergorabi. 2022a. Application of starch-based coatings incorporated with antimicrobial agents for preservation of fruits and vegetables: A review. *Progress in Organic Coatings*, *166*, 106800. https://doi.org/10.1016/j.porgcoat.2022.106800.

Palou, L., and J. L. Smilanick (Eds.). 2019. *Postharvest Pathology of Fresh Horticultural Produce*. CRC Press. https://doi.org/10.1201/9781315209180.

Perez-Nadales, E., Nogueira, M. F. A., Baldin, C., Castanheira, S., El Ghalid, M., Grund, E., . . . J. Wendland. 2014. Fungal model systems and the elucidation of pathogenicity determinants. *Fungal Genetics and Biology*, *70*, 42–67. https://doi.org/10.1016/j.fgb.2014.06.011.

Rao, V., Rao, L., Ahiduzzaman, M., and A. A. Islam (Eds.). 2021. *Nuts and Nut Products in Human Health and Nutrition*. BoD–Books on Demand. https://doi.org/10.5772/intechopen.92282.

Rodriguez, A., Rodriguez, M., Andrade, M. J., and J. J. Cordoba. 2015. Detection of filamentous fungi in foods. *Current Opinion in Food Science*, *5*, 36–42. https://doi.org/10.1016/j.cofs.2015.07.003.

Shi, C., and M. Maktabdar. 2022. Lactic acid bacteria as biopreservation against spoilage molds in dairy products–A review. *Frontiers in Microbiology*, *12*, 819684. https://doi.org/10.3389/fmicb.2021.819684.

Snyder, A. B., and R. W. Worobo. 2018a. Fungal spoilage in food processing. *Journal of Food Protection*, *81*(6), 1035–1040. https://doi.org/10.4315/0362-028X.JFP-18–031.

Snyder, A. B., and R. W. Worobo. 2018b. The incidence and impact of microbial spoilage in the production of fruit and vegetable juices as reported by juice manufacturers. *Food Control*, *85*, 144–150. https://doi.org/10.1016/j.foodcont.2017.09.033.

Souza, L. V., Rodrigues, R. D. S., Fusieger, A., da Silva, R. R., de Jesus Silva, S. R., Martins, E., . . . de A. F. Carvalho. 2022. Diversity of filamentous fungi associated with dairy processing environments and spoiled products in Brazil. *Foods*, *12*(1), 153. https://doi.org/10.3390/foods12010153.

Steinberg, G., and S. J. Gurr. 2020. Fungi, fungicide discovery and global food security. *Fungal Genetics and Biology*, *144*, 103476. https://doi.org/10.1016/j.fgb.2020.103476.

Taylor, N. P., and N. J. Cunniffe. 2023. Optimal resistance management for mixtures of high-risk fungicides: Robustness to the initial frequency of resistance and pathogen sexual reproduction. *Phytopathology®*, *113*(1), 55–69. https://doi.org/10.1094/PHYTO-02-22-0050-R.

Tiwari, R. K., Kumar, R., Sharma, S., Sagar, V., Aggarwal, R., Naga, K. C., . . . M. Kumar. 2020. Potato dry rot disease: Current status, pathogenomics and management. *3 Biotech, 10*, 1–18. https://doi.org/10.1007/s13205-020-02496-8.

Udoh, I. P., Eleazar, C. I., Ogeneh, B. O., and M. E. Ohanu. 2015. Studies on fungi responsible for the spoilage/deterioration of some edible fruits and vegetables. *Advances in Microbiology, 5*(04), 285. https://doi.org/10.4236/aim.2015.54027.

Valizadeh, S., Naseri, M., Babaei, S., and S. M. H. Hosseini. 2020. Shelf-life extension of fish patty using biopolymer-coated active paper sheets. *Food Packaging and Shelf Life, 26*, 100603. https://doi.org/10.1016/j.fpsl.2020.100603.

Wang, Y., Zhang, W., and L. Fu. 2017. *Food Spoilage Microorganisms: Ecology and Control*. CRC Press. https://doi.org/10.4324/9781315368887.

Wang, Y., Zhou, A., Yu, B., and X. Sun. 2024. Recent advances in non-contact food decontamination technologies for removing mycotoxins and fungal contaminants. *Foods, 13*(14). https://doi.org/10.3390/foods13142244.

Yao, X., Guo, H., Zhang, K., Zhao, M., Ruan, J., and J. Chen. 2023. Trichoderma and its role in biological control of plant fungal and nematode disease. *Frontiers in Microbiology, 14*, 1160551. https://doi.org/10.3389/fmicb.2023.1160551.

Zambon, A., Perez, A. G., Spilimbergo, S., and P. S. Fernández Escámez. 2022. Training in tools to develop quantitative microbial risk assessment along the food chain of Spanish products. *EFSA Journal, 20*, e200903. https://doi.org/10.2903/j.efsa.2022.e200903.

7 Mycotoxins in Cheese
Types, Sources, and Controlling Strategies

Enespa and Prem Chandra

7.1 INTRODUCTION

Some filamentous fungus synthesize secondary metabolic products called mycotoxins, which can be found in food items as a by-product of fungal development. Particularly in Europe, cheese is one of the most popular dairy items eaten globally (Izzo et al., 2022). Cheeses can be made using an extensive number of approaches and a variety of milk, and they differ in their flavor, texture, and morphological characteristics. Those dairy items are susceptible to the emergence of bacteria, yeasts, and molds that are filamentous, among other harmful and spoiling microbes (Coelho et al., 2022; Gaglio et al., 2021). There are actually two possible reasons these compounds can be found in cheese commodities: (1) indirect contamination of things, which happens when infected food is consumed by nursing animals, and (2) direct contamination, which happens when mold grows on dairy products purposefully or by mistake. *Penicillium commune* and *Penicillium nalgiovense* are among the most significant spoiling organisms found in hard, semi-hard, and semisoft cheeses produced without preservatives in a number of different nations. *P. roqueforti*, *Penicillium crustosum*, *Penicillium palitans*, *Penicillium solitum*, *Penicillium discolor*, *Penicillium verrucosum*, and *Aspergillus versicolor* are among the less-significant organisms.

In many instances, cheese is purposefully contaminated by fungi, because certain types of mold are utilized during production and ripening to impart particular qualities or extend its lifespan (Kure and Skaar, 2019). The fungi varieties utilized for this task are often initiated in the genus *Penicillium*, including *P. camemberti* and *P. roqueforti*, which can be utilized to make white and blue cheeses, accordingly (Cleere et al., 2024). The mycotoxin concentrations in cheese are prejudiced by a number of variables, comprising substrate present, temperature, and water activity. For example, when the ripening temperature is high, strains of *P. roqueforti* create large levels of roquefortine C (ROQC) (Hymery et al., 2014; Fontaine et al., 2015a). This type of fungus is capable of producing PR toxin, penicillic acid, and PAT. Likewise, cyclopiazonic acid (CPA) is formed by *P. camemberti*. Among these are *Alternaria* toxins, enniatins (ENNs), beauvericin (BEA), sterigmatocystin (STC), and mycophenolic acid (MPA) (Rodríguez-Cañás et al., 2023; Chang et al., 2009). Since fungi, plants, or animals can alter molecules like these to produce potentially

 DOI: 10.1201/9781003540113-7

harmful products that are not taken into account by the law, getting exposed to fungal toxins is not limited to prescribed and emergent mycotoxins (Khan et al., 2024).

Cheese may also include mycotoxins, which is a result of using tainted milk. Animals fed feed tainted with aflatoxin B1 (AFB1), for instance, will convert this substance to produce aflatoxin M1 (AFM1), which will be eliminated in milk (Zentai et al., 2023). The European legislation sets a threshold level of 0.05 μg/kg for AFM1 in cream milk and milk used to make dairy items in order to prevent intoxications. Nevertheless, this matrix of cells may contain additional mycotoxins, such as FBs, OTA, ZEN, cyclopiazonic acid (CPA), T-2 toxin, ROQC, ENNs, and BEA. Some mycotoxins, such as OTA, CTN, STC, CPA, ROQC, MPA, and AFM1, have been identified and found in cheese (Vaz et al., 2020; Oliveira et al., 2008). However, the European Food Safety Authority (EFSA) has asked for more information regarding incidence on OTA in order to assess if the ingredient's existence in cheese could be hazardous to humans (EFSA, 2022). The only milk derivative that is actually susceptible to fungal formation is cheese. Because molds appear frequently in milk products that are fermented in to cheese, the genus *Penicillium* is believed to be the most common (Garnier et al., 2017). The proliferation of these species in or on cheese results in disagreeable variations that reduce its nutritional quality. The development of mold on the outer layers of cheeses is thought to be a sign that their contents are going bad (Lu and Wang, 2017).

Throughout many years, molds known as *Penicillium camemberti* and *Penicillium roqueforti* are used to manufacture mold-ripened cheeses that are enjoyed worldwide. *P. roqueforti* is essential to the ultimate microflora of a number of cheeses, including Danish Blue (Denmark), Gorgonzola (Italy), Roquefort (France), Stilton (United Kingdom), Tulum (Turkey), and Blauschimmelkäse (Switzerland) (Chávez et al., 2023). *P. camemberti* includes cyclopiazonic acid (CPA), but *P. roqueforti* produces a minimum of three toxins: PR toxin, roquefortine, and patulin. *Aspergillus, Penicillium, Cladosporium, Geotrichum, Mucor*, and *Trichoderma* are just a few of the species of molds that can cause problems in the dairy product cheese production (Le Bars, 1979). In addition to the substantial financial damages caused by spoilage, some fungus have the ability to create mycotoxins and therefore might jeopardize the safety of food. The likelihood of mycotoxins in the final product increases when harmful fungi like *Aspergillus* and *Penicillium* are allowed to grow throughout the production and processing of cheese (Ogunade et al., 2018; Pouris et al., 2024). The main factors contributing to the growth of fungus in cheese are their low pH, low oxygen tensions, lipolytic activity, vulnerability to weak acid stabilization agents, and ability to grow at low temperatures (Silva et al., 2023).

Although it can occur at any stage of the cheese-making procedure, one of the primary contributors of fungal contamination in aging operations is dampness (Coelho et al., 2022). Because they are commonly found in foods and other products and can have serious negative health effects on both humans and animals, including carcinogenicity, genotoxicity/teratogenicity, mutagenicity, nephrotoxicity, hepatotoxicity, and immunotoxicity, mycotoxins rank among the naturally occurring toxicants of greatest community health concern. They are also linked to unexpected incarcerating illnesses (Zain, 2011; Hussein and Brasel, 2001). When dangerous molds are let to thrive and produce mycotoxins during the production or processing of food, especially dairy products, this risk may increase (Awuchi et al., 2022). The incorporation of fungicide compounds such as sorbates, benzoates, and natamycin to prevent food against degrading due to fungi during the process of making cheese is the most often utilized business method for minimizing spoilage of food, and the related financial implications. Fortunately, buyers are asking for additional possibilities for foods that are prepared with natural ingredients and/or without contaminants (Nasrollahzadeh et al., 2022; ZamaniMazdeh et al., 2017). Particular kinds or strains of lactic acid bacteria (LAB) can be used as bioprotective chemicals because they are capable of inhibiting fungal growth by producing a variety of active animosity metabolites and including organic acids, substances that include proteins, fatty acids, and chemical substances that look like bacteriocin (Raman et al., 2022; Ibrahim et al., 2021). This chapter will go to great lengths about the types, sources, and controlling strategies of mycotoxins in cheese which are found in raw milk and unprocessed milk cheeses. Finally, methods for preventing the growth of fungi and, as a result, eliminating or reducing mycotoxin concentrations in cheese are discussed. Biological, chemical, and physical strategies are among them.

7.2 MYCOTOXINS IN CHEESE

When AFs were found in the 1960s, the first study of strategies on mycotoxins in milk and dairy products had been released. Additional study on the prevalence of CPA in cheese was carried out in the 1970s. Considering the creation of potent analytical techniques such as LC-MS/MS in 2000, a large number of studies on the subject have emerged (Ajmal et al., 2022; Hymery et al., 2014). As already pointed out, cheeses can be distinguished from various varieties of mold that correspond to ripening indigenous or infected molds that may have food mycotoxins. Many analytical techniques are being developed to detect and measure mycotoxins in foodstuffs and diets, and regulatory agencies have established acceptable levels and limits (Izzo et al., 2022; Hymery et al., 2014). Numerous mycotoxins have been discovered in cheese; those that are most well-known are mentioned here. Roquefortine C has been found in commercialized Finnish cheeses, blue cheese (3.4 µg/g), and Manchego cheese (Lopez-Diaz et al., 2000) in amounts up to 12,000 µg/kg. Using thin-layer chromatography (TLC), MPA was found in a number of blue mold cheeses at 1.5 mg/kg (Engel et al., 1982). Using LC-MS-MS, MPA was found in cheese at considerably lower levels (0.0003 µg/kg). Cheese may contain this mycotoxin because AFM1 is present in milk. AFM1 has been detected in cheese in several studies. Nevertheless, comparatively fewer investigations have examined AFG1 and AFB1 in cheese (Mudannayake et al., 2024; Trombete et al., 2014).

In these instances, mycotoxins were directly administered to cheese, and the mixture's firmness was followed over time. AFB1 and AFG1 both stayed constant in Swiss cheese. Mycotoxigenic strains of *Aspergillus parasiticus* and *A. flavus* were introduced into cheese in comparable investigations, and both mycotoxins have been restored (Lorán et al., 2022). The quantities of sterigmatocystin (STC) in the topmost layer of hard cheeses infiltrated with *A. versicolor* range from 5 to 600 µg/kg. White cheese has been determined to contain CPA at levels of 0.02 to 0.08 mg/kg and 0.5 to 1.5 µg/g (Pietri et al., 2022; Veršilovskis et al., 2009). It was demonstrated by Stott and Bullerman (1976) that TLC could not detect patulin after it had been added to cheddar cheese. However, semi-hard cheeses contain 500 and 2,620 μg/kg of patulin and ochratoxin, respectively (Gong et al., 2024).

A rotten or filamentous fungus that produces mycotoxin may contaminate cheese either through direct contact or by encounters with milk. AFM1 is the primary cause of unintentional dissemination in cheese (Hymery et al., 2014). AFM1 is present in the finished item because dairy animals that have eaten moldy agricultural commodities contaminated with AFB1 have introduced this mycotoxin to their milk. Actually, AFB1 is changed into AFM1 by enzymes that reside primarily in animal liver, and this AFM1 is then delivered into the animal's milk and urine (Giovati et al., 2015). Numerous studies have discovered AFM1 in a range of cheeses. Whenever cheese is kept at low temperatures, the level of AFM1 may decrease (Mudannayake et al., 2024). Additional mycotoxins, such as STC, fumonisins, or CPA, have also been reported to inadvertently contaminate milk (Dorner, 2004). Although this kind of infiltration in cheese is considerably more complicated, it is contingent upon the production techniques used (Lu and Wang, 2017). Direct exposure to mycotoxin might happen intentionally because of interfering molds present during the process of producing cheese or endogenously because easily accessible fungal cultures that produce mycotoxin are available (CoulombeJr, 1993). Cheeses that have unintentionally developed mold have been found to contain mycotoxins. Mold also seen on the outside of cheese may not always indicate that mycotoxins are contemporary, because mycotoxin production can differ significantly between mycotoxigenic funguses (Fontaine et al., 2015b).

7.2.1 Types of Mycotoxins in Cheese

Cheese, along with other dietary supplements, may be harmed by mycotoxins, which are toxic compounds produced by some fungal organisms (molds). Even though proper production and shipping practices often lower the fungicide content of cheese, certain varieties of mold can nonetheless produce substances that may be toxic. The main mycotoxins found in cheese are specified in Table 7.1.

7.2.1.1 Aflatoxin M1

Aflatoxin is an assortment of insecticides manufactured by *Aspergillus flavus* and *Aspergillus parasiticus* (Payne, 2016). There are four primary varieties of aflatoxins produced through these molds: B1, B2, G1, and G2. Animal feed and contaminated food usually contain the highest concentration of aflatoxin B1 (Romero-Sánchez et al., 2022). After breastfeeding animals, including cattle, consume infectious AFB1 and AFB2 substances, monohydroxylated metabolites of AFB1 and AFB2, known as aflatoxin M1 (AFM1) and M2, can be produced in and removed from their milk (Richard, 2008). AFB1 is converted to AFM1 in milk and remains relatively constant in both manufactured and unfiltered milk-derived goods. Dairy foods may contain it (Jiang et al., 2021). Consequently, cheese prepared from milk that has not been pasteurized, which includes AFM1, will likewise include the toxic material. The destiny of AFM1 in milk during the cheese-making process is influenced by the primary methods of manufacture (Sarmast et al., 2021). There have also been contradictory

TABLE 7.1
Examples of Mycotoxins Found in Cheese

Cheese	Mold Species	Secondary Metabolites	References
Blue	*P. roqueforti*	Roquefortine, PR toxin, isofumigaclavine A	García-Estrada and Martín (2016)
Blue	*P. roqueforti*	Penicillic acid (100%), PR toxin (37%), patulin	Chávez et al. (2023)
Blue-molded	*P. roqueforti*	Mycophenolic acid	Lafont et al. (1979)
Blue-molded	*P. roqueforti*	Patulin, penicillic acid	Coton et al. (2020)
Tulum cheese	*P. roqueforti*	PR toxin, roquefortine	Erdogan and Sert (2004)
Blue-veined	*P. roqueforti*	Penicillic acid	Engel et al. (1982)
Manchego	*P. roqueforti*	Roquefortine, mycophenolic acid	Lopez-Diaz et al. (1996)
Cheddar	*P. roqueforti*	Penicillic acid, patulin	Olivigni and Bullerman (1977)
Tulum cheese	*P. roqueforti*	Penicillic acid, PR toxin, patulin	Erdogan and Sert (2004)
Blue-veined	*P. roqueforti*	Mycophenolic acid	Fontaine et al. (2015b)
Blue-molded	*P. roqueforti*	Roquefortine C, mycophenolic acid	Chávez et al. (2023)
Cheese	*P. camemberti*	Cyclopiazonic acid	Maragos et al. (2023)
Gouda and Edam	*A. versicolor*	Sterigmatocystin	EFSA (2013)
Semi-hard cheese	*Penicillium* sp.	Patulin, ochratoxin	Ramos-Pereira et al. (2019)
Turkish cheese	*Aspergillus* sp.	Aflatoxin M1	Ardic et al. (2009)
Italian white cheeses	*P. camemberti*	Cyclopiazonic acid	Maragos et al. (2023)
Artisanal Minas cheese	*Debaryomyces hansenii*, *Yarrowia lipolytica*, *A. fumigatus*, *P. roqueforti*, *Penicillium* spp., *Scopulariopsis* spp.	Aflatoxins B1, e B2 (in vitro)	Martin and Cotter (2023)
Cave cheese	*Aspergillus westerdijkiae*, *P. biforme*, *P. chrysogenum*, *A. puulaauensis*	Ochratoxin A (rind cheese)	Anelli et al. (2019)

reports of M1 rehabilitation. Early studies found that AFM1 decreased in several ways during the cheese-making process. Cottage cheese made by acid aggregation of unsurprisingly adulterated milk did not contain AFM1; it was completely removed in the whey (Sarımehmetoglu et al., 2004). However, other studies have shown that AFM1 is consistent throughout the course of making cheese and that the curd contains 40 to 57% of the total M1 (Hattimare et al., 2022).

Most of this poison ought to be separated into the whey, according to the partition coefficient of AFM1 in drinking water (Mendonca and Venâncio, 2005).

Nevertheless, a higher-than-anticipated percentage of the toxin winds up in the curd, which may be a result of M1 binding to the casein. Therefore, the existence of M1 in cheese could be due to the toxin binding to casein while also retaining humidity in the curd (Kiermeier and Buchner, 1977). An analysis of many cheese varieties revealed that M1 remains highly stable throughout maturing and storage facilities. Although there may be variations in M1 concentrations during these periods, it seems that little to no M1 is lost throughout the procedure of creating cheese (Gonçalves et al., 2020).

According to the study's outcomes, the prevalence of AFM1 in Minas cheese is probably going to lead to a rise in the total amount of aflatoxins consumed in the human diet. This suggests that a tolerance limit for AFM1 in cheese has to be established in Brazil (Esam et al., 2022). Limiting M1 presence in feed is the most effective method of controlling the presence of it in milk. The European Commission has set an acceptable limit of 10 mg kg^{-1} for AFB1 in animal feed and 0.050 μg kg^{-1} for AFM1 in raw milk, high-temperature dried milk, and milk used to make milk-based products (EFSA, 2011).

Fortunately, it has been demonstrated that 50% of the AFB1 found in the unprocessed milk used to make feta cheese is completely lost when pasteurization, milk transformation, and at least 50 days of brine preservation are combined (Chandra and Enespa, 2021). An intriguing new method for decreasing AFM1 levels in milk is to vaccinate cattle with aflatoxin B1 (AnAFB1), which lowers the amount of M1 released in the vaccinated cow's milk. In fact, immunization may be one of the most successful strategies to avoid AFM1 contamination of products such as cheese and milk (Giovati et al., 2014). The subsequent investigation that showed pre-calving immunization of dairy calves with concatenated AnAFB1 supports this, which produced even higher results in immunized animals. AFs change the DNA base configurations of genes, cause interference with DNA, and have a detrimental effect on damage repair. In both human and animal communities, AF is linked to cytotoxicity and carcinogenicity (RahimzadehBarzoki et al., 2023). According to the International Agency for Research on Cancer (1993), AFB1 is a group class 1 member of the AF family that has been shown to be responsible for malignancies in humans.

7.2.1.2 Sterigmatocystin

A different approach of fungal secondary metabolic product found in cheese is called STC. *A. versicolor*, *A. chevalieri*, *A. ruber*, *A. amstelodami*, *A. aureolatus*, *A. quadrilineatus*, and *A. sydowi* are among the *Aspergillus* species that generate it (Veršilovskis and De Saeger, 2010). The xerophilic mold *A. versicolor* is the primary generator of STC. Gouda and Edam cheeses aged in Dutch warehouses have been found to contain STC at deliberations fluctuating from 5 to 600 μg kg^{-1} (Mounier et al., 2005). According to reports, STC remains constant in infested cheeses for three months at different temperatures (between −18°C and +16°C). Cold temperatures of 5 to 7°C nonetheless ought to stop *A. versicolor* from growing and from producing the toxin throughout maturation and storage facilities (Bullerman et al., 1984).

Therefore, if STC contamination is to be prevented, cheese should ripen at lower temperatures. This is because molds that do not create STC, such as *Penicillium* species, are likely to grow at these temperatures. STC was discovered in *A. versicolor*–inoculated Ras cheese in Egypt; the toxin was discovered 45 days into the ripening process and peaked 90 days later (Zain, 2011). STC was detected in 2 of the 21 specimens that were tested, with values ranging from 0.52 to 1.23 g kg–1. STC was additionally identified in a number of cheeses produced in Belgium or Latvia (Veršilovskis et al., 2009). STC relocated to 0.5 cm from the outside of the cheese after it was created, according to studies conducted on Wilstermarsch cheese following *Aspergillus nidulans* penetration and 40 days of fermentation at 15°C or at ambient temperatures (Abd-Elhamed et al., 2024). Three natively infected hard cheeses with 45–330 μg kg^{-1} in the exterior 2 cm layer showed no STC in the middle region in one migratory study; nevertheless, six Gouda cheeses that were impulsively infested with *A. versicolor* showed identical results (Salazar et al., 2018). Significant amounts of moisture (>86%) during the process of ripening, the outside temperature, and the lactose, glycerol, and fat content all encourage the formation of STC by *A. versicolor* in cheese. Additional stimuli include non-organic phosphates and citric acid cycle constituents, such as fumarates, succinates, malates, citrates, α-ketoglutarates, and acetates (Murtaza et al., 2012). In recent years, it was shown that STC is produced by plastic-coating cheese once during the first 5 to 6 weeks of ripening at 14°C and 90% moisture content. After 8 days of ripening, toxin levels might vary from 160 to 700 μg kg^{-1}, depending on the triggering factors present (Rodríguez et al., 2024). Since STC creation was significantly reduced in cheeses that were plastic-coated many times, it is anticipated that the emergence of numerous layers of plastic covering on the cheese surface during the first few weeks of maturation will have a positive impact on reducing the manufacturing of toxin (Ritschard and Schuppler, 2024). According to this study, STC manufacturing and place of storage are also more likely to occur in environments with high levels of infection quantity, such as when warehouse air becomes heavily polluted with mold and/or mold spores, or when display shelves in unsafe circumstances are deployed in conjunction with prolonged moisture storage facilities (Hymery et al., 2014).

7.2.1.3 Citrinin

This mycotoxin has been determined to be produced by *P. citrinum* and *P. verrucosum* in cheese. Goat cheese was among the cheeses in which citrinin seems to be exceptionally durable. This mycotoxin and OTA, which are both generated by *P. verrucosum*, are commonly linked (Kamle et al., 2022). Numerous *Aspergillus* species and a few strains of *P. camemberti* are among the 12 *Penicillium* species that can generate citrinin. The largest number of citrinin-producing fungus found outdoors have been identified as *P. citrinum*, *P. viridicatum*, *P. expansum*, *P. verrucosum*, *A. carneus*, *A. niveus*, and *A. terreus*; however, other species, including *Aspergillus candidus*, were additionally found to produce citrinin in lab settings (Bragulat et al., 2008). But the only known bacteria that may make citrinin in dairy products are *P. verrucosum*, *P. citrinum*, and *P. expansum*. Once created, it is extremely durable in cheese, such as goat cheese, Saint Marcellin, and Soignon; after eight days of storage, over half of the cheese's original citrinin concentration is still there. Since both are generated by *P. verrucosum*, citrinin production is usually linked to the existence of OTA (Mladenović et al., 2021).

7.2.1.4 Roquefortine C

Amongst the most well-known secondary metabolic products of *P. roqueforti* are roquefortines. The extended family of prenylated indole alkaloids consists of roquefortine C and its associated roquefortines D (3, 12-dihydroroquefortine C), 16-hydroxyroquefortine C, and roquefortine L (Martín and Liras, 2017). The antimicrobial properties and immune-suppressive metabolism product mycophenolic acid has been encountered to be present instinctively in some blue and Manchego cheeses. Roquefortine C is consistently produced in blue cheeses and is sometimes referred to as a mycotoxin; however, this assumption is not supported by any significant pharmacological data in the available research (Vallone et al., 2014). The most common metabolites generated by the 25/58 species of the *Penicillium* genus is roquefortine C. Furthermore, almost every strain of *P. roqueforti* that has been investigated produces roquefortine C and a precursor to roquefortine D (Chávez, et al., 2023).

7.2.1.5 Ochratoxin A

Many species of *Aspergillus* produce OTA, as well as *Aspergillus alliaceus*, *Aspergillus ochraceus*, *Aspergillus auricomus*, *Aspergillus carbonarius*, *Aspergillus glaucus*, *Aspergillus melleus*, and *Aspergillus niger*, in addition to *Penicillium verrucosum* (Bayman et al., 2002). Whereas the latter is more common throughout more moderate places, such as eastern and northeastern Europe, the former is the main source of OTA in tropical parts of the entire globe. *Penicillium nordicum*, one of the other strains of *Penicillium* that make OTA, is well suited to rotting in food environments like pickled meats and cheeses (Nazareth et al., 2024). Although the levels of OTA found in Gorgonzola and Roquefort cheeses were extremely low and probably will not pose a risk to consumers, researchers continue to raise concerns over the genesis of the contaminants, which the researchers hypothesize might have originated from contaminated milk processed to produce the cheese. Crop storing produces ochratoxin A, which causes a number of harmful effects in different animal species. Kidney deterioration is one with the most notable and immediate effects, but it may also have a severe impact on the immune system and fetal growth (Battacone et al., 2010).

Contrary to the obvious indications of toxicity to the kidneys and cancer that are frequently linked to ochratoxin A in animal species, it is unknown whether this is the case in humans, yet its effects on the kidney have been demonstrated. Ochratoxin A has already been demonstrated to have properties as a nephrotoxic, a liver toxin, an immune suppressor molecule, a strong teratogen, and an animal carcinogen (Bui-Klimke and Wu, 2015). The use of blood and other tissues of animals, as well as milk and pork manufactured for humans to consume, has been demonstrated to contain ochratoxin. Additionally, ochratoxin is linked to poultry mortality and illness (Vlachou et al., 2022).

7.2.1.6 Enniatin B

Because of their increased presence in feed and food products and the general concern about their being alive, ENNs—which were initially discovered to be produced by *Fusarium* spp.—belong to the so-called "emerging mycotoxins" family (Tolosa et al., 2021). ENNs are primarily found contaminating cereals, but they have also been found in other matrix structures, such as animal-derived products, as a result of the metabolites' transfer into animal tissues following the ingestion of tainted feed. According to reports, food items have a substantial amount of ENN contaminants (Abia et al., 2013). Baby nourishment, comprising infant formula types of milk, dairy items like cheese and yogurt, cereal-based baby food, fruit and vegetable compotes, and fruits and veggies puree from the Italian market, was used to quantify ENNs B, B1, B4, A, and A1. ENN B has the highest number of documented contaminated levels, with measured amounts ranging from 11.8 to 832 μg/kg (Gasparre et al., 2022). Among this category of mycotoxins, ENN B is the most common component, and its availability indicates comparatively large quantities in food and feed compromised with *Fusarium*. An effective multi-analytes technique for the identification and measurement of mycotoxins in maize silage from dairy farms in Spain was validated using multiple mycotoxin tests (Zhang et al., 2022).

Furthermore, a pilot research for the existence of fungal metabolites in sheep milk from the first springtime milking verified ENN B. Only one mycotoxin, ENN B, was found in sheep samples of milk out of 700 bacterial, fungal, and plant-derived compounds that were examined for (average concentration: 0.0078 μg/kg; range: 0.0055–0.0121 μg/kg) (Piątkowska et al., 2018). ENN B has not yet been subject to any restrictions, in contrast to other *Fusarium* mycotoxins, such deoxynivalenol, T-2, HT-2, fumonisins, and zearalenone, whose presence in food and feed has been controlled by governments (Prosperini et al., 2017).

7.2.1.7 Trichothecenes

Trichothecenes are fashioned by a quantity of fungal species, including *Fusarium*, *Stachybotrys*, *Myrothecium*, *Trichothecium*, *Trichoderma*, *Cephalosporium*, *Cylindrocarpon*, *Verticimonosporium*, and *Phomopsis*. Depending on whether a macrocyclic ester or an ester–ether bridge exists between C-4 and C-15, trichomecenes are classified as either macrocyclic or non-macrocyclic (Zhu et al., 2020). The largest profitable sources of non-macrocyclic trichothecenes are *Fusarium* molds. Different field fungi that are adapted to spread through plants such as wheat, corn, barley, oats, and forages are found in the genus *Fusarium*. *Fusarium* is primarily found in temperate regions, where contamination of meals is common. Trichothecenes are toxic to bacteria, plants, animals, and molds (Foroud and Eudes, 2009). They are also powerful barriers of protein synthesis. Trichothecenes are typically recognized as pollutants in food and feed. Contact with the substance causes dermatitis, whereas the ingestion of trichothecenes causes rapid inflammation of the intestinal mucosa, which causes vomiting, diarrhea, and alimentary bleeding. Additionally, animals' immune systems are suppressed after extensive exposure (McCormick et al., 2011).

7.2.1.8 Mycophenolic acid

If consumed repeatedly enough, MPA may weaken the body's response to infection, opening the door for bacteria to thrive. The evaluation of mutagenic activity has revealed that this mycotoxin could cause chromosome aberration, although the Ames test came out negative (Hartwig et al., 2020).

7.2.1.9 PR toxin

Blue cheeses have not been the subject of much research documenting the existence of MPA generated by *P. roqueforti* in cheeses. MPA is a phthalide that has a number of potential uses in addition to being a mycotoxin with possible hazardous repercussions (Nielsen et al., 2006). In order to prevent organ failure, it is being utilized as an immunosuppressant in patients undergoing liver, heart, and kidney transplants. Mammals show minimal toxicity: Rats have an LD50 of 2,500 mg/kg (oral pathway) and 500 mg/kg (IV pathway); mice have an LD50 of 700 mg/kg and 450 mg/kg IV (Diehl et al., 2017). In cheese-making circumstances, a different recognized mycotoxin that *P. roqueforti* produces becomes unstable and transforms into the least-lethal PR imine. When basic and neutral amino acids are present, The PR toxin can also be changed into PR amide. Other additional metabolites, including eremofortins A, B, C, and D, have also been found to be associated with the manufacture of PR toxin by *P. roqueforti* (Cakmakci et al., 2015).

7.2.1.10 Other Mycotoxins

There are not many toxicological investigations on the additional mycotoxins that *P. roqueforti* produces in cheese. *P. roqueforti* may synthesize the alkaloids festuclavine, agroclavine, and isofumigaclavines (A and B). They are members of the mycotoxin family of ergot alkaloids (Ohmomo et al., 1979). Manufactured blue cheeses have been shown to contain trace amounts of isofumigaclavines A and B. The process of the hydrolysis outcome of isofumigaclavine A is equivalent to isofumigaclavine B. In vitro and in vivo, agroclavine exhibited an immunostimulatory effect by increasing the activity of natural killer (NK) cells in non-stressful situations (Hymery et al., 2014).

7.3 SOURCES OF MYCOTOXIN CONTAMINATION

Direct or indirect contamination might result in the presence of mycotoxins in dairy products (Flores-Flores et al., 2015). At any point during the production process, toxic molds in dairy products produce mycotoxins in situ, which lead to direct contamination. This is especially true for cheese types where molds are a natural component of their microbiota and where the molds that produce mycotoxin can come from one of three sources, as listed in what follows:

- Contamination of the milk due to post-pasteurization contaminants or survivability to pasteurization (*B. fulva*). At some point throughout the manufacturing of cheese or the storing procedure, these kinds of molds have the potential to develop and generate mycotoxins (Schröder, 1984). Nevertheless, elements such as temperature, aw, pH, oxygen, and/or cheese maturing stage that are conducive to the growth of toxic molds may not prove appropriate for the formation and stability of mycotoxin (Kokkonen et al., 2005). Mycotoxins are typically produced after the mold has entered its stationary period of growth and are secondary metabolic products that are not essential for the development of the strain. For instance, it has been demonstrated that asexual spore formation and sterigmatocystin (ST) are co-regulated, necessitating the dissolution of regenerative growth (Ismaiel and Papenbrock, 2015).

- Natural variants whose development is intentionally promoted throughout the production of raw milk cheese in order to preserve its originality or satisfy the standards for cheeses granted PDO designation. Given the variety and unidentified identification of the varieties of mold and strains participating in the maturing interpret, some of which could potentially be quite toxic, these types of cheeses would be more likely to possess excessive levels of mycotoxins (Mangione et al., 2023).
- Manufactured variants that are purposefully inserted as supplementary starting cultures have the potential to develop mycotoxins in cheese under specific situations (Gruber-Dorninger et al., 2017). *P. roqueforti*, a frequent component used in the manufacturing of blue-veined cheeses (Roquefort in France, Stilton in the UK, Tulum in Turkey, Gorgonzola in Italy, Cabrales in Spain, Blauschimmelkäse in Switzerland, and Danablu in Denmark), is capable of producing a variety of mycotoxins in vitro, including roquefortine, PR toxin, penicillic acid, isofumigaclavine, mycophenolic acid, and patulin. Numerous commercialized cheese variants that depend on *P. roqueforti* as the main ingredient for their maturing have been found to contain some of the mentioned mycotoxins (Fontaine et al., 2015b).
- Similar to this, *P. camemberti* has been shown to produce cyclopiazonic acid in vitro. This mold is also accountable for the maturity and exterior outgrowth synthesis in commercial sampling of camembert and comparable cheeses, where the chemical is detected in minor quantities (Ollinger et al., 2024).

The animals' livers metabolize the aflatoxins (AFM1 and AFM2) in their milk to form the hydroxylase products derived from AFM1 and AFM2, respectively, when they consume lactate from animals that are virulent with AFB1 and AFB2 (Jiang et al., 2021). Cheese milk that has been kept from contaminated diets is a case of passive contaminating. The quantity of AFB1 in the nutrients that a lactating animal eats and the amount that is of AFM1 in the milk that she produces are correlated linearly (Kang'ethe and Lang'a, 2009). The mycotoxin is thought to be present in any milking-based product that uses contaminated dairy products as an initial ingredient since it is stable under thermal stress, especially when the milk is heated to processed and ultra-high temperatures.

It has been validated that the attentiveness of AFM1 in cheese made from contamination of milk is usually larger than that of the original milk because a sizable portion (40–100%) of AFM1 binds to the milk casein and becomes more concentrated in curd after exhaustion (Iha et al., 2014). Under typical manufacturing settings, this toxic material does not metabolize or physiologically decompose in the following manufacturing processes. AFM1 was shown to rise by a factor of 1.74–2.45 when fresh Brazilian cheese Minas frescal was processed (Corassin et al., 2022). While other cheese types had AFM1 quantities that were between 3.2 and 8.1 times higher than the milk and other dairy products used to create them, hard and moderately hard cheeses displayed the largest amount of ratios (3.7 to 8.1) (Govaris et al., 2001).

Similarly, vaporized preserved milk has significantly higher AFM1 than fresh milk due to the combined influence of the toxin's resistance to heat and the evaporating water, which produces mycotoxin concentrations. However, a number of authors noted that the simmering procedure decreased AFM1 by 12–40%, depending on the settings employed when steaming the milk (RahimzadehBarzoki et al., 2023). Although their concentrations were significantly lower compared to that of AFM1, similar transference methods were noted for additional mycotoxins, including AFB1, CPA, FB1, OTA, trichothecenes, zearalenone (ZEA), and patulin. Since these mainly affect cereal grains, forages, and feed for animals, the so-called *Fusarium* mycotoxins—such as fumonisins, zearalenone, trichothecenes, and related toxic substances—have not been more extensively investigated in milk and milk products (Zaki et al., 2012). However, given current investigations that confirm that these toxins are transferred to milk in animals fed destroyed feed and that they appear in milk and dairy products, study participants are reminded to pay acceptable and careful attention to the potential health risks that these toxins symbolize to humans through the consumption of contaminated dairy items (Jiang et al., 2021).

7.3.1 Mold Growth and Cheese Spoilage

Penicillium commune and *Penicillium nalgiovense* are the most important examples spoiling microbes found in hard, semi-hard, and semisoft cheeses produced in a number of nations despite the use of preservatives. *P. roqueforti*, *Penicillium crustosum*, *Penicillium palitans*, *Penicillium solitum*, *Penicillium discolor*, *Penicillium verrucosum*, and *Aspergillus versicolor* are less-significant species (Pitt, 2019). AFB1, ochratoxin A (OTA), AFG1, sterigmatocystin (STC), nephrotoxic (OTA, citrinin), neurotoxic (penitrem A, CPA), and teratogenicity (OTA, AFB1) mycotoxins have all been found to be present in cheese (Creppy, 2002). Others, such as penicillic acid, patulin, and PR toxin, have been documented but are known to not linger in cheese. Cheese has also been found to include a variety of additional secondary metabolic products that are produced by various *Penicillium* spp. Historically unreported from cheese, they comprise viridic acid, verrucine A, sclerotigenin, cyclopeptin, chactoglobosin A, compactin, rugulovasine A, verrucolone, PC-2, viridic acid, chactoglobosin A, and viridicatol (Vallone et al., 2014).

Interestingly, MPA, a recognized derivative of *Penicillium* linked to blue-veined cheese, is an active component of several significant medications that suppress immunity. Mycotoxin-producing molds are legally required aerobes, so it is critical to package cheese properly (Cakmakci et al., 2015). For instance, roquefortine C and CPA generation by *P. roqueforti* and *P. commune* has been prevented or substantially significantly reduced through the use of customized atmospheric packing, where environments containing 20% to 40% CO_2 and 1% O_2 limit CPA formation to extremely low amounts (Caron et al., 2021). The prevalence of mycotoxins, which in cheese contamination with *Aspergillus* species has been reported to create aflatoxin throughout the development on the surface of Cheddar cheese, has also been examined. The poisonous substance can reach an extent of up to 4 cm within the cheese. It has been additionally demonstrated that Manchego-type cheese ripening for 60 days at 15°C produces up to 130 $\mu g\ g^{-1}$ of aflatoxin (Sengun et al., 2008). After 30 days of implantation at 28°C, aflatoxin was also found in the subsequent 10 mm layer and outside the 10mm layer. The presence of mold growth on the outside of cheese does not always mean that it includes mycotoxins because mycotoxigenic molds may have significantly different water activities (aw) for proliferation and hazard development (Benkerroum, 2020).

7.4 MANAGEMENT OF CHEESE FUNGAL INFECTIONS

7.4.1 Control of Mycotoxins in Cheeses

As mentioned before, certain researchers believe that cheese is particularly vulnerable to the development of mold and the formation of mycotoxin (Sengun et al., 2008). Conversely, other scientists have pointed out that there are contradictory findings in the scientific literature about the creation of mycotoxin in cheese and milk and dairy products, and they have come to the conclusion that cheese actually serves as more suitable for mold development than for the production of mycotoxin (Hymery et al., 2014). If cheese has been stored correctly at a low temperature (5 to 7 °C), it is an inadequate platform for mycotoxin development and benign molds outgrow hazardous ones. Hazardous molds have an insignificant impact on cheese. Regardless of whether the danger is thought to be sufficiently minimal, it needs to still be managed (Bullerman, 1979). Avoiding mycotoxins in milk is the initial objective. Two other factors deserve to be taken into account throughout the manufacturing and preservation of cheese as long as it is consumed: Harmful fungal substances that have the ability to manufacture mycotoxins are unfavorable and are not recommended, and fungal the maturation cultures that are purposefully added during the cheese-making process (such as camembert-type or blue-veined cheeses) shouldn't be mycotoxigenic (Hof, 2016).

7.4.2 Rheostat of Mycotoxins in Milk

To prevent mycotoxins from appearing in food items and, eventually, in milk (mostly AFM1), good animal nutrition habits are essential. Prior to harvest, during transportation and storage, and during the conversion of plant resources into feed products, mycotoxin-producing fungi can grow on them (Awuchi et al., 2021). Mycotoxin prevention techniques over harvesting and after harvesting involve early harvesting, which is typically linked with low mycotoxin levels; high eliminating dimensions, to prevent contaminants from particles in the soil and to remove impaired cereals that may encourage fungal infections growth; low levels of water content (the primary factor to prevent fungal growth, with a vital threshold of 14% w/w); and cool room temperatures and ventilation, which are crucial additional considerations to lower the growth risk (Loi et al., 2023). It is vital to adhere to the ideal moisture content at harvest (33–35% of dried matter for the whole vegetation) and employ excellent farming techniques when refilling elevators because moisture content for fodder stays relatively high (Rasheed et al., 2022).

By implementing these recommendations, lactic fermentation occurs and anaerobic circumstances are reached, preventing the growth of fungi. Because of their antifungal abilities, LAB can be added to silage and is increasingly being utilized instead of synthetic preservation agents. The capacity of a *Lactobacillus casei* sp. *pseudoplantarum* 371 isolate from forage to suppress the growth of mold and the generation of *A. flavus* subsp. *parasiticus* NRRL 2999 AF was investigated (Liu et al., 2022). For fodder surveillance, a variety of plant-based supplements are now on the marketplace, most of which comprise LAB. Nonetheless, the two most important suggestions for standard procedure are to use propionic acid, which is renowned for its anti-mold properties, and to rapidly establish stable anaerobic circumstances (Papademas and Bintsis, 2017).

Propionic acid (E code = E280) is permitted for usage in feed in Europe with no time constraint. As a result, several acidic additives for silage that contain propionic acid are now accessible for purchase. Curative physical, chemical, and/or biological procedures to remove mycotoxins from milk are ineffective and prohibitively costly, making them unsuitable for providing workable feed purification options (EFSA et al., 2024). A small number of various physical and chemical techniques that have been suggested for decontaminating food and feed, such as treating AF with ammonia, are actually applied. Ammoniation decomposes up to 95–98% of AF, which is why it is employed in many nations to decontaminate animal feed ingredients. Animals that produce milk serve as relatively strong safeguards toward mycotoxins. The liver and rumen microbes both manufacture harmful chemicals (Méndez-Albores et al., 2007). Fortunately, since they biotransform many mycotoxins (such as zearalenone or deoxynivalenol, the primary mycotoxins generated through silage) into innocuous metabolites, the rumen microbial populations of healthy animals also help mitigate illnesses. As a result, mycotoxin content of milk is typically low. Overall, mycotoxins are durable substances. Nonetheless, it has been demonstrated that using UV radiation and heated processing is an effective way to lower the amount of AFM1 in milk (Adácsi et al., 2022).

7.4.3 Chemical Preservatives

Mycotoxins can be managed with chemical-based preservatives. For example, it has been demonstrated that pimaricin (= natamycin; E code = E235; generated by the actinobacterium *Streptomyces natalensis*) can prevent mycotoxins by delaying fungal growth. Its direct impact on cheese's mycotoxin generation is still not well understood. Both Europe and the United States now allow the use of this antifungal drug in cheese (Meena et al., 2021). The solutions of natamycin for usage during the making of cheese are sold by small enterprises. In Europe, 0.3 mg/kg of body weight is the permissible daily consumption amount. During a 5-month maturation period at 4°C, mold growth was prevented in vacuum-packed Kashar cheese that contained natamycin. Stabilizers such sorbic, benzoic, and propionic acids can stop the growth of mold, except for *P. roqueforti* (Kallinteri et al., 2013). With the exception of blue-veined cheeses, this kind of cheese is the most significant spoiling variety because it is resilient. Sorbate, also known as sorbic acid, is allowed in cheeses at concentrations below 0.3% according to the US Federal Standard of Identity. Propionates can be utilized to treat the outer coating of cheese and cheese services, including packaging substances (Aly, 1996). By reducing the amount of water required for proliferation and the production of mycotoxin, high sodium chloride contents can also be utilized to stop mold from growing on cheeses; nevertheless, in this situation, both flavor characteristics and health risks must be taken into consideration (Nájera et al., 2021).

7.4.4 Microbial Control

The proliferation of fungi and mycotoxin generation may be further restricted by contending and/or inhibitory bacteria. The subsequent maturing cultures *P. camemberti* and *G. candidum* are capable of successfully competing with pollutants when they are completely viable (Lessard et al., 2012). Although they produce organic acid, compete for vitamins and nutrients, and produce other hostile substances, species in the *Lactococcus* and *Lactobacillus* genera are most suited to inhibit or stop the formation of mycotoxigenic mold. It has been demonstrated that unprocessed milk from ewes, cows, and goats is a prolific repository for antifungal bacteria, the majority of which are *Lactobacillus* (Coelho et al., 2022). *Debaryomyces hansenii, Kluyveromyces lactis, Kluyveromyces marxianus, Penicillium brevicompactum, Rhodotorula mucilaginosa*, and *Yarrowia lipolytica* were among the 11 *Lactobacillus* spp. that were shown to have antifungal properties in buttermilk with strain-dependent activation patterns. In yogurt, *L. harbinensis* had a potent antifungal capability, suppressing every examined fungus (Mayoral et al., 2005). It has been demonstrated that *Penicillium expansum, P. roqueforti, Aspergillus niger, Aspergillus fumigatus*, and *Fusarium culmorum* are inhibited by *Lactobacillus amylovorus* DSM 19280. Despite causing any discernible degradation in cheese effectiveness, cheddar cheeses exposed to naturally occurring airborne fungi profited from a 6-day grace period for the development of filaments on the outermost layer when the bacterial supplement was present (Leyva Salas et al., 2017). It is widely recognized that dairy propionibacteria, which are often used in the manufacturing of Swiss-style cheese, can ferment glycerol, lactic acid, and glucose to produce carbon dioxide, acetic acid, and propionic acid. Glycerol was used as a carbon source for bacterial growth in order to investigate the antifungal effect produced by dairy propionibacteria (Piwowarek et al., 2018). The molds *Penicillium commune* and *Penicillium roqueforti*, as well as the yeast *Rhodotorula mucilaginosa*, have been investigated toward five types of propionibacteria. While yeast was fewer-impacted, an increase in glycerol contents improved the suppression of the fungus. Propionic acid generation and the medium's concurrent pH decrease were found to be the causes of that impact (Lind et al., 2010).

7.5 CONCLUSION

Mold growth is far more probable to evolve in cheese than carcinogenic development. It is uncommon for cheese to include some toxins such as MPA, CPA, roquefortine C, penicillic acid, STC, PR toxin, aflatoxin, and citrinin despite the fact that it is not a particularly tempting habitat for mold that generates mycotoxin. A shortage of conclusive toxicological findings with regard the potential adverse effects of mycotoxin usage in human beings, mainly in connection to exposure to relatively small amounts of various types of toxins, across a prolonged amount of time, should be kept in mind, even though there is currently not any evidence regarding human toxic effects resulting from the consumption of mold-ripened cheese.

ACKNOWLEDGMENTS

Dr. Enespa thanks the Department of Plant Pathology, Sri Mahesh Prasad Degree College, Lucknow, and Dr. Prem Chandra would like to acknowledge the School of Agriculture, IGNOU, New Delhi, for assistance in the preparation of this chapter.

REFERENCES

Abd-Elhamed, E. Y., El-Bassiony, T. A. E. R., Elsherif, W. M. & Shaker, E. M. 2024. Enhancing Ras cheese safety: Antifungal effects of nisin and its nanoparticles against *Aspergillus flavus. BMC Veterinary Research*, 20(1): 493.

Abia, W. A., Warth, B., Sulyok, M., Krska, R., Tchana, A. N., Njobeh, P. B. & Moundipa, P. F. 2013. Determination of multi-mycotoxin occurrence in cereals, nuts and their products in Cameroon by liquid chromatography tandem mass spectrometry (LC-MS/MS). *Food Control*, 31(2): 438–453.

Adácsi, C., Kovács, S., Pócsi, I. & Pusztahelyi, T. 2022. Elimination of deoxynivalenol, aflatoxin b1, and zearalenone by gram-positive microbes (firmicutes). *Toxins,* 14(9): 591.

Ajmal, M., Bedale, W., Akram, A. & Yu, J. H. 2022. Comprehensive review of aflatoxin contamination, impact on health and food security, and management strategies in Pakistan. *Toxins,* 14(12): 845.

Aly, M. E. 1996. Prolongation of the keeping quality of Mozzarella cheese by treatment with sorbate. *Food/Nahrung,* 40(4): 194–200.

Anelli, P., Haidukowski, M., Epifani, F., Cimmarusti, M. T., Moretti, A., Logrieco, A. & Susca, A. 2019. Fungal mycobiota and mycotoxin risk for traditional artisan Italian cave cheese. *Food Microbiology*, 78: 62–72.

Ardic, M., Karakaya, Y., Atasever, M. & Adiguzel, G. 2009. Aflatoxin M1 levels of Turkish white brined cheese. *Food Control,* 20(3): 196–199.

Awuchi, C. G., Ondari, E. N., Nwozo, S., Odongo, G. A., Eseoghene, I. J., Twinomuhwezi, H. & Okpala, C. O. R. 2022. Mycotoxins' toxicological mechanisms involving humans, livestock and their associated health concerns: A review. *Toxins,* 14(3): 167.

Awuchi, C. G., Ondari, E. N., Ogbonna, C. U., Upadhyay, A. K., Baran, K., Okpala, C. O. R. & Guiné, R. P. 2021. Mycotoxins affecting animals, foods, humans, and plants: Types, occurrence, toxicities, action mechanisms, prevention, and detoxification strategies—A revisit. *Foods,* 10(6): 1279.

Battacone, G., Nudda, A. & Pulina, G. 2010. Effects of ochratoxin A on livestock production. *Toxins,* 2(7): 1796–1824.

Bayman, P., Baker, J. L., Doster, M. A., Michailides, T. J. & Mahoney, N. E. 2002. Ochratoxin production by the *Aspergillus ochraceus* group and *Aspergillus alliaceus. Applied and Environmental Microbiology*, 68(5): 2326–2329.

Benkerroum, N. 2020. Aflatoxins: Producing-molds, structure, health issues and incidence in Southeast Asian and Sub-Saharan African countries. *International Journal of Environmental Research and Public Health*, 17(4): 1215.

Bragulat, M. R., Abarca, M. L. & Cabanes, F. J. 2008. Low occurrence of patulin-and citrinin-producing species isolated from grapes. *Letters in Applied Microbiology*, 47(4): 286–289.

Bui-Klimke, T. R. & Wu, F. 2015. Ochratoxin A and human health risk: A review of the evidence. *Critical Reviews in Food Science and nnutri*tion, 55(13): 1860–1869.

Bullerman, L. B. 1979. Significance of mycotoxins to food safety and human health. *Journal of Food Protection*, 42(1): 65–86.

Bullerman, L. B., Schroeder, L. L. & Park, K. Y. 1984. Formation and control of mycotoxins in food. *Journal of Food Protection*, 47(8): 637–646.

Cakmakci, S., Gurses, M., Hayaloglu, A. A., Cetin, B., Sekerci, P. & Dagdemir, E. 2015. Mycotoxin production capability of *Penicillium roqueforti* in strains isolated from mould-ripened traditional Turkish civil cheese. *Food Additives & Contaminants: Part A*, 32(2): 245–249.

Caron, T., Le Piver, M., Péron, A. C., Lieben, P., Lavigne, R., Brunel, S. & Chassard, C. 2021. Strong effect of *Penicillium roqueforti* populations on volatile and metabolic compounds responsible for aromas, flavor and texture in blue cheeses. *International Journal of Food Microbiology*, 354: 109174.

Chandra, P. & Enespa. 2021. Aflatoxins: Food safety, human health hazards and their prevention. In *Aflatoxins-Occurrence, Detoxification, Determination and Health Risks* (pp. 1–21). Intech Open.

Chang, P. K., Ehrlich, K. C. & Fujii, I. 2009. Cyclopiazonic acid biosynthesis of *Aspergillus flavus* and *Aspergillus oryzae*. *Toxins*, 1(2): 74–99.

Chávez, R., Vaca, I. & García-Estrada, C. 2023. Secondary metabolites produced by the blue-cheese ripening mold *Penicillium roqueforti*; Biosynthesis and regulation mechanisms. *Journal of Fungi*, 9(4): 459.

Cleere, M. M., Novodvorska, M., Geib, E., Whittaker, J., Dalton, H., Salih, N. & Dyer, P. S. 2024. New colours for old in the blue-cheese fungus *Penicillium roqueforti*. *Science of Food*, 8(1): 3.

Coelho, M. C., Malcata, F. X. & Silva, C. C. 2022. Lactic acid bacteria in raw-milk cheeses: From starter cultures to probiotic functions. *Foods*, 11(15): 2276.

Corassin, C. H., Borowsky, A., Ali, S., Rosim, R. E. & de Oliveira, C. A. F. 2022. Occurrence of Aflatoxin M1 in milk and dairy products traded in São Paulo, Brazil: An update. *Dairy*, 3(4): 842–848.

Coton, E., Coton, M., Hymery, N., Mounier, J. & Jany, J. L. 2020. *Penicillium roqueforti*: An overview of its genetics, physiology, metabolism and biotechnological applications. *Fungal Biology Reviews*, 34(2): 59–73.

CoulombeJr, R. A. 1993. Biological action of mycotoxins. *Journal of Dairy Science*, 76(3): 880–891.

Creppy, E. E. 2002. Update of survey, regulation and toxic effects of mycotoxins in Europe. *Toxicology Letters*, 127(1–3): 19–28.

Diehl, R., Ferrara, F., Müller, C., Dreyer, A. Y., McLeod, D. D., Fricke, S. & Boltze, J. 2017. Immunosuppression for in vivo research: State-of-the-art protocols and experimental approaches. *Cellular & Molecular Immunology*, 14(2): 146–179.

Dorner, J. W. 2004. Biological control of aflatoxin contamination of crops. *Journal of Toxicology: Toxin Reviews*, 23(2–3): 425–450.

EFSA Panel on Additives and Products or Substances used in Animal Feed (FEEDAP), Villa, R. E., Azimonti, G., Bonos, E., Christensen, H., Durjava, M. & Amaduzzi, A. 2024. Assessment of the feed additive consisting of propionic acid for all terrestrial animal species for the renewal of its authorization (Eastman Chemical BV, Perstorp AB, Dow Europe GmbH, BASF SE). *EFSA Journal*, 22(10): e9020.

EFSA Panel on Contaminants in the Food Chain (CONTAM). 2013. Scientific Opinion on the risk for public and animal health related to the presence of sterigmatocystin in food and feed. *EFSA Journal*, 11(6): 3254.

Engel, G., von Milczewski, K. E., Prokopek, D. & Teuber, M. 1982. Strain-specific synthesis of mycophenolic acid by *Penicillium roqueforti* in blue-veined cheese. *Applied and Environmental Microbiology*, 43(5): 1034–1040.

Erdogan, A. & Sert, S. 2004. Mycotoxin-forming ability of two *Penicillium roqueforti* strains in blue moldy Tulum cheese ripened at various temperatures. *Journal of Food Protection*, 67(3): 533–535.

Esam, R. M., Hafez, R. S., Khafaga, N. I. M., Fahim, K. M. & Ahmed, L. I. 2022. Assessment of aflatoxin M1 and B1 in some dairy products with referring to the analytical performances of enzyme-linked immunosorbent assay in comparison to high-performance liquid chromatography. *Veterinary World*, 15(1): 91.

European Food Safety Authority. 2011. Use of the EFSA comprehensive European food consumption database in exposure assessment. *EFSA Journal*, 9(3): 2097.

European Food Safety Authority (EFSA), Carrasco Cabrera, L. & Medina Pastor, P. 2022. The 2020 European Union report on pesticide residues in food. *EFSA Journal*, 20(3): e07215.

Flores-Flores, M. E., Lizarraga, E., de Cerain, A. L. & González-Peñas, E. 2015. Presence of mycotoxins in animal milk: A review. *Food Control*, 53: 163–176.

Fontaine, K., Hymery, N., Lacroix, M. Z., Puel, S., Puel, O., Rigalma, K. & Mounier, J. 2015a. Influence of intraspecific variability and abiotic factors on mycotoxin production in Penicillium roqueforti. *International Journal of Food Microbiology*, 215: 187–193.

Fontaine, K., Passeró, E., Vallone, L., Hymery, N., Coton, M., Jany, J. L. & Coton, E. 2015b. Occurrence of roquefortine C, mycophenolic acid and aflatoxin M1 mycotoxins in blue-veined cheeses. *Food Control*, 47: 634–640.

Foroud, N. A. & Eudes, F. 2009. Trichothecenes in cereal grains. *International Journal of Molecular Sciences*, 10(1): 147–173.

Gaglio, R., Todaro, M. & Settanni, L. 2021. Improvement of raw milk cheese hygiene through the selection of starter and non-starter lactic acid bacteria: The successful case of PDO Pecorino Siciliano cheese. *International Journal of Environmental Research and Public Health,* 18(4): 1834.

García-Estrada, C. & Martín, J. F. 2016. Biosynthetic gene clusters for relevant secondary metabolites produced by *Penicillium roqueforti* in blue cheeses. *Applied Microbiology and Biotechnology,* 100(19): 8303–8313.

Garnier, L., Valence, F. & Mounier, J. 2017. Diversity and control of spoilage fungi in dairy products: An update. *Microorganisms,* 5(3): 42.

Gasparre, N., Mefleh, M. & Boukid, F. 2022. Nutritional facts and health/nutrition claims of commercial plant-based infant foods: Where do we stand? *Plants,* 11(19): 2531.

Giovati, L., Gallo, A., Masoero, F., Cerioli, C., Ciociola, T., Conti, S. & Polonelli, L. 2014. Vaccination of heifers with an aflatoxin improves the reduction of aflatoxin B1 carry over in milk of lactating dairy cows. *PLoS One*, 9(4): e94440.

Giovati, L., Magliani, W., Ciociola, T., Santinoli, C., Conti, S. & Polonelli, L. 2015. AFM1 in milk: Physical, biological, and prophylactic methods to mitigate contamination. *Toxins*, 7(10): 4330–4349.

Gonçalves, B. L., Muaz, K., Coppa, C. F. S. C., Rosim, R. E., Kamimura, E. S., Oliveira, C. A. F. & Corassin, C. H. 2020. Aflatoxin M1 absorption by non-viable cells of lactic acid bacteria and *Saccharomyces cerevisiae* strains in Frescal cheese. *Food Research International*, 136: 109604.

Gong, Q., Zhang, Z., Huang, P., Wang, B. & Zheng, X. 2024. Assessment of fungal and contamination of ochratoxin A and patulin in foods susceptible to contamination in the Yangzhou Market, China. *Foods,* 13(19): 3205.

Govaris, A., Roussi, V., Koidis, P. A. & Botsoglou, N. A. 2001. Distribution and stability of aflatoxin M1 during processing, ripening and storage of Telemes cheese. *Food Additives & Contaminants*, 18(5): 437–443.

Gruber-Dorninger, C., Novak, B., Nagl, V. & Berthiller, F. 2017. Emerging mycotoxins: Beyond traditionally determined food contaminants. *Journal of Agricultural and Food Chemistry*, 65(33): 7052–7070.

Hartwig, A., Arand, M., Epe, B., Guth, S., Jahnke, G., Lampen, A. & Eisenbrand, G. 2020. Mode of action-based risk assessment of genotoxic carcinogens. *Archives of Toxicology,* 94: 1787–1877.

Hattimare, D., Shakya, S., Patyal, A., Chandrakar, C. & Kumar, A. 2022. Occurrence and exposure assessment of Aflatoxin M1 in milk and milk products in India. *Journal of Food Science and Technology*, 59(6): 2460–2468.

Hof, H. 2016. Mycotoxins in milk for human nutrition: Cow, sheep and human breast milk. *GMS Infectious Diseases*, 4: 1–5.

Hussein, H. S. & Brasel, J. M. 2001. Toxicity, metabolism, and impact of mycotoxins on humans and animals. *Toxicology*, 167(2): 101–134.

Hymery, N., Vasseur, V., Coton, M., Mounier, J., Jany, J. L., Barbier, G. & Coton, E. 2014. Filamentous fungi and mycotoxins in cheese: A review. *Comprehensive Reviews in Food Science and Food Safety*, 13(4): 437–456.

Ibrahim, S. A., Ayivi, R. D., Zimmerman, T., Siddiqui, S. A., Altemimi, A. B., Fidan, H. & Bakhshayesh, R. V. 2021. Lactic acid bacteria as antimicrobial agents: Food safety and microbial food spoilage prevention. *Foods,* 10(12): 3131.

Iha, M. H., Barbosa, C. B., Heck, A. R. & Trucksess, M. W. 2014. Aflatoxin M1 and ochratoxin A in human milk in RibeirãoPreto-SP, Brazil. *Food Control*, 40: 310–313.

Ismaiel, A. A. & Papenbrock, J. 2015. Mycotoxins: Producing fungi and mechanisms of phytotoxicity. *Agriculture,* 5(3): 492–537.

Izzo, L., Mikušová, P., Lombardi, S., Sulyok, M. & Ritieni, A. 2022. Analysis of mycotoxin and secondary metabolites in commercial and traditional Slovak cheese samples. *Toxins,* 14(2): 134.

Jiang, Y., Ogunade, I. M., Vyas, D. & Adesogan, A. T. 2021. Aflatoxin in dairy cows: Toxicity, occurrence in feedstuffs and milk and dietary mitigation strategies. *Toxins,* 13(4): 283.

Kallinteri, L. D., Kostoula, O. K. & Savvaidis, I. N. 2013. Efficacy of nisin and/or natamycin to improve the shelf-life of Galotyri cheese. *Food Microbiology*, 36(2): 176–181.

Kamle, M., Mahato, D. K., Gupta, A., Pandhi, S., Sharma, N., Sharma, B. & Kumar, P. 2022. Citrinin mycotoxin contamination in food and feed: Impact on agriculture, human health, and detection and management strategies. *Toxins,* 14(2): 85.

Kang'ethe, E. K. & Lang'a, K. 2009. Aflatoxin B1 and M1 contamination of animal feeds and milk from urban centers in Kenya. *African Health Sciences*, 9(4): 218–226.

Khan, R., Anwar, F. & Ghazali, F. M. 2024. A comprehensive review of mycotoxins: Toxicology, detection, and effective mitigation approaches. *Heliyon*, 10(8): e28361.

Kiermeier, F. & Buchner, M. 1977. Distribution of aflatoxin M 1 in whey and curd during cheese processing. *ZeitschriftfuerLebensmittel-Untersuchung und Forschung*, 164: 82–86.

Kokkonen, M., Jestoi, M. & Rizzo, A. 2005. The effect of substrate on mycotoxin production of selected *Penicillium* strains. *International Journal of Food Microbiology*, 99(2): 207–214.

Kure, C. F. & Skaar, I. 2019. The fungal problem in cheese industry. *Current Opinion in Food Science*, 29: 14–19.

Lafont, P. H. I. L. I. P. P. E., Debeaupuis, J. P., Gaillardin, M. I. C. H. E. L. & Payen, J. 1979. Production of mycophenolic acid by *Penicillium roqueforti* strains. *Applied and Environmental Microbiology*, 37(3): 365–368.

Le Bars, J. O. S. E. P. H. 1979. Cyclopiazonic acid production by *Penicillium camemberti* Thom and natural occurrence of this mycotoxin in cheese. *Applied and Environmental Microbiology*, 38(6): 1052–1055.

Lessard, M. H., Bélanger, G., St-Gelais, D. & Labrie, S. 2012. The composition of Camembert cheese-ripening cultures modulates both mycelial growth and appearance. *Applied and Environmental Microbiology*, 78(6): 1813–1819.

Leyva Salas, M., Mounier, J., Valence, F., Coton, M., Thierry, A. & Coton, E. 2017. Antifungal microbial agents for food biopreservation—A review. *Microorganisms,* 5(3): 37.

Lind, H., Broberg, A., Jacobsson, K., Jonsson, H. & Schnürer, J. 2010. Glycerol enhances the antifungal activity of dairy propionibacteria. *International Journal of Microbiology,* 2010(1): 430873.

Liu, A., Xu, R., Zhang, S., Wang, Y., Hu, B., Ao, X. & Liu, S. 2022. Antifungal mechanisms and application of lactic acid bacteria in bakery products: A review. *Frontiers in Microbiology*, 13: 924398.

Loi, M., Logrieco, A. F., Pusztahelyi, T., Leiter, É., Hornok, L. & Pócsi, I. 2023. Advanced mycotoxin control and decontamination techniques in view of an increased aflatoxin risk in Europe due to climate change. *Frontiers in Microbiology*, 13: 1085891.

Lopez-Diaz, T. M., Alonso, C., Roman, C., Garcıa-Lopez, M. L. & Moreno, B. 2000. Lactic acid bacteria isolated from a hand-made blue cheese. *Food Microbiology*, 17(1): 23–32.

Lopez-Diaz, T. M., Roman-Blanco, C., Garcia-Arias, M. T., García-Fernández, M. C. & García-López, M. L. 1996. Mycotoxins in two Spanish cheese varieties. *International Journal of Food Microbiology*, 30(3): 391–395.

Lorán, S., Carramiñana, J. J., Juan, T., Ariño, A. & Herrera, M. 2022. Inhibition of *Aspergillus parasiticus* growth and aflatoxins production by natural essential oils and phenolic acids. *Toxins*, 14(6): 384.

Lu, M. & Wang, N. S. 2017. Spoilage of milk and dairy products. In *The Microbiological Quality of Food. Foodborne Spoilers* (pp. 151–178). Cambridge: Wood head Publishing.

Mangione, G., Caccamo, M., Natalello, A. & Licitra, G. 2023. Graduate Student Literature Review: History, technologies of production, and characteristics of ricotta cheese. *Journal of Dairy Science*, 106(6): 3807–3826.

Maragos, C. M., Probyn, C., Proctor, R. H. & Sieve, K. K. 2023. Cyclopiazonic acid in soft-ripened and blue cheeses marketed in the USA. *Food Additives & Contaminants: Part B,* 16(1): 14–23.

Martín, J. F. & Liras, P. 2017. Secondary metabolites in cheese fungi. In *Fungal Metabolites* (pp. 29–3315). Springer, Cham.

Martin, J. G. P. & Cotter, P. D. 2023. Filamentous fungi in artisanal cheeses: A problem to be avoided or a market opportunity. *Heliyon*, 9(4): e15110

Mayoral, M. B., Martín, R., Sanz, A., Hernández, P. E., González, I. & García, T. 2005. Detection of Kluyveromyces marxianus and other spoilage yeasts in yoghurt using a PCR-culture technique. *International Journal of Food Microbiology*, 105(1): 27–34.

McCormick, S. P., Stanley, A. M., Stover, N. A. & Alexander, N. J. 2011. Trichothecenes: From simple to complex mycotoxins. *Toxins*, 3(7): 802–814.

Meena, M., Prajapati, P., Ravichandran, C. & Sehrawat, R. 2021. Natamycin: A natural preservative for food applications—a review. *Food Science and Biotechnology*, 30: 1481–1496

Méndez-Albores, A., Del Río-García, J. C. & Moreno-Martinez, E. 2007. Decontamination of aflatoxin duckling feed with aqueous citric acid treatment. *Animal Feed Science and Technology,* 135(3–4): 249–262.

Mendonca, C. & Venâncio, A. 2005. Fate of aflatoxin M1 in cheese whey processing. *Journal of the Science of Food and Agriculture,* 85(12): 2067–2070.

Mladenović, K. G., Grujović, M. Ž., Kocić-Tanackov, S. D., Bulut, S., Iličić, M., Degenek, J. & Semedo-Lemsaddek, T. 2021. Serbian traditional goat cheese: Physico-chemical, sensory, hygienic and safety characteristics. *Microorganisms,* 10(1): 90.

Mounier, J., Gelsomino, R., Goerges, S., Vancanneyt, M., Vandemeulebroecke, K., Hoste, B. & Cogan, T. M. 2005. Surface microflora of four smear-ripened cheeses. *Applied and Environmental Microbiology*, 71(11): 6489–6500.

Mudannayake, A., Karunarathne, S., Jayasooriya, P. W., Nanayakkara, D., Abesooriya, A., Silva, S. & Fernando, R. 2024. Occurrence of aflatoxin M1 in cheese products commonly available in Sri Lankan market. *Heliyon,* 10(15): e35155.

Murtaza, M. A., Rehman, S. U., Anjum, F. M., Huma, N., Tarar, O. M. & Mueen-Ud-Din, G. 2012. Organic acid contents of buffalo milk cheddar cheese as influenced by accelerated ripening and sodium salt. *Journal of Food Biochemistry*, 36(1): 99–106.

Nájera, A. I., Nieto, S., Barron, L. J. R. & Albisu, M. 2021. A review of the preservation of hard and semi-hard cheeses: Quality and safety. *International Journal of Environmental Research and Public Health*, 18(18): 9789.

Nasrollahzadeh, A., Mokhtari, S., Khomeiri, M. & Saris, P. E. 2022. Antifungal preservation of food by lactic acid bacteria. *Foods,* 11(3): 395.

Nazareth, T. D. M., Soriano Pérez, E., Luz, C., Meca, G. & Quiles, J. M. 2024. Comprehensive review of aflatoxin and ochratoxin A dynamics: Emergence, toxicological impact, and advanced control strategies. *Foods,* 13(12): 1920.

Nielsen, K. F., Sumarah, M. W., Frisvad, J. C. & Miller, J. D. 2006. Production of metabolites from the *Penicillium roqueforti* complex. *Journal of Agricultural and Food Chemistry,* 54(10): 3756–3763.

Ogunade, I. M., Martinez-Tuppia, C., Queiroz, O. C. M., Jiang, Y., Drouin, P., Wu, F. & Adesogan, A. T. 2018. Silage review: Mycotoxins in silage: Occurrence, effects, prevention, and mitigation. *Journal of Dairy Science*, 101(5): 4034–4059.

Ohmomo, S., Ōhashi, T. & Abe, M. 1979. On the mechanism of the formation of indole alkaloids in *Penicillium roqueforti. Agricultural and Biological Chemistry*, 43(10): 2035–2038.

Oliveira, C. A. F. D., Sebastião, L. S., Fagundes, H., Rosim, R. E. & Fernandes, A. M. 2008. Aflatoxins and cyclopiazonic acid in feed and milk from dairy farms in Sao Paulo, Brazil. *Food additives and Contaminants*, 1(2): 147–152.

Olivigni, F. J. & Bullerman, L. B. 1977. Simultaneous production of penicillic acid and patulin by a *Penicillium* species isolated from Cheddar cheese. *Journal of Food Science,* 42(6): 1654–1657.

Ollinger, N., Malachová, A., Schamann, A., Sulyok, M., Krska, R. & Weghuber, J. 2024. Limited effectiveness of *Penicillium camemberti* in preventing the invasion of contaminating molds in camembert cheese. *Foods*, 13(18): 2865.

Papademas, P. & Bintsis, T. 2017. *Global Cheese Making Technology: Cheese Quality and Characteristics* (pp. 120–156). John Wiley & Sons, Ltd.

Payne, G. A. 2016. Mycotoxins and product safety. In *Peanuts* (pp. 347–361). Champaign: AOCS Press.

Piątkowska, M., Sulyok, M., Pietruszka, K. & Panasiuk, Ł. 2018. Pilot study for the presence of fungal metabolites in sheep milk from first spring milking. *Journal of Veterinary Research*, 62(2): 167–172.

Pietri, A., Leni, G., Mulazzi, A. & Bertuzzi, T. 2022. Ochratoxin A and sterigmatocystin in long-ripened Grana cheese: Occurrence, wheel rind contamination and effectiveness of cleaning techniques on grated products. *Toxins*, 14(5): 306.

Pitt, J. I. 2019. Penicillium. In *Foodborne Disease Handbook* (pp. 581–592). Boca Raton: CRC Press.

Piwowarek, K., Lipińska, E., Hać-Szymańczuk, E., Kieliszek, M. & Ścibisz, I. 2018. *Propionibacterium* spp.—source of propionic acid, vitamin B12, and other metabolites important for the industry. *Applied Microbiology and Biotechnology*, 102: 515–538.

Pouris, J., Kolyva, F., Bratakou, S., Vogiatzi, C. A., Chaniotis, D. & Beloukas, A. 2024. The Role of Fungi in Food Production and Processing. *Applied Sciences*, 14(12): 5046.

Prosperini, A., Berrada, H., Ruiz, M. J., Caloni, F., Coccini, T., Spicer, L. J. & Lafranconi, A. 2017. A review of the mycotoxin enniatin B. *Frontiers in Public Health,* 5: 304.

RahimzadehBarzoki, H., Faraji, H., Beirami, S., Keramati, F. Z., Nayik, G. A., Izadi-Yazdanaabadi, Z. & MozaffariNejad, A. S. 2023. Seasonal study of aflatoxin M1 contamination in cow milk on the retail dairy market in Gorgan, Iran. *Dairy*, 4(4): 571–580.

Raman, J., Kim, J. S., Choi, K. R., Eun, H., Yang, D., Ko, Y. J. & Kim, S. J. 2022. Application of lactic acid bacteria (LAB) in sustainable agriculture: Advantages and limitations. *International Journal of Molecular Sciences*, 23(14): 7784.

Ramos-Pereira, J., Mareze, J., Patrinou, E., Santos, J. A. & López-Díaz, T. M. 2019. Polyphasic identification of *Penicillium* spp. isolated from Spanish semi-hard ripened cheeses. *Food Microbiology*, 84: 103253.

Rasheed, M. W., Tang, J., Sarwar, A., Shah, S., Saddique, N., Khan, M. U. & Sultan, M. 2022. Soil moisture measuring techniques and factors affecting the moisture dynamics: A comprehensive review. *Sustainability*, 14(18): 11538.

Richard, J. L. 2008. Discovery of aflatoxins and significant historical features. *Toxin Reviews*, 27(3–4): 171–201.

Ritschard, J. S. & Schuppler, M. 2024. The microbial diversity on the surface of smear-ripened cheeses and its impact on cheese quality and safety. *Foods*, 13(2): 214.

Rodríguez, A., Magan, N. & Delgado, J. 2024. Exploring a cheese ripening process that hinders ochratoxin A production by *Penicillium nordicum* and *Penicilliumverrucosum*. *Biology*, 13(8): 582.

Rodríguez-Cañás, I., González-Jartín, J. M., Alvariño, R., Alfonso, A., Vieytes, M. R. & Botana, L. M. 2023. Detection of mycotoxins in cheese using an optimized analytical method based on a QuEChERS extraction and UHPLC-MS/MS quantification. *Food Chemistry*, 408: 135182.

Romero-Sánchez, I., Ramírez-García, L., Gracia-Lor, E. & Madrid-Albarrán, Y. 2022. Simultaneous determination of aflatoxins B1, B2, G1 and G2 in commercial rices using immunoaffinity column clean-up and HPLC-MS/MS. *Food Chemistry*, 395: 133611.

Salazar, J. K., Carstens, C. K., Ramachandran, P., Shazer, A. G., Narula, S. S., Reed, E. & Schill, K. M. 2018. Metagenomics of pasteurized and unpasteurized gouda cheese using targeted 16S rDNA sequencing. *BMC Microbiology*, 18: 1–13.

Sarımehmetoglu, B., Kuplulu, O. & Celik, T. H. 2004. Detection of aflatoxin M1 in cheese samples by ELISA. *Food Control*, 15(1): 45–49.

Sarmast, E., Fallah, A. A., Jafari, T. & Khaneghah, A. M. 2021. Impacts of unit operation of cheese manufacturing on the aflatoxin M1 level: A global systematic review and meta-analysis. *LWT*, 148: 111772.

Schröder, M. J. 1984. Origins and levels of post pasteurization contamination of milk in the dairy and their effects on keeping quality. *Journal of Dairy Research*, 51(1): 59–67.

Sengun, I., Yaman, D. B. & Gonul, S. 2008. Mycotoxins and mould contamination in cheese: A review. *World Mycotoxin Journal*, 1(3): 291–298.

Silva, S. P., Teixeira, J. A. & Silva, C. C. 2023. Prevention of fungal contamination in semi-hard cheeses by whey–gelatin film incorporated with *Levilactobacillus brevis* SJC120. *Foods*, 12(7): 1396.

Stott, W. T. & Bullerman, L. B. 1976. Instability of patulin in cheddar cheese. *Journal of Food Science*, 41(1): 201–203.

Tolosa, J., Rodríguez-Carrasco, Y., Ruiz, M. J. & Vila-Donat, P. 2021. Multi-mycotoxin occurrence in feed, metabolism and carry-over to animal-derived food products: A review. *Food and Chemical Toxicology*, 158: 112661.

Trombete, F. M., Castro, I. M. D., Teixeira, A. D. S., Saldanha, T. & Fraga, M. E. 2014. Aflatoxin M1 contamination in grated parmesan cheese marketed in Rio de Janeiro-Brazil. *Brazilian Archives of Biology and Technology*, 57: 269–273.

Vallone, L., Giardini, A. & Soncini, G. 2014. Secondary metabolites from *Penicillium roqueforti*, a starter for the production of Gorgonzola cheese. *Italian Journal of Food Safety*, 3(3): 2118.

Vaz, A., Cabral Silva, A. C., Rodrigues, P. & Venâncio, A. 2020. Detection methods for aflatoxin M1 in dairy products. *Microorganisms*, 8(2): 246.

Veršilovskis, A. & De Saeger, S. 2010. Sterigmatocystin: Occurrence in foodstuffs and analytical methods–an overview. *Molecular Nutrition & Food Research*, 54(1): 136–147.

Veršilovskis, A., Van Peteghem, C. & De Saeger, S. 2009. Determination of sterigmatocystin in cheese by high-performance liquid chromatography-tandem mass spectrometry. *Food Additives and Contaminants*, 26(1): 127–133.

Vlachou, M., Pexara, A., Solomakos, N. & Govaris, A. 2022. Ochratoxin A in slaughtered pigs and pork products. *Toxins,* 14(2): 67.

Zain, M. E. 2011. Impact of mycotoxins on humans and animals. *Journal of Saudi Chemical Society*, 15(2): 129–144.

Zaki, M. M., El-Midany, S. A., Shaheen, H. M. & Rizzi, L. 2012. Mycotoxins in animals: Occurrence, effects, prevention and management. *Journal of Toxicology and Environmental Health Sciences,* 4(1): 13–28.

ZamaniMazdeh, F., Sasanfar, S., Chalipour, A., Pirhadi, E., Yahyapour, G., Mohammadi, A. & Hajimahmoodi, M. 2017. Simultaneous determination of preservatives in dairy products by HPLC and chemometric analysis. *International Journal of Analytical Chemistry*, 2017(1): 3084359.

Zentai, A., Jóźwiak, Á., Süth, M. & Farkas, Z. 2023. Carry-over of aflatoxin B1 from feed to cow milk—a review. *Toxins,* 15(3): 195.

Zhang, D., Zhao, L., Chen, Y., Gao, H., Hua, Y., Yuan, X. & Yang, H. 2022. Mycotoxins in maize silage from China in 2019. *Toxins*, 14(4): 241.

Zhu, M., Cen, Y., Ye, W., Li, S. & Zhang, W. 2020. Recent advances on macrocyclic trichothecenes, their bioactivities and biosynthetic pathway. *Toxins,* 12(6): 417.

8 Conventional Techniques of Food Preservation

Muhammad Wasim Haider

8.1 INTRODUCTION

Food preservation is important for food security as it ensures food storage for long-term use (Haider et al., 2025). Food preservation techniques like canning, pickling, drying, refrigeration, and freezing have been in practice since ancient times to reduce post-harvest losses and increase the shelf life of produce (Adegoke and Olapade, 2012). These techniques originated from very basic concepts of moisture elimination, temperature control, and making solutions to restrict the activities of enzymes and the growth of spoiling agents (Benyam et al., 2021).

The importance of food preservation can be predicted from the rate of post-harvest losses occurring on a global scale (Hasan et al., 2021). According to an estimate, almost one-third of all food produced for human consumption is lost up to the point between production and processing (FAOSTAT, 2019). This amounts to about 1.3 billion tons, a notable loss of resources, including capital, energy, labor, land, and water (Ishangulyyev et al., 2019). The extent of post-harvest losses varies between 20 and 50% worldwide and depends on the region, as well as food type (Porat et al., 2018). In developing countries, a significant percentage of these losses are attributed to an insufficient infrastructure with regards to storage capabilities and transport (Kumar and Kalita, 2017; O'Connor et al., 2021). Alternatively, food waste is more prevalent at the consumer level in developed countries and typically a result of over-purchase and improper storage (Yahia and Mourad, 2019; Schneider and Eriksson, 2020). For example, fruits and vegetables are particularly susceptible to spoilage (Figure 8.1).

The conventional techniques of food preservation are vital in mitigating these losses (Figure 8.2). By understanding and implementing these techniques, we can not only extend the availability of seasonal produce but also contribute to reducing the environmental impact of food production and waste. Furthermore, preserving food through conventional techniques plays a crucial role in maintaining food quality, nutritional value, and safety, thus supporting both food security and public health. The usage and improvement of these preservation techniques have become increasingly important as food demand has risen with the expansion of the world's population. Preserving food effectively ensures that more of the food we produce reaches consumers, thus reducing waste and contributing to sustainable food systems.

DOI: 10.1201/9781003540113-8

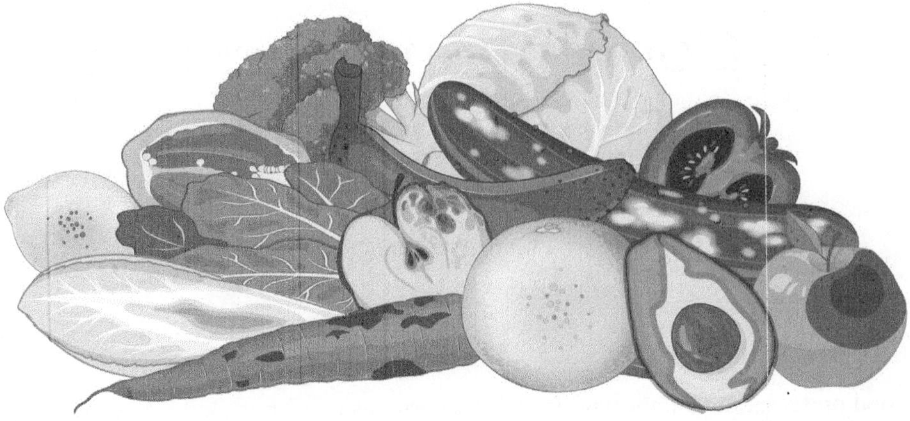

FIGURE 8.1 Post-harvest spoilage in fruits and vegetables.

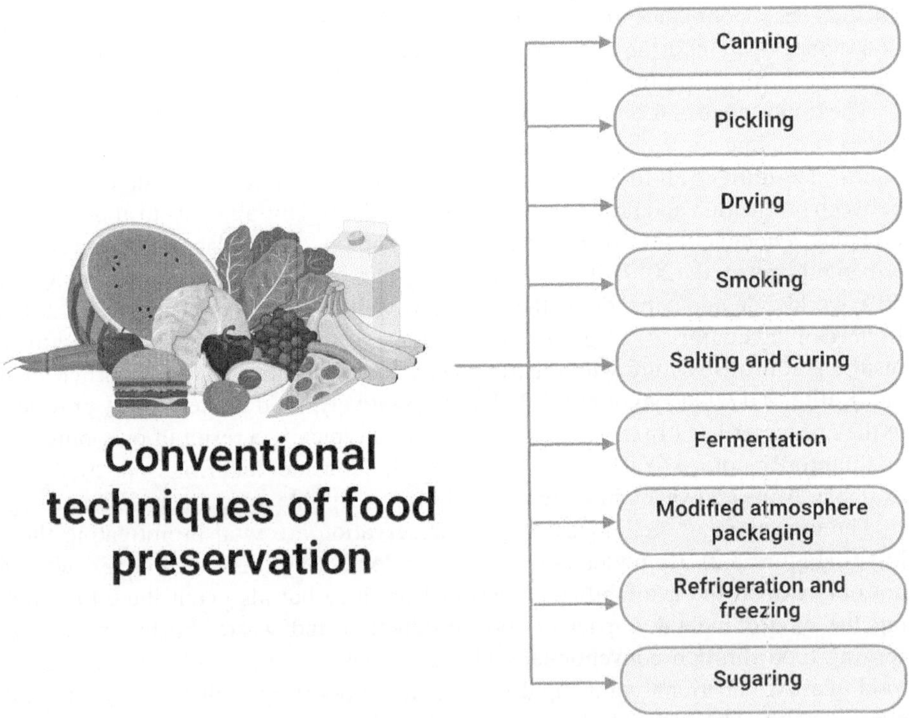

FIGURE 8.2 The most common conventional techniques of food preservation.

The most common conventional techniques of food preservation include drying, canning, pickling, smoking, salting, fermenting, refrigerating and freezing, sugaring, heating, and modified atmosphere packaging, each method effectively prolonging the storability of foods by hindering spoilage and growth of microbes.

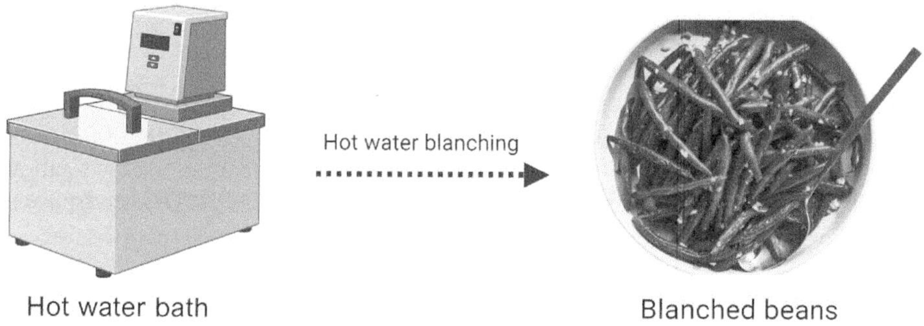

FIGURE 8.3 Hot water blanching in green beans.

8.2 CANNING

Canning is the food preservation technique of placing commodities in containers that are tightly closed and then subjecting them to heat to kill the undesirable microbes that cause decay or foodborne problems (Tumuluru, 2023). The principles of canning rely on creating a vacuum seal that prevents new microorganisms from entering the container while also eliminating oxygen, which many bacteria need to survive (Owusu-Apenten and Vieira, 2022). The process typically involves applying heat to kill molds, yeasts, bacteria and enzymes that can cause food to deteriorate (Naczk and Artyukhova, 2020). For example, fresh peaches are peeled, sliced, and packed in syrup or juice before being sealed in jars and processed in a boiling water bath. This method preserves both the flavor and texture of the peaches, providing a sweet treat that can be enjoyed year-round. Likewise, green beans are blanched, packed into jars, and processed in a pressure canner to ensure they are safely preserved (Figure 8.3). This method retains the color of beans and their nutritional value, making them a convenient side dish. Tuna fish is cooked, packed in oil or water, and then sealed in cans before being sterilized in a pressure canner. Canned tuna is a popular source of protein that has a long shelf life and is widely used in salads, sandwiches, and casseroles.

8.2.1 Steps of the Canning Process

8.2.1.1 Preparation

The first step in canning is to select and prepare the food (Rajput et al., 2022). Fresh, high-quality fruits, vegetables, or meats are washed, peeled, chopped, or otherwise prepped according to the specific canning recipe (Dar et al., 2022). This ensures that the final product is both safe and of high quality.

8.2.1.2 Blanching

Blanching is the process of immersing prepared food in ice water after it has been short-boiled (Xin et al., 2015). This step helps preserve color, texture, and nutritional value by inactivating enzymes that can cause spoilage (Xiao et al., 2017). It also helps in loosening the skins of some fruits and vegetables, making them easier to peel (Sumonsiri and Barringer, 2014).

8.2.1.3 Filling

After blanching, food is packed into sterilized jars or cans (Ramesh, 2007). The containers are typically filled to a certain level, leaving headspace at the top to allow for the expansion of food during processing (Miranda et al., 2019). The headspace is crucial for creating a vacuum seal during the canning process.

8.2.1.4 Sealing

Once the food is packed, the containers are sealed with lids that are designed to form an airtight seal (Rooney, 2005). This step is critical as it prevents any microorganisms from entering the container after processing. Proper sealing also creates the vacuum necessary to preserve the food over time.

8.2.1.5 Sterilization

The sealed containers are then subjected to heat treatment, typically in a water bath canner or a pressure canner (Owusu-Apenten and Vieira, 2022). The type of canner used depends on the acidity of the food being preserved. Acidic foods like fruits and pickles are canned using boiling water baths, while foods with low acid content, like meats and vegetables requiring higher temperatures, are canned using pressure canners to ensure safety.

8.2.1.6 Cooling

After sterilization, the cans or jars are removed from the canner and allowed to cool (Ramesh, 2020). During cooling, a vacuum seal is formed as the content of the jar contract. Once cooled, the seals are checked to ensure they are airtight, and any jars that did not seal properly are either reprocessed or refrigerated for immediate use.

8.2.2 Types of Canned Foods

Canned foods are categorized based on their pH value and the method of canning.

8.2.2.1 High-pH Foods

Vegetables, meats, poultry, seafood, and soups fall into this category and require pressure canning to ensure safety.

8.2.2.2 Low-pH Foods

These include fruits, tomatoes, pickles, and jams, which can be safely processed using a boiling water bath.

8.2.2.3 Mixed Foods

Foods that combine ingredients from both high- and low-pH food categories, such as stews or sauces, are typically processed using pressure canning.

8.2.3 Safety Considerations

Canning is a safe method of food preservation when done correctly, but there are important safety considerations.

8.2.3.1 Acidity Levels

Understanding the acidity of the food is essential to determining the correct canning method (boiling water bath or pressure canning).

8.2.3.2 Seal Integrity

The vacuum seal must be intact to prevent contamination. Any jars that do not seal properly should be refrigerated and consumed promptly.

8.2.3.3 Proper Sterilization

Ensuring that the food reaches the necessary temperature during canning is crucial to killing harmful microorganisms, including *Clostridium botulinum*, which can cause botulism.

8.2.3.4 Retention of Nutrients

Many vitamins and minerals are well-preserved during canning, especially if the process is done quickly after harvesting.

8.2.3.5 Nutritional Impact

Canning can affect the nutritional content of foods, although the impact varies.

8.2.3.6 Loss of Heat-Sensitive Nutrients

Certain heat-sensitive nutrients, like vitamin C and several B group vitamins, can be affected during canning. Nevertheless, the extent of nutrient loss is generally minimal and can be offset by the convenience and extended shelf life of canned foods.

8.2.4 Advantages of Canning

- Canned foods are ready to use and require no further preparation, making them a convenient option for quick meals.
- When done correctly, canning effectively kills harmful microorganisms, making food safe to eat.
- Canning also has the advantage of greatly increasing the storability of foods for months or even years without refrigeration.

8.2.5 Limitations of Canning

- The texture of some foods may change, becoming softer due to the high heat used in canning.
- Certain vitamins, like vitamin C, may be destroyed during the process of canning due to their heat-sensitive nature.

- The equipment needed for home canning, such as pressure canners and jars, can be expensive to acquire.
- The process of preparing, packing, and processing foods for canning is labor-intensive and time-consuming.

8.3 PICKLING

Pickling is a preservation technique that entails soaking food in an acidic solution, generally vinegar, or allowing it to ferment in brine (Lee and Kang, 2004). The acidic or saline environment inhibits the growth of harmful bacteria, molds, and yeasts, effectively preserving the food and extending its shelf life (Peréz-Dıaz et al., 2013). Pickling not only is used in food storage but also offers an extra flavor, making it a popular technique for enhancing the taste of various fruits, vegetables, and even meats (Behera et al., 2020). A classic example of pickled foods is mango pickle in vinegar (Figure 8.4). Other examples are dill pickle (made from cucumbers and flavored with dill, garlic, and spices), kimchi (a traditional Korean dish), hot and sweet peppers pickles, and pickled eggs and meats.

8.3.1 Types of Pickling

8.3.1.1 Vinegar Pickling

Vinegar pickling, also known as quick pickling, involves submerging food in a solution of vinegar, water, salt, and often sugar (Giudici et al., 2017). The vinegar's acetic acid creates an environment where bacteria cannot survive, thus preserving the food (Sengun and Karabiyikli, 2011). Vinegar pickles are usually ready to eat within a few hours to a few days, depending on the recipe, and can be stored for extended periods.

8.3.1.2 Fermentation Pickling

In fermentation pickling, food is placed in a brine (saltwater solution) and left to ferment naturally over time (Alan, 2019). The salt encourages the growth of beneficial

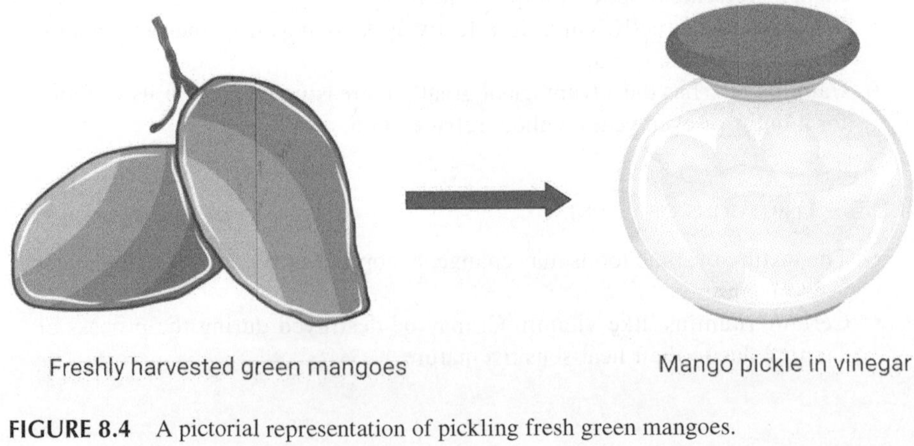

FIGURE 8.4 A pictorial representation of pickling fresh green mangoes.

lactic acid bacteria (LAB), which produce lactic acid as they consume sugars in the food (Chen, 2021). This lactic acid acts as a preservative, lowering the pH and creating a tangy flavor (Sionek et al., 2023). Fermentation pickles take longer to develop their flavor, often requiring several weeks, but they offer probiotic benefits due to the presence of live bacteria (Beena Divya et al., 2012).

8.3.2 Steps of Pickling Process

The process of pickling involves several key steps, which can vary slightly depending on the type of pickling, either vinegar pickling or fermenting.

8.3.2.1 Preparation

The first step in pickling is selecting fresh, high-quality produce. The produce is thoroughly washed, peeled, sliced, or chopped as needed. Spices, herbs, and other flavorings are also prepared at this stage.

8.3.2.2 Brining or Salting

In some cases, particularly for fermentation pickling, the produce is salted or brined to draw out excess moisture. This step helps create the right conditions for fermentation or to ensure that the pickling liquid will adequately penetrate the food.

8.3.2.3 Packing

The prepared produce is packed into sterilized jars. For vinegar pickling, a hot vinegar-based solution is poured over the produce, ensuring it is fully submerged. In fermentation pickling, the produce is covered with brine, often weighted down to keep it below the liquid surface.

8.3.2.4 Sealing and Storing

The jars are sealed with airtight lids. For vinegar pickles, the jars are often processed in a boiling water bath to ensure a strong seal and extend shelf life. Fermented pickles are typically stored in a cool, dark place to allow the fermentation process to occur over several weeks.

8.3.2.5 Aging

Pickles are usually allowed to age to develop their flavors fully. Vinegar pickles might be ready in a few days, while fermented pickles can take several weeks. The aging process allows the flavors to meld and the preservation process to complete.

8.3.3 Advantages of Pickling

- Increase the shelf life of perishable foods, allowing them to be stored for months or even years.
- Introduce a tangy, savory, or sweet flavor that enhances the taste of food.
- Fermented pickles, in particular, offer probiotic benefits that support digestive health.
- Important culinary tradition in many cultures, contributing to regional cuisines and diets.

8.3.4 Limitations of Pickling

- Some vitamins, particularly water-soluble ones, like vitamin C, can be lost during the pickling process.
- Many pickles, particularly fermented ones, can be high in salt, which may be a concern for individuals on low-sodium diets.
- Fermentation pickling requires time for the flavors to develop, which might not be ideal for those seeking quick preservation methods.
- The high acidity of pickled foods might not be suitable for individuals with certain digestive issues, such as acid reflux.

8.4 DRYING

Drying is a process that involves the removal of moisture from materials, typically using heat (Musielak et al., 2016). The primary goal is to reduce the water content to a level that prevents microbial growth, enzymatic reactions, and other spoilage mechanisms (Rawat, 2015). The process relies on the principles of heat transfer and mass transfer, where heat is applied to a substance, causing the moisture within to evaporate (Saravacos, 2005). This moisture is then removed from the material's surface, typically by air or another gas (Wang et al., 2011). A common example of drying in the food industry is the preparation of raisins from grapes. The grapes are typically dried in the sun, reducing their moisture content and concentrating their sugars (Figure 8.5). This simple yet effective process transforms a perishable fruit into a long-lasting, sweet snack with multiple culinary uses. Similarly, tomatoes are often dried using air-drying or oven-drying methods. This process removes most of the moisture, concentrating the flavor and preserving the tomatoes for use in various dishes like pastas and sauces. Apples can be sliced and dehydrated using a food dehydrator or oven-drying method. Dried slices of apples are a nutritious, portable snack with a long shelf life commonly used in baking.

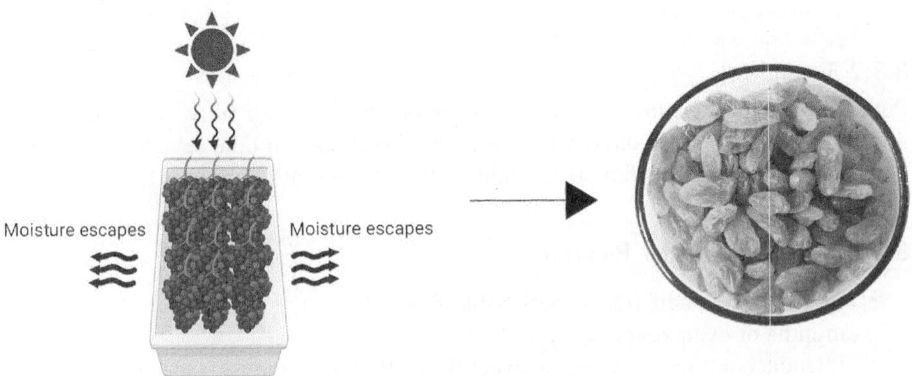

FIGURE 8.5 A pictorial representation of drying perishable grapes into lifelong raisins.

8.4.1 Methods of Drying

Various methods are employed for drying, depending on the nature of the material and the desired end product. The most common drying methods involve the following.

8.4.1.1 Sun Drying

Sun drying is known to be one of the most primitive and simplest techniques of drying, especially for agricultural products like fruits, vegetables, and grains (Janjai and Bala, 2012). Material is spread out under direct sunlight. This method is cost-effective and environmentally friendly but is limited by weather conditions, and there is a risk of contamination from dust, insects, and animals (Lingayat et al., 2021).

8.4.1.2 Oven Drying

Oven drying uses controlled heat in an enclosed environment, allowing for precise control over temperature and drying time (Moses et al., 2014). It is commonly used for small batches or materials that require a specific drying profile (Zambrano et al., 2019). Oven drying is efficient and produces consistent results but can be energy-intensive and may not be suitable for large-scale operations (Bramsiepe et al., 2012).

8.4.1.3 Dehydration

Dehydration typically involves the use of specialized equipment, such as dehydrators or freeze dryers, to remove the moisture (Rehman and Awan, 2012). This method can reduce moisture content to very low levels, extending the shelf life of the product (Mishra et al., 2017). Dehydration preserves the nutritional content and flavor of foods better than other drying methods but requires a significant initial investment in equipment (Qiu et al., 2019).

8.4.1.4 Air-Drying

Air-drying involves passing air, typically at ambient temperature, over or through the material to be dried (Chua and Chou, 2003). This method can be passive, relying on natural air currents, or active, using fans to enhance air circulation (Chan et al., 2010). Air-drying is less energy-intensive than other methods but is generally slower and less efficient in removing moisture (Inyang et al., 2017).

8.4.2 Applications

Drying has a wide range of applications across various industries.

8.4.2.1 Agriculture

Drying grains and seeds ensures their longevity and viability for planting or consumption.

8.4.2.2 Food Industry

Drying is used to preserve fruits, vegetables, spices, and meats.

8.4.2.3 Pharmaceuticals

Drying is crucial in the preparation of powders and granules for medication.

8.4.2.4 Textiles

Drying is essential in processes like dyeing and finishing fabrics.

8.4.3 Advantages

- Prolongs the storage of perishables by lowering their moisture level, hence inhibiting microbial growth.
- Reduces the weight and volume of materials, lowering storage and transportation costs.
- Retains the nutritional value and flavor of foods, particularly with dehydration.
- Provides a method for preserving seasonal products for year-round availability.

8.4.4 Limitations

- Drying processes can be energy-intensive, particularly oven drying and dehydration.
- Some methods, like sun drying, are highly dependent on weather conditions and can lead to inconsistent results.
- The initial cost of specialized drying equipment can be high.
- Over-drying can lead to loss of texture, color, and nutritional value in food products.

8.5 SMOKING

Smoking is one of the earliest food storage methods, having originated thousands of years ago. Initially, smoking was used by ancient cultures as a means to preserve meat and fish during times when refrigeration and other preservation methods were not available. The food is subjected to smoke produced from burning wood or any other material, to gain color, flavor, and preservative effect (Nizio et al., 2023). The smoke contains compounds that inhibit the growth of bacteria and fungi, thereby extending the shelf life of food (Mei et al., 2019). Over time, smoking evolved from a purely preservative technique to one that is also valued for the unique flavors it imparts to food. For instance, the gentle smoking of salmon fish imparts a delicate flavor while preserving the fish's tender texture (Figure 8.6).

8.5.1 Types of Smoking Technique

8.5.1.1 Hot Smoking

Hot smoking, on the other hand, exposes food to both smoke and heat, with temperatures typically ranging between 49 and 107°C (120–225°F) (Ledesma et al., 2017). This method not only imparts a smoky flavor but also cooks the food. Hot smoking

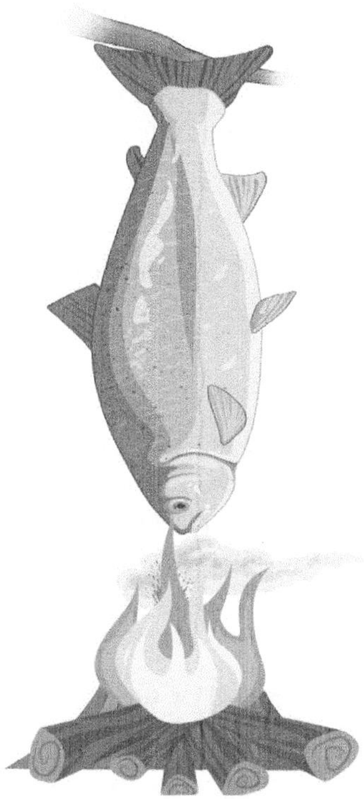

FIGURE 8.6 Traditional smoking of salmon over an open bonfire smoke.

is faster than cold smoking, often taking just a few hours, and is commonly used for meats, poultry, and fish. The higher temperatures ensure that the food is cooked through, making it safe to eat immediately after smoking (Suleman et al., 2020).

8.5.1.2 Cold Smoking

Cold smoking is a process where food is exposed to smoke at temperatures typically ranging between 20 and 30°C (68–86°F) (Ledesma et al., 2016). The lower temperature allows the food to absorb the smoke flavor without cooking it (Belichovska et al., 2019). Cold smoking is often used for foods like salmon, cheese, and cured meats. This method can take several days or even weeks, depending on the desired level of smokiness and the type of food being smoked. Cold smoking requires careful control of temperature to prevent spoilage, as the food remains raw throughout the process (Løvdal, 2015).

8.5.2 Smoking Process

The smoking process involves several steps, whether using cold or hot smoking methods.

8.5.2.1 Preparation

The first step in the smoking process is preparing the food. For meats and fish, this may involve curing with salt, sugar, or spices to enhance flavor and further preserve the food. The curing process can last from a few hours to several days, depending on the recipe and the type of food.

8.5.2.2 Drying

After curing, the food is usually air-dried to form a pellicle, a tacky surface that helps smoke adhere to the food. This step is essential for achieving an even smoke coating and enhancing the flavor.

8.5.2.3 Smoking

The prepared food is placed in a smoker or smokehouse, where it is exposed to smoke generated by burning wood chips, sawdust, or other materials. The type of wood used, such as hickory, mesquite, or applewood, can significantly affect the flavor profile of the smoked food. During this stage, the temperature is carefully monitored and controlled to ensure proper smoking, whether the goal is cold or hot smoking.

8.5.2.4 Aging (Optional)

In some cases, smoked foods are aged or further cured after smoking to develop deeper flavors. This is common with smoked sausages and some types of fish.

8.5.3 Types of Smoked Foods

Smoking is applied to a wide variety of foods, with some of the most common types including the following.

8.5.3.1 Cheese

Certain cheeses, such as Gouda and cheddar, are smoked to develop a unique, robust flavor. Smoked cheese is often used in sandwiches, appetizers, and gourmet dishes.

8.5.3.2 Vegetables and Nuts

While less common, smoking is also used for vegetables and nuts, imparting a smoky flavor that complements their natural taste.

8.5.3.3 Fish and Seafood

Smoking is a popular method for preserving and flavoring fish, especially salmon, mackerel, and trout. Smoked fish can be served as is or used in various dishes.

8.5.3.4 Meats

Beef, pork, poultry, and sausages are often smoked to enhance flavor and preserve them for longer periods. Examples include smoked ham, bacon, and smoked brisket.

8.5.4 Health Implications

While smoking food offers the benefits of preservation and flavor, it also has some health implications that should be considered.

8.5.4.1 Formation of Harmful Compounds

The process of smoking may generate potentially hazardous compounds, such as polycyclic aromatic hydrocarbons (PAHs) and heterocyclic amines (HCAs), both of which have been related to an elevated risk of cancer (Bulanda and Janoszka, 2022). These compounds form when meat is smoked at high temperatures or for prolonged periods.

8.5.4.2 High Sodium Content

Many smoked foods, particularly those that are cured before smoking, can be high in sodium (Arvanitoyannis and Kotsanopoulos, 2011). Excessive sodium intake is associated with health risks, such as hypertension and cardiovascular disease (Whelton et al., 2012).

8.5.4.3 Moderation

Despite these concerns, smoked foods can be enjoyed safely in moderation as part of a balanced diet (Haider, personal communication). Choosing lean cuts of meat, avoiding excessive smoking times, and incorporating a variety of other foods can help mitigate potential health risks (Kushi et al., 2006).

8.6 SALTING AND CURING

Salting is another traditional and quite effective way of food preservation. It involves the application of salt to food, which creates an environment that is inhospitable to bacteria, molds, and other microorganisms that cause spoilage (Kaczmarek et al., 2019). The process works through osmosis, where salt draws moisture out of the food and the microorganisms, effectively dehydrating them and inhibiting their growth (Taormina, 2010). This approach not only extends the shelf life of the food but also improves its texture and flavor. Salted pistachios are a well-known example of salting and curing (Figure 8.7). Another example is corned beef, which is made by brining beef in a mixture of water, salt, and spices. The wet curing process gives the beef its characteristic salty flavor and tender texture, making it a popular choice for dishes like corned beef and cabbage.

8.6.1 Methods of Curing

Curing is a broader category that includes salting but also incorporates other processes, like smoking and drying to preserve food. There are several methods of curing, each with its specific techniques and applications.

8.6.1.1 Dry Curing

Dry curing is a method where salt, often mixed with sugar and spices, is rubbed directly onto the surface of the food (Rysová and Šmídová, 2021). The food is then left to cure for a period from several days to weeks depending on the type of food and the desired outcome. This method is commonly used for meats. The salt draws moisture out of the food, creating a dry environment that prevents bacterial growth and extends the shelf life of the product.

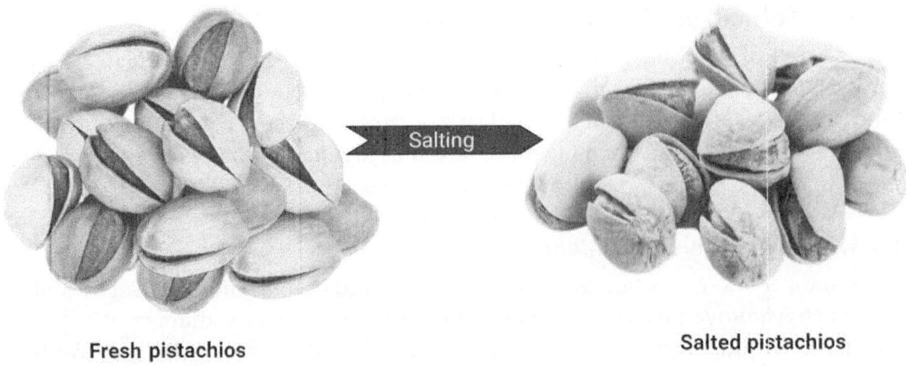

FIGURE 8.7 Preservation of freshly harvested pistachios through salting.

8.6.1.2 Wet Curing

Wet curing, also known as brining, involves immersing the food in a solution of water, salt, and sometimes sugar, spices, and other flavorings (Joardder and Masud, 2019). The food absorbs the salt from the brine, which helps preserve it while also imparting flavor. Wet curing is typically faster than dry curing and is used for products like pickles, corned beef, and certain types of fish (Sindelar and Houser, 2009). The brining process can last from several hours to several days.

8.6.1.3 Combination Curing

Combination curing is a method that uses both dry and wet curing techniques (Hall, 2010). It typically involves rubbing the food with a dry cure mixture and then submerging it in brine (Martin, 2001). This method is often used for large cuts of meat that require deep penetration of the curing agents. Combination curing ensures thorough preservation and flavor distribution, making it ideal for products like meats and large fish.

8.6.2 Commonly Salted and Cured Foods

Salting and curing are applied to a wide variety of foods, with some of the most common including the following.

8.6.2.1 Vegetables

Some vegetables, like cucumbers and cabbages, are salted or brined as part of the pickling process, which preserves them and adds a tangy flavor. Sauerkraut is an example of cured vegetable.

8.6.2.2 Fish

Fish such as salmon is often salted or cured to extend their shelf life and enhance their flavor.

8.6.2.3 Cheese

Certain cheeses are cured with salt to develop their flavor and texture. Examples include feta and blue cheese, which are often stored in brine.

8.6.3 Advantages

- Salting and curing significantly extend the shelf life of perishable foods, making them suitable for long-term storage without refrigeration.
- Imparts unique and complex flavors to the food, which are highly valued in many culinary traditions.
- Curing can be applied to a wide range of foods, from meats and fish to vegetables and cheeses, allowing for diverse culinary applications.
- Many cured foods are integral to specific cultural cuisines, contributing to regional diets and food traditions.

8.6.4 Limitations

- If not carefully monitored, the curing process can lead to overly salty or tough textures, which may be undesirable in some cases.
- Some nutrients, particularly water-soluble vitamins like vitamin C, can be lost during the salting and curing process.
- Curing, especially dry curing, can be a time-consuming process that requires careful attention and patience.

8.7 FERMENTATION

Fermentation is a natural metabolic process where microorganisms such as bacteria, yeast, or fungi convert organic compounds, typically sugars, into other substances, like acids, gases, or alcohol (Binood et al., 2013). This process occurs under anaerobic conditions (in the absence of oxygen) and is used to store food and improve its flavor and nutritional profile. Fermentation creates an environment that inhibits the growth of spoilage-causing microorganisms while encouraging the proliferation of beneficial microbes. These beneficial microbes contribute to the safety, taste, and shelf life of fermented foods. Yogurt is a popular product produced through lactic acid fermentation of milk by bacteria such as *Lactobacillus bulgaricus* and *Streptococcus thermophilus* (Figure 8.8). The process thickens the milk and makes a nutritious dairy product.

8.7.1 Methods of Fermentation

8.7.1.1 Acetic Acid Fermentation

Acetic acid fermentation is the process by which ethanol is converted into acetic acid by acetic acid bacteria (AAB). This method is used to produce vinegar from alcoholic liquids like wine, cider, or beer. The acetic acid gives vinegar its characteristic sharp flavor and serves as a natural preservative. The fermentation process involves

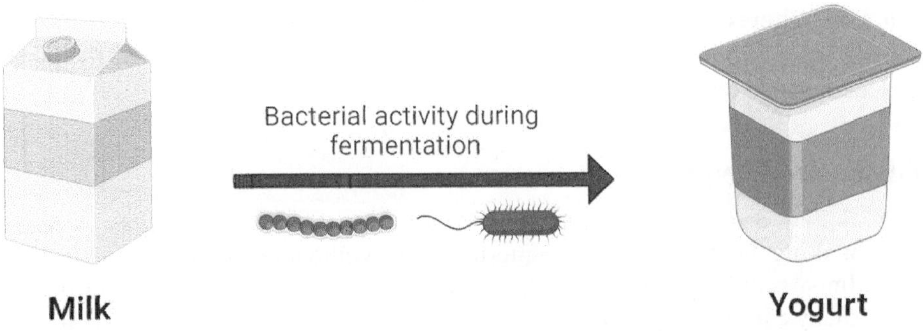

FIGURE 8.8 A visual illustration of milk fermentation into fresh yogurt.

exposing the alcohol to oxygen, which allows AAB to oxidize the ethanol into acetic acid.

8.7.1.2 Alcoholic Fermentation

Alcoholic fermentation is carried out by yeast, particularly the species *Saccharomyces cerevisiae*, which converts sugars into ethanol (alcohol) and carbon dioxide (Walker and Stewart, 2016). This process is the basis for producing alcoholic beverages like beer, wine, and spirits (Benito, 2020). In bread-making, alcoholic fermentation occurs when yeast ferments the sugars in the dough, producing carbon dioxide, which causes the dough to rise, and a small amount of alcohol that evaporates during baking (Heitmann et al., 2018).

8.7.1.3 Lactic Acid Fermentation

Lactic acid fermentation is one of the most common forms of fermentation, where LAB convert sugars into lactic acid (Von Wright and Axelsson, 2019). This type of fermentation is responsible for the sour taste in foods like yogurt, sauerkraut, kimchi, and pickles (Mani, 2018). The lactic acid slows down the rate of deterioration of food as the food becomes more acidic and therefore unfit for proliferation of hazardous bacteria. Lactic acid fermentation is also used in the production of sourdough bread, where it enhances the flavor and texture of the dough.

8.7.2 Commonly Fermented Foods

Fermentation is used to produce a wide variety of foods across different cultures. Some of the most common examples include the following.

8.7.2.1 Grains

Sourdough bread and certain fermented grain porridges or beverages are produced through lactic acid fermentation, which improves their flavor and nutritional content.

8.7.2.2 Vegetables

Sauerkraut, kimchi, and pickles are popular examples of fermented vegetables. The lactic acid fermentation process preserves the vegetables and gives them their tangy flavor.

8.7.2.3 Dairy Products

Yogurt, kefir, and certain cheeses are produced through lactic acid fermentation. The process enhances the flavor, texture, and digestibility of dairy products.

8.7.2.4 Beverages

Beer, wine, kombucha, and cider are produced through alcoholic and acetic acid fermentation. These beverages are enjoyed for their unique flavors and potential health benefits.

8.7.3 Advantages

- Extends the shelf life of perishable foods by creating an acidic or alcoholic environment that inhibits the growth of harmful bacteria and molds.
- Produces complex flavors and aromas that are highly valued in many culinary traditions.
- Increases the bioavailability of nutrients, produces beneficial probiotics, and reduces antinutritional factors in foods.
- Fermented foods are often easier to digest because the fermentation process breaks down complex carbohydrates, proteins, and fats.

8.7.4 Limitations

- *Fermentation* is a natural process that can be difficult to control, leading to variations in flavor, texture, and safety. Improper fermentation can result in spoilage or the growth of harmful microorganisms.
- The fermentation process can take days or weeks, requiring patience and careful monitoring.
- Fermentation requires specific conditions, such as temperature, pH, and the presence of certain microorganisms, making it challenging to achieve consistent results without proper knowledge and equipment.

8.8 MODIFIED ATMOSPHERE PACKAGING

Modified atmosphere packaging (MAP) is a food preservation technique that involves altering the atmosphere surrounding the food inside a package to extend its shelf life (Kargwal et al., 2020). The principle behind MAP is to reduce the oxygen content and increase the levels of gases like carbon dioxide and nitrogen within the package (Priyadarshi et al., 2020). This altered atmosphere slows down the growth of aerobic microorganisms (those that require oxygen) and delays the ripening, spoilage, and oxidation processes in foods. By modifying the internal atmosphere, MAP helps maintain the quality, freshness, and safety of perishable foods over a longer period (Figure 8.9).

8.8.1 Methods of Modified Atmosphere Packaging

8.8.1.1 Vacuum Packaging

Vacuum packaging involves removing air from the package before sealing it. This method significantly reduces the oxygen content around the food, creating a

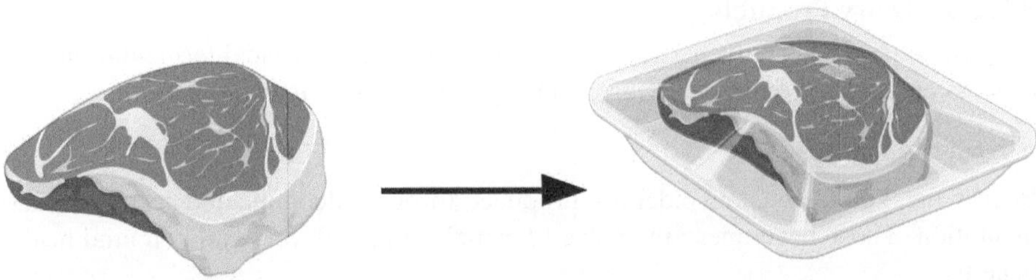

FIGURE 8.9 The use of modified atmosphere packaging in the preservation of meat.

near-vacuum environment (Mullan and McDowell, 2003). The absence of oxygen inhibits the growth of aerobic bacteria and molds, which are primary causes of spoilage (Taniwaki et al., 2009). Vacuum packaging is widely used for meats, cheeses, and other perishable products to extend their shelf life. The vacuum-sealed packaging also helps retain the moisture and flavor of the food.

8.8.1.2 Gas Flushing

Gas flushing, also known as gas replacement, involves replacing the air inside the package with a specific gas mixture, usually a combination of carbon dioxide, nitrogen, and sometimes oxygen (Lee et al., 2022). The choice of gases depends on the type of food being preserved. For example, a high concentration of carbon dioxide is used to inhibit the growth of microorganisms, while nitrogen acts as an inert filler that displaces oxygen and prevents oxidation. Gas flushing is commonly used for fresh produce, baked goods, and snacks to maintain freshness and prevent spoilage.

8.8.1.3 Barrier Packaging

Barrier packaging refers to the use of materials that prevent or slow down the transmission of gases, moisture, and other environmental factors into and out of the package (Rooney, 1995). These materials include plastic films, laminates, and coated papers that act as barriers to oxygen, carbon dioxide, and water vapor (Tyagi et al., 2021). Barrier packaging is often combined with vacuum or gas flushing methods to create an optimal environment for preserving food (Ahn and Min, 2007). It is used for products like coffee, snacks, and ready-to-eat meals where maintaining the integrity of the modified atmosphere is crucial for extending shelf life.

8.8.2 Mechanism of Action

The effectiveness of MAP relies on its ability to control the internal gas composition of the package (Boz et al., 2018). By reducing the oxygen levels and increasing carbon dioxide and nitrogen concentrations, MAP creates an environment that is unfavorable for the growth of aerobic microorganisms and slows down the metabolic processes of the food (Czerwiński et al., 2021). For example, in fresh fruits and vegetables, MAP reduces the respiration rate, delaying ripening and senescence. In meats, lowering oxygen levels prevents oxidation, which can cause rancidity and color changes. The specific gas mixture used in MAP is tailored to the type of food and its preservation needs, ensuring that the food remains fresh, safe, and appealing for a longer time.

8.8.3 Applications of Modified Atmosphere Packaging in Food Preservation

Modified atmosphere packaging is used across various food categories to extend shelf life, enhance safety, and maintain quality:

8.8.3.1 Fruits and Vegetables

Fruits and vegetables are commonly packaged using MAP to slow down respiration and prevent spoilage, allowing for longer storage and transportation times.

8.8.3.2 Dairy Products

Cheeses and dairy-based products are often vacuum-packed or gas-flushed to prevent spoilage and extend shelf life.

8.8.3.3 Bakery Products

Breads, cakes, and pastries benefit from MAP by maintaining freshness and preventing mold growth, especially when gas flushing is used to replace oxygen with carbon dioxide.

8.8.3.4 Meat and Poultry

MAP is used to preserve fresh meats and poultry by inhibiting microbial growth and oxidation, which helps retain color, texture, and flavor.

8.8.3.5 Ready-to-Eat Meals

Prepared meals, salads, and sandwiches are packaged using MAP to keep them fresh and safe for consumption over several days.

8.8.4 Advantages

- MAP significantly prolongs the shelf life of perishable foods by creating an environment that inhibits spoilage and microbial growth.
- By controlling oxygen levels and other gases, MAP helps preserve the texture, flavor, and nutritional value of foods, ensuring they remain appealing to consumers.
- A longer shelf life leads to less food spoilage and waste, which is beneficial for both consumers and retailers.
- MAP can be used to preserve a wide variety of food commodities, including fresh vegetables, dairy, meat, and baked goods, making it a versatile and extensively adopted preservation technique.

8.8.5 Limitations

- The equipment and materials required for MAP can be expensive, making it less accessible for small-scale producers.
- The process of modifying the atmosphere and selecting the appropriate gas mixture can be complex and requires specialized knowledge and technology.
- While MAP extends shelf life, it does not prevent spoilage indefinitely. Over time, the protective atmosphere may degrade, leading to reduced effectiveness.
- The use of specialized packaging materials, such as plastics and laminates, can contribute to environmental waste if not properly managed and recycled.

8.9 REFRIGERATION AND FREEZING

Refrigeration and freezing are two fundamental food storage techniques that work by minimizing the food temperature, thereby slowing down the activity of microbes and the rate of chemical reactions that lead to spoilage (Amit et al., 2017). Refrigeration typically involves cooling food to temperatures between 0 and 4°C (32–39°F), which slows down bacterial growth without freezing the food (Chaves and Zaritzky, 2018). Whereas freezing is the process of decreasing the food temperature beyond its freezing point, generally to −18°C (0°F) or lower, which not only halts bacterial growth but also significantly slows down enzymatic and oxidative reactions (Rehman et al., 2021).

In refrigeration, the cold temperature slows down the metabolic rates of microorganisms, reducing their ability to multiply and cause spoilage. In freezing, the water content in the food turns into ice, making it unavailable for microorganisms and enzymatic reactions, thus preserving the food for extended periods. The use of household refrigerators and freezers is a common example of preserving various kinds of food products including fruits, vegetables, milk, and meat (Figure 8.10).

8.9.1 Methods of Refrigeration and Freezing

8.9.1.1 Domestic Refrigeration

Domestic refrigeration involves the cooling and freezing of food in household fridges and freezers. These appliances are capable of maintaining a low temperature to preserve perishable produce for an extended period. The cold storage is usually in the range of 0–5°C, while the freezing capacity functions at temperatures below −18°C (Chen et al., 2021). Domestic refrigeration is ideal for preserving fresh produce, dairy products, meats, and leftovers, ensuring that they remain safe to eat and retain their nutritional value.

8.9.1.2 Cryogenic Freezing

Cryogenic freezing is a method that involves freezing food at extremely low temperatures using cryogenic gases, such as liquid nitrogen or carbon dioxide (Goswami, 2010). The food is exposed to these gases, which rapidly lower the temperature to as

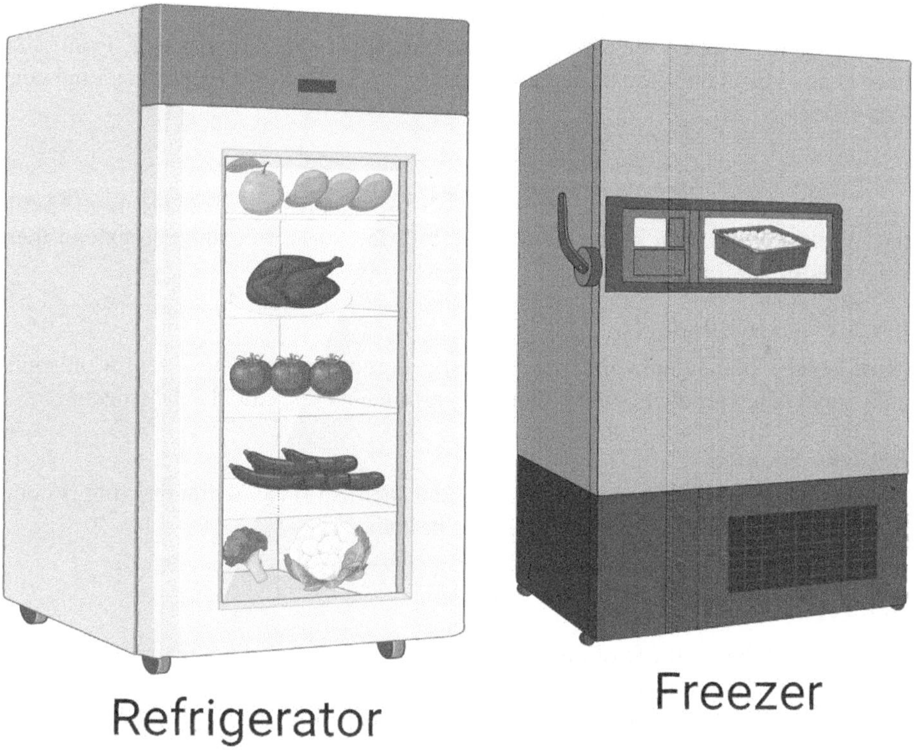

FIGURE 8.10 The use of refrigerator and freezer to preserve food.

low as −196°C (−320°F) (Sanjeev and Ramesh, 2006). Cryogenic freezing is highly effective in preserving the texture, flavor, and nutritional content of food because it prevents the formation of large ice crystals (James et al., 2015). This method is often used for high-value products like gourmet foods, delicate seafood, and specialty items that require exceptional quality preservation.

8.9.1.3 Commercial Freezing

Commercial freezing is used on a larger scale in the food industry to preserve food products for extended periods (Zhao et al., 2021). This method involves freezing food quickly at very low temperatures, often using blast freezers, which circulate cold air rapidly around the food. Quick freezing helps to minimize the formation of large ice crystals, which can damage the texture and quality of food (Zhu et al., 2019). Commercial freezing is commonly used for meats, seafood, vegetables, fruits, and ready-to-eat meals, allowing them to be stored and transported over long distances without compromising quality.

8.9.2 Commonly Preserved Foods by Refrigeration and Freezing

Refrigeration and freezing are used to preserve a wide range of foods, ensuring they remain safe and nutritious over time.

8.9.2.1 Fruits and Vegetables

Fresh produce is often refrigerated to maintain its crispness and flavor. Fruits like berries and vegetables like peas are frequently frozen to retain their nutritional content and taste.

8.9.2.2 Meat and Poultry

Both refrigeration and freezing are commonly used to preserve fresh meats and poultry. While refrigeration keeps them fresh for a few days, freezing can extend their shelf life for several months.

8.9.2.3 Dairy Products

Milk, cheese, butter, and yogurt are typically stored in refrigerators to prevent spoilage. Some dairy products, like butter and certain cheeses, can also be frozen.

8.9.2.4 Seafood

Refrigeration is essential for keeping fish and shellfish fresh, while freezing is commonly used to preserve seafood for longer periods.

8.9.3 Advantages

- Extend the shelf life of perishable foods, reducing food waste and ensuring a consistent food supply.
- Preserve the nutritional content of foods, including vitamins and minerals, by halting enzymatic activity.
- Provide a convenient way to store food, allowing for bulk purchasing and meal preparation in advance.
- Effectively slow down or stop the growth of harmful bacteria, reducing the risk of foodborne illnesses.

8.9.4 Limitations

- Refrigeration and freezing require a constant energy supply, making them costly and environmentally impactful, especially in regions with limited access to electricity.
- Freezing can alter the texture of certain foods, particularly those with high water content, leading to a loss of quality upon thawing.
- While freezing preserves most nutrients, some, particularly water-soluble vitamins, like vitamin C, can be lost during the freezing process.
- Both refrigeration and freezing require significant storage space, which can be a limitation for households or businesses with limited capacity.

8.10 SUGARING

Sugaring is a traditional method of food preservation that involves using sugar to inhibit the growth of microorganisms (Tucker, 2011). Sugar acts as a preservative by creating an environment with high osmotic pressure, which draws water out of

FIGURE 8.11 Food preservation through sugaring.

the cells of microorganisms through a process called osmosis (Rastogi, 2020). This dehydration prevents bacteria, yeast, and molds from growing and spoiling the food. The principle behind sugaring is similar to salting, where a high concentration of solute (in this case, sugar) helps preserve food by reducing the available water content, which is essential for microbial activity. Sugaring is commonly used to preserve fruits, as the sweet taste of sugar complements the natural flavors of these foods. It also helps in retaining the color and texture of the preserved food, making it a popular choice for jams, jellies, and candied fruits (Figure 8.11).

8.10.1 Methods of Sugaring

8.10.1.1 Dry Sugaring

Dry sugaring involves coating or mixing food with granulated sugar or powdered sugar (Broomfield, 1996). This method is often used for preserving fruits like strawberries, blueberries, and cherries. The food is either rolled in sugar or layered with sugar in a container. Over time, the sugar draws out moisture from the food, creating a syrup that further aids in preservation (Ergun et al., 2010). The food is then stored in an airtight container to maintain its quality. Dry sugaring can also be used to make candied fruits, where the fruit is slowly dried and crystallized with sugar.

8.10.1.2 Sugar Syrup Preservation

Sugar syrup preservation involves immersing food, typically fruits, in a concentrated sugar solution or syrup (Rux et al., 2017). The syrup is usually made by dissolving sugar in water and sometimes adding lemon juice or other flavorings. The food is cooked in the syrup, which helps preserve it by creating a barrier against microbial growth (Ahmed et al., 2016). This method is commonly used for making jams, jellies, marmalades, and fruit preserves. The high sugar content, combined with the cooking process, ensures that the food is preserved for an extended period, often in sterilized jars, to further enhance shelf life.

8.10.2 Commonly Sugared Foods

Sugaring is primarily used to preserve fruits, but it can also be applied to certain vegetables and even some meats. Common examples include the following.

8.10.2.1 Fruits

Strawberries, cherries, apricots, and plums are often sugared to make jams, jellies, and preserves. These fruits are selected for their natural sweetness, which is enhanced by the sugaring process.

8.10.2.2 Candied Fruits

Fruits like citrus peels, ginger, and pineapple are commonly candied by boiling in sugar syrup and then drying to create a sweet, chewy treat that can be stored for long periods.

8.10.2.3 Confectioneries

Sugaring is used to make various confectioneries, such as fruit preserves, sugar-coated nuts, and crystallized ginger, all of which are popular for their extended shelf life and rich flavor.

8.10.3 Advantages

- Sugaring can be used for a wide range of foods, from fruits and vegetables to nuts and certain meats, offering flexibility in preservation techniques.
- Extends the shelf life of perishable foods by reducing moisture content, which inhibits microbial growth.
- Enhances the natural sweetness of food, making it a popular method for creating sweet preserves, candies, and desserts.
- Retains the vibrant color and texture of fruits, making them visually appealing and enjoyable to eat even after preservation.

8.10.4 Limitations

- Foods preserved by sugaring tend to have a high sugar content, which may not be suitable for individuals with dietary restrictions, such as diabetes, or those seeking to reduce sugar intake.

- The sugaring process, particularly when combined with cooking, can lead to the loss of heat-sensitive nutrients, like vitamin C.
- Sugared foods are often calorie-dense, which can contribute to weight gain if consumed in large quantities.
- In some cases, sugaring can alter the texture of food, making it softer or chewier, which may not be desirable for all types of food.

8.11 CONCLUSIONS AND FUTURE RECOMMENDATION

Traditional methods of food preservation are important to retain quality and prolong the shelf life of perishable and non-perishable commodities. The traditional base in these techniques will continue to be upgraded by modern needs and requirements toward achieving food security and reducing losses. In fact, these methods should be upgraded with advanced technologies in the near future. Further research in the future may be related to increasing efficiency and sustainability without affecting nutritional value. Integrating these techniques with emerging preservative technologies could lead to even more efficient food preservation solutions.

REFERENCES

Adegoke, G.O. and A.A. Olapade. 2012. Preservation of plant and animal foods: An overview. In *Progress in food preservation*, eds. R. Bhat, A.K. Alias and G. Paliyath, 603–611. Wiley-Blackwell, Chichester, West Sussex, UK. [Google Scholar]

Ahmed, I., I.M. Qazi and S. Jamal. 2016. Developments in osmotic dehydration technique for the preservation of fruits and vegetables. *Innovative Food Science and Emerging Technologies* 34:29–43. https://doi.org/10.1016/j.ifset.2016.01.003

Ahn, D.U. and B. Min. 2007. Packaging and storage. In *Handbook of Fermented Meat and Poultry*, ed. T. Fidel, 289–300. Blackwell, Oxford, UK. [Google Scholar]

Alan, Y. 2019. Culture fermentation of Lactobacillus in traditional pickled gherkins: Microbial development, chemical, biogenic amine and metabolite analysis. *Journal of Food Science and Technology* 56:3930–3939. https://doi.org/10.1007/s13197-019-03866-8

Amit, S.K., M.M. Uddin, R. Rahman, S.M.R. Islam and M.S. Khan. 2017. A review on mechanisms and commercial aspects of food preservation and processing. *Agriculture and Food Security* 6:1–22. https://doi.org/10.1186/s40066-017-0130-8

Arvanitoyannis, I.S. and K.V. Kotsanopoulos. 2011. Smoking of fish and seafood: History, methods and effects on physical, nutritional and microbiological properties. *Food and Bioprocess Technology* 5:831–853. https://doi.org/10.1007/s11947-011-0690-8

Beena Divya, J., K. Kulangara Varsha, K. Madhavan Nampoothiri, B. Ismail and A. Pandey. 2012. Probiotic fermented foods for health benefits. Engineering in Life Sciences 12:377–390. https://doi.org/10.1002/elsc.201100179

Behera, S.S., A.F. El Sheikha, R. Hammami and A. Kumar. 2020. Traditionally fermented pickles: How the microbial diversity associated with their nutritional and health benefits? *Journal of Functional Foods* 70:103971. https://doi.org/10.1016/j.jff.2020.103971

Belichovska, K., D. Belichovska and Z. Pejkovski. 2019. Smoke and smoked fish production. *Meat Technology* 60:37–43. https://doi.org/10.18485/meattech.2019.60.1.6

Benito, S. 2020. Combined use of *Lachanceathermotolerans* and *Schizosaccharomyces pombe* in winemaking: A review. *Microorganisms 8*:655. https://doi.org/10.3390/microorganisms8050655

Benyam, A.A., T. Soma and E. Fraser. 2021. Digital agricultural technologies for food loss and waste prevention and reduction: Global trends, adoption opportunities and barriers. *Journal of Cleaner Production* 323:129099. https://doi.org/10.1016/j.jclepro.2021.129099

Binood, P., R. Sindhu and A. Pandey. 2013. The alcohol fermentation step: The most common ethanologenic microorganisms among yeasts, bacteria and filamentous fungi. In *Lignocellulose Conversion*, ed. V. Faraco, 131–149. Springer, Berlin/Heidelberg, Germany. [Google Scholar]

Boz, Z., B.A. Welt, J.K. Brecht et al. 2018. Review of challenges and advances in modification of food package headspace gases. *Journal of Applied Packaging Research* 10:5.[Google Scholar]

Bramsiepe, C., S. Sievers, T. Seifert et al. 2012. Low-cost small scale processing technologies for production applications in various environments–Mass produced factories. *Chemical Engineering and Processing* 51:32–52. https://doi.org/10.1016/j.cep.2011.08.005

Broomfield, R.W. 1996. The manufacture of preserves, flavourings and dried fruits. In *Fruit Processing*, 165–195. Springer International Publishing, Boston, MA, USA. [Google Scholar]

Bulanda, S. and B. Janoszka. 2022. Consumption of thermally processed meat containing carcinogenic compounds (polycyclic aromatic hydrocarbons and heterocyclic aromatic amines) versus a risk of some cancers in humans and the possibility of reducing their formation by natural food additives–A literature review. *International Journal of Environmental Research and Public Health* 19:4781. https://doi.org/10.3390/ijerph19084781

Chan, H.Y., S.B. Riffat and J. Zhu. 2010. Review of passive solar heating and cooling technologies. *Renewable and Sustainable Energy Reviews* 14:781–789. https://doi.org/10.1016/j.rser.2009.10.030

Chaves, A. and N. Zaritzky. 2018. Cooling and freezing of fruits and fruit products. In *Fruit Preservation*, eds. A. Rosenthal, R. Deliza, J. Welti-Chanes and G.V. Barbosa-Cánovas, 127–180. Springer, Berlin/Heidelberg, Germany. [Google Scholar]

Chen, P. 2021. Lactic acid bacteria in fermented food. In *Advances in Probiotics*, eds. D. Dhanasekaran and A. Sankaranarayanan, 397–416. Elsevier, Amsterdam, The Netherlands. [Google Scholar]

Chen, X., J. Wu, X. Cai and S. Wang. 2021. Production, structure-function relationships, mechanisms, and applications of antifreeze peptides. *Comprehensive Reviews in Food Science and Food Safety* 20:542–62. https://doi.org/10.1111/1541-4337.12655

Chua, K.J. and S.K. Chou. 2003. Low-cost drying methods for developing countries. *Trends in Food Science and Technology* 14:519–528. https://doi.org/10.1016/j.tifs.2003.07.003

Czerwiński, K., T. Rydzkowski, J. Wróblewska-Krepsztul and V.K. Thakur. 2021. Towards impact of modified atmosphere packaging (MAP) on shelf-life of polymer-film-packed food products: Challenges and sustainable developments. *Coatings* 11:1504. https://doi.org/10.3390/coatings11121504

Dar, A.H., N. Kumar, S. Shah, R. Shams and M.B. Aga. 2022. Processing of fruits and vegetables. In *Agro-Processing and Food Engineering*, 535–579. Springer International Publishing, New York, USA. [Google Scholar]

Ergun, R., R. Lietha and R.W. Hartel. 2010. Moisture and shelf life in sugar confections. *Critical Reviews in Food Science and Nutrition* 50(2):162–192. https://doi.org/10.1080/10408390802248833

FAOSTAT. 2019. *Food and Agriculture Organization of the United Nations Database*. Available at: http://faostat.fao.org/.

Giudici, P., L. De Vero and M. Gullo. 2017. Vinegars. In *Acetic Acid Bacteria: Fundamentals and Food Applications*, 1st ed., ed. I.Y. Sengun, 261–287. CRC Press, Boca Raton, FL, USA. [Google Scholar]

Goswami, T. 2010. Role of cryogenics in food processing and preservation. *International Journal of Food Engineering* 6:15. https://doi.org/10.2202/1556-3758.1771

Haider, M.W., M. Nafees, M. Valipour, H.U. Asad and R.A. Marc. 2025. Ultraviolet (UV) light technology for postharvest fruits and vegetables. In *Sustainable Postharvest*

Technologies for Fruits and Vegetables, eds. S. Ali, S.A. Mir, B.N. Dar and S. Ejaz, 137–148. CRC Press, Boca Raton, FL, USA. [Google Scholar]

Hall, G.M. 2010. Preservation by curing (drying, salting and smoking). In *Fish Processing*, ed. G.M. Hall, 51–76. Wiley-Blackwell, Chichester, West Sussex, UK. https://doi.org/10.1002/9781444328585.ch3

Hasan, M.U., A.U. Malik, R. Anwar, A.S. Khan, M.W. Haider, R. Riaz, S. Ali, R.N.U. Rehman and K. Ziaf. 2021. Postharvest Aloe vera gel coating application maintains the quality of harvested green chillies during cold storage. *Journal of Food Biochemistry* 45:13682. https://doi.org/10.1111/jfbc.13682

Heitmann, M., E. Zannini and E. Arendt. 2018. Impact of *Saccharomyces cerevisiae* metabolites produced during fermentation on bread quality parameters: A review. *Critical Reviews in Food Science and Nutrition* 58:1152–1164. https://doi.org/10.1080/10408398.2016.1244153

Inyang, U., I. Oboh and B. Etuk. 2017. Drying and the different techniques. *International Journal of Food Nutrition and Safety* 8:45–72.[Google Scholar]

Ishangulyyev, S., S. Kim and S.H. Lee. 2019. Understanding food loss and waste-why are we losing and wasting food? *Foods* 8:297. https://doi.org/10.3390/foods8080297

James, C., G. Purnell and S.J. James. 2015. A review of novel and innovative food freezing technologies. *Food and Bioprocess Technology* 8:1616–1634. https://doi.org/10.1007/s11947-015-1542-8

Janjai, S. and B.K. Bala. 2012. Solar drying technology. *Food Engineering Reviews* 4:16–54. https://doi.org/10.1007/s12393-011-9044-6

Joardder, M.U.H. and M.H. Masud. 2019. Food preservation techniques in developing countries. In *Food Preservation in Developing Countries: Challenges and Solutions*, 67–125. Springer International Publishing, New York, USA. https://doi.org/10.1007/978-3-030-11530-2_4

Kaczmarek, M., S.V. Avery and I. Singleton. 2019. Microbes associated with fresh produce: Sources, types and methods to reduce spoilage and contamination. *Advances in Applied Microbiology* 107:29–82. https://doi.org/10.1016/bs.aambs.2019.02.001

Kargwal, R., M.K. Garg, V.K. Singh, R. Garg and N. Kumar. 2020. Principles of modified atmosphere packaging for shelf life extension of fruits and vegetables: An overview of storage conditions. *International Journal of Chemical Studies* 8:2245–2252. [Google Scholar]

Kumar, D. and P. Kalita. 2017. Reducing postharvest losses during storage of grain crops to strengthen food security in developing countries. *Foods* 6:8. https://doi.org/10.3390/foods6010008

Kushi, L.H., T. Byers, C. Doyle, E.V. Bandera, M. McCullough, T. Gansler, K.S. Andrews and M.J. Thun. 2006. American Cancer Society Guidelines on Nutrition and Physical Activity for cancer prevention: Reducing the risk of cancer with healthy food choices and physical activity. *CA: A Cancer Journal for Clinicians* 56:254–281. https://doi.org/10.3322/canjclin.56.5.254

Ledesma, E., M. Rendueles and M. Díaz. 2016. Contamination of meat products during smoking by polycyclic aromatic hydrocarbons: Processes and prevention. *Food Control* 60:64–87. https://doi.org/10.1016/j.foodcont.2015.07.016

Ledesma, E., M. Rendueles and M. Díaz. 2017. Smoked food. In *Current Developments in Biotechnology and Bioengineering*, eds. A. Pandey, S. Negi and C.R. Soccol, 201–243. Elsevier, Amsterdam, The Netherlands. [Google Scholar]

Lee, D.S., H.J. Wang, C. Jaisan and D.S. An. 2022. Active food packaging to control carbon dioxide. *Packaging Technology and Science* 35:213–227. [Google Scholar]

Lee, S. Y and D.H. Kang. 2004. Microbial safety of pickled fruits and vegetables and hurdle technology. *Internet Journal of Food Safety* 4:21–32.[Google Scholar]

Lingayat, A., R. Balijepalli and V. Chandramohan. 2021. Applications of solar energy based drying technologies in various industries–A review. *Solar Energy* 229:52–68. https://doi.org/10.1016/j.solener.2021.05.058

Løvdal, T. 2015. The microbiology of cold smoked salmon. *Food Control* 54:360–373. https://doi.org/10.1016/j.foodcont.2015.02.025

Mani, A. 2018. Food preservation by fermentation and fermented food products. *International Journal of Academic Research and Development* 1:51–57. [Google Scholar]

Martin, M. 2001. Meat curing technology. In *Meat Science and Applications*, eds. Y.H. Hui, W.K. Nip, R.W. Rogers and A.O. Young, 491–508. Marcel Dekker, New York, USA. [Google Scholar]

Mei, J., X. Ma and J. Xie. 2019. Review on natural preservatives for extending fish shelf life. *Foods* 8:490. https://doi.org/10.3390/foods8100490

Miranda, G., A. Berna and A. Mulet. 2019. Dried-fruit storage: An analysis of package headspace atmosphere changes. *Foods* 8:56. https://doi.org/10.3390/foods8020056

Mishra, B., J. Mishra, P. Pati and P. Rath. 2017. Dehydrated meat products: A review. *International Journal of Livestock Research* 7:10–22. [Google Scholar]

Moses, J.A., T. Norton, K. Alagusundaram and B. Tiwari. 2014. Novel drying techniques for the food industry. *Food Engineering Reviews* 6:43–55. https://doi.org/10.1007/s12393-014-9078-7

Mullan, M. and D. McDowell. 2003. Modified atmosphere packaging. In *Food Packaging Technology*, eds. R. Coles, D. McDowell and M. Kirwan, 303–324. Blackwell Publishing, London, UK. [Google Scholar]

Musielak, G., D. Mierzwa and J. Kroehnke. 2016. Food drying enhancement by ultrasound–A review. *Trends in Food Science and Technology* 56:126–141. https://doi.org/10.1016/j.tifs.2016.08.003

Naczk, M. and A.S. Artyukhova. 2020. Canning. In *Seafood*, ed. Z.E. Sikorski, 181–198. CRC Press, Boca Raton, FL, USA. [Google Scholar]

Nizio, E., K. Czwartkowski and G. Niedbała. 2023. Impact of smoking technology on the quality of food products: Absorption of polycyclic aromatic hydrocarbons (PAHs) by food products during smoking. *Sustainability* 15:16890. https://doi.org/10.3390/su152416890

O'Connor, J., S.A. Hoang, L. Bradney, S. Dutta, X. Xiong, D.C. Tsang, K. Ramadass, A. Vinu, M.B. Kirkham and N.S. Bolan. 2021. A review on the valorisation of food waste as a nutrient source and soil amendment. *Environmental Pollution* 272:115985. https://doi.org/10.1016/j.envpol.2020.115985

Owusu-Apenten, R. and E. Vieira. 2022. Thermal processing and canning. In *Elementary Food Science*, ed. V. Richard, 253–288.Cham, Springer International Publishing, New York, USA. https://doi.org/10.1007/978-3-030-65433-7

Peréz-Dıaz, I.M., F. Breidt, R.W. Buescher, F.N. Arroyo-López, R. Jiménez-Dıaz, A. Garrido-Fernández and S. Johanningsmeire. 2013. Fermented and acidified vegetables. In *Compendium of Methods for the Microbiological Examination of Foods*, 4th ed., eds. Y. Salfinger and M.L. Tortorello, 521–532.American Public Health Association, Washington, DC, USA. [Google Scholar]

Porat, R., A. Lichter, L.A. Terry, R. Harker and J. Buzby. 2018. Postharvest losses of fruit and vegetables during retail and in consumers' homes: Quantifications, causes, and means of prevention. *Postharvest Biology and Technology* 139:135–149. https://doi.org/10.1016/j.postharvbio.2017.11.019

Priyadarshi, R., F. Deeba, Sauraj and Y.S. Negi. 2020. Modified atmosphere packaging development. In *Processing and Development of Polysaccharide-Based Biopolymers for Packaging Applications*, ed. Y. Zhang, 261–280. Elsevier, Amsterdam, The Netherlands. [Google Scholar]

Qiu, L., M. Zhang, J. Tang, B. Adhikari and P. Cao. 2019. Innovative technologies for producing and preserving intermediate moisture foods: A review. *Food Research International* 116:90–102. https://doi.org/10.1016/j.foodres.2018.12.055

Rajput, H., D. Goswami, M. Arya and A. Randhawa. 2022. Technology for canning. *Global Hi-Tech Horticulture* 6:135–151.[Google Scholar]

Ramesh, M.N. 2007. Canning and sterilization of foods. In *Handbook of Food Preservation*, 2nd ed., ed. M.S. Rahman, 584–587. CRC Press, Boca Ratón, FL, USA. [Google Scholar]

Ramesh, M.N. 2020. Canning and Sterilization of Foods. In *Handbook of Food Preservation*, 609–636. CRC Press, Boca Raton, FL, USA. [Google Scholar]

Rastogi, N.K. 2020. 1-Forward osmosis: Principles, applications, and recent developments. In *Current Trends and Future Developments on (Bio-) Membranes*, eds. A. Basile, A. Cassano, N.K. Rastogi, 3–35. Elsevier, Amsterdam, Netherlands. [Google Scholar]

Rawat, S. 2015. Food Spoilage: Microorganisms and their prevention. *Asian Journal of Plant Science and Research* 5:47–56. [Google Scholar]

Rehman, R.N.U., S. Ali, M.U. Hasan, R. Anwar, M.W. Haider, A.S. Khan, A.U. Malik and L. Pengmin. 2021. Low temperature and hypoxic conditions induce flavonoids biosynthesis and enhances antioxidant potential of crabapple (*Malus profusion*) fruits. *Acta Physiologiae Plantarum* 43:131. https://doi.org/10.1007/s11738-021-03302-5

Rehman, S.U. and J.A. Awan. 2012. Dehydration of fruit and vegetables in tropical regions. In *Progress in Food Preservation*, eds. R. Bhat, A.K. Alias and G. Paliyath, 191–209. Wiley-Blackwell, Chichester, West Sussex, UK. [Google Scholar]

Rooney, M.L. 1995. *Active Food Packaging*. Springer, Boston, MA, USA, p. 260. [Google Scholar]

Rooney, M.L. 2005. Introduction to active food packaging technologies. In *Innovations in Food Packaging*, 63–79. Elsevier Ltd. https://doi.org/10.1016/b978-012311632-1/50037-1

Rux, G., O.J. Caleb, A. Fröhling, W.B. Herppich and P.V. Mahajan. 2017. Respiration and storage quality of fresh-cut apple slices immersed in sugar syrup and orange juice. *Food and Bioprocess Technology* 10:2081–2091. https://doi.org/10.1007/s11947-017-1980-6

Rysová, J. and Z. Šmídová. 2021. Effect of salt content reduction on food processing technology. *Foods* 10:2237. https://doi.org/10.3390/foods10092237

Sanjeev, K. and M.N. Ramesh. 2006. Low oxygen and inert gas processing of foods. *Critical Reviews in Food Science and Nutrition* 46:423–451. https://doi.org/10.1080/10408390500215670

Saravacos, G.D. 2005. Mass transfer properties of foods. In *Engineering Properties of Foods*, eds. M.A. Rao and S.S.H. Rizvi, 89–132. Marcel Dekker Inc., New York, USA. [Google Scholar]

Schneider, F. and M. Eriksson. 2020. Food waste (and loss) at the retail level. In *Routledge Handbook of Food Waste*. Routledge, Oxfordshire, UK. [Google Scholar]

Sengun, Y.I. and S. Karabiyikli. 2011. Importance of acetic acid bacteria in food industry. *Food Control* 22:647–656. https://doi.org/10.1016/j.foodcont.2010.11.008

Sindelar, J.J. and T.A. Houser. 2009. Alternative curing systems. In *Ingredients in Meat Products: Properties, Functionality and Applications*, ed. R. Tarté, 379–405. Springer, New York, USA. [Google Scholar]

Sionek, B., A. Szydłowska, K. Küçükgöz and D. Kołożyn-Krajewska. 2023. Traditional and new microorganisms in lactic acid fermentation of food. *Fermentation* 9:1019. https://doi.org/10.3390/fermentation9121019

Suleman, R., Z. Wang, R.M. Aadil, T. Hui, D.L. Hopkins and D. Zhang. 2020. Effect of cooking on the nutritive quality, sensory properties and safety of lamb meat: Current challenges and future prospects. *Meat Science* 167:108172. https://doi.org/10.1016/j.meatsci.2020.108172

Sumonsiri, N. and S.A. Barringer. 2014. Fruits and vegetables—processing technologies and applications. In *Food Processing*, 363–381. John Wiley & Sons, Hoboken, NJ, USA. [Google Scholar]

Taniwaki, M.H., A.D. Hocking, J.I. Pitt and G.H. Fleet. 2009. Growth and mycotoxin production by food spoilage fungi under high carbon dioxide and low oxygen atmospheres.

International Journal of Food Microbiology 132:100–108. https://doi.org/10.1016/j.ijfoodmicro.2009.04.005

Taormina, P.J. 2010. Implications of salt and sodium reduction on microbial food safety. *Critical Reviews in Food Science and Nutrition* 50:209–227. https://doi.org/10.1080/10408391003626207

Tucker, G.S. 2011. Food biodeterioration and methods of preservation. In *Food and Beverage Packaging Technology*, eds. R. Coles and M. Kirwan, 31–57. Wiley-Blackwell, Chichester, West Sussex, UK. [Google Scholar]

Tumuluru, J.S. 2023. Introductory chapter: Food processing, preservation, and packaging: A brief overview. *Food Processing and Packaging Technologies-Recent Advances*. https://doi.org/10.5772/intechopen.110229

Tyagi, P., K.S. Salem, M.A. Hubbe and L. Pal. 2021. Advances in barrier coatings and film technologies for achieving sustainable packaging of food products—A review. *Trends in Food Science and Technology* 115:461–485. https://doi.org/10.1016/j.tifs.2021.06.036

Von Wright, A. and L. Axelsson. 2019. Lactic acid bacteria: An introduction. In *Lactic Acid Bacteria: Microbiological and Functional Aspects*, eds. S. Lahtinen, A.C. Ouwehand, S. Salminen and A. von Wright, 1–16. CRC Press, Boca Raton, FL, USA.[Google Scholar]

Walker, G.M. and G.G. Stewart. 2016. *Saccharomyces cerevisiae* in the production of fermented beverages. *Beverages* 2:30. https://doi.org/10.3390/beverages2040030

Wang, T., K.S. Lackner and A. Wright. 2011. Moisture swing sorbent for carbon dioxide capture from ambient air. *Environmental Science and Technology* 45:6670–6675. https://doi.org/10.1021/es201180v

Whelton, P.K., L.J. Appel, R.L. Sacco et al. 2012. Sodium, blood pressure, and cardiovascular disease: Further evidence supporting the American heart association sodium reduction recommendations. *Circulation* 126:2880–2889. https://doi.org/10.1161/CIR.0b013e318279acbf

Xiao, H.W., Z. Pan, L.Z. Deng, H.M. El-Mashad, X.H. Yang, A.S. Mujumdar, Z.J. Gao and Q. Zhang. 2017. Recent developments and trends in thermal blanching—a comprehensive review. *Information Processing in Agriculture* 4:101–127. https://doi.org/10.1016/j.inpa.2017.02.001

Xin, Y., M. Zhang, B. Xu, B. Adhikari and J. Sun. 2015. Research trends in selected blanching pretreatments and quick freezing technologies as applied in fruits and vegetables: A review. *International Journal of Refrigeration* 57:11–25. https://doi.org/10.1016/j.ijrefrig.2015.04.015

Yahia, E.M. and M. Mourad. 2019. Food waste at the consumer level. In *Preventing Food Losses and Waste to Achieve Food Security and Sustainability*, ed. E.M. Yahia, 9–11. Burleigh Dodds Science Publishing, Cambridge, UK. [Google Scholar]

Zambrano, M.V., B. Dutta, D.G. Mercer, H.L. MacLean and M.F. Touchie. 2019. Assessment of moisture content measurement methods of dried food products in small-scale operations in developing countries: A review. *Trends in Food Science and Technology* 88: 484–496. https://doi.org/10.1016/j.tifs.2019.04.006

Zhao, Y., M.J. Powell-Palm, J. Wang, C. Bilbao-Sainz, T. McHugh and B. Rubinsky. 2021. Analysis of global energy savings in the frozen food industry made possible by transitioning from conventional isobaric freezing to isochoric freezing. *Renewable and Sustainable Energy Reviews* 151:111621. https://doi.org/10.1016/j.rser.2021.111621

Zhu, Z., Q. Zhou and D.W. Sun. 2019. Measuring and controlling ice crystallization in frozen foods: A review of recent developments. *Trends in Food Science and Technology* 90:13–25. https://doi.org/10.1016/j.tifs.2019.05.012

9 Recent Developments in Food Preservation Techniques

Milad Tavassoli, Reza Abedi-Firoozjah, Mohammad Hashemi, Ehsan Sadeghi, and Seyyed Mohammad Ali Noori

9.1 INTRODUCTION

Food preservation has always been crucial in human history. Early humans used fermentation, salting, and sun drying. As society progressed, the need for greater amounts and a higher quality of processed food increased, leading to the rise of the food preservation industry. This industry aims to provide economical, nutritious, and satisfying food while reducing food degradation and promoting efficient food utilization through both traditional and modern preservation methods, thereby minimizing waste production and fostering a sustainable food industry (Priyanka, Arun Kumar, Moses, & Anandharamakrishnan, 2024). The FAO estimates that the global value of food lost or wasted amounts to US$1 trillion per year, representing a significant economic loss. By minimizing the production of hazardous substances, wastewater, fossil energy use, and the consumption of water and solvents, new techniques for processing, preservation, and extraction can improve production efficiency while also supporting environmental sustainability. Thermal processing remains a significant aspect of the food processing industry. It is crucial to employ a variety of commercial and innovative preservation techniques, or a combination thereof, to minimize waste and preserve human health (Boateng, 2024; Gyawali, Mahapatra, Bardsley, & Niemira, 2024; Xie et al., 2024).

Developing and implementing more efficient, energy- and resource-saving technologies globally is crucial for meeting the increasing demand for food, reducing environmental impact, and promoting sustainability. The United Nations estimates that around 2 billion tons of globally produced food is lost or wasted for a variety of reasons. Efficient, sustainable food preservation technologies are essential in minimizing food spoilage. Inadequate food safety and quality process programs cost the industry millions of dollars annually and can lead to food losses. Consumers play a significant role in food waste, and secure global food systems are crucial for minimizing all food losses (Lobatón-García, López-Mejía, & Cruz-Guayacundo, 2023).

The realm of food items, including meat and meat products, dairy products, fruits, vegetables, and other foods like flours, sugar, fats, coffee, and grains, is a rich

DOI: 10.1201/9781003540113-9

tapestry woven with a diverse range of nutrients, flavors, and compounds. Before reaching consumers, these products undergo various processing and preservation methods to prepare them for consumption and to extract essential food ingredients. Traditional techniques such as frying, drying, filtering, and cooking have been relied upon for these processes (Allai, Azad, Mir, & Gul, 2023). Despite their benefits, these methods may pose several challenges, such as diminished nutritional value, inefficient production rates, and excessive resource use. In response to these challenges, innovative and sustainable methods have been developed for extraction, processing, and pasteurization (Figure 9.1). These modern methods, like cold plasma, ultrasonication, microwave processing, supercritical fluid extraction, pulse electric field, and controlled pressure drop process, are designed to minimize resource usage and energy consumption (Chiozzi, Agriopoulou, & Varzakas, 2022).

Considerable research efforts have been directed towards these greener food processing techniques, which aim not only to boost production efficiency but also to reduce environmental impact. The ongoing shift towards unconventional food processing underpins the rapidly advancing field of food technology. Adopting alternative approaches for processing, preservation, and extraction could enhance production efficiency and aid in environmental conservation by decreasing water and solvent consumption, lowering energy use, and preventing the creation of hazardous substances. "Green food processing," guided by principles of green chemistry and green engineering, calls for technical processes that emphasize energy and water conservation, facilitate the recycling of food by-products via bio-refining, and ensure the safety and quality of these products (Novickij et al., 2020).

In this chapter, a comprehensive examination of cutting-edge processing techniques at an industrial level is conducted. The focus is on the advanced development

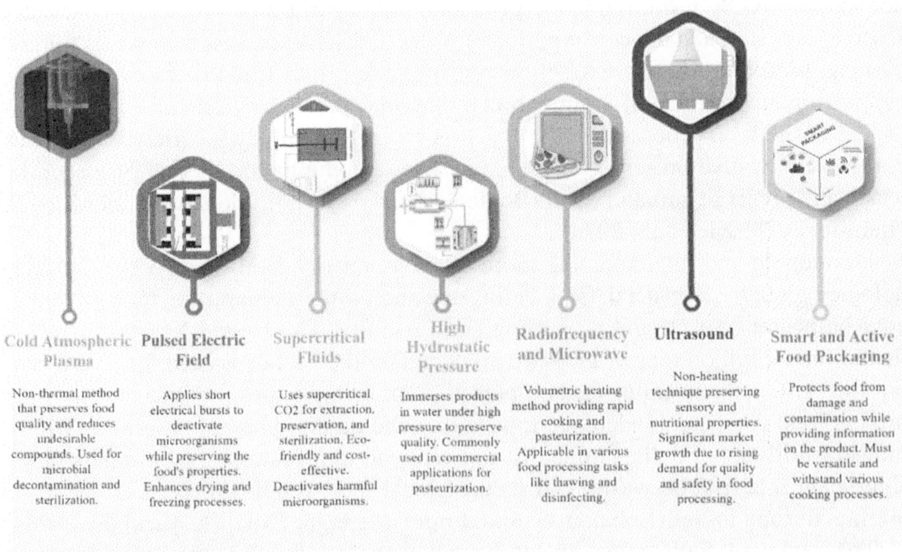

FIGURE 9.1 Advanced food processing technologies.

of modern techniques aimed at enhancing and streamlining existing processes, utilizing versatile equipment, and introducing innovation in procedures. The review encompasses an analysis of the challenges, limitations, and safety considerations associated with these techniques.

9.2 ADVANCED FOOD PROCESSING TECHNOLOGIES

Preserving food involves a range of techniques to maintain its quality and nutritional value. This includes cultivation, gathering, preparation, packaging, and distribution of food products. The primary goals of food preservation include addressing inadequate agricultural planning, creating value-added products, and introducing dietary variety. Traditional methods like drying, chilling, freezing, and pasteurization have evolved into advanced technologies, such as cold atmospheric plasma, irradiation, pulsed electric field, high-pressure processing, ultrasound, and supercritical fluids. These methods aim to prevent the chemical and microbial deterioration of food items. Table 9.1 provides a comprehensive overview of the main highlights concerning advanced food processing technologies.

9.2.1 Cold Atmospheric Plasma

Recently, the food industry has witnessed an increasing focus on non-thermal processing methods, especially for food components that are sensitive to heat. These innovative technologies offer numerous benefits, including preserving food quality, minimizing energy consumption, and reducing the production of undesirable compounds (Chao et al., 2024; Karthik, Mavelil-Sam, Thomas, & Thomas, 2024). Cold plasma technology, originally used in other industries, is gaining traction in the food industry for microbial decontamination, toxin degradation, enzyme inactivation, food packaging, and sterilization. The nutritional value and sensory qualities of food can be adversely affected by the elevated temperatures used in conventional thermal processing (Birania, Attkan, & Ghanghas, 2024). Figure 9.2 illustrates the essential components of cold atmospheric plasma (CAP) technology and its applications in food processing.

Non-thermal processing methods are gaining popularity for preserving the nutritional and functional aspects of food. Some techniques include ozone processing, HHP, pulsed electric field, ultrasonication process, and cold atmospheric plasma. These methods show promise in maintaining the stability of processed foods and have antimicrobial properties (Kulawik et al., 2023).

An innovative approach in the food industry, known as cold plasma (CP) processing or non-thermal plasma, is employed for purposes such as microbial decontamination, functional modification, and enzyme inactivation. As an alternative to conventional sterilization methods, this technique utilizes ionized gas to disinfect food. CP treatment involves highly energetic ionic and reactive species, which bring about beneficial changes to the food matrix and impact food quality. This method is cost-effective, uses atmospheric air as gas, and operates at room temperature and atmospheric pressure. The complex interaction between small organic compounds and plasma reactive species has puzzled researchers due to the dynamic and energetic

TABLE 9.1
Summary of Key Points about Advanced Food Processing Technologies

Technology	Description	Food-Related Applications	Key Benefits	Considerations	Ref.
CAP	Non-thermal process preserving food quality and preventing degradation of nutrients	Microbial decontamination, surface treatment, toxin degradation, food packaging, sterilization	Reduces energy consumption, high efficacy	Effectiveness can vary by food type	(Abedi-Firoozjah, Ghasempour, Khorram, Khezerlou, & Ehsani, 2021; Sriraksha, Ayenampudi, Noor, Raghavendra, & Chakka, 2023; Zhang et al., 2023)
PEF	Short bursts of electricity that deactivate microorganisms while preserving food characteristics	Liquid food products, drying, freezing	Maintains sensory and nutritional qualities	Requires precise control of electric fields	(Abedi-Firoozjah et al., 2021; Alkanan et al., 2024; Roy, Manna, Gayen, Bandyopadhyay, & Maiti, 2024; Zare, Ghasemi, Bansal, & Hosano, 2023)
Supercritical fluids	Uses supercritical CO_2 for extraction and preservation without solvents	Food, flavor extraction, sterilization	Eco-friendly, cost-effective, and versatile	High pressure may not be feasible for all products	(Rojas et al., 2024)
HHP	Immerses products in water under high pressure to achieve pasteurization	Preservation of meats, juices, and ready-to-eat meals	Extends shelf life while retaining nutritional content	Requires specialized packaging and equipment	(Chiozzi et al., 2022; Guillou & Membré, 2019; Kulawik, Rathod, Ozogul, Ozogul, & Zhang, 2023)

RF and microwave	Electromagnetic energy providing rapid, volumetric heating for various food processing tasks	Bread heating, thawing, pasteurization	Fast and efficient, reduces traditional heating limitations	Must manage heating rates to ensure quality	(Du et al., 2024; Geng et al., 2024)
Ultrasound	Non-heating method that uses high-frequency sound waves to preserve food properties	Deactivating micro-organisms, extending shelf life	Maintains quality while improving safety	Reliability depends on design and application specifics	(Mukherjee, 2024; Nabi et al., 2024) (Khadhraoui, Fabiano-Tixier, Robinet, Imbert, & Chemat, 2019; Mahmoud, Fagiry, Davidson, & Abdelbasset, 2022)
Smart and active food packaging	Enhances food safety through innovative packaging that provides information and protects against spoilage	Transport and storage of various food products	Conveys nutritional info, improves handling and safety	Must accommodate different food types and processes	(Bangar, Whiteside, Kajla, & Tavassoli, 2023; Khezerlou, Alizadeh Sani, et al., 2023; Khezerlou, Zolfaghari, et al., 2023; Yekta, Abedi-Firoozjah, Azimi Salim, Khezer-lou, & Abdolmaleki, 2023)

Note: HHP = high hydrostatic pressure; CAP = cold atmospheric plasma; RF = radio frequency; PEF = pulsed electric field.

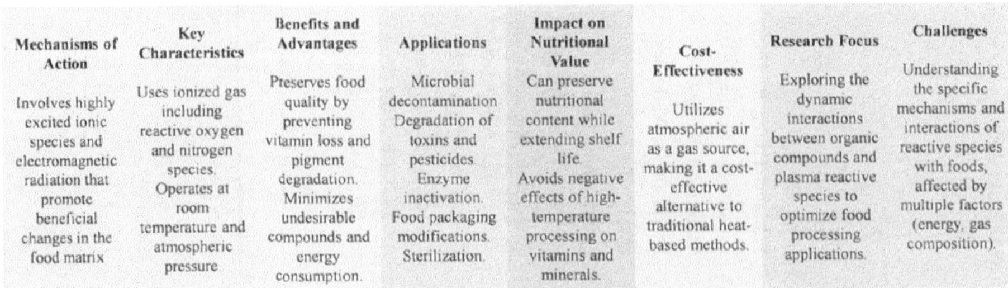

FIGURE 9.2 Key aspects of CAP technology in food processing.

nature of plasma. Understanding this interaction is crucial for using plasma as a tool in food processing (Sriraksha et al., 2023; Zhang et al., 2023).

The impact of CP treatment on phenolic compounds arises from the presence of different reactive species. CP processing has been remarkably recognized to enhance the phenolic content, stimulate the generation of catechin polymerization products, and increase hydroxycinnamic acids and chalcones. CP treatment disrupts the cell membrane, improves mass transfer, and enhances phenolic extraction. Additionally, it results in the release of flavonoids and other compounds from cells (Birania et al., 2024; Limnaios et al., 2021; Quintana, Llalla, García-Risco, & Fornari, 2021).

A unique process involving the generation of ozone as a reactive species was observed, in which small organic molecules were broken down when exposed to CP. This resulted in the degradation of an unstable ozonide and the breaking of double bonds in organic compounds. CP processing was indicated to affect the content of vitamins in food: The free-form vitamin A was increased, the enzyme dehydroascorbate reductase was activated, and the overall ascorbic acid content in the plasma-treated sample was assessed to be increased (Birania et al., 2024; Limnaios et al., 2021). CP treatment has a substantial effect on the quality of food, with notable implications. These affect factors such as pH levels and phenolic substances in various foods. Cold plasma processing can lead to changes in pH, impacting texture, taste, and shelf life. It can also influence the quantity of phenolic substances in foods. Factors such as gas choice, flow rate, exposure time, and power intensity influence the effects of cold plasma treatment on food (Zhang et al., 2023).

Almeida et al. (2015) conducted a study on the effects of atmospheric cold plasma on various fruit juices and observed interesting findings (Almeida et al., 2015). They found that prebiotic orange juice preserved 76% of its initial phenolic content after 60 seconds of indirect exposure to plasma. This preservation was attributed to the sensitivity of phenolic compounds to ozone attack, with significant changes occurring only after the 60-second exposure. Interestingly, blueberry juice, pomegranate juice, and cashew apple juice experienced an increase in phenolic content after plasma processing. This increase was linked to the activation of phenylalanine ammonia-lyase, an essential enzyme in synthesizing phenolic compounds. Furthermore, the treatment was noted to facilitate the better extraction of phenolic compounds by depolymerizing and suspending cell wall polysaccharides.

On the other hand, Charoux et al. (2020) found minimal changes in phenolic compounds in plasma-treated black pepper grains (Charoux et al., 2020). This was attributed to the structural differences in these foods, which limited the penetration of plasma reactive species. The findings suggest that exceeding the capacity to prevent oxidative reactions during treatment may have deleterious effects on phenolic compounds. It is also important to note that optimized treatment conditions, including gas composition and process conditions, should be considered for meaningful comparisons between studies.

CP has become a hot topic for deactivating food enzymes. A variety of enzymes, such as polyphenol oxidase, peroxidase, trypsin, and lysozyme, can be effectively inactivated. The impact of CP on these enzymes is significant due to its ability to cause disruptions in the 3D protein structure. Factors such as input power, gas type, exposure duration, enzyme structure, and stability determine the degree of enzyme deactivation. Understanding how enzymes and CP interact is essential for targeted application in food processing (Chao et al., 2024; Karthik et al., 2024). Table 9.2 presents a range of advanced food processing technologies and their potential applications.

Many protein sources found in food can lead to negative reactions in the body. Antinutrients are substances that may reduce the nutritional benefits of food and hinder the absorption of essential nutrients, while non-toxic reactions triggered by the immune system are known as food allergies. However, food processing techniques can be used to modify proteins and reduce the risk of food allergies. Research studies have provided evidence of the performance of CP processing in minimizing allergenic properties in common food allergens, such as soybeans, shrimp, peanuts, and milk. Nonetheless, more comprehensive studies are required to fully understand and control the adverse impact of CP treatment on other antinutrients found in different food sources. The use of cold plasma to process food products has gained interest in the past decade due to its potential in preserving agricultural produce. While cold plasma shows promise in eliminating bacteria, there is a lack of research on how it affects crucial components and properties of food. Optimizing the parameters of cold plasma treatment can be beneficial in preserving food without compromising its quality attributes.

9.2.2 Pulsed Electric Field

The *pulsed electric field* (PEF) technology involves applying short bursts of electrical treatments with varying durations and electric field strengths. It can be used to deactivate harmful microorganisms and enzymes in food while maintaining the sensory, nutritional, and health-promoting characteristics of liquid food products. PEF can also electrically perforate biological membranes to potentially allow for the selective recovery of valuable compounds from various materials and enhance drying and freezing processes (Alkanan et al., 2024; Roy et al., 2024; Zare et al., 2023).

TABLE 9.2
Some Example of the Application of Advanced Food Processing Technologies

Food Processing Technologies	Research Purpose	Remarks	Food Sample	Ref.
CAP	Effect on the quality of orange juice containing prebiotic oligosaccharides	Plasma treatments partially degrade oligosaccharides while preserving their prebiotic properties, phenolic content, and antioxidant activity.	Orange juice	(Almeida et al., 2015)
	Microbial inactivation	Reduced *Bacillus subtilis* growth, achieving a 2.92 log reduction while showing no significant impact on total phenolic content. Further, reactive species, atomic oxygen, and nitrogen were identified as contributors to microbial cell death.	Black pepper grains	(Charoux et al., 2020)
	Ochratoxin A detoxification	Effectively inhibited mycotoxigenic fungi and reduced ochratoxin A levels in coffee, achieving a complete fungal kill and 50% toxin reduction after 30 minutes.	Roasted coffee	(Casas-Junco et al., 2019)
	Inactivation of *S. Typhimurium*, *L. monocytogenes*, and *E. coli*	30 seconds of plasma treatment eliminated bacteria, while 300 seconds reduced biofilm levels by up to 5 log^{10} CFU/sample.	Lettuce broth	(Ziuzina, Han, Cullen, & Bourke, 2015)
PEF	Effect on the onion drying efficiency and quality	PEF pre-treatment significantly enhances onion drying efficiency and quality, achieving a 32% reduction in drying time and a 47% improvement in rehydration coefficient at the 85/65°C profile.	Onion	(Jalté, Lanoisellé, Lebovka, & Vorobiev, 2009)
	Effect on the *Lacticaseibacillus rhamnosus* and *Lacticaseibacillus paracasei*	The enhancement of lactic acid production and the release of proteins due to sub-lethal PEF treatment are particularly noteworthy.	Not stated	(Djukić-Vuković, Meglič, Flisar, Mojović, & Miklavčič, 2021)

	Reduction in trichothecenes, zearalenone, enniatins, beauvericin, and tentoxin	Reductions in several mycotoxins of up to 84%, while some showed increased extractability into water, suggesting enhanced extraction.	Wheat	(Stranska et al., 2023)
Supercritical fluids	Antimicrobial food packaging PLA nanocomposite foams	A quasi-Fickian diffusion release mechanism with release rate constants up to 3.6 times lower than those of pure eugenol, effectively inhibiting foodborne pathogens like *Listeria* and *Salmonella*.	Not used	(Rojas et al., 2024)
	Pasteurization process	Maintained microbiological stability, color, and total phenols over 28 days, while antioxidant activity decreased significantly for all methods.	Pomegranate juice	(Bertolini et al., 2020)
	Pasteurization process	Achieving a 4 log CFU/g reduction of various microorganisms with a 15-minute treatment at 45°C and 12 MPa. Further, the treatment did not affect hardness.	Fresh-cut coconut	(Ferrentino, Balzan, Dorigato, Pegoretti, & Spilimbergo, 2012)
HHP	Effect on *E. coli* O157:H7, *S. enterica*, and *L. monocytogenes*	600 MPa for 1.5, 0.9, and 1.5 minutes, respectively, achieves 5-log reductions, with the Weibull model providing a better fit for inactivation kinetics.	Acidified juices and beverages	(Usaga, Acosta, Churey, Padilla-Zakour, & Worobo, 2021)
	Inactivation microorganisms	HHP processing at 600 MPa for 8 minutes effectively reduced *L. monocytogenes* by 3.2 log^{10} CFU/g in Iberian chorizo, with significant reductions in yeast and mold counts.	**Dry-cured sausage**	(Cava, García-Parra, & Ladero, 2020)

(Continued)

TABLE 9.2 (*Continued*)
Some Example of the Application of Advanced Food Processing Technologies

Food Processing Technologies	Research Purpose	Remarks	Food Sample	Ref.
	Improvement of safety and quality	HPP effectively inactivated pathogens in fresh and frozen coconut water, resulting in <1 CFU/mL for inoculated samples, while non-HPP controls showed gradual pathogen decline and spoilage within weeks.	Raw coconut water	(Raghubeer et al., 2020)
RF and microwave	Improvement of uniform heating and sterilization	The best heating process for a large amount of black fungus and achieved ideal heating uniformity and sterilization effects.	Dried black fungus	(Geng et al., 2024)
	Inactivation bacteria and extended shelf life	Deactivated bacteria and extended the shelf life of pre-made carrots to 42 days while preserving their nutritional content.	Pre-made carrots	(Du et al., 2024)
	Impact on grain quality	Effectively kills the stored grain pest *Rhyzopertha dominica*, achieving 100% mortality at 58°C, with faster mortality rates noted at 50–52°C; heating time varies by grain type, influenced by kernel size and moisture content.	Grain	(Raghubeer et al., 2020)
Ultrasound	Impact on microbiological quality	Decreased the growth of lactic acid, mesophilic, and psychrophilic bacteria in chicken meat, indicating its potential for preserving chicken meat due to its bactericidal properties.	Chicken meat	(Pinon et al., 2018)

	Impact on microbiological quality	Decreased the number of *enterobacteria*, mesophilic, lactic acid, and psychrotrophic bacteria. However, the treatment did not significantly impact quality parameters, such as lipid and protein oxidation, shear force, anaerobic glycolysis, and muscle structure.	Pre-chill chicken carcasses	(Cichoski et al., 2019)
	Impact on the nutritional and sensory quality of milk	There is increasing interest in non-thermal preservation techniques, such as ultrasound, to deactivate microorganisms and enzymes while preserving the quality of the products.	Milk and milk products	(Górska-Warsewicz, Rejman, Laskowski, & Czeczotko, 2019)
Smart and active food packaging	Extending the shelf life of strawberries	The packaging included a UV light barrier and prolonged bacterial inhibition, extending the shelf life of strawberries at RT. This packaging shows promising potential for use in food packaging materials.	Strawberries	(Fu et al., 2024)
	Extending the shelf life of red meat	A composite antibacterial material, ZIF-8-TA/PBQDs, was developed, showcasing 100% broad-spectrum antibacterial efficacy and a photothermal conversion efficiency of 27.92%. This multifunctional film, when integrated into chitosan, significantly extends food shelf life.	Red meat	(Wang, Zhou, Lu, Wang, & Deng, 2024)
	Pork preservation and blueberry storage	The study developed N-CDs incorporated into chitosan films, achieving high photodynamic antibacterial rates of 91.2% against *E. coli* and 99.9% against *S. aureus*, while significantly improving the preservation of pork and blueberries.	Pork and blueberry	(Chen et al., 2023)

(Continued)

TABLE 9.2 (*Continued*)
Some Example of the Application of Advanced Food Processing Technologies

Food Processing Technologies	Research Purpose	Remarks	Food Sample	Ref.
	Fruits preservation	The study developed a zein-gum arabic nanoparticle-stabilized composite essential oil emulsion (ZGCEO) that significantly enhanced the properties of film and demonstrated strong antioxidant and antimicrobial properties.	Huangguan fruits	(Pu, Chen, & Jiang, 2024)
	Freshness indicator	Anthocyanin-rich SPSP + MEO films enhanced showed improved antimicrobial properties and significant changes in physical characteristics and also provided intuitive color changes (purple → green) to indicate food freshness.	Shrimp and pork	(Yue, Zhang, Liu, & Sun, 2024)

Note: CAP = cold atmospheric plasma; PEF = pulsed electric field; HHP = high hydrostatic pressure, RF = radio frequency; RT = room temperature; ZIF-8-TA/PBQDs = polyphenol functionalized metal–organic frameworks/black phosphorus quantum dots; N-CDs = nitrogen-doped carbon dots; SPSP = steamed purple sweet potato; MEO = mandarin essential oil.

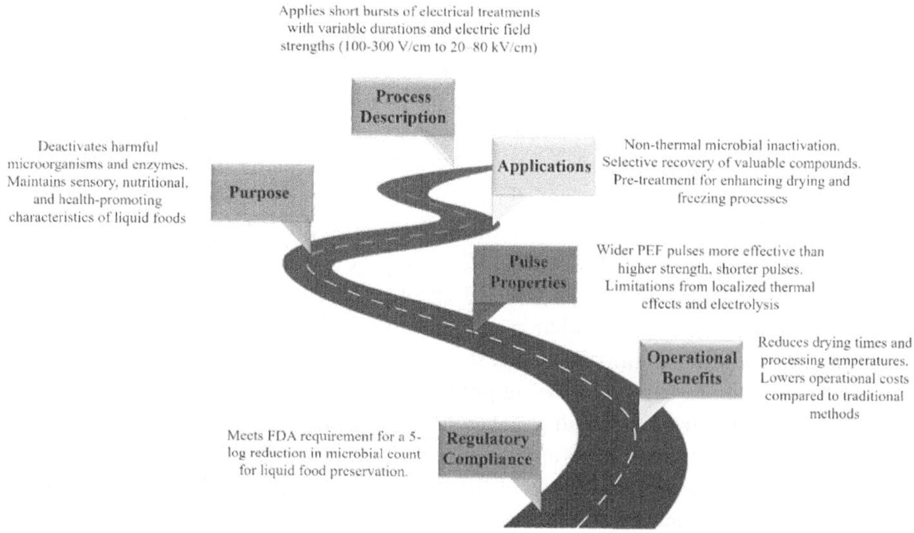

FIGURE 9.3 Organized information about pulsed electric fields (PEF) technology.

Figure 9.3 illustrates organized information about the principles, applications, and advantages of PEF technology. It provides an in-depth analysis of the mechanisms involved, the potential areas of use, and the benefits of employing PEF technology in various industrial and scientific contexts.

When biological cells are exposed to an applied voltage, a process known as electroporation occurs, which creates an electric field of a magnitude larger than that of the critical transmembrane potential. This leads to the formation of temporary or permanent pores in the cell membrane. In the food processing industry, PEF treatment employs electroporation and is commonly used for extracting valuable compounds from plant-based sources, non-thermal microbial inactivation, or toxins degradation, and serving as a pre-treatment to enhance freezing and drying methods. It can also be used to induce stress responses in cells and introduce functional materials or medications in biological tissues. Irreversible electroporation has been extensively researched for enzyme deactivation, pasteurization process for liquid foods, and pre-treatment of biomass products (Kulawik et al., 2023; Reza & Milad, 2024).

The use of PEF technology offers a promising alternative to traditional pasteurization methods for preserving liquid food products while meeting FDA requirements for microbial reduction. PEF treatments have been shown to effectively reduce microbial counts in products such as milk, orange juice, and apple juice (Timmermans et al., 2019).

Studies have shown that broader PEF treatment pulses are more efficient compared to shorter pulse widths with higher pulse strength. PEF technology has various applications in solid food, such as improving structure, altering physical properties, extracting bioactive compounds, and preserving food. It has a lethal effect on microorganisms and brings about beneficial structural changes. PEF also reduces the required processing temperature and shortens drying times in food processing (Ostermeier, Giersemehl, Siemer, Töpfl, & Jäger, 2018).

The process of freezing food involves turning water into ice to preserve the food by reducing water content without increasing glucose levels. Freezing food quickly is important to prevent the spread of salt and avoid decomposition. The duration of freezing plays an important and undeniable role in assessing the quality of frozen foods. Research has shown that PEF treatment can accelerate the freezing process, reduce production costs, and improve the cellular structure of animal products, such as meat, leading to greater tenderness. However, there are conflicting reports on the effect of PEF on meat tenderness, with different research showing negative or neutral effects. Nonetheless, recent evidence suggests that PEF treatment can enhance meat tenderization, leading to improved digestibility and nutritional quality (Lammerskitten et al., 2019). For example, treating potatoes with a specific PEF reduced freezing time by 65%, while treating fresh apples with PEF before freezing helped them retain a texture similar to that of fresh apples after defrosting (Jalté et al., 2009).

Before large-scale industrial use of this innovative food preservation method can be attempted, its economic viability compared to pasteurization methods must be proven. This involves examining operational and investment costs, ensuring product quality, and assessing consumer acceptance. It will also be necessary to design impulse generation systems that can provide sufficient electrical field strength, power, and repetition rates at a reasonable cost. Besides, although consumer acceptance is vital, the legal implications of this method are not yet very clear. Further, new processing techniques must demonstrate meeting conventional standards (Mahendran et al., 2019; Mukhtar et al., 2022). Concerns about electrochemical reactions at electrode–medium interfaces indicate a need to explore alternative materials for electrodes or modify pulse generator systems to mitigate these reactions. Before industrial-scale use, potential toxicity risks and the production of harmful compounds resulting from PEF treatments must be addressed. PEF-based food processing generates less industrial waste than traditional processes and contributes to energy conservation, water preservation, and minimal emission of harmful gases. An increased collaboration among scientists and engineers is necessary to advance PEF technology and optimize economic factors for its implementation in the food processing industry.

9.2.3 Supercritical Fluids

Supercritical fluids (SF), such as carbon dioxide (CO_2), are increasingly used in green food processing. They efficiently extract food components without using organic solvents, help preserve food by deactivating microbes and enzymes, and are cost-effective. When a fluid becomes supercritical, its temperature and pressure exceed the critical point, eliminating the distinct boundary between liquid and gas phases. In the food industry, supercritical CO_2's solvent power and diffusivity make it suitable for sterilization and food preservation. This method effectively deactivates microbial spores and ensures food safety without using heat (Bertolini et al., 2020).

Supercritical fluid processing replaces traditional solvents with substances that exhibit unique properties. These fluids are ideal for extracting compounds and are environmentally friendly. Carbon dioxide is widely used in food applications due to its mild critical conditions. Despite its advantages, it is considered a weak solvent for some compounds, so polar cosolvents are often added to enhance solubility and recovery rates. Ethanol, due to its low toxicity, is considered a better modifier compared to others (Gyawali et al., 2024; Xie et al., 2024).

Supercritical fluid sterilization is based on the unique physical and chemical characteristics of the fluid, particularly when it is near its critical point. SF processing, particularly CO_2, has the remarkable ability to deactivate spores due to their unique combination of solvent-like power and gas-like diffusivity. CO_2 can effectively penetrate through the various layers of spores, and there are several proposed mechanisms for inactivating microbes and spores using supercritical CO_2. Supercritical carbon dioxide ($scCO_2$) technology offers an environmentally friendly approach to producing solvent-free materials and is advantageous for industrial applications, such as supercritical solvent impregnation (SSI) used in diverse industries like food packaging, pharmaceuticals, and textiles (Boateng, 2024; Lobatón-García et al., 2023). Beyond SSI, $scCO_2$ is also used as a blowing agent in polymer processing to produce foam. This technology shows promise for creating environmentally friendly and versatile materials for various industries. In 2024, Rojas and their team developed novel eugenol extended-release poly(lactic acid) nanocomposite foams by employing supercritical fluid technology and cocrystallization techniques. They prepared eugenol–phenazine cocrystals and incorporated them into PLA foams with varying amounts of Cloisite 30B. The research explored how the cocrystallization method and the amount of Cloisite 30B influenced the release kinetics of eugenol, as well as its antimicrobial properties. The modified foam samples effectively inhibited the adherence of *Listeria monocytogenes* and *Salmonella enteritidis*, demonstrating sustained antimicrobial effectiveness in broth culture against both bacteria. This research proposes a novel approach for developing extended-release antimicrobial materials for food packaging (Rojas et al., 2024). Figure 9.4 outlines the important characteristics of SF treatment, with a focus on $scCO_2$ as they relate to food processing.

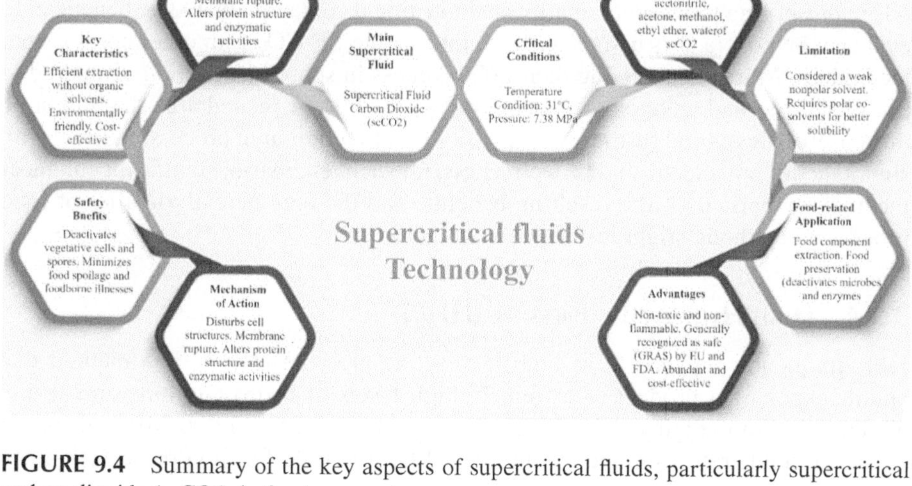

FIGURE 9.4 Summary of the key aspects of supercritical fluids, particularly supercritical carbon dioxide ($scCO_2$), in food processing.

When extracting essential oils, the most common method is hydrodistillation. However, supercritical carbon dioxide extraction has gained attention for its ability to preserve the plant's natural composition. A study comparing the two methods found that supercritical fluid extraction, especially when using ethanol as a cosolvent with CO_2, resulted in higher yields. The extracts were used to create chitosan-based coatings for strawberries, with supercritical extracts showing reduced fungal decay and hydrodistillation extracts better maintaining phenolic compounds in the fruit (Quintana et al., 2021).

The success of $scCO_2$ pasteurization relies on optimizing process parameters. Researchers are exploring a novel approach to pasteurize pomegranate juice at lower temperatures using $scCO_2$, aiming to maintain microbial stability and preserve the juice's characteristics. Experiments have fine-tuned time, temperature, and pressure to determine optimal conditions for microbial inactivation. The findings suggest that $scCO_2$-treated juice remained free from harmful microorganisms throughout its shelf life and retained its color and overall microbial count. Although antioxidant activity decreased during storage, this decline was comparable to other pasteurization methods. Overall, the $scCO_2$ process shows great potential for producing highly nutritious juices (Bertolini et al., 2020).

The use of SF-CO_2 will reduce the harmful effects on the nutritional quality of foodstuffs, their texture, and their flavor compared to the conventional methods of thermal processing. The SF-CO_2 efficiently diffuses through and disperses on coatings, cell walls, and spore membranes, while disruption of these structures causes loss by degradation within the spore core through dispersion of the lipids in them. The assessment and clarification of the inactivation mechanisms were also emphasized. The mechanisms by which SF-CO_2 spreads and permeates were classified into three main categories: (1) disruption and rupture of cell walls and membranes, (2) breakdown of cellular proteins, and (3) inhibition of enzyme functionality and performance. The addition of a cosolvent was observed to enhance the efficacy of SF-CO_2 in inactivating microbial cells and spores. This goes to highlight further research in applying SF-CO_2 to inactivate microbial cells and spores within food and pharmaceutical products. Given the benefits of non-thermal food processing methods in preserving food safety quality and nutritional content, it would be beneficial to evaluate the cost factors associated with integrating SF-CO_2 with other non-thermal techniques. More so, assessing their effectiveness in spore inactivation compared to traditional thermal technologies would be worthwhile. A related area that might be most usefully explored further is the possibility of reduction in processing time, with simultaneous enhancement of effectiveness. Of such a scenario, significant enhancements in overall costs, the resulting benefits, and the commercial viability of food processing methods might result.

9.2.4 High Hydrostatic Pressure (HHP)

HPP, frequently referred to as high hydrostatic pressure (HHP), is a method that entails exposing a product to extremely high levels of hydrostatic pressure within a water medium (Figure 9.5). Pressure ranges from 100 to 1,000 MPa (megapascals), and temperatures range from 0 to 120°C. It is used for food preservation and

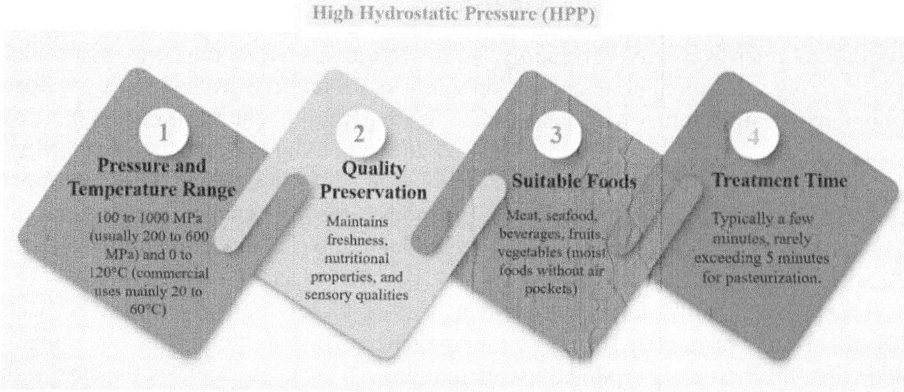

FIGURE 9.5 Some key aspects of HPP technology.

extending shelf life while retaining quality. HHP is suitable for various foods and is commonly used for meat, seafood, beverages, fruits, and vegetables (Chiozzi et al., 2022; Guillou & Membré, 2019; Kulawik et al., 2023).

Compared to other food preservation methods, HHP preservation is relatively costly, ranging from 0.1 to 0.3 EUR/kg. It is used for preserving high-quality foods while maintaining their freshness and nutritional properties. Studies have shown that HHP does not significantly change the fundamental structure of food components and can be used to intentionally modify food compositions (Zhu et al., 2024). Health foods have become popular due to a growing interest in preventive health care as the global population ages. There is a need to offer healthy foods in diverse forms, and one promising approach is integrating them into a balanced diet without major changes to eating habits. HPP offers a practical and economical approach to improving the health-beneficial elements found in foods. This method makes these enhanced products more widely available, ensuring that a diverse array of consumers can access healthier food options. This includes pressure-tolerant probiotics used in various HPP products and natural health-promoting materials derived from medicinal plants, herbs, and animals (Allai et al., 2023; Kulawik et al., 2023).

The application of HHP poses challenges in ensuring product safety and expanding its potential uses. Achieving uniformity in the process and temperature distribution is critical, especially for high-pressure thermal processes and spore deactivation. Specialized temperature sensors are needed to detect the HP vessel's flow profiles, distribution, and temperature (Obileke et al., 2022). Challenges in mechanical engineering, material science, and microbiology must be addressed to optimize HPP technology for food. There is still limited understanding of microbial deactivation, virus inactivation, and chemical reactions under HPP. The effectiveness of HPP treatment is influenced by pressure, temperature, pH, and other factors. Additional comprehensive studies are required on the impact of HPP on food storage stability, packaging materials, and allergenic potential.

9.2.5 Radio Frequency and Microwave

Similar to microwave (MW) energy, *radio frequency* (RF) refers to a form of non-ionizing electromagnetic energy, though it has a longer wavelength. It provides volumetric heating rather than just surface heating, and recent research has focused on its food-related applications for tasks like bread heating, pasteurization, thawing, drying, and blanching. RF heating has emerged as a viable alternative to traditional cooking methods in the processing of fresh foods, offering benefits such as microorganism reduction, disinfection, thawing, and blanching (Du et al., 2024; Geng et al., 2024).

The principal problem of inconsistent heating affects both RF and MW heating; this results principally from the inhomogeneous distribution of the electromagnetic field inside the material and its surroundings. This is particularly the case with moisture-rich products, such as fresh foods. To improve thermal uniformity, surface heating methods—such as hot air or water—can be combined with RF heating, and the process can be further enhanced by moving or mixing the samples. More so, computer simulation can be used to predict and adjust the electromagnetic field to improve heating uniformity. Their RF and MW range from 3 kHz to 3 GHz, where their frequencies have been legally divided by the US Federal Communications Commission for each of the following: industrial, scientific, and medical purposes. Compared to the complex and uneven electromagnetic field found in a microwave oven, RF heating typically generates a more straightforward and uniform field (Costa & Marra, 2024; Gao et al., 2023; Muñoz-Almagro, Morales-Soriano, Villamiel, & Condezo-Hoyos, 2021).

Black fungus, also known as *Auricularia auriculata*, is a valued edible fungus with medicinal and culinary properties found in countries such as Japan, China, South Korea, and Thailand. Fresh black fungus is harmful and must be dried before consumption to ensure safety. Dried black fungus is at risk of contamination by pathogenic bacteria, emphasizing the need for sterilization before sale. Hot air and MW heating are the main techniques used for sterilizing and drying black fungus, but they have drawbacks, such as declining product quality and uneven drying. RF heating offers more uniform heating and improved sterilization efficiency, and a study developed a high-capacity RF rotation heating system to sterilize and heat-dry black fungus, achieving ideal heating uniformity and sterilization effects (Geng et al., 2024).

Belonging to the annual and biennial plant family, the carrot is categorized into yellow, orange, and red varieties according to its carotenoid production pathway. These colorful vegetables are rich in beneficial compounds, including quercetin, carbohydrates, vitamin C, naphthol, carotenoids, and various other bioactive substances. Processed carrots are famous for their vibrant color, crunchy texture, juiciness, and versatility for raw or cooked consumption. Recent studies suggest that RF pasteurization combined with nisin can enhance the preservation of processed carrots. Du and colleagues (2024) successfully optimized a combination of nisin and RF to treat pre-made carrots. The results demonstrated that this method effectively deactivated bacteria, leading to an impressive shelf life extension of 42 days while maintaining the carrots' nutritional integrity. Furthermore, the treatment addressed unpleasant odors and enhanced the overall appearance of the carrots. These findings hold significant promise for elevating the quality of pre-made carrots and similar food products within the food processing industry (Du et al., 2024).

The food industry has been slow to adopt RF technology primarily because of the difficulties associated with designing effective RF heating systems. Future research should focus on implementing digital technology, computer simulation, and parameter evaluation to enhance heating uniformity and reduce costs. Efforts should be directed toward integrating RF processing systems into renewable energy grids and simplifying the structural design of RF generation to reduce industrial costs. Additionally, attention should be given to dynamic impedance matching and addressing the impact of sample volume changes on heating uniformity (López-Lluch et al., 2020; Ptak, Żarski, & Kapuśniak, 2020). Product inconsistency can be a drawback for food safety, leading to microorganism growth. Research on RF heating as a promising alternative for the food industry has focused on process optimization, equipment designs, and simulations. RF technology has been used for various foods and is environmentally friendly for pest disinfestation and microorganism inactivation while preserving food quality. Further exploration of RF technology at different frequencies could be beneficial, using computer-aided food engineering methodology for designing optimal processes and equipment.

9.2.6 Ultrasound

The use of ultrasonic treatment for food preservation has become widely adopted for heat-sensitive foods. This method preserves food's properties, extends shelf life, and ensures safety. The global market for food ultrasound is projected to reach a value of USD 204.9 million by 2028, with a CAGR of 7.4%. Factors driving market growth include stringent food safety regulations, increased investments in the food processing industry, high demand for processed food, and emphasis on minimizing food waste. Indeed, by deactivating microorganisms and enzymes, ultrasonic technology helps maintain the quality of food (Khadhraoui et al., 2019; Mahmoud et al., 2022).

Ultrasound is a form of sound wave that operates at a frequency higher than the 20 kHz threshold for human hearing. In relation, classification based on frequency and intensity can be done with ultrasound. Low-frequency waves lie between 20 and 100 kHz, while high-intensity ultrasound is classified to be above 1 W/cm^2. On the other hand, low-intensity ultrasound uses frequencies above 100 kHz and below 1 W/cm^2. Applications for both types of ultrasound are found in various industries, such as food processing as well as safety-oriented industries. In quality control and processing stages, low-intensity ultrasound serves as an analytical instrument, enabling non-destructive testing. On the other hand, high-intensity ultrasound alters the characteristics of the medium in various applications, including emulsification, defoaming, and textural modification of fatty products. Additionally, it plays a role in sonocrystallization and the functional properties of food proteins (Bariya et al., 2023).

Ultrasound treatment is increasingly important for enhancing the quality of food products. Low-intensity ultrasound improves taste, tenderness, and overall food quality. High-intensity ultrasound can be used in fresh meat processing for various applications, such as tenderizing, cooking, brining, thawing, freezing, and inhibiting bacterial growth. Recent studies focus on the impact of high-intensity ultrasound on bovine muscles, particularly semitendinosus and longissimus, including microscopic changes, salt distribution, water content, texture, oxidative stability, sensory attributes, and color of meat products (Bariya et al., 2023; Mahmoud et al., 2022). Research suggests that ultrasound waves can be useful in the meat industry. Pinon et al. suggest that high-intensity ultrasound with a frequency above 20 kHz and an intensity of 27.6 W/cm^2 can reduce the growth of bacteria in chicken meat. This indicates that ultrasound may help preserve chicken meat due to its ability to kill bacteria (Pinon et al., 2018). Cichoski and co-workers used ultrasound with slightly acidic electrolyzed water (SAEW) at 10°C for 10 minutes to pre-chill chicken carcasses. The combination significantly reduced certain bacteria but did not affect quality parameters. The study suggests that this technology is feasible for maintaining carcass quality during pre-chilling (Cichoski et al., 2019). Poultry meat processing involves stages such as the cooling process, which can potentially cause cross-contamination. Rossi and co-workers demonstrated the effectiveness of ultrasound alone or in combination with chlorine dioxide in eliminating bacteria in poultry processing chiller tanks. The study found that ultrasound treatment reduced *Salmonella* and *E. coli* activity, and chlorine dioxide did not decrease activity for either microorganism when subjected to agitation (Paula Rossi et al., 2021). Milk and dairy products are known to be a valuable source of nutrients; however, they are also prone to microbial spoilage, as noted by Górska-Warsewicz and her team (Górska-Warsewicz et al., 2019). The dairy industry is exploring non-thermal preservation methods like ultrasound to address spoilage and contamination challenges. A study by Ragab and colleagues examined the effects of combining pasteurization and ultrasound on raw goat milk. They found that thermosonication at 300 W and 400 W power levels significantly reduced bacterial count without affecting the milk's color or nutritional content (Ragab et al., 2019). The storage of seafood is crucial due to its quick spoilage. Various methods, like using chemicals and applying heat, can improve safety and extend shelf life, but they may affect taste, texture, and appearance. As the demand for fresh and minimally processed foods increases, interest in non-heat treatments, such as ultrasound, is rising. Further investigation is required to explore the application of ultrasound for disinfecting seafood in industrial settings. Pedrós-Garrido et al. found notable decreases in bacterial counts for salmon and mackerel with ultrasound treatment, but white fish showed lower reductions potentially due to their irregular surfaces protecting bacteria (Pedrós-Garrido et al., 2017).

Concerns about food safety have led to increased interest in using ultrasonic technology to ensure a safe and quality food production. Ultrasound can deactivate harmful microorganisms and enzymes, extending shelf life and preserving food properties. However, unresolved issues need to be addressed, and combining ultrasound with other techniques for comprehensive food safety is essential. Further research is required to create automated systems for industrial use and better understand ultrasound's effects on food quality.

9.3 SMART AND ACTIVE FOOD PACKAGING

Maintaining the quality and safety of food products relies heavily on effective food packaging. It protects food from physical damage, harmful light, and gases, while also preventing contamination and conveying important information to consumers. Packaging materials must be versatile, be capable of being shaped into various sizes and forms, and possess a wide range of functional qualities (Bangar et al., 2023; Khezerlou, Alizadeh Sani, et al., 2023; Khezerlou, Zolfaghari, et al., 2023). The food industry is using bioactive films and advanced methods for packaging. Plastic films are popular due to their lightweight nature and cost-effectiveness, and they help protect food from various elements. The development of innovative barrier films aimed at prolonging the shelf life of food has seen significant advancements lately, like in innovative, active, and intelligent packaging. However, synthetic plastics contribute to environmental pollution and pose risks to human health due to their slow biodegradation rate. To address these issues, manufacturers are exploring eco-friendly, sustainable packaging materials, including biopolymer-based options derived from proteins and/or polysaccharides sourced from animals or plants (Noori et al., 2024; Tavassoli et al., 2023). Bionanocomposites combine biopolymers with nanosized particles for various applications, offering large surface areas and improved physico-mechanical, biodegradable, and thermal properties. Noori et al. (2024) developed a film using barberry anthocyanins fixed on cobalt-based metal–organic framework nanoparticles with biodegradable polymers, suitable for maintaining the freshness of red meat. The film exhibited enhanced antibacterial effectiveness, UV protection, water vapor resistance, and suitability for monitoring red meat spoilage, making it an effective food packaging material (Noori et al., 2024). The exploration of active packaging films, which include antibacterial and antioxidant attributes, is on the rise. Scientists are examining how MOF additives can enhance the effectiveness of (bio)polymer-based food packaging systems. MOFs can act as nanocarriers for delivering bioactive compounds, offering antibacterial, antifungal, and antioxidant activities in packaging films or coatings. They show promise in combating mold, yeast, and bacteria and can enhance the properties of polymer matrices. The incorporation of MOF nanoparticles into different polymer matrices has demonstrated significant improvements in both mechanical and antibacterial effects (Tavassoli, Bahramian, et al., 2024). Tavassoli and colleagues created active packaging systems by combining methylcellulose (MC) and ChNF with silver metal organic framework (Ag-MOF) nanoparticles loaded with lactoferrin (LAC). The nanoparticles improve mechanical properties, thermal stability, moisture content, solubility, swelling power, as well as gas barrier and UV-shielding properties. The films containing Ag-MOF-LAC2% exhibited strong and prolonged antibacterial effects against both *E. coli* and *S. aureus*. It was demonstrated that these composite films can keep apples fresh for at least 7 days. While this breakthrough holds promise for application in active packaging for fresh fruits and vegetables, additional studies are required to confirm their safety and economic viability (Tavassoli, Khezerlou, et al., 2024). Strawberries are widely recognized for their abundant bioactive compounds and micronutrients. While China holds the lead in cultivation and production, strawberries are susceptible to spoilage. Numerous preservation techniques are employed, although some may pose risks to the environment and human health. The development of sustainable and innovative natural resources is seen as a viable approach to preserving fruits and vegetables. In line with this, Fu and collaborators (2024) have devised a multifaceted food packaging system using a layer-by-layer electrostatic self-assembly technique to safeguard strawberries. This packaging incorporates a UV light barrier and extends bacterial inhibition, thereby prolonging the shelf life of strawberries at room temperature. Such packaging exhibits promising potential for integration into food packaging materials (Fu et al., 2024). The development of advanced food packaging films has focused on incorporating antibacterial and antioxidant properties using metal nanoparticles, essential oils, and natural polyphenols. For example, nano-ZnO has strong antibacterial qualities and has been approved by the FDA for antimicrobial applications. Studies have shown that incorporating nano-ZnO into the film matrix can inhibit the growth of bacteria. Additionally, natural anthocyanins found in various fruits and vegetables have been identified as effective antioxidants. Researchers created food packaging films with nano-ZnO and purple potato anthocyanins to preserve chilled hairtail pieces, showing promising results in prolonging shelf life and maintaining quality (Liu et al., 2024).

Developing biodegradable food packaging using natural ingredients has gained attention as a sustainable alternative to traditional plastics. Research is ongoing to create edible packaging materials with desired attributes, including active and intelligent packaging, to enhance food preservation. Challenges remain, such as encapsulating essential oils effectively and expanding intelligent packaging applications beyond meat and seafood to fruits and dairy products. The future of intelligent food packaging envisions real-time data recording and transmission for food traceability through smart devices. However, scalability, cost-effectiveness, and functional performance under practical conditions need further assessment for commercial applications.

9.4 CHALLENGES, LIMITATIONS, AND SAFETY ASPECTS

Techniques for preservation, such as ultrasound processing, MW treatment, supercritical fluid methods, instant controlled pressure drop, and PEF present benefits, including decreased extraction time, lower energy consumption, and minimized need for water or solvents. Overcoming the challenge of the restricted diffusion of water or solvent into biomass involves enhancing diffusion and disrupting cell walls. Ultrasound and electric pulse fields effectively disrupt cells, while MW heating facilitates mass and heat transfer, leading to cell destruction. Education, expanding operational methods, and establishing regulations are essential for the widespread adoption of non-thermal technologies. However, scaling up these innovative methods to an industrial level can be costly due to current small-scale usage. It is essential to carefully consider the significant investment costs for long-term implementation and utilization of non-thermal technologies. Specialized facilities are needed to enhance operational methods and cost-effectiveness. Further research into non-thermal technologies and their interaction with food structure is necessary to better understand the process and product parameters. Plasma technology offers advantages but raises

TABLE 9.3
Summary of the Challenges, Limitations, and Safety Aspects of Food Preservation Technologies

Aspect	Challenges	Potential Solutions
Consumer awareness	Low consumer knowledge about safety and benefits of non-thermal technologies.	Increase education and establish food safety regulations.
Scaling up	High initial investment and the need for specialized facilities can impede widespread use.	Assess long-term feasibility and investment costs.
Efficacy variability	Lack of standard protocols for different products hinders effective application.	Conduct tailored studies for each product to ensure efficacy.
Decision-making tools	Balancing efficiency, cost, and safety can be challenging.	Create decision aids for selecting preservation methods based on material type.
Environmental concerns	Meeting sustainability goals in food production remains a challenge.	Innovate and implement technologies that support environmental sustainability.
Product safety	Optimizing conditions for treatment while preserving nutrients is complex.	Integrate cooling systems to manage temperature during processing.
Cost assessments	Insufficient data on costs and process parameters can hinder adoption.	Conduct detailed cost assessments for initiating and maintaining technology processes.

Note: Table categorized based on the different aspects, the challenges, and their potential solutions.

concerns about the generation of free radicals and oxidative species when in contact with food components, especially fatty products. The effectiveness of plasma treatment is influenced by food surface topography and constituents, such as polyphenols. Potential downsides may include altered sensory properties. The food industry faces challenges, like meeting the demand for safe, top-notch quality and nutrient-enriched and health-optimized food products while minimizing waste and addressing climate change. Table 9.3 provides an overview of the challenges, limitations, and safety considerations associated with food preservation technologies. When considering implementing emerging food powder decontamination technologies on an industrial scale, it is important to address the high initial costs and the lack of sufficient data on process control parameters. Optimization of treatment conditions, product preservation, and integration of advanced technologies for minimizing temperature increases during scaling up are key considerations. Industry reports show potential for enhancing microbial safety at a lower operational cost, but a comprehensive assessment of all associated costs is essential for practical implementation.

9.5 CONCLUSION AND FUTURE PROSPECTIVE

The advancement of the human civilization has been significantly influenced by food preservation, which has allowed communities to thrive and settle in one place. However, the ongoing challenge of prolonging the shelf life of food items while keeping their original qualities intact remains. As an organic and perishable material, food is vulnerable to deterioration from a range of physical, chemical, and microbial factors. Throughout history, diverse traditional preservation methods, including freezing, drying, cooling, and fermentation, have been developed to maintain the nutritional value and texture of food. These ancient techniques have evolved and advanced over time. Modern innovations, including cold plasma, irradiation, MW heating, HHP, and PEF treatment, have been introduced to effectively maintain food quality and prolong its shelf life. Despite the rapid growth of the food preservation and processing industry driven by the rising demand for preserved foods, the inclusion of chemical additives and preservatives has sparked worries about possible health risks. It is essential for professionals and researchers in the food processing and food safety fields to grasp the mechanisms of food spoilage and preservation techniques in order to ensure food safety and longevity. This comprehensive review encompasses a wide array of food categories, spoilage mechanisms, and advanced preservation methods and is intended to serve as a valuable resource for professionals and researchers aiming to develop effective food preservation strategies.

The integration of technology with a focus on design and cost-effectiveness is essential. Ongoing research aims to identify natural preservatives with powerful antioxidant and antimicrobial properties to replace processed food, ensuring its safety for consumption. In order to satisfy consumer demands for a more efficient food system, researchers are investigating hurdle technology, which encompasses various methods for monitoring parameters such as moisture levels, water activity, temperature, pH, as well as enzyme activity. More so, there is a drive for advancements in smart and active packaging for foods. However, there are ongoing discussions and debates regarding research on various nanomaterials, addressing concerns about their toxicity, safety for consumers, and genetic factors. Further progress is necessary in the areas of data visualization, artificial intelligence, Internet of Things, and machine learning, all of which have the potential to bring about significant changes in the food and agricultural industry through digitalization, particularly in the domain of functional foods and crops. Understanding the intricate composition of food and the microbial environment it harbors is essential for creating new food products and investigating preservation methods. The interplay between food components and innovative preservation techniques may have little to no influence on the final food items. Furthermore, the adoption of new preservation methods often necessitates approval from regulatory agencies, which can lead to a lengthy review process. Nonetheless, additional research is essential to clarify and demonstrate how emerging preservation technologies affect the safety and shelf life of food items.

REFERENCES

Abedi-Firoozjah, R., Ghasempour, Z., Khorram, S., Khezerlou, A., & Ehsani, A. (2021). Non-thermal techniques: A new approach to removing pesticide residues from fresh products and water. *Toxin Reviews*, *40*(4), 562–575.

Alkanan, Z. T., Altemimi, A. B., Younis, M. I., Ali, M. R., Cacciola, F., & Abedelmaksoud, T. G. (2024). Trends, recent advances, and application of pulsed electric field in food processing: A review. *ChemBioEng Reviews*, *11*(4). https://doi.org/10.1002/cben.202300078

Allai, F. M., Azad, Z. R. A. A., Mir, N. A., & Gul, K. (2023). Recent advances in non-thermal processing technologies for enhancing shelf life and improving food safety. *Applied Food Research*, *3*(1). https://doi.org/10.1016/j.afres.2022.100258

Almeida, F. D. L., Cavalcante, R. S., Cullen, P. J., Frias, J. M., Bourke, P., Fernandes, F. A. N., & Rodrigues, S. (2015). Effects of atmospheric cold plasma and ozone on prebiotic orange juice. *Innovative Food Science & Emerging Technologies*, *32*, 127–135. https://doi.org/10.1016/j.ifset.2015.09.001

Bangar, S. P., Whiteside, W. S., Kajla, P., & Tavassoli, M. (2023). Value addition of rice straw cellulose fibers as a reinforcer in packaging applications. *International Journal of Biological Macromolecules*, *243*, 125320. https://doi.org/10.1016/j.ijbiomac.2023.125320

Bariya, A. R., Rathod, N. B., Patel, A. S., Nayak, J. K. B., Ranveer, R. C., Hashem, A., . . . Rocha, J. M. (2023). Recent developments in ultrasound approach for preservation of animal origin foods. *Ultrasonics Sonochemistry*, *101*, 106676. https://doi.org/10.1016/j.ultsonch.2023.106676

Bertolini, F. M., Morbiato, G., Facco, P., Marszałek, K., Pérez-Esteve, É., Benedito, J., . . . Spilimbergo, S. (2020). Optimization of the supercritical CO2 pasteurization process for the preservation of high nutritional value of pomegranate juice. *The Journal of Supercritical Fluids*, *164*, 104914. https://doi.org/10.1016/j.supflu.2020.104914

Birania, S., Attkan, A. K., & Ghanghas, S. (2024). Cold plasma: Principles and applications. In *Nonthermal Food Engineering Operations* (pp. 141–162). Wiley Online Library. https://doi.org/10.1002/9781119776468

Boateng, I. D. (2024). Recent processing of fruits and vegetables using emerging thermal and non-thermal technologies. A critical review of their potentialities and limitations on bioactives, structure, and drying performance. *Critical Reviews in Food Science and Nutrition*, *64*(13), 4240–4274. https://doi.org/10.1080/10408398.2022.2140121

Casas-Junco, P. P., Solís-Pacheco, J. R., Ragazzo-Sánchez, J. A., Aguilar-Uscanga, B. R., Bautista-Rosales, P. U., & Calderón-Santoyo, M. (2019). Cold plasma treatment as an alternative for ochratoxin A detoxification and inhibition of mycotoxigenic fungi in roasted coffee. *Toxins*, *11*(6), 337.

Cava, R., García-Parra, J., & Ladero, L. (2020). Effect of high hydrostatic pressure processing and storage temperature on food safety, microbial counts, colour and oxidative changes of a traditional dry-cured sausage. *LWT*, *128*, 109462. https://doi.org/10.1016/j.lwt.2020.109462

Chao, H., Hu, W., Li, Y., Gui, H., Tantai, S., Yu, Y., . . . Li, L. (2024). Effects of cold plasma treatment on reactive oxygen metabolism and storage quality of Brassica chinensis. *Innovative Food Science and Emerging Technologies*, *92*. https://doi.org/10.1016/j.ifset.2024.103574

Charoux, C. M. G., Free, L., Hinds, L. M., Vijayaraghavan, R. K., Daniels, S., O'Donnell, C. P., & Tiwari, B. K. (2020). Effect of non-thermal plasma technology on microbial inactivation and total phenolic content of a model liquid food system and black pepper grains. *LWT*, *118*, 108716. https://doi.org/10.1016/j.lwt.2019.108716

Chen, S., Zeng, Q., Tan, X., Ye, M., Zhang, Y., Zou, L., . . . Hu, K. (2023). Photodynamic antibacterial chitosan/nitrogen-doped carbon dots composite packaging film for food preservation applications. *Carbohydrate Polymers*, *314*, 120938. https://doi.org/10.1016/j.carbpol.2023.120938

Chiozzi, V., Agriopoulou, S., & Varzakas, T. (2022). Advances, applications, and comparison of thermal (pasteurization, sterilization, and aseptic packaging) against non-thermal (ultrasounds, UV radiation, ozonation, high hydrostatic pressure) technologies in food processing. *Applied Sciences (Switzerland)*, *12*(4). https://doi.org/10.3390/app12042202

Cichoski, A. J., Flores, D. R. M., De Menezes, C. R., Jacob-Lopes, E., Zepka, L. Q., Wagner, R., . . . Campagnol, P. C. B. (2019). Ultrasound and slightly acid electrolyzed water application: An efficient combination to reduce the bacterial counts of chicken breast during pre-chilling. *International Journal of Food Microbiology*, *301*, 27–33. https://doi.org/10.1016/j.ijfoodmicro.2019.05.004

Costa, J. M., & Marra, F. (2024). Advances in food processing through radio frequency technology: Applications in pest control, microbial and enzymatic inactivation. *Food Engineering Reviews*, 1–19.

Djukić-Vuković, A., Meglič, S. H., Flisar, K., Mojović, L., & Miklavčič, D. (2021). Pulsed electric field treatment of Lacticaseibacillus rhamnosus and Lacticaseibacillus paracasei, bacteria with probiotic potential. *LWT*, *152*, 112304. https://doi.org/10.1016/j.lwt.2021.112304

Du, T., Jiang, J., Suo, M., Adhikari, B., Lv, W., & Xu, J. (2024). Nisin combined with radio frequency: Effectiveness on microbial growth, physiological quality, and structure of pre-made carrots. *Food Control*, *160*, 110355. https://doi.org/10.1016/j.foodcont.2024.110355

Ferrentino, G., Balzan, S., Dorigato, A., Pegoretti, A., & Spilimbergo, S. (2012). Effect of supercritical carbon dioxide pasteurization on natural microbiota, texture, and microstructure of fresh-cut coconut. *Journal of Food Science*, *77*(5), E137–E143.

Fu, X., Chang, X., Xu, S., Xu, H., Ge, S., Xie, Y., . . . Ding, S. (2024). Development of a chitosan/pectin-based multi-active food packaging with both UV and microbial defense functions for effectively preserving of strawberry. *International Journal of Biological Macromolecules*, *254*, 127968. https://doi.org/10.1016/j.ijbiomac.2023.127968

Gao, J., Wu, M., Du, S., Zhang, H., Wang, S., & Ling, B. (2023). Recent advances in food processing by radio frequency heating techniques: A review of equipment aspects. *Journal of Food Engineering*, *357*, 111609.

Geng, Z., Xie, Y., Ye, P., Sun, Y., Fu, H., Chen, X., . . . Wang, Y. (2024). Effect of hot air-assisted radio frequency rotation heating system on improving heating uniformity of dried black fungus (Auricularia auricula). *Food and Bioproducts Processing*, *147*, 384–392. https://doi.org/10.1016/j.fbp.2024.07.022

Górska-Warsewicz, H., Rejman, K., Laskowski, W., & Czeczotko, M. (2019). Milk and dairy products and their nutritional contribution to the average polish diet. *Nutrients*, *11*(8). https://doi.org/10.3390/nu11081771

Guillou, S., & Membré, J. M. (2019). Inactivation of Listeria monocytogenes, Staphylococcus aureus, and Salmonella enterica under High Hydrostatic Pressure: A quantitative analysis of existing literature data. *Journal of Food Protection*, *82*(10), 1802–1814. https://doi.org/10.4315/0362-028X.JFP-19–132

Gyawali, R., Mahapatra, A. K., Bardsley, C. A., & Niemira, B. A. (2024). Nonthermal techniques, antimicrobial agents, and packaging methods to improve the microbial safety of nuts. *Trends in Food Science and Technology*, *146*. https://doi.org/10.1016/j.tifs.2024.104363

Jalté, M., Lanoisellé, J.-L., Lebovka, N. I., & Vorobiev, E. (2009). Freezing of potato tissue pre-treated by pulsed electric fields. *LWT—Food Science and Technology*, *42*(2), 576–580. https://doi.org/10.1016/j.lwt.2008.09.007

Karthik, C., Mavelil-Sam, R., Thomas, S., & Thomas, V. (2024). Cold plasma technology based eco-friendly food packaging biomaterials. *Polymers*, *16*(2). https://doi.org/10.3390/polym16020230

Khadhraoui, B., Fabiano-Tixier, A.-S., Robinet, P., Imbert, R., & Chemat, F. (2019). 2—Ultrasound technology for food processing, preservation, and extraction. In F. Chemat & E. Vorobiev (Eds.), *Green Food Processing Techniques* (pp. 23–56). Academic Press.

Khezerlou, A., Alizadeh Sani, M., Tavassoli, M., Abedi-Firoozjah, R., Ehsani, A., & McClements, D. J. (2023). Halochromic (pH-responsive) indicators based on natural anthocyanins for monitoring fish freshness/spoilage. *Journal of Composites Science*, *7*(4). https://doi.org/10.3390/jcs7040143

Khezerlou, A., Zolfaghari, H., Forghani, S., Abedi-Firoozjah, R., Alizadeh Sani, M., Negahdari, B., . . . McClements, D. J. (2023). Combining non-thermal processing techniques with edible coating materials: An innovative approach to food preservation. *Coatings*, *13*(5), 830.

Kulawik, P., Rathod, N. B., Ozogul, Y., Ozogul, F., & Zhang, W. (2023). Recent developments in the use of cold plasma, high hydrostatic pressure, and pulsed electric fields on microorganisms and viruses in seafood. *Critical Reviews in Food Science and Nutrition*, *63*(29), 9716–9730. https://doi.org/10.1080/10408398.2022.2077298

Lammerskitten, A., Mykhailyk, V., Wiktor, A., Toepfl, S., Nowacka, M., Bialik, M., . . . Parniakov, O. (2019). Impact of pulsed electric fields on physical properties of freeze-dried apple tissue. *Innovative Food Science & Emerging Technologies*, *57*, 102211. https://doi.org/10.1016/j.ifset.2019.102211

Limnaios, A., Pathak, N., Grossi Bovi, G., Fröhling, A., Valdramidis, V. P., Taoukis, P. S., & Schlüter, O. (2021). Effect of cold atmospheric pressure plasma processing on quality and shelf life of red currants. *LWT*, *151*. https://doi.org/10.1016/j.lwt.2021.112213

Liu, J., Huang, J., Jiang, L., Lin, J., Ge, Y., & Hu, Y. (2024). Chitosan/polyvinyl alcohol food packaging incorporated with purple potato anthocyanins and nano-ZnO: Application on the preservation of hairtail (Trichiurus haumela) during chilled storage. *International Journal of Biological Macromolecules*, *277*, 134435. https://doi.org/10.1016/j.ijbiomac.2024.134435

Lobatón-García, H. F., López-Mejía, N., & Cruz-Guayacundo, W. (2023). CFD-response surface methodology to optimize the effective thermal conductivity and homogeneity in tray dryer. *Revista U.D.C.A Actualidad and Divulgacion Cientifica*, *26*(2). https://doi.org/10.31910/rudca.v26.n2.2023.2241

López-Lluch, D. B., Cano-Lamadrid, M., Hernández, F., Zimmer, A., Lech, K., Figiel, A., . . . Wojdyło, A. (2020). Hydroxycinnamic acids and carotenoids of dried loquat fruit cv. 'Algar' affected by freeze-, convective-, vacuum-microwave- And combined-drying methods. *Molecules*, *25*(16). https://doi.org/10.3390/molecules25163643

Mahendran, R., Ramanan, K. R., Barba, F. J., Lorenzo, J. M., López-Fernández, O., Munekata, P. E. S., . . . Tiwari, B. K. (2019). Recent advances in the application of pulsed light processing for improving food safety and increasing shelf life. *Trends in Food Science and Technology*, *88*, 67–79. https://doi.org/10.1016/j.tifs.2019.03.010

Mahmoud, M. Z., Fagiry, M. A., Davidson, R., & Abdelbasset, W. K. (2022). The benefits, drawbacks, and potential future challenges of the most commonly used ultrasound-based hurdle combinations technologies in food preservation. *Journal of Radiation Research and Applied Sciences*, *15*(1), 206–212. https://doi.org/10.1016/j.jrras.2022.03.006

Mukherjee, A. K. (2024). Role of ultrasound-assisted mechanical shear force in the preservation of foods. In *Advanced Research Methods in Food Processing Technologies: Technology for Sustainable Food Production* (pp. 197–213). Apple Academic Press. https://doi.org/10.1201/9781003369394

Mukhtar, K., Nabi, B. G., Arshad, R. N., Roobab, U., Yaseen, B., Ranjha, M. M. A. N., . . . Ibrahim, S. A. (2022). Potential impact of ultrasound, pulsed electric field, high-pressure processing and microfludization against thermal treatments preservation regarding sugarcane juice (Saccharum officinarum). *Ultrasonics Sonochemistry*, *90*. https://doi.org/10.1016/j.ultsonch.2022.106194

Muñoz-Almagro, N., Morales-Soriano, E., Villamiel, M., & Condezo-Hoyos, L. (2021). Hybrid high-intensity ultrasound and microwave treatment: A review on its effect on quality and bioactivity of foods. *Ultrasonics Sonochemistry*, *80*. https://doi.org/10.1016/j.ultsonch.2021.105835

Nabi, B. G., Mukhtar, K., Ansar, S., Hassan, S. A., Hafeez, M. A., Bhat, Z. F., . . . Aadil, R. M. (2024). Application of ultrasound technology for the effective management of waste from fruit and vegetable. *Ultrasonics Sonochemistry*, *102*. https://doi.org/10.1016/j.ultsonch.2023.106744

Noori, S. M. A., Khezerlou, A., Hashemi, M., Alizadeh-Sani, M., Firoozy, S., Khodaian, F., . . . Tavassoli, M. (2024). Polyvinyl alcohol/chitosan nanofiber-based films incorporated with barberry anthocyanin-loaded CO-MOF as multifunctional performance for red meat sample packaging. *Food and Bioprocess Technology*. https://doi.org/10.1007/s11947-024-03548-0

Novickij, V., Stanevičienė, R., Staigvila, G., Gruškienė, R., Sereikaitė, J., Girkontaitė, I., . . . Servienė, E. (2020). Effects of pulsed electric fields and mild thermal treatment on antimicrobial efficacy of nisin-loaded pectin nanoparticles for food preservation. *LWT*, *120*. https://doi.org/10.1016/j.lwt.2019.108915

Obileke, K., Onyeaka, H., Miri, T., Nwabor, O. F., Hart, A., Al-Sharify, Z. T., . . . Anumudu, C. (2022). Recent advances in radio frequency, pulsed light, and cold plasma technologies for food safety. *Journal of Food Process Engineering*, *45*(10). https://doi.org/10.1111/jfpe.14138

Ostermeier, R., Giersemehl, P., Siemer, C., Töpfl, S., & Jäger, H. (2018). Influence of pulsed electric field (PEF) pre-treatment on the convective drying kinetics of onions. *Journal of Food Engineering*, *237*, 110–117. https://doi.org/10.1016/j.jfoodeng.2018.05.010

Paula Rossi, A., Lahis Kalschne, D., Paula Iglikowski Byler, A., Lisandro de Moraes Flores, E., Donizeti Leite, O., dos Santos, D., . . . Canan, C. (2021). Effect of ultrasound and chlorine dioxide on Salmonella Typhimurium and Escherichia coli inactivation in poultry chiller tank water. *Ultrasonics Sonochemistry*, *80*, 105815. https://doi.org/10.1016/j.ultsonch.2021.105815

Pedrós-Garrido, S., Condón-Abanto, S., Beltrán, J. A., Lyng, J. G., Brunton, N. P., Bolton, D., & Whyte, P. (2017). Assessment of high intensity ultrasound for surface decontamination of salmon (S. salar), mackerel (S. scombrus), cod (G. morhua) and hake (M. merluccius) fillets, and its impact on fish quality. *Innovative Food Science & Emerging Technologies*, *41*, 64–70. https://doi.org/10.1016/j.ifset.2017.02.006

Pinon, M., Alarcon-Rojo, A., Paniwnyk, L., Mason, T., Luna, L., & Renteria, A. (2018). Ultrasound for improving the preservation of chicken meat. *Food Science and Technology*, *39*(Suppl. 1), 129–135.

Priyanka, S., Arun Kumar, E., Moses, J. A., & Anandharamakrishnan, C. (2024). Food preservation and hurdle technology. In *Emerging Technologies for the Food Industry: Volume 1 Fundamentals of Food Processing Technology* (pp. 39–84). Apple Academic Press. https://doi.org/10.1201/9781003413592

Ptak, S., Żarski, A., & Kapuśniak, J. (2020). Technological, economic and health aspects of application of microwave radiation in food processing. *Zywnosc. Nauka. Technologia. Jakosc/Food. Science Technology. Quality*, *27*(1), 47–62. https://doi.org/10.15193/zntj/2020/122/321

Pu, Y., Chen, L., & Jiang, W. (2024). Antimicrobial guar gum films optimized with Pickering emulsions of zein-gum arabic nanoparticle-stabilized composite essential oil for food preservation. *International Journal of Biological Macromolecules*, *278*, 134911. https://doi.org/10.1016/j.ijbiomac.2024.134911

Quintana, S. E., Llalla, O., García-Risco, M. R., & Fornari, T. (2021). Comparison between essential oils and supercritical extracts into chitosan-based edible coatings on strawberry quality during cold storage. *The Journal of Supercritical Fluids*, *171*, 105198. https://doi.org/10.1016/j.supflu.2021.105198

Ragab, E. S., Lu, J., Pang, X. Y., Nassar, K. S., Yang, B. Y., Zhang, S. W., & Lv, J. P. (2019). Effect of thermosonication process on physicochemical properties and microbial load of goat's milk. *Journal of Food Science and Technology*, *56*(12), 5309–5316. https://doi.org/10.1007/s13197-019-04001-3

Raghubeer, E. V., Phan, B. N., Onuoha, E., Diggins, S., Aguilar, V., Swanson, S., & Lee, A. (2020). The use of High-Pressure Processing (HPP) to improve the safety and quality of raw coconut (Cocos nucifera L) water. *International Journal of Food Microbiology, 331*, 108697. https://doi.org/10.1016/j.ijfoodmicro.2020.108697

Reza, A.-F., & Milad, T. (2024). Functionality of food additives. In L. Dr. Vasiliki (Ed.), *Chemistry of Food Additives* (Ch. 4). Rijeka: IntechOpen.

Rojas, A., Misic, D., Zizovic, I., Dicastillo, C. L. d., Velásquez, E., Rajewska, A., . . . Galotto, M. J. (2024). Supercritical fluid and cocrystallization technologies for designing antimicrobial food packaging PLA nanocomposite foams loaded with eugenol cocrystals with prolonged release. *Chemical Engineering Journal, 481*, 148407. https://doi.org/10.1016/j.cej.2023.148407

Roy, A., Manna, A. S., Gayen, S., Bandyopadhyay, D., & Maiti, D. K. (2024). Sustainable novel food coating of chitosan-acetylated gallic acid nanoparticles with pulsed light treatment: A promising edible food preservative. *ACS Food Science and Technology, 4*(6), 1527–1543. https://doi.org/10.1021/acsfoodscitech.4c00153

Sriraksha, M. S., Ayenampudi, S. B., Noor, M., Raghavendra, S. N., & Chakka, A. K. (2023). Cold plasma technology: An insight on its disinfection efficiency of various food systems. *Food Science and Technology International, 29*(4), 428–441. https://doi.org/10.1177/10820132221089169

Stranska, M., Prusova, N., Behner, A., Dzuman, Z., Lazarek, M., Tobolkova, A., . . . Hajslova, J. (2023). Influence of pulsed electric field treatment on the fate of Fusarium and Alternaria mycotoxins present in malting barley. *Food Control, 145*, 109440. https://doi.org/10.1016/j.foodcont.2022.109440

Tavassoli, M., Bahramian, B., Abedi-Firoozjah, R., Ehsani, A., Phimolsiripol, Y., & Bangar, S. P. (2024). Application of lactoferrin in food packaging: A comprehensive review on opportunities, advances, and horizons. *International Journal of Biological Macromolecules, 273*, 132969. https://doi.org/10.1016/j.ijbiomac.2024.132969

Tavassoli, M., Khezerlou, A., Moghaddam, T. N., Firoozy, S., Bakhshizadeh, M., Sani, M. A., . . . Lorenzo, J. M. (2023). Sumac (Rhus coriaria L.) anthocyanin loaded-pectin and chitosan nanofiber matrices for real-time monitoring of shrimp freshness. *International Journal of Biological Macromolecules, 242*, 125044. https://doi.org/10.1016/j.ijbiomac.2023.125044

Tavassoli, M., Khezerlou, A., Sani, M. A., Hashemi, M., Firoozy, S., Ehsani, A., . . . McClements, D. J. (2024). Methylcellulose/chitosan nanofiber-based composites doped with lactoferrin-loaded Ag-MOF nanoparticles for the preservation of fresh apple. *International Journal of Biological Macromolecules, 259*, 129182. https://doi.org/10.1016/j.ijbiomac.2023.129182

Timmermans, R. A. H., Mastwijk, H. C., Berendsen, L. B. J. M., Nederhoff, A. L., Matser, A. M., Van Boekel, M. A. J. S., & Nierop Groot, M. N. (2019). Moderate intensity Pulsed Electric Fields (PEF) as alternative mild preservation technology for fruit juice. *International Journal of Food Microbiology, 298*, 63–73. https://doi.org/10.1016/j.ijfoodmicro.2019.02.015

Usaga, J., Acosta, Ó., Churey, J. J., Padilla-Zakour, O. I., & Worobo, R. W. (2021). Evaluation of high pressure processing (HPP) inactivation of Escherichia coli O157:H7, Salmonella enterica, and Listeria monocytogenes in acid and acidified juices and beverages. *International Journal of Food Microbiology, 339*, 109034. https://doi.org/10.1016/j.ijfoodmicro.2020.109034

Wang, J., Zhou, S., Lu, F., Wang, S., & Deng, Q. (2024). Polyphenols functionalized MOF encapsulated BPQDs for synergistic photothermal/photodynamic antibacterial properties and multifunctional food preservation. *Food Chemistry, 451*, 139451. https://doi.org/10.1016/j.foodchem.2024.139451

Xie, Y., Zhang, J., Zhang, P., Regenstein, J. M., Liu, D., & Zhou, P. (2024). Improving the microbiological safety and quality of aquatic products using nonthermal processing. *Comprehensive Reviews in Food Science and Food Safety*, *23*(3). https://doi.org/10.1111/1541-4337.13368

Yekta, R., Abedi-Firoozjah, R., Azimi Salim, S., Khezerlou, A., & Abdolmaleki, K. (2023). Application of cellulose and cellulose derivatives in smart/intelligent bio-based food packaging. *Cellulose*, *30*(16), 9925–9953.

Yue, R., Zhang, Y., Liu, J., & Sun, J. (2024). Preparation of steamed purple sweet potato-based films containing mandarin essential oil for smart packaging. *Molecules*, *29*(10), 2314.

Zare, F., Ghasemi, N., Bansal, N., & Hosano, H. (2023). Advances in pulsed electric stimuli as a physical method for treating liquid foods. *Physics of Life Reviews*, *44*, 207–266. https://doi.org/10.1016/j.plrev.2023.01.007

Zhang, Q., Yang, W., Zhang, S., Tang, J., Shi, X., Qin, S., . . . Xiao, H. (2023). Enhancing the applicability of gelatin-carboxymethyl cellulose films by cold plasma modification for the preservation of fruits. *LWT*, *178*. https://doi.org/10.1016/j.lwt.2023.114612

Zhu, Y., Gu, M., Su, Y., Li, Z., Xiao, Z., Lu, F., & Han, C. (2024). Recent advances in spoilage mechanisms and preservation technologies in beef quality: A review. *Meat Science*, *213*. https://doi.org/10.1016/j.meatsci.2024.109481

Ziuzina, D., Han, L., Cullen, P. J., & Bourke, P. (2015). Cold plasma inactivation of internalised bacteria and biofilms for Salmonella enterica serovar Typhimurium, Listeria monocytogenes and Escherichia coli. *International Journal of Food Microbiology*, *210*, 53–61. https://doi.org/10.1016/j.ijfoodmicro.2015.05.019

10 Microbial Dynamics
Classification and Functional Roles in Fermented Foods

Amit Patnaik, Titirsha Kayal, Soumya Basu, Ponoop L. Kumar Patro, and Duryodhan Sahu

10.1 INTRODUCTION

From the tangy zest of yogurt to the effervescent jubilance of champagne, fermented foods represent a cornerstone of culinary heritage, offering a diverse tapestry of flavors and textures that tantalize our palates (MansukhlalGondaliya et al., 2024). However, the true magic of these culinary creations lies concealed within a vibrant, microscopic world, a bustling metropolis of microbial life. Bacteria and yeasts, the often-unsung heroes of fermentation, orchestrate a complex biochemical symphony, transforming simple raw ingredients into gastronomic delights imbued with enhanced flavor, altered texture, and even a plethora of potential health benefits (Ramírez Rojas et al., 2022; Sharma et al., 2020).

This chapter embarks on a captivating exploration of the microbial dynamics underpinning this ancient culinary art, unveiling the intricate interplay of these microscopic powerhouses and their profound influence on the characteristics of fermented foods.

Our journey begins with a systematic classification of the key microbial players in this fascinating world, dissecting the distinct roles of lactic acid bacteria (LAB) and yeasts. LAB, the diligent workhorses of many fermentations, diligently convert sugars into lactic acid, a natural preservative that not only extends the shelf life of food products but also imparts the characteristic tanginess so often associated with fermented delicacies (Abedi & Hashemi, 2020; Zapaśnik et al., 2022). Yeasts, in contrast, contribute significantly to the production of alcohol and carbon dioxide, lending their effervescent touch to beverages like beer and wine and providing the leavening power essential for the airy texture of certain breads (Liszkowska & Berlowska, 2021; Maicas, 2020).

But the microbial story does not end there. Beyond these central protagonists, a supporting cast of diverse microbial actors further enriches the fermentation drama. Filamentous fungi, commonly referred to as molds, contribute their unique enzymatic repertoire to specific cheese varieties, shaping texture and driving characteristic flavor development (Gomes et al., 2018; Yassunaka Hata et al., 2023).

DOI: 10.1201/9781003540113-10

Acetic acid bacteria, with their remarkable ability to oxidize alcohol into acetic acid, impart the distinctive acidic bite to fermented products like sauerkraut and kombucha. This complex microbial consortium, working in harmonious synergy, breathes life into fermented foods, each member contributing distinct nuances to the final product's unique sensory profile.

The functional significance of these microbial communities extends far beyond their role in shaping flavor and texture. Their metabolic activities confer a multitude of benefits, including enhanced preservation through the creation of an acidic environment that inhibits the growth of spoilage and pathogenic microorganisms (Agriopoulou et al., 2020; Zapaśnik et al., 2022). Moreover, fermentation often unlocks enhanced nutritional value, improving digestibility and potentially increasing the bioavailability of essential vitamins and minerals (Dimidi et al., 2019; Sharma et al., 2020). Notably, some fermented foods serve as rich reservoirs of beneficial live bacteria, known as probiotics, which contribute to a balanced gut microbiota and potentially bolster immune function (Leeuwendaal et al., 2022; Valentino et al., 2024).

The story of fermented foods is not simply a tale of the final product; it is a dynamic narrative of microbial succession and metabolic transformation. The composition and activity of microbial communities shift and evolve throughout the fermentation process, with different stages favoring the proliferation of specific microbial guilds. Understanding these intricate ecological dynamics is crucial to appreciating the complex interplay between microbes and their environment, ultimately paving the way for a more profound understanding and appreciation of this time-honored culinary art.

10.2 CLASSIFICATION OF MICROBES IN FERMENTED FOODS: A STARRING CAST AND SUPPORTING PLAYERS

The microbial world within fermented foods is a vibrant ecosystem teeming with diverse life-forms. While the specific players can vary depending on the type of fermentation, some key groups take center stage, each with a crucial role to play.

10.2.1 The Main Players

Lactic acid bacteria (LAB). These are the undisputed champions of many fermentation processes. LAB, a diverse group encompassing genera like *Lactobacillus*, *Streptococcus*, *Leuconostoc*, and *Pediococcus*, possess the remarkable ability to convert sugars (often lactose in dairy products) into lactic acid (Abedi & Hashemi, 2020; Bintsis, 2018). Lactic acid plays a multifaceted role:

Preservation: It creates an acidic environment that inhibits the growth of spoilage and pathogenic bacteria, extending the shelf life of the food significantly (Sionek et al., 2023; Zapaśnik et al., 2022).

Flavor: Lactic acid contributes to the characteristic tangy flavor profile of many fermented products, from yogurt and kimchi to sauerkraut and sourdough bread (Dimidi et al., 2019).

Yeasts. Yeasts, particularly *Saccharomyces cerevisiae* (baker's yeast), are another fundamental player (Leeuwendaal et al., 2022). They are masters of sugar fermentation, converting them into thus:

Alcohol (ethanol): This is the foundation for alcoholic beverages like beer, wine, and some ciders (Iyabo Christianah, 2019).

Carbon dioxide (CO_2): CO_2 plays a vital role in the leavening of dough during bread fermentation, creating the characteristic rise and airy texture (Srivastava et al., 2022).

Flavor and aroma: Certain yeast strains contribute additional volatile compounds that enhance the flavor and aroma profile of fermented products (Zhang et al., 2023).

10.2.2 Metabolic Pathways and Microbial Interactions

The biochemical transformations orchestrated by microbes during fermentation are complex and multifaceted. LAB, through glycolysis and subsequent fermentation pathways, primarily produce lactic acid, lowering the pH and creating an environment hostile to many spoilage organisms. This acidification also influences the solubility and bioavailability of minerals, contributing to the nutritional enhancement observed in many fermented foods. Furthermore, LAB produce *bacteriocins*, antimicrobial peptides that provide an additional layer of protection against competing microorganisms. Yeasts, while primarily known for alcoholic fermentation, also contribute a range of flavor compounds, including esters, higher alcohols, and aldehydes, adding complexity and depth to the sensory profile of fermented beverages and baked goods. The metabolic interplay between LAB and yeasts can be synergistic or competitive, depending on factors like substrate availability and environmental conditions. For example, in sourdough bread production, LAB create an acidic environment that favors yeast strains tolerant to low pH, resulting in a unique flavor profile not achievable with either microbe alone (Leeuwendaal et al., 2022).

10.2.3 Environmental Influences on Microbial Communities

The environment within a fermenting food matrix is dynamic and highly influential in shaping microbial community structure and function (Hao et al., 2021). Temperature plays a crucial role, impacting both the growth rate and the metabolic activity of different microbes. *Psychrophilic* microbes thrive at lower temperatures, while *thermophilic* microbes prefer warmer conditions. Oxygen availability also exerts a profound influence, differentiating aerobic fermentations, like kombucha production (Gonzalez & Aranda, 2023), from anaerobic processes, like the fermentation of kimchi or sauerkraut. The availability of nutrients, including carbohydrates, proteins, and minerals, directly impacts microbial growth and metabolic output, affecting the final product's characteristics. Salt concentration, often used as a preservative, plays a selective role, inhibiting the growth of salt-sensitive microbes while favoring halotolerant species. Furthermore, the physical structure of the food matrix, such as the surface area available for microbial colonization or the presence of biofilms, can significantly impact microbial activity and community dynamics (Shen et al., 2021; Wolfe & Dutton, 2015).

10.2.4 Controlling and Manipulating Fermentations

Understanding the factors influencing microbial communities empowers food producers to control and manipulate fermentation processes to achieve desired product attributes (Sawant et al., 2025). *Backslopping*, the practice of inoculating a new batch of fermentation with a portion of a previously successful batch, provides a controlled introduction of desirable microbial communities. Starter cultures, containing carefully selected strains of LAB, yeasts, or other microbes, offer a more standardized approach, ensuring consistency in product quality and characteristics. Furthermore, manipulating environmental parameters like temperature, oxygen availability, and salt concentration provides fine-tuned control over microbial growth and metabolic activity (Siddiqui et al., 2023). Modern fermentation technologies, such as bioreactors, allow for the precise control of these parameters, optimizing fermentation processes for enhanced efficiency and product consistency (Gonzalez & Aranda, 2023).

10.2.5 The Future of Fermented Foods

As our understanding of microbial ecology deepens, the future of fermented foods holds immense potential. Metagenomics and other molecular techniques offer unprecedented insights into the complex microbial communities driving these processes, allowing for the identification of novel microbial species and the elucidation of their functional roles (Leech et al., 2020). This knowledge can be harnessed to develop innovative fermentation strategies, leading to the creation of new fermented food products with enhanced nutritional value, improved functionality, and novel flavor profiles. Furthermore, the potential of fermented foods to contribute to human health is an area of ongoing research, with studies exploring the role of probiotics and other bioactive compounds produced during fermentation in promoting gut health, boosting immunity, and even mitigating chronic diseases (Caffrey et al., 2024; Leech et al., 2020; Yu et al., 2022).

10.2.6 The Supporting Cast

While LAB and yeasts are the stars of the show, other microbes play crucial supporting roles in specific fermented foods:

Molds. Certain mold species, like *Penicillium* and *Aspergillus*, are essential for the development of specific textures and flavors in some cheeses. They contribute by breaking down complex molecules in the cheese matrix, leading to the characteristic creaminess, sharpness, or blue veining found in different varieties (Martin & Cotter, 2023; Zheng et al., 2021).

Acetic acid bacteria. These bacteria specialize in the oxidation of ethanol into acetic acid. Their presence is particularly important in products like sauerkraut, kombucha, and vinegar, where the characteristic vinegar tang is a defining feature (Gomes et al., 2018; Yassunaka Hata et al., 2023).

10.2.7 Beyond the Classification

It is important to understand that the microbial landscape in fermented foods is not static. Different stages of the fermentation process favor the growth of specific microbes. For example, in the early stages, yeasts might dominate due to their ability to readily utilize simple sugars. Later, as the environment becomes more acidic due to lactic acid production, LAB become more prominent. This dynamic interplay between microbial populations is a key factor in achieving the desired characteristics in the final fermented product Alekseeva et al., 2021; Dimidi et al., 2019; Sharma et al., 2020).

10.2.8 The Takeaway

An overall scheme has been given in **Figure 10.1.** The classification of microbes in fermented foods reveals a fascinating cast of characters, each with a unique contribution. Understanding the roles played by LAB, yeasts, molds, and acetic acid bacteria allows us to appreciate the intricate dance of life within the fermenting vessel. This knowledge is not just for scientific curiosity; it empowers producers to control the fermentation process and create delicious and nutritious fermented foods for our enjoyment (Dimidi et al., 2019; Sharma et al., 2020; Voidarou et al., 2020).

10.3 FUNCTIONAL ROLES OF MICROBES IN FERMENTATION: A SYMPHONY OF METABOLISM, ENERGY TRANSFER, AND PRESERVATION

The magic of fermented foods lies not just in their unique flavors and textures but also in the unseen symphony orchestrated by microbes. These tiny organisms are not simply passengers on a culinary journey; they are the driving force behind the metabolic transformations that define fermentation. Let us delve into the functional roles of microbes, exploring how they manipulate energy transfer and conservation to create the foods we love.

10.3.1 Metabolic Transformations: A Tale of Two Pathways

Fermentation, at its core, is a metabolic strategy employed by microorganisms to extract energy from organic substrates in the absence of oxygen or when oxygen utilization is suppressed, even if present. This process can be broadly classified into two main categories based on the presence or absence of oxygen: aerobic and anaerobic fermentation. While both strategies ultimately serve the purpose of energy generation, their metabolic pathways, end products, and ecological implications differ significantly.

10.3.1.1 Aerobic Fermentations: Harnessing the Power of Oxygen

While the term "aerobic fermentation" may appear counterintuitive, it refers to a specific set of metabolic processes where microorganisms utilize oxygen, albeit incompletely, to oxidize organic substrates. In these scenarios, even though oxygen is present, the full respiratory pathway is not employed, and the final electron acceptor

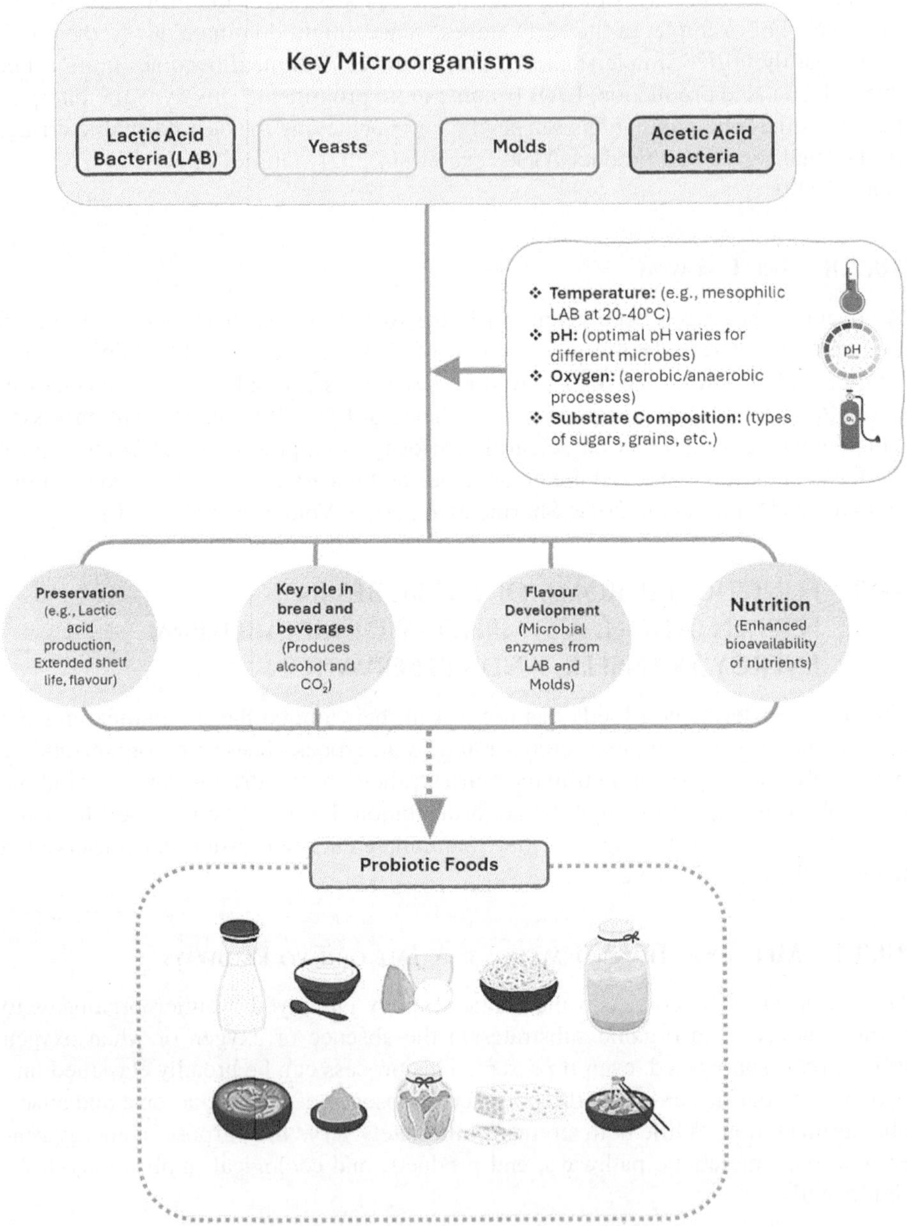

FIGURE 10.1 Microbes in fermentation and their basic attributes.

is not oxygen. Instead, organic molecules like acetic acid or gluconic acid are produced. This distinguishes aerobic fermentation from true aerobic respiration, where oxygen serves as the terminal electron acceptor, leading to the complete oxidation of substrates to carbon dioxide and water. Aerobic fermentations are crucial in the production of several important food products (Hagman & Piškur, 2015; Tsuchiya et al., 2021). For instance, *Acetobacter* species, key players in vinegar production, utilize oxygen to convert ethanol to acetic acid. This reaction is responsible for the characteristic sour taste and pungent aroma of vinegar. Similarly, some yeasts employed in kombucha production utilize oxygen to produce gluconic acid, contributing to the beverage's unique flavor profile. Another example is the production of certain fermented dairy products like kefir, where yeasts can contribute to the production of acids and other flavor compounds under partially aerobic conditions. While aerobic fermentations generate less ATP compared to full respiration, they still yield significantly more energy than anaerobic fermentations, enabling faster growth rates and higher biomass production (Hagman & Piškur, 2015; Møller et al., 2002; Tsuchiya et al., 2021).

10.3.1.2 Anaerobic Fermentations: Thriving in Oxygen's Absence

In contrast to aerobic fermentations, anaerobic fermentations occur strictly in the absence of oxygen. This is often the case in submerged fermentations, such as the production of sauerkraut, kimchi, yogurt, or certain types of cheese. In these oxygen-depleted environments, microorganisms must rely on alternative metabolic pathways to generate ATP. LAB, the quintessential anaerobic fermenters, employ a variety of pathways, the most common being homolactic fermentation. In this pathway, glucose is converted to pyruvate through glycolysis, and the pyruvate is subsequently reduced to lactic acid (Buckel, 2021). This process regenerates NAD+, a crucial coenzyme required for glycolysis to continue. Some LAB utilize the heterolactic fermentation pathway, producing not only lactic acid but also ethanol and carbon dioxide (Jiang et al., 2008). This pathway is employed by species like *Leuconostoc mesenteroides*, which contributes to the characteristic flavor and texture of sauerkraut. Yeasts, while capable of aerobic respiration, can also thrive under anaerobic conditions, employing alcoholic fermentation. In this pathway, glucose is converted to pyruvate through glycolysis, and the pyruvate is then decarboxylated to acetaldehyde, which is subsequently reduced to ethanol (Møller et al., 2002). This process, employed in the production of beer and wine, generates the characteristic alcohol content and contributes to the complex flavor profiles of these beverages. While anaerobic fermentations yield significantly less ATP compared to aerobic respiration or even aerobic fermentation, they are essential for microbial survival in oxygen-limited environments (Hagman et al., 2013). Moreover, the diverse array of end products generated during anaerobic fermentations contribute significantly to the unique flavors, textures, and nutritional profiles of a wide range of fermented foods (da Silva Vale et al., 2022).

The interplay between aerobic and anaerobic metabolisms, along with the diverse metabolic pathways employed by different microbial species, underscores the complexity and richness of the microbial world driving the art of fermentation. This intricate metabolic dance is further influenced by a multitude of environmental factors, shaping the unique characteristics of each fermented food product.

10.3.2 Energy Transfer and Conservation

The metabolic transformations driven by microbes involve the transfer of energy stored in the chemical bonds of sugars (like glucose) to ATP, which fuels various cellular processes. This energy transfer occurs through a series of enzyme-mediated reactions:

- *Substrate-level phosphorylation.* This is a key mechanism employed by LAB during fermentation. Here, a phosphate group is directly transferred from a high-energy intermediate molecule (like phosphoenolpyruvate) to ADP (adenosine diphosphate), generating ATP (Bonora et al., 2012).
- *Electron transport chain.* Though not the primary energy source in anaerobic fermentations, some microbes, like certain yeasts, might utilize a simplified electron transport chain to generate a small amount of ATP. Here, electrons are transferred through a series of protein complexes, ultimately leading to the generation of a proton gradient across the cell membrane. This gradient can then be used to drive ATP synthesis through a mechanism called chemiosmosis (Müller et al., 2012).

10.3.3 Preservation Powerhouse

The metabolic by-products of microbial activity play a crucial role in food preservation. Lactic acid, the primary product of LAB fermentation, creates an acidic environment that inhibits the growth of spoilage and pathogenic bacteria (Ibrahim et al., 2021; Zapaśnik et al., 2022). This acidic environment disrupts essential cellular processes for these undesirable microbes, effectively extending the shelf life of fermented foods (Zapaśnik et al., 2022).

10.4 BEYOND PRESERVATION: A SPECTRUM OF BENEFITS

The functional roles of microbes extend far beyond food preservation. Fermentation processes can:

- *Enhance nutrient availability.* Microbial activity can break down complex carbohydrates and proteins into simpler forms, making them more readily digestible and bioavailable for human consumption (Nkhata et al., 2018; Oliphant & Allen-Vercoe, 2019; Sharma et al., 2020).
- *Boost vitamin content.* LAB, for instance, have the ability to synthesize certain B vitamins during fermentation, potentially enriching the nutritional profile of the food (LeBlanc et al., 2011).
- *Develop flavor and aroma.* The diverse range of metabolites produced by microbes, such as organic acids, alcohols, and esters, contributes significantly to the characteristic flavor and aroma profiles of fermented foods (Zhang et al., 2023).
- *Promote gut health.* Some fermented foods harbor live beneficial bacteria (probiotics) that can improve gut microbiota composition and potentially

contribute to gut health and immunity (Leeuwendaal et al., 2022; Valentino et al., 2024).

While the specific figures can vary depending on the fermentation process and the microbial strains involved, here are some illustrative figures to provide a sense of scale for the functional roles of microbes in fermentation:

- *Energy production.* During anaerobic fermentation with LAB, the ATP yield can be around 2–3 ATP molecules per molecule of glucose fermented. This is significantly lower compared to the theoretical maximum of around 36 ATP molecules possible through cellular respiration under aerobic conditions (Rabinowitz & Enerbäck, 2020).
- *Lactic acid production.* LAB fermentations can result in the production of lactic acid in the range of 0.5% to 3% (w/v) in the final fermented product. This amount is sufficient to create a significantly acidic environment (pH below 4.5) that inhibits the growth of most spoilage and pathogenic bacteria (Lee et al., 2021).
- *Vitamin B12 production.* The nutritional enhancement of food products through fermentation extends beyond improved digestibility and mineral bioavailability. Certain lactic acid bacteria (LAB) strains, notably those belonging to the genus *Lactobacillus*, possess the remarkable ability to synthesize vitamin B12 (cobalamin) de novo, a capability not found in plants or animals. This microbial biosynthesis of B12 can significantly enrich the vitamin content of fermented foods, offering a valuable dietary source of this essential nutrient, particularly for individuals following vegetarian or vegan diets who may be at risk of B12 deficiency. The specific B12 vitamers produced by LAB can vary, with some strains producing pseudocobalamin, an inactive analogue of B12 that can interfere with B12 uptake. However, other strains, such as *Lactobacillus reuteri*, are known to produce active forms of B12, including adenosylcobalamin and methylcobalamin, which are directly bioavailable (Molina et al., 2009; Santos et al., 2008). The amount of B12 synthesized during fermentation is influenced by a multitude of factors, including the specific LAB strain employed, substrate composition, fermentation temperature, pH, and oxygen availability. For instance, studies have demonstrated that the addition of cobalt, a crucial component of the B12 molecule, to the fermentation medium can enhance B12 production by certain LAB strains (De Angelis et al., 2014). Furthermore, the presence of prebiotic compounds, such as fructooligosaccharides, can stimulate the growth and metabolic activity of B12-producing LAB, leading to increased B12 synthesis during fermentation. Quantifying the precise amount of B12 produced during fermentation can be challenging due to the complex matrix of the fermented food and the presence of various B12 analogues. However, advanced analytical techniques, such as high-performance liquid chromatography (HPLC) coupled with mass spectrometry, allow for accurate quantification of different B12 vitamers, providing valuable insights into the B12-biosynthetic capacity of LAB and the potential of fermented foods

as a source of this essential vitamin (Chamlagain et al., 2015). While the contribution of fermented foods to dietary B12 intake is promising, further research is needed to optimize fermentation conditions for maximal B12 production and to fully understand the bioavailability of B12 from different fermented food matrices.

- *Probiotic content.* The probiotic potential of fermented foods, their ability to deliver beneficial live microorganisms to the human gut, is a key aspect of their health-promoting properties. However, the number of viable probiotic bacteria present in these foods is not uniform and can fluctuate significantly depending on a multitude of factors. These factors encompass the specific type of fermented food, the starter cultures or spontaneous fermentation processes employed, the processing methods used post-fermentation (e.g., pasteurization, heating, drying), and the storage conditions (e.g., temperature, duration). While some fermented foods, such as live-culture yogurt, kefir, and kimchi, are renowned for their rich probiotic content, others may contain significantly lower numbers of viable probiotic bacteria due to processing or storage conditions that negatively impact microbial viability. For instance, heat treatments applied to some fermented products for safety or shelf life extension can drastically reduce or eliminate the presence of live probiotics. Studies have shown that probiotic counts in fermented foods like yogurt, kefir, and kimchi can range from 10^{-7} to 10^{-10} CFU (colony-forming units) per gram or milliliter (Marco et al., 2021; Rezac et al., 2018). This wide range highlights the variability inherent in fermented food production and the importance of careful quality control to ensure consistent probiotic delivery. The specific probiotic strains present also vary depending on the product. Yogurt typically contains *Lactobacillus delbrueckii* subsp. *bulgaricus* and *Streptococcus thermophilus*, while kefir may harbor a more diverse community, including *Lactobacillus acidophilus*, *Bifidobacterium bifidum*, and various yeasts (Dimidi et al., 2019). Kimchi, a traditional Korean fermented cabbage dish, typically contains *Lactobacillus plantarum*, *Leuconostoc mesenteroides*, and *Weissella koreensis*, among other LAB species (Jung et al., 2014). The probiotic benefits associated with these different strains are not identical, with each strain exhibiting specific properties, such as adhesion to intestinal cells, immunomodulatory effects, or the production of antimicrobial compounds (Hill et al., 2014). It is crucial to note that the mere presence of high probiotic counts does not guarantee health benefits. The probiotic strains must survive transit through the gastrointestinal tract and reach the gut in sufficient numbers to exert their beneficial effects. Furthermore, the specific health benefits associated with probiotic consumption are strain-specific and depend on the individual's gut microbiota composition and overall health status (Saarela et al., 2000). Therefore, characterizing the probiotic content of fermented foods requires not only quantifying viable cell numbers but also identifying the specific strains present and evaluating their potential health benefits through in vitro and in vivo studies.

10.5 CONCLUSION: A FUTURE FERMENTED WITH POSSIBILITIES

The exploration of fermented foods transcends mere culinary appreciation; it unveils a captivating intersection of microbiology, biochemistry, and human ingenuity. By delving into the intricate microbial ecosystems that drive the transformation of raw ingredients, we gain a profound understanding of the complex biological processes underpinning this ancient culinary art. This knowledge empowers us not only to preserve and optimize traditional fermentation practices, ensuring the continuation of culturally significant food traditions, but also to harness the vast, untapped potential of microbial communities for the development of novel food products with enhanced flavor, improved functionality, and augmented nutritional value. The remarkable diversity of fermented foods reflects the remarkable metabolic versatility of the microbial world. From the tangy complexity of kimchi, shaped by the intricate interplay of LAB and yeasts, to the creamy texture and nuanced flavors of cheese, influenced by the proteolytic activities of specific bacterial and fungal communities, each fermented food represents a unique microbial signature. This diversity is further amplified by the influence of environmental factors, including temperature, oxygen availability, and substrate composition, which shape the microbial landscape and dictate the trajectory of biochemical transformations. For instance, the production of natto, a traditional Japanese fermented soybean product, relies on the specific activity of *Bacillus subtilis* var. *natto*, which produces a unique array of enzymes, including nattokinase, a fibrinolytic enzyme with potential health benefits. Similarly, the fermentation of cocoa beans, a crucial step in chocolate production, relies on a complex succession of yeast and bacterial species, each contributing to the development of the characteristic chocolate flavor profile. As we continue to unravel the secrets of these microbial artisans, the future of fermented foods promises to be as diverse and delicious as the cultures that have nurtured them for millennia. Metagenomics and other advanced molecular techniques offer unprecedented insights into the complex composition and functional dynamics of microbial communities within fermented food ecosystems. This knowledge is not only expanding our understanding of the microbial ecology of fermentation but also providing a powerful toolbox for manipulating and optimizing these processes. For example, the identification of specific microbial strains with desirable metabolic properties, such as the production of specific flavor compounds or bioactive molecules, allows for the development of targeted starter cultures, enhancing the consistency and functionality of fermented foods. Furthermore, the application of high-throughput screening methods and metabolic engineering techniques holds immense potential for the development of novel microbial strains with tailored metabolic capabilities, paving the way for the creation of entirely new fermented food products with enhanced nutritional and functional attributes. However, the journey into the future of fermented foods also presents important challenges. Ensuring food safety remains paramount, requiring rigorous monitoring and control of fermentation processes to prevent the growth of pathogenic microorganisms. The increasing demand for standardized and consistent fermented food products necessitates the development of robust and reliable fermentation technologies, particularly for traditionally spontaneous fermentations. Furthermore, addressing consumer acceptance of novel fermented foods and navigating regulatory frameworks for these products are crucial steps in realizing the full potential of this burgeoning field. As we embark on this exciting frontier, continued research and interdisciplinary collaboration will be essential to overcome these challenges and unlock the full potential of microbial communities to revolutionize the food landscape, creating a future fermented with possibilities.

REFERENCES

Abedi, E., & Hashemi, S. M. B. (2020). Lactic acid production—producing microorganisms and substrates sources-state of art. *Heliyon*, *6*(10), e04974. https://doi.org/10.1016/j.heliyon.2020.e04974

Agriopoulou, S., Stamatelopoulou, E., Sachadyn-Król, M., & Varzakas, T. (2020). Lactic acid bacteria as antibacterial agents to extend the shelf life of fresh and minimally processed fruits and vegetables: Quality and safety aspects. *Microorganisms*, *8*(6), 952. https://doi.org/10.3390/microorganisms8060952

Alekseeva, A. Y., Groenenboom, A. E., Smid, E. J., & Schoustra, S. E. (2021). Eco-evolutionary dynamics in microbial communities from spontaneous fermented foods. *International Journal of Environmental Research and Public Health*, *18*(19), 10093. https://doi.org/10.3390/ijerph181910093

Bintsis, T. (2018). Lactic acid bacteria as starter cultures: An update in their metabolism and genetics. *AIMS Microbiology*, *4*(4), 665–684. https://doi.org/10.3934/microbiol.2018.4.665

Bonora, M., Patergnani, S., Rimessi, A., De Marchi, E., Suski, J. M., Bononi, A., Giorgi, C., Marchi, S., Missiroli, S., Poletti, F., Wieckowski, M. R., & Pinton, P. (2012). ATP synthesis and storage. *Purinergic Signalling*, *8*(3), 343–357. https://doi.org/10.1007/s11302-012-9305-8

Buckel, W. (2021). Energy conservation in fermentations of anaerobic bacteria. *Frontiers in Microbiology*, *12*. https://doi.org/10.3389/fmicb.2021.703525

Caffrey, E. B., Olm, M. R., Kothe, C. I., Evans, J., & Sonnenburg, J. L. (2024). MiFoDB, a workflow for microbial food metagenomic characterization, enables high-resolution analysis of fermented food microbial dynamics. *BioRxiv*. https://doi.org/10.1101/2024.03.29.587370

Chamlagain, B., Edelmann, M., Kariluoto, S., Ollilainen, V., & Piironen, V. (2015). Ultra-high performance liquid chromatographic and mass spectrometric analysis of active vitamin B12 in cells of Propionibacterium and fermented cereal matrices. *Food Chemistry*, *166*, 630–638. https://doi.org/10.1016/j.foodchem.2014.06.068

da Silva Vale, A., Balla, G., Rodrigues, L. R. S., de Carvalho Neto, D. P., Soccol, C. R., & de Melo Pereira, G. V. (2022). Understanding the effects of self-induced anaerobic fermentation on coffee beans quality: Microbiological, metabolic, and sensory studies. *Foods*, *12*(1), 37. https://doi.org/10.3390/foods12010037

De Angelis, M., Bottacini, F., Fosso, B., Kelleher, P., Calasso, M., Di Cagno, R., Ventura, M., Picardi, E., van Sinderen, D., & Gobbetti, M. (2014). Lactobacillus rossiae, a vitamin B12 producer, represents a metabolically versatile species within the Genus Lactobacillus. *PloS One*, *9*(9), e107232. https://doi.org/10.1371/journal.pone.0107232

Dimidi, E., Cox, S. R., Rossi, M., & Whelan, K. (2019). Fermented foods: Definitions and characteristics, impact on the gut microbiota and effects on gastrointestinal health and disease. *Nutrients*, *11*(8). https://doi.org/10.3390/nu11081806

Gomes, R. J., Borges, M. de F., Rosa, M. de F., Castro-Gómez, R. J. H., & Spinosa, W. A. (2018). Acetic acid bacteria in the food industry: Systematics, characteristics and applications. *Food Technology and Biotechnology*, *56*(2). https://doi.org/10.17113/ftb.56.02.18.5593

Gonzalez, J. M., & Aranda, B. (2023). Microbial growth under limiting conditions-future perspectives. *Microorganisms*, *11*(7), 1641. https://doi.org/10.3390/microorganisms11071641

Hagman, A., & Piškur, J. (2015). A study on the fundamental mechanism and the evolutionary driving forces behind aerobic fermentation in yeast. *PLOS ONE*, *10*(1), e0116942. https://doi.org/10.1371/journal.pone.0116942

Hagman, A., Säll, T., Compagno, C., & Piskur, J. (2013). Yeast "Make-Accumulate-Consume" life strategy evolved as a multi-step process that predates the whole genome duplication. *PLoS ONE*, *8*(7), e68734. https://doi.org/10.1371/journal.pone.0068734

Hao, F., Tan, Y., Lv, X., Chen, L., Yang, F., Wang, H., Du, H., Wang, L., & Xu, Y. (2021). Microbial community succession and its environment driving factors during initial fermentation of Maotai-flavor baijiu. *Frontiers in Microbiology*, *12*. https://doi.org/10.3389/fmicb.2021.669201

Hill, C., Guarner, F., Reid, G., Gibson, G. R., Merenstein, D. J., Pot, B., Morelli, L., Canani, R. B., Flint, H. J., Salminen, S., Calder, P. C., & Sanders, M. E. (2014). Expert consensus document. The International Scientific Association for Probiotics and Prebiotics consensus statement on the scope and appropriate use of the term probiotic. *Nature Reviews. Gastroenterology & Hepatology*, *11*(8), 506–514. https://doi.org/10.1038/nrgastro.2014.66

Ibrahim, S. A., Ayivi, R. D., Zimmerman, T., Siddiqui, S. A., Altemimi, A. B., Fidan, H., Esatbeyoglu, T., & Bakhshayesh, R. V. (2021). Lactic acid bacteria as antimicrobial agents: Food safety and microbial food spoilage prevention. *Foods*, *10*(12), 3131. https://doi.org/10.3390/foods10123131

Iyabo Christianah, O. (2019). Alcoholic beverages and human health: An overview. In *Frontiers and New Trends in the Science of Fermented Food and Beverages*. IntechOpen. https://doi.org/10.5772/intechopen.81054

Jiang, H., Guan, W., Pinney, D., Wang, W., & Gu, Z. (2008). Relaxation of yeast mitochondrial functions after whole-genome duplication. *Genome Research*, *18*(9), 1466–1471. https://doi.org/10.1101/gr.074674.107

Jung, J. Y., Lee, S. H., & Jeon, C. O. (2014). Kimchi microflora: History, current status, and perspectives for industrial kimchi production. *Applied Microbiology and Biotechnology*, *98*(6), 2385–2393. https://doi.org/10.1007/s00253-014-5513-1

LeBlanc, J. G., Laiño, J. E., del Valle, M. J., Vannini, V., van Sinderen, D., Taranto, M. P., de Valdez, G. F., de Giori, G. S., & Sesma, F. (2011). B-Group vitamin production by lactic acid bacteria—current knowledge and potential applications. *Journal of Applied Microbiology*, *111*(6), 1297–1309. https://doi.org/10.1111/j.1365-2672.2011.05157.x

Lee, S.-J., Jeon, H.-S., Yoo, J.-Y., & Kim, J.-H. (2021). Some important metabolites produced by lactic acid bacteria originated from kimchi. *Foods*, *10*(9), 2148. https://doi.org/10.3390/foods10092148

Leech, J., Cabrera-Rubio, R., Walsh, A. M., Macori, G., Walsh, C. J., Barton, W., Finnegan, L., Crispie, F., O'Sullivan, O., Claesson, M. J., & Cotter, P. D. (2020). Fermented-food metagenomics reveals substrate-associated differences in taxonomy and health-associated and antibiotic resistance determinants. *MSystems*, *5*(6). https://doi.org/10.1128/mSystems.00522-20

Leeuwendaal, N. K., Stanton, C., O'Toole, P. W., & Beresford, T. P. (2022). Fermented foods, health and the gut microbiome. *Nutrients*, *14*(7), 1527. https://doi.org/10.3390/nu14071527

Liszkowska, W., & Berlowska, J. (2021). Yeast fermentation at low temperatures: Adaptation to changing environmental conditions and formation of volatile compounds. *Molecules*, *26*(4), 1035. https://doi.org/10.3390/molecules26041035

Maicas, S. (2020). The role of yeasts in fermentation processes. *Microorganisms*, *8*(8), 1142. https://doi.org/10.3390/microorganisms8081142

MansukhlalGondaliya, S., Sheth, J., & Shah, S. (2024). Exploring the culinary and health significance of fermented foods: A comprehensive review. *World Journal of Advanced Research and Reviews*, *21*(3), 2609–2616. https://doi.org/10.30574/wjarr.2024.21.3.1000

Marco, M. L., Sanders, M. E., Gänzle, M., Arrieta, M. C., Cotter, P. D., De Vuyst, L., Hill, C., Holzapfel, W., Lebeer, S., Merenstein, D., Reid, G., Wolfe, B. E., & Hutkins, R. (2021). The International Scientific Association for Probiotics and Prebiotics (ISAPP) consensus statement on fermented foods. *Nature Reviews. Gastroenterology & Hepatology*, *18*(3), 196–208. https://doi.org/10.1038/s41575-020-00390-5

Martin, J. G. P., & Cotter, P. D. (2023). Filamentous fungi in artisanal cheeses: A problem to be avoided or a market opportunity? *Heliyon*, *9*(4), e15110. https://doi.org/10.1016/j.heliyon.2023.e15110

Molina, V. C., Médici, M., Taranto, M. P., & Font de Valdez, G. (2009). Lactobacillus reuteri CRL 1098 prevents side effects produced by a nutritional vitamin B deficiency. *Journal of Applied Microbiology*, *106*(2), 467–473. https://doi.org/10.1111/j.1365-2672.2008.04014.x

Møller, K., Christensen, B., Förster, J., Piškur, J., Nielsen, J., & Olsson, L. (2002). Aerobic glucose metabolism of *Saccharomyces kluyveri:* Growth, metabolite production, and quantification of metabolic fluxes. *Biotechnology and Bioengineering*, *77*(2), 186–193. https://doi.org/10.1002/bit.10122

Müller, M., Mentel, M., van Hellemond, J. J., Henze, K., Woehle, C., Gould, S. B., Yu, R.-Y., van der Giezen, M., Tielens, A. G. M., & Martin, W. F. (2012). Biochemistry and evolution of anaerobic energy metabolism in eukaryotes. *Microbiology and Molecular Biology Reviews*, *76*(2), 444–495. https://doi.org/10.1128/MMBR.05024-11

Nkhata, S. G., Ayua, E., Kamau, E. H., & Shingiro, J. (2018). Fermentation and germination improve nutritional value of cereals and legumes through activation of endogenous enzymes. *Food Science & Nutrition*, *6*(8), 2446–2458. https://doi.org/10.1002/fsn3.846

Oliphant, K., & Allen-Vercoe, E. (2019). Macronutrient metabolism by the human gut microbiome: Major fermentation by-products and their impact on host health. *Microbiome*, *7*(1), 91. https://doi.org/10.1186/s40168-019-0704-8

Rabinowitz, J. D., & Enerbäck, S. (2020). Lactate: The ugly duckling of energy metabolism. *Nature Metabolism*, *2*(7), 566–571. https://doi.org/10.1038/s42255-020-0243-4

Ramírez Rojas, A. A., Swidah, R., & Schindler, D. (2022). Microbes of traditional fermentation processes as synthetic biology chassis to tackle future food challenges. *Frontiers in Bioengineering and Biotechnology*, *10*. https://doi.org/10.3389/fbioe.2022.982975

Rezac, S., Kok, C. R., Heermann, M., & Hutkins, R. (2018). Fermented foods as a dietary source of live organisms. *Frontiers in Microbiology*, *9*, 1785. https://doi.org/10.3389/fmicb.2018.01785

Saarela, M., Mogensen, G., Fondén, R., Mättö, J., & Mattila-Sandholm, T. (2000). Probiotic bacteria: Safety, functional and technological properties. *Journal of Biotechnology*, *84*(3), 197–215. https://doi.org/10.1016/s0168-1656(00)00375-8

Santos, F., Vera, J. L., van der Heijden, R., Valdez, G., de Vos, W. M., Sesma, F., & Hugenholtz, J. (2008). The complete coenzyme B12 biosynthesis gene cluster of Lactobacillus reuteri CRL1098. *Microbiology*, *154*(1), 81–93. https://doi.org/10.1099/mic.0.2007/011569-0

Sawant, S. S., Park, H.-Y., Sim, E.-Y., Kim, H.-S., & Choi, H.-S. (2025). Microbial fermentation in food: Impact on functional properties and nutritional enhancement—A review of recent developments. *Fermentation*, *11*(1), 15. https://doi.org/10.3390/fermentation11010015

Sharma, R., Garg, P., Kumar, P., Bhatia, S. K., & Kulshrestha, S. (2020). Microbial fermentation and its role in quality improvement of fermented foods. *Fermentation*, *6*(4), 106. https://doi.org/10.3390/fermentation6040106

Shen, D., Shen, H., Yang, Q., Chen, S., Dun, Y., Liang, Y., Zheng, J., & Zhao, S. (2021). Deciphering succession and assembly patterns of microbial communities in a two-stage solid-state fermentation system. *Microbiology Spectrum*, *9*(2). https://doi.org/10.1128/Spectrum.00718-21

Siddiqui, S. A., Erol, Z., Rugji, J., Taşçı, F., Kahraman, H. A., Toppi, V., Musa, L., Di Giacinto, G., Bahmid, N. A., Mehdizadeh, M., & Castro-Muñoz, R. (2023). An overview of fermentation in the food industry—looking back from a new perspective. *Bioresources and Bioprocessing*, *10*(1), 85. https://doi.org/10.1186/s40643-023-00702-y

Sionek, B., Szydłowska, A., Küçükgöz, K., & Kołożyn-Krajewska, D. (2023). Traditional and new microorganisms in lactic acid fermentation of food. *Fermentation*, *9*(12), 1019. https://doi.org/10.3390/fermentation9121019

Srivastava, S., Kollemparembil, A. M., Zettel, V., Claßen, T., Mobarak, M., Gatternig, B., Delgado, A., Jekle, M., & Hitzmann, B. (2022). An innovative approach in the baking of bread with CO2 gas hydrates as leavening agents. *Foods*, *11*(22), 3570. https://doi.org/10.3390/foods11223570

Tsuchiya, Y., Nakamura, T., Izumi, Y., Okazaki, K., Shinano, T., Kubo, Y., Katsuhara, M., Sasaki, T., & Yamamoto, Y. (2021). Physiological role of aerobic fermentation constitutively expressed in an aluminum-tolerant cell line of tobacco (*Nicotiana tabacum*). *Plant and Cell Physiology*, *62*(9), 1460–1477. https://doi.org/10.1093/pcp/pcab098

Valentino, V., Magliulo, R., Farsi, D., Cotter, P. D., O'Sullivan, O., Ercolini, D., & De Filippis, F. (2024). Fermented foods, their microbiome and its potential in boosting human health. *Microbial Biotechnology*, *17*(2). https://doi.org/10.1111/1751-7915.14428

Voidarou, C., Antoniadou, M., Rozos, G., Tzora, A., Skoufos, I., Varzakas, T., Lagiou, A., & Bezirtzoglou, E. (2020). Fermentative foods: Microbiology, biochemistry, potential human health benefits and public health issues. *Foods*, *10*(1), 69. https://doi.org/10.3390/foods10010069

Wolfe, B. E., & Dutton, R. J. (2015). Fermented foods as experimentally tractable microbial ecosystems. *Cell*, *161*(1), 49–55. https://doi.org/10.1016/j.cell.2015.02.034

Yassunaka Hata, N. N., Surek, M., Sartori, D., Serrato, R. V., & Spinosa, W. A. (2023). Role of acetic acid bacteria in food and beverages. *Food Technology and Biotechnology*, *61*(1), 85–103. https://doi.org/10.17113/ftb.61.01.23.7811

Yu, Y., Li, L., Xu, Y., Li, H., Yu, Y., & Xu, Z. (2022). Metagenomics reveals the microbial community responsible for producing biogenic amines during mustard [Brassica juncea (L.)] fermentation. *Frontiers in Microbiology*, *13*. https://doi.org/10.3389/fmicb.2022.824644

Zapaśnik, A., Sokołowska, B., & Bryła, M. (2022). Role of lactic acid bacteria in food preservation and safety. *Foods*, *11*(9), 1283. https://doi.org/10.3390/foods11091283

Zhang, K., Zhang, T.-T., Guo, R.-R., Ye, Q., Zhao, H.-L., & Huang, X.-H. (2023). The regulation of key flavor of traditional fermented food by microbial metabolism: A review. *Food Chemistry: X*, *19*, 100871. https://doi.org/10.1016/j.fochx.2023.100871

Zheng, X., Shi, X., & Wang, B. (2021). A review on the general cheese processing technology, flavor biochemical pathways and the influence of yeasts in cheese. *Frontiers in Microbiology*, *12*. https://doi.org/10.3389/fmicb.2021.703284

11 Food Factors Influencing Microbial Development

Puja Bhoi and Narendra Kumar

11.1 INTRODUCTION TO MICROBIAL DEVELOPMENT IN FOOD

The growth of microbes in food results in metabolic activity that alters the food's properties to varied degrees. Some of these alterations that occur in fermentation are advantageous, while others, which cause food contamination and spoiling, are undesirable (Hamad 2012). When food products get contaminated and inhabited by bacteria, they transform into a microbial ecosystem. While processed foods correlate to a worse environment for development, lowering the natural microbial population associated with raw food, fresh meals allow for fast microbial growth because of their high nutritional content. Food may get contaminated from external sources during manufacture, processing, storage, transportation, and distribution, in addition to the natural microbiota associated with its place of origin and environmental circumstances. Therefore, some of the main causes of food spoiling are the development and actions of microorganisms like bacteria, yeasts, and molds (Sherawat et al. 2021). While many microbes are helpful in causing desired changes during food fermentation, very few are pathogens. If there are enough nutrients available, a lot of bacteria can develop in food at once. The study of the types of microorganisms present (diversity and structure), their frequency of occurrence, their activities (functionality), and their interactions with their surroundings and one another (microbial communities) is known as microbial ecology of food. Since ecophysiological characteristics dictate the activities within individual cells, the kind of microbes that can develop in a certain food habitat is controlled by these combined impacts (McMeekin et al. 2010). Many of the parameters that influence growth are interdependent, so the successful control of multiplication is difficult. Some of those growth parameters include the nutrient content of the food, the preservatives used, water activity, pH, temperature, and atmosphere in which the food is exposed (Preetha and Narayanan 2020). Additionally, bacteriostatic agents in foods that use sequestering agents to limit the bacterial response to an antimicrobial agent are possibly limited. Additional growth factors may still be present in the system.

The exploration of food factors influencing microbial development is a critical area of research, given the implications for food safety and public health. The purpose of this introductory chapter is to reveal a complex interplay of environmental conditions, intrinsic food properties, and preservation techniques that collectively shape microbial behavior in food products. Collectively, this chapter underscores the multifaceted nature of microbial development in food, highlighting the critical

 DOI: 10.1201/9781003540113-11

factors that influence microbial behavior and the ongoing need for innovative preservation methods to ensure food safety. Future research plans will be designed to provide useful information for the food industry.

11.2 LEADING FACTORS OF MICROBIAL GROWTH

The factors influencing microbial development in food and the effects of food components on growth are of inherent interest for a variety of reasons. The factors have a bearing on questions of prime importance in matters of nutrition, food safety, the enhancement of flavor, and maintaining the quality of the product (Rolfe and Daryaei 2020). For these reasons, the general subject forms the basis for many of the scientific efforts of the food industry and institutional and private research organizations around the world. The environment, composition of foods, processing methods used, and food product's storage conditions throughout time all affect the microbes' ability to proliferate and persist within it. The qualities that are innate to the food matrix are known as intrinsic factors. Extrinsic variables, on the other hand, are characteristics of the surrounding environment, particularly during processing and storage. Each microbe has a minimum and maximum threshold as well as ideal growth conditions for each of the intrinsic and extrinsic factors in order to tolerate these changes (Banerjee et al. 2019). In order to thrive in a food environment, a microbe has to be able to get the nutrients required for cellular biosynthesis and energy generation. Adequate air and temperature conditions are also crucial. The development of pathogenic and spoiling microorganisms can be inhibited or inactivated by a combination of many factors (Davidson et al. 2012). A product's overall quality and safety may be improved by employing both internal and extrinsic elements, whether they are produced intentionally or organically. The shelf life of food can be increased, or microbial population development and growth can be prevented or minimized, by adjusting these characteristics (Preetha and Narayanan 2020). The factors listed in what follows are inherent to the food product itself (Figure 11.1).

11.2.1 INTRINSIC FACTORS

Intrinsic factors are those that are present in foods by origin. These are the biological, chemical, and physical characteristics of the food matrix that can arise spontaneously or intentionally. Water activity, pH levels, nutrient content, oxidation-reduction (redox) potential, and natural antimicrobial and biological constituents are a few examples of these factors (Banerjee et al. 2019; Rolfe and Daryaei 2020). Each of these is covered in detail in the following.

11.2.1.1 Water Activity

Microorganisms in food items need water in an accessible form to flourish. The water activity (*aw*) of food or the environment is used to define a microorganism's water needs. In order to adequately characterize the behavior of microorganisms in environments with lower moisture content, the term "water activity" was developed (Breene et al. 1988). Water activity significantly influences microorganism

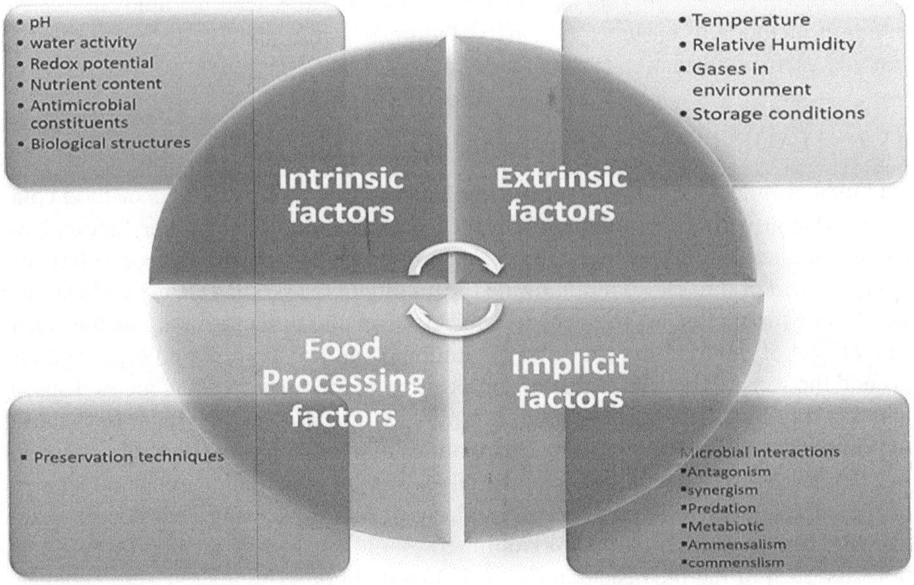

FIGURE 11.1 Different factors affecting the development of microorganisms in food.

development. *Water activity* is the ratio of the food's water vapor pressure to that of pure water at the same temperature (Chichester et al. 1963). It can be mathematically written as:

$$\mathbf{aw} = \frac{\boldsymbol{P}}{\boldsymbol{P}^{\circ}}$$

Here, *aw* denotes water activity, *P* denotes the vapor pressure of the food, and P° denotes the vapor pressure of pure water.

Completely dehydrated food has a water activity of 0, while pure water has a water activity of 1.00. The equilibrium relative humidity above the food on a scale of 0 to 100% is correlated with the meal's *aw* on this scale, which ranges from 0.00 to 1.00.

According to Labuza et al. (1972), microbial spoiling may be avoided by lowering water activity below 0.7.

In general, microorganisms have minimal and maximum aw levels for growth based on other environmental growth variables. Based on its aw range, the food is divided into three groups:

1. High-moisture foods (aw >0.85)
2. Intermediately moist foods (aw 0.60–0.85)
3. Low-moisture foods (aw <0.60)

Foods that are filled with moisture and have a water activity of 0.85 or higher are considered extremely perishable due to their susceptibility to the growth of harmful

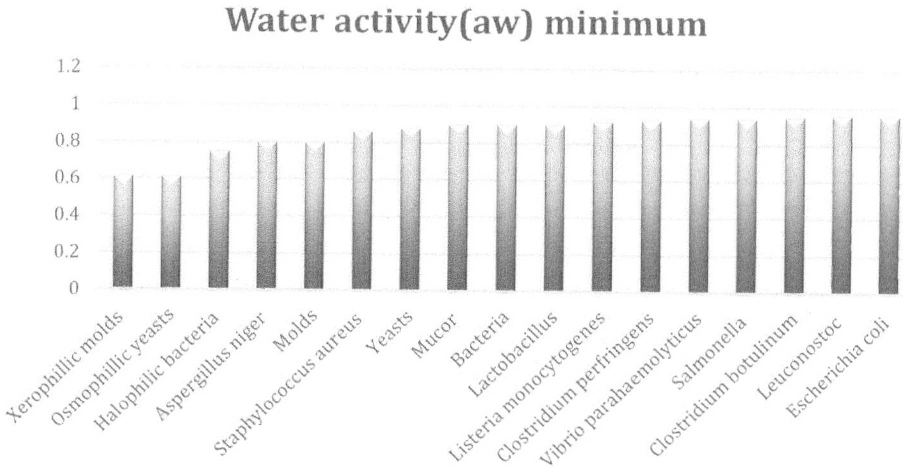

FIGURE 11.2 Minimum aw range required for specific foodborne microbes to grow.

microbes, particularly bacteria that multiply quickly (Figure 11.2). Yeasts and molds are the primary culprits for the comparatively gradual microbial deterioration of foods with intermediate moisture content (Maltinia et al. 2003). Microorganisms cannot contaminate low-moisture foods unless their moisture content is increased in any way. Although microbial development terminates when water activity falls below the minimum threshold, the microorganisms may persist in the food for extended periods of time in a latent state (Hamad 2012)

11.2.1.2 pH Levels

The pH of food is a primary factor influencing the nature and activity of microorganisms. pH is measured on a scale ranging from 0 (most acidic) to 14 (most alkaline), with a pH of 7 representing neutrality. In the region of 0 to 7, acidity is related directly to concentration levels of hydrogen ions, with the greater concentration indicating increasing acidity. Conversely, in the region of 7 to 14, alkalinity or basicity is related to concentration levels of hydroxyl ions, with the greater concentration indicating increasing alkalinity. As a general rule, spoilage is usually due to bacteria, fungi, and yeasts found to be growing at levels of pH greater than 4 to 4.5. Additionally, certain types of bacteria can survive and grow at pH levels below 4. Other types, such as the general food spoilage bacteria, are able to survive in acidic food environments (Kim et al. 2018). The approximate pH ranges of specific organisms that are pertinent to food for their development in lab conditions are listed in Table 11.1.

The pH level of a food product also influences spore germination and outgrowth in addition to the rate of bacterial growth. For example, when canning tomatoes or vegetable products using an acidified recipe, sterile ingredients before canning and proper reduction of the pH to 4.6 or below are important to restrict the growth of spoilage organisms. Past studies have shown that foods with a pH less than 4.6 are antagonistic to microbiologic pathogens, including bacteria, viruses, and parasites (Breidt et al. 2014). However, spoilage bacteria or other nonpathogenic microorganisms are

TABLE 11.1
Approximate pH Values for the Development of Certain Microorganisms in Food

Bacteria	Optimum	Minimum	Maximum
Bacillus cereus	6.0–7.0	4.9	9.3
Clostridium perfringens	7.2	5.5–5.8	8.0–9.0
Vibrio vulnificus	7.8	5.0	10.2
Salmonella	6.0 –7.5	4.5–5.0	8.0–9.0
Escherichia coli	6.0–8.0	4.3–4.4	9.0–10.0
Aspergillus niger	3.0–6.0	1.2	–
Penicillium	4.5–6.7	1.9	9.3

often capable of growing in acidic and acidified environments, close to or below the 4.6 threshold, and many have been further studied in neutral, acidic, and acidified foods. To regulate hazardous activities by these spores and other bacteria, it is essential to consider the pH level, acidulant presence, and other factors. For this, producers must employ knowledge of prepared product formulation principles, food safety regulations and guidelines, and specialized recipes and process calculations to ensure food safety, minimal food waste that results from spoilage, and maximum shelf life (Beales 2004).

11.2.1.3 Nutrient Content

Microorganisms require a few basic nutrients, such as phosphorus, iron, magnesium, sulfur, manganese, calcium, and potassium, to thrive and maintain their metabolic processes. Depending on the microbe, different types and amounts of nutrients are needed, such as nitrogen, water, vitamins, minerals, and an energy source (Jay 2000). Available nutrients include small reactive molecules, large molecules that may be broken down to simpler components by enzymes produced by microorganisms, and exogenous molecules, such as inorganic salts and organic vitamins. It is this group of exogenous molecules—vitamins, minerals, and trace elements—which is essential for the activity and development of microorganisms (Emerson and Tang 2007). Alcohols, amino acids, and carbohydrates can all be used as sources of energy by foodborne microbes. Most microbes metabolize glucose, but some may also be able to metabolize more complex carbohydrates, like cellulose or starch, which are found in plant meals, or glycogen, which is found in foods that are used by muscles. Although a very limited number of microorganisms in foods target these chemicals, certain microbes also utilize fat as an energy source (Kim and Gadd 2008).

Generally, just trace quantities of these minerals are needed, and a variety of foods are an excellent source of them. Numerous bacterial species, particularly pathogenic spore formers, like *Clostridium botulinum*, and Gram-negative bacteria, like *Pseudomonas*, *Acinetobacter*, *Moraxella*, *Shewanella*, and *Aeromonas*, are proteolytic and can thrive in protein-rich foods, quickly spoiling them or causing illness (André et al. 2017). Microorganisms that ferment carbohydrates, especially molds and yeasts, can contaminate foods high in carbohydrates. Bacterial species belonging to the following genera are capable of degrading carbohydrates: *Aeromonas*, *Pseudomonas*, *Leuconostoc*, *Enterobacter*, *Bacillus*, and *Clostridium* (Ray 2004). Gram-positive bacteria are often the least artificial. Therefore, in order for them to grow, they need to be given one or more of these substances. It is possible to find Gram-negative bacteria and molds flourishing on diets deficient in B vitamins since they are able to synthesize most of or all their needs. For growth to occur, all microbes need sources of electrons and energy (Yoshii et al. 2019). Based on how well they meet each of these criteria, microorganisms may be categorized into nutritional groups. Organisms may obtain their energy from just two sources: light energy and energy obtained by the oxidation of organic or inorganic molecules (Mamatha 2021).

11.2.1.4 Oxidation-Reduction Potential (Redox Potential)

The ratio of a substance's total oxidizing (electron-accepting) power to its total reducing (electron-donating) power is known as its *oxidation-reduction* or *redox* potential. The measurement of a substance's electron uptake or loss is called its redox potential. Millivolts are used to measure *Eh*. The Eh depends on the substrate's pH; typically, it is measured at 7.0 (Beuchat 1974). Anaerobes require a negative redox potential (a reduced state) in order to develop, while aerobic creatures require food in order to have a positive redox potential. Since other molecules may receive electrons, oxygen is not a necessary component for oxidation-reduction processes.

In general, the following Eh ranges are suitable for the growth of certain microorganisms: facultative anaerobes +300 to −100 mV, aerobes +500 to +300 mV, and anaerobes +100 to less than −250 mV. Depending on their oxygen (O_2) needs, microorganisms can be classified as aerobes, anaerobes, or facultative anaerobes. Eh decreases as a result of the aerobes' reduction of the medium's O_2. With oxidized substrates, the medium gets worse, and with reduced substrates, it gets better. When electrons are transferred between atoms or molecules, an oxidation-reduction (redox) process takes place. The substrate that contributes an electron is reduced, while the substrate that receives the electrons is oxidized. Both hydrogen and electron transfer processes are crucial components of metabolism and energy production through oxidative (substrate) phosphorylation in living cells (Wimpenny 1969). For example, *C. botulinum* is a stringent anaerobe that needs an Eh of less than +60 mV to grow; however, greater Eh values might cause slower development. The formation of anaerobes, which are contaminants in the food, may begin when the Eh of the meal approaches negative values as a result of the growth of aerobes, further spoiling the food (Figure 3) (Alwazeer 2020). The most significant reducing agents (antioxidants) found naturally in plant and animal tissues include reducing sugars, enzymes (catalase, superoxide dismutase, peroxidase), thiols (mercaptans), ascorbic acid, and polyphenols (tannins, lignin, flavonoids) (Hamad 2012). Possessing food capacity is still another crucial need. This feature is influenced by the food's active respiratory enzyme systems as well as its oxidizing and reducing ingredients. Because their respiratory enzyme systems are active, muscle foods and meats continue to breathe, which results in low Eh values (Figure 11.3) (Morris 2000). For livestock products

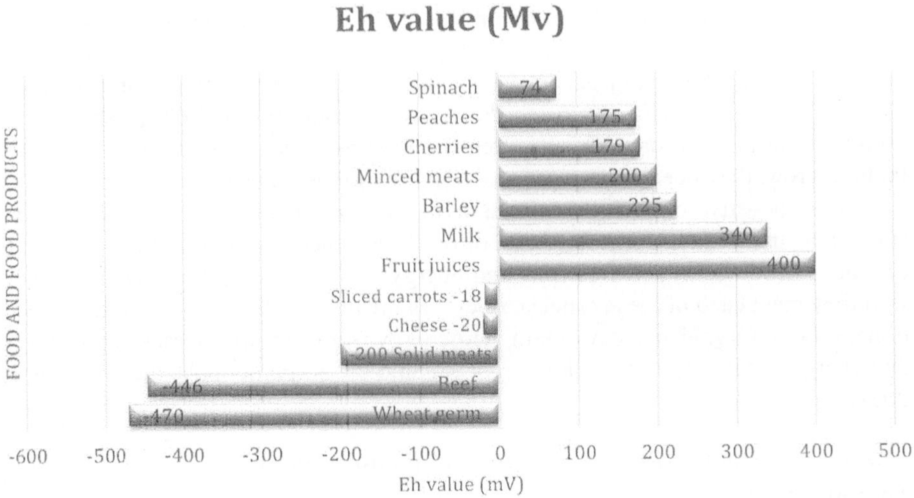

FIGURE 11.3 Eh value of various foods and food items.

with one or more components, determining a product's redox potential is a rather simple process.

11.2.1.5 Presence of Natural Antimicrobials

Natural antimicrobial compounds found in some meals will have an inhibiting effect on the development of microorganisms. These might occur during food preparation, be naturally existing in the food ingredient, or be purposefully added (Mahmud and Khan 2018). Among the many forms of antimicrobial effects are:

1. Substances that damage a cell's wall, cytoplasmic membrane, or ribosome structure or function
2. Substances (oxidative agents, chelating agents, heavy metals, and anti-metabolites) that impact microbial enzymes
3. Chemical molecules that react with DNA (cytostatics, alkylating or deaminating agents, and chemical mutations)

The antimicrobials found naturally in food, that is, essential oils from specific plant species, are recognized to have antibacterial properties. These include eugenol in cloves, allicin in garlic, eugenol and cinnamic aldehyde in cinnamon, eugenol and thymol in sage, carvacrol (isothymol) and thymol in oregano, and allyl isothiocyanate in mustard (Shelef 1984). Milk and egg whites also contain natural antimicrobials. The lactoperoxidase enzyme system, lactoferrin, and conglutinin proteins found in cow's milk all have antibacterial properties (Jay et al. 2008). There are common antibacterial agents, such as bacteriocins, organic acids, carbon dioxide, unrefined and refined herbal extracts, essential oils, chitosan, sodium carbonate, electrolyzed water, and hydrogen peroxide. The form of antibacterial agents and their antimicrobial concentrations are influenced by multiple environmental factors in food (Savvaidis and Ayala-Zavala 2021).

11.2.1.6 Biological Constituents

A natural biological structure of some kind is often present in the majority of raw food that come from plants or animals, which may prevent microbes from entering the food's cells and tissues (Jay 2000). Seed testa, fruit outer layer, nutshell, animal skin, and eggshell are examples of structures that fall within this group.

11.2.2 Extrinsic Factors

Extrinsic factors are defined as those that are governed by outside circumstances. These pertain to the factors of food preparation and storage. Food processing procedures, gaseous conditions, temperature, relative humidity, and the presence of other microorganisms are a few examples (Rolfe and Daryaei 2020).

11.2.2.1 Temperature

Temperature is the most important factor affecting microbial development in food. Depending on their resistant temperatures and relation to growth temperature, microbes are categorized into three groups: psychrotrophs (optimum range of 20–30°C), mesophiles (optimum range of 30–40°C), and thermophiles (optimum range of 40–70°C) (Table 11.2). As the temperature rises over the optimal range, the cytoplasmic membrane breaks down and enzymes and proteins become irreversibly denatured, which causes the growth rate to drop much more precipitously (Herendeen et al. 1979). At temperatures higher than the upper limit, these modifications are enough to eradicate the microbe. Growth rates decrease as the temperature decreases below the ideal range because the cell's enzyme processes slow down. Psychrophiles often contain larger quantities of unsaturated fatty acids than mesophiles, and many bacteria thrive at lower temperatures by increasing the amount of unsaturated fatty acids in their membrane lipids. The melting point of fatty acids lowers as their degree of unsaturation increases (Gill and Suisted 1978). At lower temperatures, membranes with increased unsaturated fatty acid content will continue to be fluid and functioning. However, in mesophiles, thermoduric bacteria and nutrient transport through the membrane are decreased, as is the fluidity of the cytoplasmic membrane at low temperature (Chintalapati et al. 2004).

An organism's heat sensitivity is also influenced by its growth temperature. The D value of *Listeria monocytogenes* in inoculated sausages increases by 2.4 times at 64°C (147°F) compared to 48°C (118°F). For example, all through a broad

TABLE 11.2
Temperature Variations for the Microorganism Groups

Microorganisms	Temperature
Psychophiles	Subzero–20°C
Psychotrophs	0–20°C
Mesophiles	20–45°C
Thermophiles	55–65°C
Hyperthermophiles	80–110°C

temperature range, *Clostridium perfringens* grows much more slowly at pH 5.8 than pH 7.2 (ICMSF 1980). Toxin generation generally occurs slowly below 20°C. For instance, *Clostridium perfringens* grows considerably more slowly at pH 5.8 than pH 7.2 throughout a wide temperature range. Below 20°C, toxin production usually happens gradually (ICMSF 1980).

Heat damages membranes, depletes nutrients and ions, causes ribosomes to aggregate, breaks DNA strands, inactivates vital enzymes, and causes proteins to coagulate. Enzymes, cytochromes, and permeases mediate membrane-associated metabolic activities, which will slow down when the temperature drops and membrane fluidity decreases. These occurrences will cause the fatty acyl composition to shift in a compensatory manner, increasing the fluidity of the membrane. Low temperatures can cause the commencement of protein synthesis to be blocked, ribosome structure to be disrupted, and secondary structures to develop, all of which the organism has a cellular response to (Mendoza 2014).

Many studies have shown that microflora grown at different initial temperatures can be clearly distinguished. Even with incubation at 7°C, the succession of the bacterial community can also be observed with the difference in storage time of leaf mustard, beef, pork, and straw mushrooms, which further provides the foundation for preventing microbial spoilage effectively (Su Wang et al. 2024).

11.2.2.2 Relative Humidity

Relative humidity (RH) can affect the water activity (aw) level, particularly on the food surface, and consequently affect the development of microorganisms. Temperature and relative humidity are related to each other. Relative humidity decreases with increasing temperature, and vice versa. As a general rule, molds require high relative humidity values for growth relative to the growth rate of bacteria. With very low moisture retention capacity, there is little difference between a few percentage points of moisture content in dried fruits, sugar, and honey compared with the major changes occurring in water activity (Peleg et al. 2015). As in the case of salt ions, a reducing effect exists on the water activity of foods by adding osmophilic agents to reduce the water activity of a food, as the proportion of water in the solution's gases and liquids, the use of glycerol based in meat, and the vacuum dehydration of concentrating fruit juices or brines, saline solution sprays, paste fills, and immersed foods with vegetables, fruits, or shellfish. For the latter use, the osmophilic activity of the salt does not necessarily preserve the food. Furthermore, after disinfection processes, air and surface dehumidification devices are frequently employed in industrial sanitation approaches to manage relative humidity (Esbelin et al. 2018).

11.2.2.3 Storage Conditions

Storage conditions such as temperature, moisture level, composition of gas phase, and light intensity can influence metabolic activity, reproductive capacity, cell exudation, and extracellular product formation of microorganisms. Time and temperature, together with the relative humidity (RH) to which the food or packing material may be exposed, are crucial factors to take into account (Baines and Borradale 2020) when evaluating the development rates of microbial pathogens while preserving food. Foods kept in a refrigerator have a shorter shelf life when the storage temperature rises. However, for safety, items that have been cooked or reheated and are served or kept hot might need to be properly regulated in terms of temperature. The food's susceptibility to the effects of RH is also significantly influenced by its packaging. Temperature changes in the surroundings have been shown to cause moisture to condense on the surface of some meals with low aw (Ray 2004). The development of spoilage and potentially harmful microbes will be encouraged by this surface water. Therefore, the product ought to be packed in a way that prevents ambient moisture from negatively altering the product's quality.

Furthermore, many preservative technologies for smoke-borne, thermal, and radiation treatment food are also affected by storage conditions. Frequently, different storage conditions must be used depending on pre-storage (Long et al. 2020).

11.2.2.4 Gases Present in the Environment

Gases present in the environment influence the growth, death, and development of microorganisms. Since oxygen is essential for most microorganisms, CO_2 may be detrimental or useful to them, and N_2 may affect the growth or detection of certain bacteria. Carbon dioxide is toxic for a large number of microorganisms, but due to its constant low concentration in the atmosphere, it is only a disadvantage to the microorganisms in the product headspace (Ray and Bhunia 2007).

Aerobic microorganisms need a positive redox value that is oxidized to be able to grow. Under aerobic respiration, the final electron acceptor that is used is oxygen. Anaerobic microorganisms need a negative value of redox—that means reduced. In anaerobic respiration, any other molecule than oxygen can be used as the final electron acceptor (Kornacki 2010). The majority of bacterial metabolic products, particularly those of the Gram-negative kind, can be utilized as a terminal electron acceptor for the organism's anaerobic respiration, which might stop or slow development or even cause CO_2-sensitive germs to become inactive or die. Aerotolerant and facultative anaerobes are two types of anaerobic microbes. Microorganisms that can thrive in anaerobic environments but grow more when oxygen is present are known as facultative anaerobes. *Lactobacillus* and *Pediococcus* species, as well as *Salmonella*, *E. coli*, and *Listeria enteric* pathogens, are examples of facultative anaerobes (Table 11.3). *Aerotolerant* anaerobes, like *Enterococcus* species, are those that can withstand oxygen in their surroundings but do not use it (André et al. 2021). *Microaerophiles*, or organisms that need less oxygen than atmospheric circumstances, are another type. According to Jay et al. (2008), *Campylobacter* species are microaerophiles. According to Ray (2004), facultative anaerobes grow in Eh ranges of +100 to +300 mV, obligatory aerobes grow at Eh ranges of +500 to +300 mV, and obligate anaerobes grow at Eh ranges of +100 to ~250 mV or below.

An alteration of the environment inside the package of food can stop bacteria from growing there. The product's environment can be modified to incorporate particular concentrations of the target microorganism of concern and the product—carbon dioxide (CO_2), nitrogen (N_2), and oxygen (O_2). This method, known as modified atmosphere packaging (MAP), enables the management and avoidance of the development of disease-causing and spoiling microorganisms while permitting the growth of fermenting bacteria (Czerwiński et al. 2021).

TABLE 11.3
Types of Microbes Based on Adaptation to Oxygen Concentrations

Terminology	Definition	Examples
Obligate aerobe	Entirely dependent on oxygen from the atmosphere to flourish	*Pseudomonas*, *Mycobacterium*, *Micrococcus luteus*, and majority of fungus, algae, and protozoa
Obligate anaerobe	Intolerant to oxygen and perishes when exposed to it	*Methanobacterium*, *Clostridium*, *Bacteroides*, and *Trepomonas agilis*
Facultative anaerobe	Grows better in the presence of oxygen but does not need it for growth	*Saccharomyces cerevisiae*, *Enterococcus*, and *Escherichia*
Microaerophile	Needs oxygen levels between 2 and 10% to grow, and is damaged by ambient oxygen (20%)	*Treponema pallidum*, *Spirillum volutans*, and *Campylobacter*
Aerotolerant anaerobe	May grow just as effectively with or without oxygen	*Streptococcus pyogenes*

11.2.3 Implicit Factors

Implicit factors are the third crucial component in identifying the kind of microbial development in food. The characteristics of the organisms themselves and their reactions to these settings are determined by this component. Microorganisms have the ability to either promote or impede each other's growth. Dominance in the flora is a dynamic process as metabolically active species continue to interact with one another. These interactions can be classified as antagonistic or synergistic depending on whether they encourage or inhibit growth. Microbial interactions like predation, parasitism, commensalism, amensalism, allotropy, asymbiosis, and neutrality are a few different behaviors and reactions that might be advantageous or detrimental to microorganisms (Preetha and Narayanan 2020). The organisms may create compounds like hydrogen peroxide, bacteriocins, antibiotics, and organic acids that either inhibit or kill other species. Moreover, metals and chelators are also supposed to influence the perceptive development of microorganisms, as with other catabolic compounds in toxic concentrations; they may serve as bacteriostatic agents and prevent spoilage (Timothy et al. 2021).

11.2.3.1 Microbial Interaction

The growth of a microbe in and on food may be influenced by its direct and indirect interaction with other species of the microflora. Antagonistic and synergistic interactions between individual microorganisms can significantly affect the behavior of the total microflora at particular points in the product (Gall 1970).

11.2.3.2 Antagonisms

The majority of relationships between microorganisms are antagonistic, and antagonism occurs when the development of one organism suppresses or otherwise hinders the development of the second. Numerous microbes produce alcohols and organic acids that hinder some of their rivals. Some create bacteriocins or antibiotics, which have extremely selective antimicrobial action and are frequently directed against closely related species. By taking advantage of or storing a vital mineral or vitamin that its rivals require, some bacteria can obtain a competitive edge. Some pseudomonads inhibit the development of rivals that need iron by producing siderophores, an iron-chelating chemical (Gram et al. 2002).

11.2.3.3 Synergisms

When two or more microorganisms grow together, they form a synergistic connection that produces an effect that none of the individual germs could achieve on their own. There are not many reported instances of true synergism. It has long been known that only when *Pseudomonas syncyanea* and *Lactococcus lactis* grow together will milk become blue (Little et al. 2021). The generation of a distinctive taste in a mixed starter (including *Streptococcus lactis* and *Leuconostoc* spp., such as *Leucodextranicum* and *L. citrovorum*) is only feasible at low pH levels because *Leuconostocs* convert citrate to volatile chemicals (Starrenburg and Hugenholtz 1991). *S. lactis* causes pH to drop because it produces lactic acid. Lactic acid bacteria are necessary for the "yeast cream" deficiency in cream that is brought on by yeasts (*Candida pseudotropicalis, Torulopsis sphaerica*) (Nannen and Hutkins 1991), because the distinctive foaming caused by gas generation then by yeasts requires the coagulation of milk by lactic acid bacteria (Gram et al. 2002).

11.2.3.4 Metabiosis

These relationships are basically "sequential synergisms," by which the development of one microbe results in good environmental conditions for the development of another, which can then result in favorable conditions for a third, and so forth. An excellent illustration of prolonged metabiosis may be found in raw milk (Blasche et al. 2021). *Lactococcus lactis* and a few coliforms are the first microorganisms to appear in raw milk. They produce lactic acid, which gives aciduric *lactobacilli* a suitable habitat. Oxidative molds and yeasts start to grow and oxidize lactic acid when the milk's built-up acidity inhibits the growth of *lactobacilli*. This raises the milk's pH and permits the growth of proteolytic bacteria (Quigley et al. 2013). A simple spoiling route will serve as an example of a metabiotic relationship. Lactic acid bacteria may break down the amino acid arginine into ornithine, which enteric bacteria can then break down into the odorous amine putrescine (Barbieri et al. 2019).

11.2.3.5 Commensalism

Commensalism has no effect on the host but is advantageous to the symbiont. It is the interaction of two populations in which one benefits while the other is left unchanged. Commensalism, which occurs when metabolic products from one microbial population may be utilized by other microbial communities without causing any particular benefit to the original population, is mostly associated with nutrition in the microbial world. One example is the generation of reduced redox-sensitive species, which can support the expansion of other microbial communities upon oxidation. Accordingly, methanogens generate methane, which methanotrophs may oxidize, and sulfate reducers generate sulfide, which a range of sulfide-oxidizing species can oxidize (Canfield et al. 2005).

11.2.3.6 Predation

A *predator* is a creature that consumes other organisms, and the main predators of prokaryotes in nature are a wide variety of protozoans. Protists aggressively consume other food sources, such as prokaryotes (particles), via a procedure called phagocytosis, when a food particle (prokaryotic, for example) is "consumed" at the cell surface in specialized feeding organelles. The protist receives prokaryotes by passive diffusion, direct interference, or filter feeding. Filter feeding is achieved by actively moving water through cilia, which are stiff tentacles on the protist's surface that filter out tiny cells and other food items from the surrounding environment. Particles are transported to the feeding protist's surface by fluid movement inside the medium during direct interception. Following ingestion, food particles develop a vacuole that joins membrane-bound enzyme sacs known as lysosomes to complete the particle's digestion (Canfield et al. 2005).

11.2.3.7 Amensalism

A relationship in which one population's actions are detrimental to another is known as *amensalism*. This usually happens to microorganisms when the by-products of one metabolism are harmful to another. For example, manufacturing oxygen by cyanobacteria, blocking the growth of anaerobic organisms, or generating chemical compounds that hinder substances as by-products of metabolism. For instance, many microbes are inhibited by ethanol, a fermentation product, especially when it is present in greater amounts. A wide range of microorganisms are inhibited by the harsh environment created by the acid produced during sulfide oxidation, especially in surface sediments where the pH can be pushed extremely low (Canfield et al. 2005). The reverse of commensalism, known as amensalism, occurs when one organism suffers harm while the other is unaffected.

11.2.4 Processing Factors

Food composition and the types and number of microorganisms present are all affected by a range of processing variables, including drying, heating, and cooling. Structural damage or physiological dysfunctions, like rupture of envelopes, changes in DNA conformation, or protein aggregation, are frequently linked to microbial cell death (Gould 1996). Multiple injuries occurring at the same time make it difficult for researchers to attribute cell viability to a single event. Protein coagulation, ribosome aggregation, DNA strand breakage, membrane damage, loss of nutrients and ions, and inactivation of vital enzymes are all brought on by heat. Bacterial cells are affected by ionizing radiation in two ways: directly through radiation absorption, and indirectly through interactions between reactive species (Russell 2003).

The transition of the membrane from liquid crystalline to gel section is facilitated by high pressure, which tightens the packing of acyl chains inside phospholipid bilayers. Bacterial resistance to hydrogen peroxide (HHP) is influenced by the makeup and condition of the bacterial cell membrane prior to pressure treatment (Ulmer et al. 2002). Although it is unclear how this makes them more resistant, cells with a more fluid membrane are more barotolerant. There have also been reports of cytoplasmic membrane disruption following pressurization, osmotic responsiveness loss, intracellular material loss, and the development of lipid-based buds and vesicles (Mañas and Mackey 2004). Membrane damage is one of the main targets of HHP inactivation, which is multitarget. Bacterial spores may tolerate up to 1,000 MPa for extended periods of treatment, demonstrating their strong resistance to HHP (Sarker et al. 2015).

11.2.4.1 Preservation Techniques

The primary purpose of food preservation is to stop or exclude microbial activity using a variety of techniques. These techniques include destroying microorganisms, excluding or eliminating them, and preventing their development or survival. A hostile environment is produced by chemical food additives including salts, sugars, and organic acids, but many plant- and animal-based diets naturally contain antibacterial substances. Through both direct and indirect harmful effects, gases like carbon dioxide, ozone, and oxygen can also prevent bacteria from growing and proliferating. Antimicrobial compounds are created in food by food processing methods, including fermentation, smoking, and heating. One of the earliest methods of food processing that can inhibit the development and survival of foodborne spoilage microbes is food fermentation (Ibrahim et al. 2021).

Lactic acid bacteria (LAB) are utilized as the foundation for the fermentation of milk, meats, cereals, and vegetables because of their distinct metabolic activities (Bintsis 2018). Lactic, acetic, and propionic acids; ethanol; hydrogen peroxide; diacetyl; and bacteriocins—protein or peptide antimicrobial compounds that render other bacteria inactive—can all be produced by LAB. Another obstacle to managing the natural population of microorganisms is the production of antimicrobial compounds by these organisms (Şanlıbaba and Güçer 2015).

11.3 CONCLUSION AND FUTURE PROSPECTS

The growth of the microbe is therefore greatly influenced by all the aforementioned factors (intrinsic, extrinsic, implicit, and processing) to varying degrees, and as a result, controlling these aspects is crucial. However, microorganism development and resistance in food items are influenced by prominently both extrinsic and intrinsic factors. The main factor influencing shelf life is often, but not always, microbial proliferation. Several bacteria tend to outcompete other microbes and take over at the end of their shelf life because their minimal growth temperatures are at or below the temperatures at which perishable goods are stored. The harmless development of homofermentative lactic acid bacteria in some food products treated in specific ways is not associated with sensory loss. While there are more quick and accurate methods that are utilized in industrial research and specific projects, standard plate count procedures are frequently employed for routine shelf life assessment. The methods for detecting chemical by-products of microbial metabolism can be used to predict when food will reach the end of its shelf life. However, harmful bacteria may be controlled by modifying their energy sources and metabolic needs. Food quality and safety may be preserved throughout storage by adjusting these factors. In order to better forecast products and stop microorganism development in food systems, further research employing physiological, molecular, and genetic methods might enhance our understanding of microorganism reactions.

REFERENCES

Alwazeer, Duried. "Importance of consideration of oxidoreduction potential as a critical quality parameter in food industries." *Food Research International* 132 (2020): 109108.

André, Antonin C., Lorine Debande, and Benoit S. Marteyn. "The selective advantage of facultative anaerobes relies on their unique ability to cope with changing oxygen levels during infection." *Cellular Microbiology* 23, no. 8 (2021): e13338.

André, Stéphane, Tatiana Vallaeys, and Stella Planchon. "Spore-forming bacteria responsible for food spoilage." *Research in Microbiology* 168, no. 4 (2017): 379–387.

Baines, Janis, and David Borradale. "Food Safety: Microbiological Risks." In *Food and Nutrition*, pp. 123–138. Routledge, 2020.

Banerjee, Dipak, Arun Kumar Das, Neelam Thakur, Sharif Uddin Talukder, Animesh Chandra Das, J Sonowal, Brown Paul, Rajendra Kumar Khare, and Dt Sakunde. "Factors affecting microbial growth in livestock products: A review." *International Journal of Chemical Studies* 7 (2019): 4017–4022.

Barbieri, Federica, Chiara Montanari, Fausto Gardini, and Giulia Tabanelli. "Biogenic amine production by lactic acid bacteria: A review." *Foods* 8, no. 1 (2019): 17.

Beales, Nikki. "Adaptation of microorganisms to cold temperatures, weak acid preservatives, low pH, and osmotic stress: A review." *Comprehensive Reviews in Food Science and Food Safety* 3, no. 1 (2004): 1–20.

Beuchat, L. R. "Combined effects of water activity, solute, and temperature on the growth of Vibrio parahaemolyticus." *Applied Microbiology* 27, no. 6 (1974): 1075–1080.

Bintsis, Thomas. "Lactic acid bacteria as starter cultures: An update in their metabolism and genetics." *AIMS Microbiology* 4, no. 4 (2018): 665.

Blasche, Sonja, Yongkyu Kim, Ruben A. T. Mars, Daniel Machado, Maria Maansson, Eleni Kafkia, Alessio Milanese, Georg Zeller, Bas Teusink, Jens Nielsen, Vladimír Beneš, Rute Neves, Uwe H. Sauer, and Kiran Raosaheb Patil. "Metabolic cooperation and spatiotemporal niche partitioning in a kefir microbial community." *Nature Microbiology* 6 (2021): 196–208.

Breene, W. M., P. S. Taoukis, and T. P. Labuza. "The role of water activity in foods." *Advanced Cereal Science Technology* 9 (1988): 91–94.

Breidt, Frederick, Kathryn Kay, Jason A. Osborne, Barbara H. Ingham, and Fletcher M. Arritt. "Thermal processing of acidified foods with pH 4.1 to pH 4.6." *Food Protection Trends* 34, no. 3 (2014): 132–138.

Canfield, Donald E., Erik Kristensen, and Bo Thamdrup. "The methane cycle". In *Advances in Marine Biology*, vol. 48, pp. 383–418. Academic Press, 2005.

Chichester, C. O. et al. "Post drying." *Microbiological Quality of Foods* (1963): 224–225.

Chintalapati, Suresh K., Mayuresh Dilip Kiran, and Sisinthy Shivaji. "Role of membrane lipid fatty acids in cold adaptation." *Cellular and Molecular Biology* 50, no. 5 (2004): 631–642.

Czerwiński, Krzysztof, Tomasz Rydzkowski, Jolanta Wróblewska-Krepsztul, and Vijay Kumar Thakur. "Towards impact of modified atmosphere packaging (MAP) on shelf-life of polymer-film-packed food products: Challenges and sustainable developments." *Coatings* 11, no. 12 (2021): 1504.

Davidson, P. Michael, and Faith Critzer. "Interventions to Inhibit or Inactivate Bacterial Pathogens in Foods." In *Microbial Food Safety: An Introduction*, pp. 189–202. New York, NY: Springer New York, 2011.

Emerson, David, and Jane S. Tang. "Nutrition and media." *Methods for General and Molecular Microbiology* (2007): 200–214.

Esbelin, Julia, Tiago Santos, and Michel Hébraud. "Desiccation: An environmental and food industry stress that bacteria commonly face." *Food Microbiology* 69 (2018): 82–88.

Gall, Lorraine S. "Significance of microbial interactions in control of microbial ecosystems." *Biotechnology and Bioengineering* 12 (1970).

Gill, Colin O., and J. R. Suisted. "The effects of temperature and growth rate on the proportion of unsaturated fatty acids in bacterial lipids." *Journal of General Microbiology* 104, no. 1 (1978): 31–36.

Gould, Grahame W. "Methods for preservation and extension of shelf life." *International Journal of Food Microbiology* 33, no. 1 (1996): 51–64.

Gram, Lone, Lars Ravn, Maria Rasch, Jesper Bartholin Bruhn, Allan B. Christensen, and Michael Givskov. "Food spoilage—interactions between food spoilage bacteria." *International Journal of Food Microbiology* 78, no. 1–2 (2002): 79–97.

Hamad, Siddig Hussein. "Factors affecting the growth of microorganisms in food." *Progress in Food Preservation* (2012): 405–427.

Herendeen, S. L., Ruth A. Vanbogelen, and Frederick C. Neidhardt. "Levels of major proteins of Escherichia coli during growth at different temperatures." *Journal of Bacteriology* 139 (1979): 185–194.

Ibrahim, Salam A., Raphael D. Ayivi, Tahl Zimmerman, Shahida Anusha Siddiqui, Ammar B. Altemimi, Hafize Fidan, Tuba Esatbeyoglu, and Reza VaseghiBakhshayesh. "Lactic acid bacteria as antimicrobial agents: Food safety and microbial food spoilage prevention." *Foods* 10, no. 12 (2021): 3131.

International Commission on Microbiological Specifications for Foods. *Factors Affecting Life and Death in Microorganisms: Microorganisms in Foods 3*. Vol. 1. Academic Press, 1980.

Jay, James M. *Modern Food Microbiology*. 6th ed., Gaithersburg, MD: Aspen Publishers, 2000.

Jay, James M., Martin J. Loessner, and David A. Golden. *Modern Food Microbiology*. Springer Science & Business Media, 2008.

Kim, Byung Hong, and Geoffrey Michael Gadd. "Heterotrophic Metabolism on Substrates Other than Glucose." Chapter. In *Bacterial Physiology and Metabolism*, pp. 202–251. Cambridge: Cambridge University Press, 2008.

Kim, C., K. Wilkins, M. Bowers, C. Wynn, and E. Ndegwa. "Influence of pH and temperature on growth characteristics of leading foodborne pathogens in a laboratory medium and select food beverages." *Austin Food Sciences* 3, no. 1 (2018): 1031.

Kornacki, Jeffrey L. "What factors are required for microbes to grow, survive, and die?" *Principles of Microbiological Troubleshooting in the Industrial Food Processing Environment* (2010): 103–115.

Labuza, T. P., L. McNally, Denise Gallagher, J. Hawkes, and F. Hurtado. "Stability of intermediate moisture foods. 1. Lipid oxidation." *Journal of Food Science* 37, no. 1 (1972): 154–159.

Little, William, Caroline Black, and Allie Clinton Smith. "Clinical implications of polymicrobial synergism effects on antimicrobial susceptibility." *Pathogens* 10, no. 2 (2021): 144.

Long, Yanzhen, Min Zhang, Sakamon Devahastin, and Ping Cao. "Progresses in processing technologies for special foods with ultra-long shelf life." *Critical Reviews in Food Science and Nutrition* 62 (2020): 2355–2374.

Mahmud, Jumana, and Ruhul A. Khan. "Characterization of natural antimicrobials in food system." *Advances in Microbiology* 8, no. 11 (2018): 894.

Maltinia, E., D. Torreggianib, E. Venira, and G. Bertolob. "Water activity and the preservation of plant foods." *Food Chemistry* 82, no. 1 (2003): 79–86.

Mamatha, Dereddy. "Microbial nutrition and growth." *Journal of Microbial & Biochemical Technology* 13 (2021): 1–1.

Mañas, Pilar, and Bernard M. Mackey. "Morphological and physiological changes induced by high hydrostatic pressure in exponential- and stationary-phase cells of Escherichia coli: Relationship with cell death." *Applied and Environmental Microbiology* 70 (2004): 1545–1554.

McMeekin, T. A., C. Hill, M. Wagner, A. Dahl, and Thomas Ross. "Ecophysiology of food-borne pathogens: Essential knowledge to improve food safety." *International Journal of Food Microbiology* 139 (2010): S64-S78.

Mendoza, Diego de. "Temperature sensing by membranes." *Annual Review of Microbiology* 68, no. 1 (2014): 101–116.

Morris, J. Gareth. "The effect of redox potential." *The Microbiological Safety and Quality of Food* 1 (2000): 235–250.

Nannen, Nancy L., and Robert Hutkins. "Intracellular pH effects in lactic acid bacteria." *Journal of Dairy Science* 74 (1991): 741–746.

Peleg, M., M. G. Corradini, and M. D. Normand. "On modeling the effect of water activity on microbial growth and mortality kinetics." *Water Stress in Biological, Chemical, Pharmaceutical and Food Systems* (2015): 263–278.

Preetha, S. S., and Rita Narayanan. "Factors influencing the development of microbes in food." *Shanlax International Journal of Arts, Science and Humanities* 7, no. 3 (2020): 57–77.

Quigley, Lisa, Orla O'Sullivan, Catherine Stanton, Tom P. Beresford, R. Paul Ross, Gerald F. Fitzgerald, and Paul D. Cotter. "The complex microbiota of raw milk." *FEMS Microbiology Reviews* 37, no. 5 (2013): 664–698.

Ray, B. "Fundamental food microbiology." 3rd ed., *Fundamental Food Microbiology*, p. 608. Boca Raton, FL: CRC Press, 2004.

Ray, Bibek, and Arun Bhunia. *Fundamental Food Microbiology*. CRC Press, 2007.

Rolfe, Catherine, and Hossein Daryaei. "Intrinsic and extrinsic factors affecting microbial growth in food systems." In *Food Engineering Series*, chap. 1, pp. 3–24. Cham: Springer International Publishing, 2020.

Russell, A. D. "Lethal effects of heat on bacterial physiology and structure." *Science Progress* 86 (2003): 115–137.

Şanlıbaba, Pınar, and Yalçın Güçer. "Antimicrobial activity of Lactic acid Bacteria." *Agriculture and Food* 3 (2015): 451–457.

Sarker, Mahfuzur R., Saeed Akhtar, J. Antonio Torres, and Daniel Paredes-Sabja. "High hydrostatic pressure-induced inactivation of bacterial spores." *Critical Reviews in Microbiology* 41 (2015): 18–26.

Savvaidis, Ioannis, and Jesús Fernando Ayala-Zavala. "Editorial for special issue of food natural antimicrobials." *International Journal of Food Microbiology* 358 (2021): 109414.

Shelef, L. A. "Antimicrobial effects of spices 1." *Journal of Food Safety* 6, no. 1 (1984): 29–44.

Sherawat, Megha, Ravi Kant Rahi, Varsha Gupta, Deepesh Kumar Neelam, and Devki Sain. "Prevention and control of food spoilage: An overview (review article)." *International Journal of Pharmacy and Biological Sciences* 11, no. 1 (2021): 124–130.

Starrenburg, Marjo, and Jeroen Hugenholtz. "Citrate fermentation by *Lactococcus* and *Leuconostoc spp*." *Applied and Environmental Microbiology* 57 (1991): 3535–3540.

Timothy, Bamgbose, Attahiru Iliyasu, and Anupkumar R. Anvikar. "Bacteriocins of lactic acid bacteria and their industrial application." *Current Topic in Lactic Acid Bacteria and Probiotics* 7, no. 1 (2021): 1–13. https://doi.org/10.35732/ctlabp.2021.7.1.1

Ulmer, Helge M., Heinz Herberhold, S Fahsel, Michael G. Gänzle, Roland H. A. Winter, and Rudi F. Vogel. "Effects of pressure-induced membrane phase transitions on inactivation of HorA, an ATP-dependent multidrug resistance transporter, in *Lactobacillus plantarum*." *Applied and Environmental Microbiology* 68 (2002): 1088–1095.

Wang, Su, Dequan Zhang, Qingfeng Yang, Xiangyuan Wen, Xin Li, Tongjing Yan, Rui Zhang, Wei Wang, Kumayl Hassan Akhtar, Caiyan Huang, and Chengli Hou. "Effects of different cold chain logistics modes on the quality and bacterial community succession of fresh pork." *Meat Science* 213 (2024): 109502.

Wimpenny, J. W. T. "The effect of Eh on regulatory processes in facultative anaerobes." *Biotechnology and Bioengineering* 11, no. 4 (1969): 623–629.

Yoshii, Ken, Koji Hosomi, Kento Sawane, and Jun Kunisawa. "Metabolism of dietary and microbial vitamin B family in the regulation of host immunity." *Frontiers in Nutrition* 6 (2019): 48.

12 Combating Foodborne Fungal Infections

Diagnostics, Control Measures, and Future Directions

Mayuri Bhagawati, Chayan Dhar, Sukanya Sonowal, Kabyashree Buragohain, Badal Kumar Datta, and Ratul Nath

12.1 INTRODUCTION

The relationship between food consumption and foodborne diseases was recognized early on, with Hippocrates (460 BC) being the first to highlight the strong connection between the food we eat and the illnesses we suffer from (Hutt and Hutt, 1984). Foodborne pathogens cause a major public health threat, causing illness and death in all populations of all ages. Mead et al. (1999) reported more than 200 different foodborne diseases have been identified. The most serious cases usually occur in the elderly, young children, people with compromised immune systems, and also in healthy individuals who are exposed to a very high dose of the pathogen. Common diseases related with foodborne pathogens include botulism, cholera, typhoid, dysentery, listeriosis, and many more.

Foodborne pathogens, including viruses, bacteria, and parasites, are biological agents that can lead to foodborne illnesses. A *foodborne disease outbreak* is defined as the occurrence of two or more individuals developing the same illness after consuming the same contaminated food. Foodborne illness occurs when a pathogen is consumed with food and establishes itself in the human body or when a toxin-producing pathogen contaminates food and generates a toxin that is then consumed by the host (Bintsis, 2017). Foodborne illness, commonly referred to as foodborne disease or food poisoning, includes any illness caused by consuming contaminated food. This contamination can occur through harmful bacteria, viruses, parasites, or some natural toxins, like those found in mushrooms. Over 250 different foodborne diseases (FBDs) have been recorded, with most of them being infections caused by various bacteria, viruses, and parasites (Hassanain et al., 2013).

Foodborne illness is typically categorized into two types: (1) foodborne infections and (2) foodborne intoxications. With foodborne infections, the incubation period

DOI: 10.1201/9781003540113-12

results in a longer delay between consuming contaminated food and the appearance of symptoms, compared to foodborne intoxications. The symptoms of foodborne diseases vary depending on the type of pathogen responsible. Many different organisms can lead to similar symptoms, especially abdominal cramps, diarrhea, and nausea. Hassanain et al. (2013) reported that the most frequently identified foodborne infections are those caused by bacteria (such as *Campylobacter* sp., *Salmonella* sp., *E. coli*, and *L. monocytogenes*), viruses (including hepatitis A and E and rotavirus), marine biotoxins, mycotoxins, and parasites (such as *Taenia solium*, *Trichinella spiralis*, *Fasciola* sp., *T. saginata*, *Echinococcus* sp., *Cryptosporidium parvum*, *Entamoeba histolytica*, and *Toxoplasma gondii*).

Diarrheal diseases, which are mainly caused by food- or waterborne microbial pathogens, are a major contributor to illness and death in developing countries, resulting in the loss of an estimated 1.9 million lives each year worldwide. These pathogens are responsible for a range of sporadic infections and chronic health issues, as well as large-scale and terrible outbreaks both within individual countries and across international borders (Heredia and Garcia, 2018). In industrialized countries, one-third of the population annually may experience microbiological foodborne diseases. An increasing number of these pathogens causing these significant problems are now identified as zoonotic, with a noticeable rise in certain zoonotic diseases in recent years. Raw or undercooked foods, such as meat, unpasteurized milk, shellfish, poultry, raw eggs, and rice, are the leading sources of foodborne illnesses.

EFSA and ECDC in 2016 reported that the primary food sources associated with foodborne outbreaks were mainly of animal origin, like eggs and egg products, as well as pig meat, accounting for 10% of all confirmed outbreaks each. Broiler chicken meat followed at 9%, while cheese was involved in 8%. Fish and fish products contributed to 7%, followed by milk and dairy products at 5%, beef at 4%, and crustaceans were responsible for 3%. Bacterial pathogens were behind most of the foodborne outbreaks reported in 2015, accounting 33.7% of all cases. Among these, *Salmonella* sp. was the cause of 21.8%, and *Campylobacter* sp. was responsible for 8.9% of outbreaks. In 2015, it was reported that *Salmonella* sp. in eggs were responsible for the highest number of foodborne diseases and that it ranked among the top five food–pathogen pairs in terms of total illness cases and hospitalizations. However, over the past five years, the frequency of outbreaks linked to *Salmonella* sp. and the consumption of "eggs and egg products" has been declining. Bacterial toxins were the second most common cause, accounting for 19.5% of all reported food- and waterborne outbreaks. By contrast, in 2024, viruses had been the most frequently identified agents, which were linked to only 9.2% of outbreaks in 2015. Parasites and other factors, including histamine, were implicated in fewer than 3% of the outbreaks. The unidentified causative agent was about 34% of the reported incidents. In foodborne outbreaks with strong evidence, *Salmonella* sp. were the most frequently reported pathogens in private households. In contrast, bacterial toxins (excluding *Clostridium botulinum*), viruses, and other agents were more commonly identified in public places, including canteens, restaurants, workplace cafeterias, and pubs (Bintsis, 2017).

Food spoilage is a metabolic process that makes food unsuitable for human consumption due to alterations in its sensory properties. While spoiled food may not be harmful if it does not contain any pathogens or toxins, the alterations in taste, texture, odor, flavor, and appearance often make it rejected. Some degree of spoilage is unavoidable, and several factors contribute to the deterioration of food. Damage caused by rodents and insects can provide pathways for microbes to invade and begin multiplying. Food spoilage is often recognized by unusual colors, bad odors, the softening of fruits and vegetables, and the presence of slime. However, the microbial breakdown of food starts long before these visible signs appear, as microbes begin decomposing food molecules for their metabolic needs. Initially, sugars and easily digestible carbohydrates are consumed first, followed by the breakdown of plant pectins. Next, proteins are targeted, resulting in the production of volatile compounds which produce distinct and unpleasant smells (Jabeen and Shahzad, 2019).

Fungi were previously placed in the plant kingdom, but they do not contain chlorophyll, are chemoheterotrophs, and do not have the ability to move. Their cells are eukaryotic, with membrane-bound organelles. Fungi reproduce by mitosis and have ergosterol in their cell membranes, along with 80S ribosomes. As osmotrophic organisms, they absorb nutrients from their surroundings, either as saprophytes or as parasites. Fungi generally reproduce via producing spores, either through asexual or sexual means. Growth occurs either by budding during reproductive processes or by elongation of hyphal tips during vegetative growth. Fungi store energy mainly in the form of lipids and glycogen. To acquire nutrients, they release digestive enzymes onto their substrate and absorb the resulting breakdown products.

The numerous genera and species of fungi found in food can be categorized into bacteria, molds, and yeasts. The changes that occur in food are driven by various microorganisms that utilize the food as a source of carbon and energy. These microorganisms include prokaryotes (bacteria), which are single-celled organisms without defined nuclei or other cell organelles, as well as eukaryotes, such as yeasts (single-celled) and molds (multicellular) having nuclei and other cell organelles (Pouris et al., 2024).

Yeasts belong to the broader group of fungi, which also includes molds and mushrooms. These are single-celled, colorless organisms which reproduce by budding. Their cells are round or oval in shape and able to grow even in the absence of air. Yeasts found in food are classified into two main classes: ascomycetes and fungi imperfecti. These fungal strains are often responsible for the spoilage of different foodstuffs. In addition, yeast plays a crucial role in the fermentation process, which is essential for alcoholic beverage production (Maicas, 2020). Zygosaccharomyces and related genera act as common spoilage organisms in foodstuffs like dry fruits, honey, jams, etc. These yeasts typically grow slowly, producing unpleasant odors and flavors, as well as carbon dioxide, which can lead to swelling or bursting of food containers (Escott et al., 2018). Saccharomyces species are widely recognized for their involvement in making bread and wine. Some species also thrive on fruits, including those found in fruit-flavored yogurt. *Candida* and its related genera can cause infections in humans. These yeasts are also involved in the spoilage of fruits, certain vegetables, and dairy items (Andrade et al., 2022). Molds are essential for decomposing dead plant and animal matter in nature, but they also contaminate a variety of foods and other human resources. They are easily recognized by their fuzzy or cotton-like appearance. They are a common cause of food spoilage and grow best at room temperature, typically between 25 and 30°C. Because they require free oxygen for growth, molds usually tend to develop on the surface of contaminated foods. Zygomycetes are regarded as one of the more primitive groups of fungi and widely found in nature. *Mucor* and *Rhizopus* are the most common species responsible for food spoilage. *Rhizopus nigricans*, often commonly known as bread mold, plays a role in the spoilage of a variety of foods, including berries, fruits, vegetables, etc. *Penicillium* and related genera are valuable to humans for their roles in producing antibiotics and blue cheese. However, many species also cause spoilage, and some produce toxic mycotoxins, such as patulin, ochratoxin, citreoviridin, and penitrem. *Penicillium* sp. are responsible for the visible decay on citrus, pears, and apples. *Aspergillus* produces harmful mycotoxins, including aflatoxins, ochratoxin, territrems, and cyclopiazonic acid. This genus is responsible for spoilage in a variety of materials, both food-related and non-food items. However, *Aspergillus* is most commonly associated for spoiling grains, dried beans, peanuts, tree nuts, etc. (Jabeen and Shahzad, 2019).

Foodborne pathogens that cause sporadic illnesses and outbreaks are typically identified through targeted laboratory tests. The primary aim of investigating these outbreaks is to manage and prevent further occurrences. The effective prevention of foodborne diseases generally involves the use of surveillance systems to track these illnesses, as well as the implementation of the Hazard Analysis Critical Control Point (HACCP) system across the entire food production chain, from farm to consumer, alongside import and export regulations. To reduce microbiological contamination, it is important to maintain proper environmental cleanliness, use high-quality feed for livestock, vaccinate animals, mandatorily inspect livestock, and enforce sanitary standards in slaughterhouses and meat processing facilities. Food safety can further be improved by enhanced proper personal hygiene, proper food handling, and regular cleaning of facilities and kitchen utensils. Additionally, routine health and laboratory checks for food handlers are necessary. Health education programs for both food handlers and consumers play a crucial role in preventing the spread of foodborne illnesses (Hassanain et al., 2013).

This chapter reviews the properties and diseases of foodborne fungal pathogens, their toxicological effects, the advancements in diagnostic techniques associated with foodborne fungal pathogens, the mechanism through which these foodborne pathogens cause disease, their control and prevention strategies, past outbreaks, and future implications.

12.2 FOODBORNE FUNGAL PATHOGENS

Fungi are highly diverse organisms that play numerous crucial roles in Earth's ecosystems and are found throughout the environment (Casadevall et al, 2008). Many fungi are vital to biotechnology, with industrial uses in the production of various medicines, food, and beverages. However, some fungi pose a serious risk to human health. Of the estimated 1.5 million fungal species on Earth, around 300 are known to cause diseases, which can range from allergic reactions to potentially fatal invasive infections (Hawksworth, 2001). Globally, approximately 600 million cases of foodborne illnesses occur annually (World Health Organization, 2015), some of

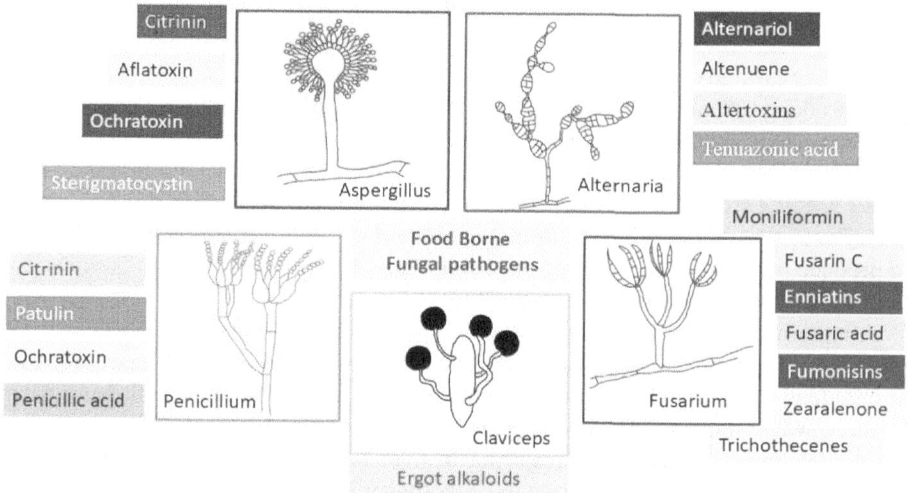

FIGURE 12.1 Types of foodborne fungal pathogens.

which may be attributed to fungi or their toxic metabolites, such as mushroom poisoning or mycotoxins. Pathogenic fungi that contaminate or spoil food include species like *Aspergillus*, *Alternaria*, *Fusarium*, *Cladosporium*, *Penicillium*, *Mucor*, and *Rhizopus* (Mateus et al., 2021). A report has also shown that some major filamentous fungi, for example, *Aspergillus*, *Fusarium*, and *Mucor*, that have contaminated food consumed or inhaled can cause localized infections in the sinuses, lungs, and other areas, even in immunocompetent people (Paterson and Lima, 2017).

Invasive fungal infections (IFIs) can be extremely severe and lead to significant mortality, although accurately determining their overall public health impact remains a challenge (Vallabhaneni et al., 2016). Among the most toxic eukaryotic foodborne pathogens, next to toxin-producing microalgae, are mycotoxin-producing fungi (Giacometti et al., 2013). These fungi, mainly from the *Aspergillus*, *Penicillium*, and *Fusarium* genera, along with certain other filamentous molds, frequently infect edible plants and animal feed. This means humans may be exposed to these toxic compounds not only by consuming contaminated food but also through the consumption of dairy products and other animal-based foods (Stoev, 2013). However, some of the compounds synthesized by these fungal members are utilized for their beneficial properties. For example, *Claviceps* spp.–synthesized ergot alkaloids have a great demand in the pharmaceutical industry for developing anti-migraine drugs, prolactin inhibitors, and medications for Parkinson's disease (Hulvova et al., 2013).

12.2.1 Major Fungal Pathogens

The following list offers a comprehensive overview of some major fungal pathogens associated with foodborne diseases.

12.2.1.1 *Aspergillus flavus*

Aspergillus flavus are multicellular microbes and a typical mold featuring structures such as hyphae and conidiophores consisting a stalk, sterigmata, vesicle, and conidia. They produce several mycotoxins, which are collectively referred to as aflatoxins that exist in various chemical variants, including B1, B2, G1, G2, and M1, where "B" and "G" denote, respectively, blue and green fluorescence (Moi et al., 2022). *Aspergillus flavus* species are present in a broad range of human environments. Several strains of *A. flavus* that commonly contaminate food and agricultural products are also able to produce mycotoxins, which can pose significant toxicity risks to both humans and animals (Gourama and Bullerman, 1995).

Aflatoxins are commonly found in nuts, spices, and figs, particularly when stored under warm and humid environments. The consumption of aflatoxin-contaminated animal feed can lead to high mortality in livestock. In humans, aflatoxins cause harm in three ways: mutagenic, causing genetic mutations (such as substitutions, frame shifts, deletions, or additions), resulting in liver damage or hepatotoxicity; teratogenic, which cause birth defects; and carcinogenic effects, leading to permanent alterations in cell function, abnormal cell growth, and the development of cancer or tumors (Mahato et al., 2019). Aspergillosis is primarily caused by *A. flavus*, especially in individuals with compromised immune systems or ongoing lung conditions. *A. flavus* accounts for approximately 30% of all aspergillosis infections (Jabeen and Shahzad, 2019).

12.2.1.2 *Aspergillus ochraceus*

Aspergillus ochraceus are microscopic fungal molds that have structures such as hyphae, conidiophores, and conidia. They produce ochratoxin, a type of mycotoxin which can be found in a wide range of foods, including wheat, corn, soybeans, oats, barley, coffee beans, meats, cheese, etc. Ochratoxins are reported to have noxious effect on both liver and kidneys and is also a potent teratogen and carcinogen. The main health problems associated with ochratoxin exposure include kidney damage and nephropathy. They interfere with cellular processes by blocking the formation of phenylalanine tRNA complex and ATP synthesis. They also trigger the lipid peroxidation process (Cousin et al., 2005). Ochratoxin A (OTA) is mainly produced by fungi such as *Aspergillus ochraceus* (FAO, 2004), which has become a global public health concern. *A. ochraceus* is a filamentous fungus characterized by biseriate conidiophores and is commonly found in various environments, including soil and agricultural products such as cereal grains, peanuts, cottonseed, and fruits (D'Mello, 2001). The OTA produced by *A. ochraceus* is noted to be stable under acidic conditions and can survive typical cooking temperatures. Furthermore, OTA is recognized as a potent carcinogen (Al Khalaileh, 2018). The fungus *A. ochraceus* readily colonizes animal feed and forage during post-harvest handling and storage. Thus, OTA negatively affects animal productivity, as the toxin can potentially remain in meat and meat-based products. In addition, this mycotoxin has also been proven to be nephrotoxic, hepatotoxic, and teratogenic in both animals as well as humans (Abdel-Wahhab and Kholif, 2010).

12.2.1.3 *Aspergillus niger*

Aspergillus niger are filamentous fungi belonging to the phylum Ascomycota. This species is primarily known to reproduce asexually, though sexual forms have also been identified. The stipes are slightly brown, the surface is smooth, the vesicle features large serrations, and the conidia surface is irregular and rough, globose in shape. *Aspergillus niger* is a common decay pathogen reported on most fruits and is often known to act as wound pathogen (Munhuweyi et al., 2016). This fungus is the cause of black mold, a condition that affects different fruits and vegetables, such as grapes, onions, and peanuts, and is a common contaminant in food items. Due to the prevalence of *A. niger*, humans are frequently exposed to its spores and vegetative forms, which are present in the air, on food items, and in stored products, leading to allergic reactions. Also, it produces mycotoxins that are toxic to the liver, the kidneys, and the immune system. This fungus, with wide distribution, is involved in the post-harvest decay of a variety of crops, leading to a considerable economic loss. For example, *A. niger*–induced black rot of onions causes productivity loss in pre-harvest as well as post-harvest phases and even during storage (Jabeen and Shahzad, 2019).

12.2.1.4 *Penicillium citrinum*

Penicillium citrinum is a saprophytic fungus, often referred to as blue or green mold. Its vegetative structure, called mycelium, consists of highly branched with septate hyphae, which can range from unicellular to multicellular forms. This fungus reproduces via sexual, asexual, and vegetative means. During asexual reproduction, it produces non-motile conidiospores that develop within a conidiophore, as well as ascospores. This species produces a mycotoxin called citrinin which is commonly found in contaminated foods, such as rice, moldy bread, oats, rye, ham, wheat, and barley. Citrinin acts as a nephrotoxin, leading to kidney damage in animals (Bennett and Klich, 2003). Schmidt-Heydt et al. (2019) reported that this species is a food-contaminating ascomycete that consistently produces large amounts of the mycotoxin citrinin. However, citrinin also shows antibiotic activity, which could contribute to the development of antibiotic resistance. Through genome sequencing, they have identified the biosynthesis gene cluster responsible for the production of citrinin, and this cluster appears to be highly conserved within the genus *Penicillium* (Ganesan et al., 2021). In their studies, Houbraken et al. (2010) and Pitt and Hocking (2009) found that *Penicillium citrinum* occurs on salt-rich foods, citrus fruits, wheat, and other cereals, leading to food contamination. This species is closely related to *P. expansum*, *P. nordicum*, and *P. verrucosum*. Citrinin has harmful effects on the kidneys and the immune system. Additionally, in the 1970s, the first statin, mevastatin, was discovered from *P. citrinum* (Endo, 2017).

12.2.1.5 *Penicillium patulum*

Penicillium patulum is a fungus species associated with the blue mold in stored apples, which is also common in pear, cherry, and other fruits. Patulin (4-hydroxy-4H-furo[3,2-c] pyran-2(6H)-one) is a toxin produced by the mold *Penicillium patulum*, later called *Penicillium urticae* and currently known as *Penicillium griseofulvum*. During the 1960s, however, it became clear that although it did have some beneficial pharmaceutical activity, this toxin was toxic to animals and plants too. After this revelation, patulin was reclassified as a true mycotoxin. This species is involved in common foods like apple sap, apple juice, etc. (Jha, 2016).

12.2.1.6 *Penicillium expansum*

Penicillium expansum are commonly known as green mold, consisting of monoverticillate, branched-septate conidiophores with sporulating cells. They produce mycotoxins named patulin, which can be degraded by SO_2 and fermentation. According to Jackson et al. (2003), a major producer of patulin mycotoxin is a mold named *Penicillium expansum*, commonly found in soil and the major source of patulin in fruits. The fungus *P. expansum* is currently regarded as the most effective natural producer of patulin. This mycotoxin is commonly found in non-fermented juices and other food items (Pique et al., 2013), but it is not resistant to fermentation processes in cider-based products, where it is readily broken down by yeasts. This mold produces a mycotoxin patulin that affects apples after harvesting or during storage, leading to rotting and decay, causing mentionable economic losses. The mycotoxin has been noted in processed apple juice, brown rot–affected pears, grapes and in animal feed (Pleadin et al., 2016).

12.2.1.7 *Fusarium* spp.

Fusarium spp. has several distinct morphological features that aid in their identification, such as the formation of macro- and microconidia having various shapes and sizes. Also, they form chlamydospores, which are specialized spores that enable the pathogenic fungus to persist in soil as well as on plant hosts for an extended period of time (Hami et al., 2021). The anamorphic genus *Fusarium* is a large group of molds consisting of around 145 species (Moss and Thrane, 2004), and many of them become a significant risk to food safety. This risk arises because most of them are plant pathogenic in nature that infect crops included in the human diet and, hence, can easily access the human body, where they induce malfunctions as well as serious ailments. Their ability to produce mycotoxins, especially fumonisins, trichothecenes, and zearalenones, has been associated with mycotoxicoses. While *Fusarium* mycotoxins are produced in the field during the plant's growth, improper storage conditions, like high moisture content, can also lead to significant mycotoxin buildup. Some *Fusarium* species, particularly *F. graminearum*, have the ability to produce deoxynivalenol, type B trichothecenes, and zearalenone. *F. sporotrichoides* is capable of producing T2 and HT2 toxins, as well as type A trichothecenes. *F. culmorum* is known to produce deoxynivalenol, along with other types of B trichothecenes, while *F. tricinctum* can produce beauvericin and enniatins (Rychlik et al., 2014). Additionally, *F. verticillioides* (previously classified as *F. moniliforme*) and *F. proliferatum* are capable of producing fumonisins, a group of long-chain polyhydroxylamines closely related in structure to sphingosine. The identification of fumonisins B1 and B2 resulted from detailed studies on fungal isolates from moldy corn following severe cases of animal poisoning in South Africa (Marasas, 2001).

12.2.1.8 *Claviceps purpurea*

Claviceps purpurea is a microscopic saprophytic fungus with septate hyphae that produces a toxic mixture of alkaloids which, while not classified as a typical mycotoxin, can result in severe health problems. This fungus commonly grows on the heads of grasses such as wheat and rye, leading to a condition known as St. Anthony's fire, named for the painful burning sensations it induces in the limbs and extremities of those affected. There are two main forms of ergotism: gangrenous and convulsive. The gangrenous form impairs blood flow, leading to tissue damage, while in the convulsive form, it affects the central nervous system. Ergotism is a serious issue for livestock, such as cattle, pigs, sheep, and chickens, causing symptoms like convulsions, abortion, gangrene, hypersensitivity, and ataxia (Cousin et al., 2005). Lunne et al. (2021) found that *Claviceps purpurea* is an ergot fungus recognized for producing neurotropic alkaloids, which have been identified as the main cause of ergotism, a disease in both livestock and humans caused by consuming ergot-contaminated feed or food.

12.2.1.9 *Mucor circinelloides*

M. circinelloides belongs to the order Mucorales, part of the early-diverging fungal lineages known as zygomycetes, and is one of the causal agents of a serious fungal infection, mucormycosis. These fungi have eucarpic, typically coenocytic thalli that contain haploid nuclei. Asexual reproduction is marked by the formation of multiple sporangiospores within a mitosporangium. Sexual reproduction occurs when two comparable gametangia fuse to form a zygospore (Botha and Botes, 2014).

Mucor species are the second most common fungi in the order Mucorales responsible for mucormycosis, after *Rhizopus* species (Petrikkos et al., 2012). According to a US FDA report, *M. circinelloides* was implicated in a foodborne illness outbreak linked to the consumption of contaminated yogurt. The *M. circinelloides* species complex includes several different species or subspecies, such as *M. circinelloides* f. *lusitanicus*, *M. circinelloides* f. *circinelloides*, and *M. circinelloides* f. *griseocyanus* (Schipper, 1976). Isolates of *M. circinelloides* f. *circinelloides* are more frequently associated with human infections and exhibit higher virulence in murine models of mucormycosis (Li et al., 2011).

12.2.1.10 *Rhizopus stolonifer*

Rhizopus stolonifer is commonly known as a bread mold which primarily grows as mycelium, made up of long, thread-like cells (hyphae) that lack cross-walls, or septa. Spores of this species are often dispersed in hot, dry weather, and they contain proteins that can trigger allergic reactions in humans, leading to symptoms like nasal issues, coughing, and chest discomfort. It is an opportunistic fungal species that causes respiratory infections, sinusitis, and otomycosis (Maness and Zubov, 2019). This fungus is a prevalent fungus that causes post-harvest diseases in a wide range of fruits and vegetables, leading to significant losses in stored products (Jabeen and Shahzad, 2019).

12.2.2 Exploration of Ecological Characteristics and Metabolic Activities of the Diverse Fungal Species Present in Food Products

Exploring the fungal diversity within food products unwraps a dynamic and intricate microbial landscape thriving with infinite interactions and metabolic activities. It is highly essential to understand the ecological dynamics of these fungal communities so that effective strategies should be formulated in order to preserve quality, ensure safety, and mitigate the risks posed by spoilage and contamination in food materials. By illustrating the roles and interactions of specific fungal taxa within food ecosystems, we can also harness the potential of beneficial fungi while reducing the risks associated with noxious species, thereby safeguarding both human health and the integrity of our food supply (Pouris et al., 2024).

A variety of ecological adaptations are possessed by foodborne fungi that enable them to colonize and break down different food substrates. Filamentous fungi, such as *Aspergillus* and *Penicillium*, produce hyphae and mycelium that penetrate and colonize food structures, resulting in visible signs of spoilage, like mold formation and change in color. Also, they secrete several extracellular enzymes, such as amylases, proteases, and lipases, which, respectively, break down carbohydrates, proteins, and lipids in food substances, leading to alterations in organoleptic properties of food, like texture, odor, flavor, etc. (Nguyen et al., 2010; Snyman et al., 2019; Cha et al., 2023)

The spoilage of food items by fungal strains involves a complex interplay of enzymatic degradation, metabolic activity, and production of secondary metabolites called mycotoxins. These mycotoxins are extremely harmful for both humans as well as other animals as they disturb the normal body processes and can even cause death. Proteolytic enzymes are secreted by foodborne fungal strains that break down proteins in food, resulting in the generation of peptides and amino acids that contribute to unpleasant flavors and odors, inducing spoilage. Similarly, lipolytic enzymes degrade lipids, which leads to rancidity and off-flavors in fatty foods. Moreover, fungal strains also generate organic acids and volatile compounds that contribute to souring and fermentation in food materials (da Rocha et al., 2014).

12.2.3 Factors Influencing Fungal Growth and Spoilage in Different Food Systems

Fungal growth and respective fungus-induced food spoilage can be affected by a variety of factors, which could be classified as intrinsic and extrinsic factors.

The inherent properties of food matrix that influence fungal growth and spoilage are termed *intrinsic factors*. Nutrient composition and availability play an important role in supporting fungal proliferation, that is, carbohydrate- and moisture-rich foods serve as ideal substrates that support fungal growth. In addition, pH levels and water and moisture content within food items are some other important intrinsic factors that impact fungal growth (Bousset et al., 2019).

Extrinsic factors, like environmental conditions, also have significant facilitating effect on the growth of pathogenic fungal mycelium and thus support rapid spoilage of both processed as well as raw food items. Temperature is considered as one of the most important extrinsic factors affecting fungal contamination. While oxygen availability is only essential in case of aerobic fungi, it is not so in the case of anaerobic fungi, which can even survive and accomplish food spoilage even in oxygen-deprived environmental conditions. Also, storage conditions, like humidity levels, packaging materials, and exposure to light, can affect fungal growth and the associated spoilage of food materials. Sometimes, fungus-induced food spoilage often happens in combination with the actions of other microorganisms, like bacteria, fungi, etc. For example, the interactions between fungi and bacteria can have either cooperative or competitive effects, which influence the extent and type of food spoilage (Pouris et al., 2024).

12.2.4 Techniques for Fungal Detection and Characterization in Different Foods

Detecting and controlling fungal pathogens in crops is crucial for maintaining food safety and security. Fungal spoilage contributes to food loss, particularly for perishable items, like fruits, vegetables, etc., and also cause damage to processed foodstuffs. Fungal contamination can occur at different stages, namely, from pre-harvest to the post-harvest stage. Warm, humid climates or during extreme weather events are some climatic parameters that facilitate food spoilage by pathogenic fungal strains (Bhilwadikar et al., 2019). In industrialized nations, spoilage concerns contribute to the loss of up to 30% of stored food at the consumer level (FAO, 2011). Thus, techniques for the identification and characterization of pathogenic fungal strains in different food samples are highly crucial. These methodologies can be generally categorized as traditional and molecular techniques and assessments based on biochemical and morphological characteristics.

1. Traditional methods, like microscopic examination, are the earliest and simplest techniques for identifying fungal pathogens in food samples. The technique of microscopic examination, in particular, involves direct observation of fungal structures, like hyphae, spores, and conidia, under the microscope, which aids in observing valuable morphological details that can assist in species identification.
2. Other traditional techniques, like culture-based methods, include isolating fungi from food samples and cultivating them on either selective or non-selective agar media. The samples are spread onto media like Sabouraud dextrose agar (SDA) or potato dextrose agar (PDA) and then incubated under suitable conditions to promote fungal growth. Ultimately, these resulting colonies are later identified based on their morphological features, growth patterns, and biochemical properties (Pouris et al., 2024).
3. Molecular methods have revolutionized the identification and characterization methods for detecting pathogenic fungal strains in food samples, offering exceptional levels of specificity, sensitivity, and precision (Borman et al., 2008).
4. Immunological methods, such as ELISA (enzyme-linked immunosorbent assay) and lateral flow assays, rely on specific antibodies in detecting fungal antigens or toxins in food samples (Ince et al., 2023).

12.2.5 Factors Related to Increasing Foodborne Illnesses

Climate change has direct effect on food production under modified environmental conditions. Basically, changes in weather patterns and temperature variations can impact the spread and distribution of foodborne microscopic pathogens. For example, changing climate conditions can cause more frequent flooding in agricultural areas, thus increasing the risk of contaminating the food supply (Shamloo et al., 2019).

1. Due to the excessive use of pesticides in agricultural fields and the exposure to harsh climatic parameters, pathogens undergo mutation to develop resistance against these antimicrobial agents and environmental changes, like different stress conditions (Ravindhiran et al., 2023), and they increase their potential to affect humans and other animals and plants as disease-causing agents (Smith et al., 2004).
2. Rapid urbanization has resulted in overcrowded living environments and, thus, enhancing difficulties in maintaining adequate food safety standards without affecting their quality and quantity. Uncontrolled urbanization is often associated with poor-quality housing and limited access to essential services, such as clean water, proper sanitation, etc., and thus providing ideal conditions for the proliferation of disease-causing vectors (Scallan et al., 2011).
3. International travel and the cross-border movement of people and food materials (processed and unprocessed) could play a critical role in the spread of foodborne pathogens. Migration and global travel are considered key factors in food-related illness transmission (Smith and Fratamico, 2018).
4. Human behavioral adjustments and consumer behavior can also lead to increased consumption of pre-packaged and ready-to-eat food items, which may carry an enhanced risk of contamination. Furthermore, people can spread microorganisms to themselves and others through actions such as coughing, sneezing, etc. The failure to wash hands before, during, and after food preparation is a notable factor in distributing foodborne microbial infections and toxins (Tack et al., 2019).

12.2.6 Advanced Strategies to Mitigating Foodborne Fungal Diseases for Prevention and Control

Pathogens have the potency to enter the food supply chain at any point, thus possessing a continuous health risk for the consumers. To mitigate these risks, it is essential to develop preventive measures and effective practices. The implementation of good agricultural methods, such as hygiene standards, thorough sanitation procedures, and effective pest control strategies, can aid in downregulating the extent of contamination during production. Implementing rigorous quality control throughout the food supply chain, such as regular monitoring and testing for fungal contamination, can help detect and alleviate potential sources that trigger spoilage (Beelman et al., 2019; Pouris et al., 2024). The application of natural preservatives, like antimicrobial compounds isolated from plants or microorganisms, can help in mitigating fungal growth and also extend the shelf life of food (da Rocha et al., 2014).

Post-harvest fungal diseases are still one of the most significant contributors of food spoilage during their storage. Several studies were conducted earlier which aimed to develop innovative methods that will reduce these post-harvest losses. For example, biocontrol agents derived from fungi have been demonstrated to prevent post-harvest fungi diseases and to extend the shelf life of several important agricultural crops. The use of antagonistic yeasts has also been mentioned in various previous literature as an efficient option to control fruit spoilage due to mycotoxin-producing fungal strains (Miranda et al., 2022), while some other previous literature have evidence of using antagonistic yeasts for mycotoxin adsorption and detoxification metabolism.

The application of competitive fungal strains is becoming an extremely trusted and popular method of combating the growth of saprophytic fungi and breaking down toxic fungal secondary metabolites as compared to chemical methods. Thus, antagonistic yeasts and their antifungal mechanisms have attracted considerable attention in securing food safety and security (Oztekin et al., 2023).

Good manufacturing practices (GMP) are considered as another pivotal technique that can be implemented in the food industry to boost the production, storage, handling, and transportation of food items. Additionally, food safety training must be provided to employees in the food sector, and this measure acts as key to maintaining high standards of food. Similarly, consumers have also been made aware of proper food handling practices and safety precautions, and this is essential in forestalling contamination during food consumption. The Hazard Analysis Critical Control Point (HACCP) system should be applied throughout all stages of food processing, including the handling of raw materials, storage, and transportation, to maintain safety in the manufactured food (Ludlow et al., 2016).

UV-C irradiation can also be applied to combat fungal diseases in fruits (Oztekin et al., 2023). Sun et al. (2022) demonstrated that UV-C can efficiently cause mycotoxin degeneration and also enhance antioxidant enzymes, leading to quality enhancement in *Pyrus sinkiangensis* T. T. Yu fruit. Post-harvest decay in blueberries has also been reported to be controlled in another study by Sanchez et al. (2021), where the authors applied a combination of UV-C treatment with yeast (*Pichia cecembensis*) for combatting inoculated *B. cinerea* pathogen and other pathogenic native microbiota. However, a more in-depth analysis is required to elucidate the biochemical pathways involved in mycotoxin detoxification and to note the by-products generated during the detoxification process. These will aid in understanding more about the advances, and even gaps, associated with the process, thus facilitating the further development of more efficient techniques (Oztekin et al., 2023). The development of biotechnological approaches like genetic engineering leads to the production of enzymes that are cost-effective and highly capable candidates that can degrade fungal secondary metabolites (Gantenbein et al., 2023). The present study offers pivotal insights into the challenges posed by emerging foodborne microbial pathogens and provides recommendations to formulate strategies to ensure food safety by alleviating the effects of pathogenic microorganisms (Hassan et al., 2023).

12.3 MYCOTOXINS AND THEIR HEALTH RISKS

12.3.1 Mycotoxins: Overview

A variety of toxic secondary metabolites produced by several filamentous fungi, such as *Aspergillus*, *Alternaria*, *Claviceps*, *Fusarium*, *Mucor*, *Monascus*, *Myrothecium*, *Penicillium*, *Pyrenophora*, *Stachybotrys*, *Trichoderma*, and *Trichothecium*, are termed mycotoxins (Khan, 2024). A large number of toxic secondary fungal metabolic compounds, namely, above 500 compounds, belonging to about 20 families have been evaluated, and the identification of more is still in progress. Among them, some frequently occurring mycotoxins in food commodities include aflatoxins, deoxynivalenol, enniatin, fumonisins, patulin, nivalenol, ochratoxins, and zearalenone (Nguyen et al., 2024; Liu et al., 2024; Miri et al., 2024). Mycotoxins, such as aflatoxin A, ochratoxin A, trichothecene-2, fumonisin B1, fumonisin B2, ergot alkaloids, patulin, deoxynivalenol, sterigmatocystin, zearalenone, etc., have a critical role in the food industry due to their presence in both processed as well as raw food items and, hence, have great impact on the economic status of food industries, along with deteriorating impacts on consumers (Goessens et al., 2024).

12.3.1.1 Aflatoxins

Aflatoxins are derivatives of difuranocoumarin with a characteristic bifuran group and a pentanone or lactone ring which is linked to the coumarin nucleus (Vila-Donat et al., 2024; El-sayed et al., 2022). The presence of a lactone ring and double bond in the terminal furan ring is considered to be the reason behind its severely toxic and carcinogenic activity. They are produced by several members of the fungal genus *Aspergillus*, primarily *Aspergillus flavus*, *A. bombycis*, *A. pseudotamarii*, *A. nomius*, and *A. parasiticus* (Awuchi et al., 2021). These group of mycotoxins contaminate a variety of food, feed, and agricultural crops, including cereals (e.g., rice, wheat, sorghum, millet, guinea corn, and corn), oilseeds (e.g., cottonseeds, sunflower, peanut, sesame, and soybean), spices (e.g., garlic, ginger, black pepper, chili pepper, coriander, turmeric), and nuts, like almonds, walnuts, and pistachios (Awuchi et al., 2022). Also, their presence in egg, meat, milk, cheese, and dry fruits has also been noted (Zhang et al., 2020). Among all the 20 discovered aflatoxins, AFB1, AFB2, AFG1, and AFG2 are the most commonly noted types, and AFB1 is considered to be the most noxious one (Liu et al., 2024), while evidence of AFM1 and AFM2 causing toxicity is also available in literature (Santos et al., 2022; Goessens et al., 2024).

12.3.1.2 Ochratoxins

Ochratoxins are fungal secondary metabolites of the genera *Aspergillus* and *Penicillium*. Ochratoxin has more than 20 subtypes, including ochratoxin A, ochratoxin B, ochratoxin C, non-amide ochratoxin α of ochratoxin A and ochratoxin C, non-amide ochratoxin β of ochratoxin B, hydroxylated ochratoxin A and ochratoxin B, etc., among which ochratoxin A is the most noxious (Ding et al., 2023). These are polyketide derivatives from dihydrocoumarins biosynthesized by several fungal species, including *A. niger*, *A. carbonarius*, *A. ochraceus*, *P. nordicum*, *P. verrucosum*, and *P. viridicatum*. The occurrence of ochratoxin in cereals, cereal-based products, coffee, licorice, raisins, vegetables, wine, nuts, pork, and poultry is mentioned (Nguyen et al., 2024). Among the different types of ochratoxins, ochratoxin A is most common and virulent. It is a phenylalanine derivative, moderately soluble in polar solvents but relatively unstable to air and light (Afsah-Hejri et al., 2020). This is formed as a result of the linkage between L-phenylalanine and isocoumarin by an amide bond. The chlorine atom in the dihydroisocoumarin ring is considered to be the main cause of its toxicity, while in ochratoxin B, there is a hydrogen atom in the same position, which makes it a relatively less-toxic mycotoxin than ochratoxin A. Meanwhile, the other subtypes of ochratoxins, namely, ochratoxin C, ochratoxin α, and ochratoxin β, are not much known to possess any significant threat to humans and animals (Ding et al., 2023; Miri et al., 2024).

12.3.1.3 Trichothecenes

Trichothecenes is a family of mycotoxin consisting of more than 200 types of structurally related fungal secondary metabolites that are mostly biosynthesized by *Fusarium*. These naturally occurring tetracyclic sesquiterpenoids are actually derivatives of terpenes, including three isoprene units, and share a common core structure consisting of acetyl groups, an epoxide group, an olfenic group, and varying numbers of hydroxyl. These are classified into four categories on the basis of functional groups: A, B, C, and D. T-2, HT-2, nivalenol, and deoxynivalenol are some common trichothecenes that contaminate foodstuff. 3-acetyldeoxynivalenol, 15-acetyldeoxynivalenol, diacetoxyscirpenol, fusarenon X, neosolaniol, NX toxin, and verrucarin are some other trichothecenes (Kolawole et al., 2021; Goessens et al., 2024). T-2 and its deacetylated form, HT-2, as well as diacetoxyscirpenol and neosolaniol, belong to type A trichothecenes, and deoxynivalenol/vomitoxin, its acetylated derivatives 3-acetyldeoxynivalenol and 15-acetyldeoxynivalenol, nivalenol, and fusarenon-X belong to the type B category (Ekwomadu et al., 2021). A very common type B trichothecene is vomitoxin, or deoxynivalenol, primarily produced by *Fusarium* spp. (*F. graminearum* and *F. culmorum*), which contaminates ~60% of leading cereal crops, like barley, corn, rice, oat, rye, sorghum, and wheat, at a global level. This trichothecene member is a huge concern because of its increased stability and heat-resistant properties. Since they are heat-tolerant, it becomes very difficult to detoxify them from food via physical techniques, like heating, milling, and other forms of processing (Zhou et al., 2020; Liao et al., 2024).

12.3.1.4 Fumonisins

These are elongated hydrocarbon chains that are linked to various methyl and hydroxyl groups, along with one free amino and two tricarballylate functionalities (Nguyen et al., 2024). Fumonisins biosynthesized by *Fusarium* (primarily *F. verticilloides* and *F. proliferatum*), and on the basis of functional groups, are classified in four categories, namely, A, B, C, and P, having 28 structural analogs, which are FA1, FA2, FA3, FB1, FB2, FB3, FB4, FB5, FC1, FC3, FC4, FP1, FP2, FP3, PHFA3a, PHFA3b, HFA3, FAK1, FBK1, Iso-FB1, PHFB1a, PHFB1b, HFB1, N-acetyl-FC1, Iso-FC1, N-acetyl-iso-FC1, OH-FC1, and N-acetyl-OH-FC1 (Qu et al., 2022). Fumonisins B group is the most commonly observed to occur in nature, which includes fumonisin B1, fumonisin B2, fumonisin B3, and fumonisin B4. Out of these, fumonisin B1 and fumonisin B2 are the most noxious and frequently occurring member of this mycotoxin family (Liu et al., 2024). They are most commonly noted to contaminate cereals and cereal-based products, especially maize and its associated processed foods. Besides asparagus, garlic, beer, dry figs, milk, and black tea all have susceptibility toward *Fusarium* infection releasing fumonisins in them (El-Sayed et al., 2022; Goessens et al., 2024).

12.3.1.5 Patulin

Another common secondary fungal metabolite synthesized by *Penicillium*, *Aspergillus*, and *Byssochlamys* is patulin, which is a water-soluble small polyketide that consists of a cyclic γ-lactone having low molecular weight (Nguyen et al., 2024). The ability of this mycotoxin to form covalent sulfhydryl-containing compounds with its target compounds or molecules is considered its key toxicity mechanism (Liu et al., 2022). Patulin is predominantly noted in fruits and vegetables, particularly in apples and apple-based products, and also in plenty other fruits, like cherries, pears, peaches, plums, grapes, and berries, like strawberries, mulberries, blueberries. Even several evidences are present in literature about their detection from plenty of manufactured foodstuffs, such as dried fruits, like apricots, dates, raisins, figs, longans, plums, and peaches; juices of fruits, like apple, lemon, orange, blueberry, and mango; and jams, like pear, apple, blueberry, strawberry, and pineapple jams (NgolongNgea et al., 2020; Stoev et al., 2024; Liu et al., 2024).

12.3.1.6 Zearalenone

Zearalenone is a significant noxious secondary fungal metabolite consisting of a resorcylic acid lactone structure and produced by *Fusarium* fungi. This mycotoxin is a serious threat that causes post-harvest contamination in cereals (Liu et al., 2024). *Fusarium cerealis*, *F. crookwellense*, *F. culmorum*, *F. equiseti*, *F. incarnatum*, *F. graminearum*, and *F. sporotrichioides* are some common members of the genus *Fusarium* that biosynthesizes zearalenone (Zhang et al., 2020; Mischler et al., 2024). Foods or feeds contaminated with this mycotoxin cannot be eliminated by employing thermal treatments due to its heat-tolerant property (Nguyen et al., 2024).

12.3.1.7 Other Mycotoxins

In addition to trichothecenes, fumonisins, and zearalenone, *Fusarium* also produce some other toxins, like moniliformin, fusarin C, fusaric acid, and enniatins (Ekwomadu et al., 2021; Kolawole et al., 2021; Goessens et al., 2024). Citrinin is another toxic secondary metabolite produced by *Aspergillus*, *Penicillium*, and *Monascus* species that predominantly contaminate cereal crops (Zargar and Wani, 2023). Alternariol, altenuene, alternariol monomethyl ether, tenuazonic acid, and altertoxins are some pivotal mycotoxins synthesized by *Alternaria* (Wang et al., 2019). The presence of another noxious mycotoxin from lupin crops and seeds has been reported. These are phomopsins, produced by *Diaporthe toxica* and considered an emerging mycotoxin (Kunz et al., 2021). Gliotoxin is another toxic metabolite biosynthesized by *Aspergillus fumigatus* (Talib et al., 2024). Ergocornine, ergometrine, ergocristine, ergocryptine, ergotamine, and ergosine are ergot alkaloids produced by fungal genera *Claviceps* and *Epichloe* mostly in cereal foodstuffs reported by EFSA on 2024.

12.3.2 Mechanisms of Mycotoxin Production and Contamination in Food Products

12.3.2.1 Biosynthesis of Mycotoxin: Mechanism

The complex mechanism of mycotoxin production depends on varied factors, including (1) pathway-specific regulators, (2) global regulators, and (3) epigenetic modification. The genes involved in the biosynthesis of fungal secondary metabolites are located adjacent to each other in the genome in biosynthetic gene clusters (BGCs), containing key biosynthetic genes, that is, synthases and/or synthetases genes (for example, cyclases, polyketide synthases, terpene synthases, isocyanide synthases, and non-ribosomal synthetases), along with many tailoring enzymatic encoding genes (such as epimerases, methyltransferases, hydroxylases, and p450 monooxygenases). The biosynthetic genes define the structural backbone and the chemical class of the secondary metabolites, whereas the tailoring genes modify the carbon molecular structure of the mycotoxins. These key transcription factors in BCGs positively influence the process of biosynthesis. Also, the presence of some additional genes was also reported, which encode proteins that are hypothesized to be involved in the transportation of the toxic property, or in the protection from the metabolite itself. The presence of several other genes in BCGs was also estimated along with the earlier mentioned genes, but their role is not yet clear and thus require more research to exemplify their significance in the mycotoxin biosynthesis pathway (Gallo and Perrone, 2021; Wang et al., 2022). However, mycotoxin biosynthesis is controlled by an even more complex regulatory network containing connected subnetworks. BGCs in all fungal species respond to several biotic and abiotic stimuli, including the intensity of light, pH, temperature, redox status, sources of carbon and nitrogen, etc. These broad domain transcription factors are often referred to as global regulators and involve a complex regulatory network to control the production of mycotoxins, called global regulation. The velvet and LaeA proteins are the most studied transcriptional complex that is involved in the regulation of secondary metabolite synthesis in also every genus of fungi, and these complexes are controlled by light. LaeA proteins have a positive role in mycotoxin synthesis, which is evident from the fact that the deletion of this particular protein downregulates the production of mycotoxins like fumagillin, fumagatin, gliotoxin, endocrocin, helvolic acid, and sterigmatocystin within the genus *Aspergillus*. Besides these, several modifiers are there that induce the epigenetic regulation of mycotoxin biosynthesis. Generally, BGCs are subject to epigenetic regulation through the remodeling of chromatin, leading the expression of cryptic metabolic clusters. Enzymes, like methyltransferases and histone acetyltransferases, can place or remove post-translational modifications on histone tails, which regulate how tight or relaxed the chromatin will be, and directly effecting BGC transcription (Hou et al., 2024).

12.3.2.2 Mycotoxins in Food

Every year, 25% of the harvested crops are contaminated by mycotoxins, making it a topic of concern and a very significant threat toward the safety, security, and trade of food worldwide. Cereals, like rice, wheat, corn, rye, sorghum, and oat, are very common targets. Besides, other crops, including legumes, like peas, soybeans, and peanuts, are also contaminated (El-sayed et al., 2022). Mycotoxins, like aflatoxin, ergot alkaloids, trichothecenes, phomopsin, patulin, and zearalenone, are noted in the feed and fodder consumed by domestic animals, like pigs, hens, ducks, sheep, cow, etc. (Kunz et al., 2021). Thus, they get easy access to the body of these animals where they interfere with normal functioning and induce several fatal health issues. The meat, eggs, and milks of these animals are also contaminated by these mycotoxins. Not only raw food items but also processed food products have been reported to get mycotoxin contamination (De Felice et al., 2023; Miri et al., 2024). Alcoholic beverages, like beer, wine, red wine, and white wine, and nonalcoholic beverages, namely, tea and coffee, have been reported to get contaminated by different ochratoxins (Miri et al., 2024; Qu et al., 2022). These mycotoxins are highly stable in nature since they are not eliminated by processes like food processing, cooking, baking, roasting, or pasteurization. Chips, cheese, tortillas and tortilla chips, cornflakes, cornstarch, corn flour, popcorn, grits, sausages, oils, and several other snack items have also been reported to contain mycotoxins in them (Goessens et al., 2024). Fruit juices (apple juice, grape juice, berry juice, tomato juice, multifruit juice), fruits and vegetables (apple, strawberry, tangerine, pomelo, orange, grapes, capsicum, tomatoes, lemon, jujube, cherry), dried fruits (raisins, dried wolfberries, dried figs, dried jujube), and nuts (almond, hazelnut, pistachio, peanut, walnut) all have been reported to be contaminated by alternariol, a mycotoxin produced by *Alternaria* spp. Similarly, processed foods, like tomato-based products, noodles, groundnut soup, vegetable oil, tomato sauce, jams, cereal-based baby foods, biscuits, mixed spices, sunflower seeds, green tea, wheat flour, wheat (durum), olive oil, and chili powder, are also mentioned to be contaminated by the same toxin (Saleh et al., 2024).

12.3.3 Toxicological Effects and Health Risks Associated with Mycotoxins

About 60–80% of food crops are contaminated by mycotoxins, and thus, exposure of humans to these secondary metabolites is highly common. Hence, it becomes a global threat toward food safety, security, and economy as well. These compounds are found to possess high toxicity toward both humans as well as animals, but their role in triggering particular ailment is less demonstrated. While, most researchers agree that these mycotoxins can effectively cause toxicity to different parts or systems of the human body, the International Agency for Research on Cancer (IARC), an intergovernmental agency of the World Health Organization, unclassified several mycotoxins in the monograph series published by them. The mycotoxins mentioned in the recently updated IARC monographs are listed in Table 12.1.

As categorized by IARC, aflatoxin is also reported by researchers to induce carcinogenesis in other chordates such as birds, in rodents, and in primates, including humans. Also, hepatotoxicity, genotoxicity, growth retardation, and liver failure are some commonly noted effects of aflatoxins (Awuchi et al., 2021, 2022). Besides the ability of mycotoxins to induce carcinogenesis, there are various other health issues that are caused by them. Vila-Donat et al. (2024) highlighted that ochratoxins have severe degenerative effects on the nervous system, leading to ailments like Parkinson's and Alzheimer's disease. In addition, it also has immunotoxic, hepatotoxic, nephrotoxic, and teratogenic effects as well. A similar noxious effect of ochratoxins were also put forward by Miri et al. (2024). Stoev et al. (2024) mentioned the effect of mycotoxins on farm animals. They claimed that, compared to other animals, pigs are more prone to these fungal secondary metabolites, particularly to trichothecene and deoxynivalenol. Pigs and poultry are highly sensitive to fusariotoxicosis caused by Trichothecenes (T-2, HT-2, and diacetoxyscirpenol) produced by *Fusarium*. In humans, this disease is referred to as alimentary toxic aleukia. During the early stages, abdominal pain, diarrhea, nausea, tremors and reduction in weight are noted.

TABLE 12.1
Mycotoxins Classified According to IARC Monograph (2024) as Carcinogenic to Humans

Serial No.	Mycotoxin	IARC Group	Publication Year of IARC Monograph
1	Aflatoxin	1	2012
2	Fumonisin B1	2B	2002
3	Ochratoxin A	2B	1993
4	Fumonisin B2	2B	1993
5	Fusarin C	2B	1993
6	Sterigmatocystin	2B	1987
7	T-2 toxin	3	1993
8	Zearalenone	3	1993
9	Deoxynivalenol	3	1993
10	Nivalenol	3	1993
11	Fsarenone X	3	1993
12	Patulin	3	1987
13	Citrinin	3	1987
14	Penicillic acid	3	1987
15	Cyclochlorotine	3	1987
16	Luteoskyrin	3	1987
17	Rugulosin	3	1987

Note: Group 1 = the agent is carcinogenic to humans; group 2B = the agent is possibly carcinogenic to humans; group 3 = the agent is not classifiable as to its carcinogenicity to humans.

While ulcerations and necrotic changes are observed in the gastrointestinal tract, damage in the liver, heart, kidney, peripheral ganglia, and brain have also been noted. Type B trichothecene, deoxynivalenol, or vomitoxin induces a high level of toxicity to the dermal, gastrointestinal, nervous, reproductive, and immune systems. The gastrointestinal tract is highly sensitive to deoxynivalenol, as its exposure can cause malfunctioning, inflammations, lesions, and several other impairments. Anorexia, delayed growth, diarrhea, immune disorders, vomiting, and weigh loss. In animals, they can induce carcinogenesis. The mode of action of vomitoxin involves mitochondrial disruption, apoptosis, and accumulation of ROS (Liao et al., 2024). De Felice et al. (2023) highlighted that enniatin B1 can have cytotoxic, genotoxic, and negative estrogenic effects. In addition, the mycotoxin can also cause cell cycle impairment, apoptosis, oxidative stress, and changes in mitochondrial membrane permeabilization. Lupinosis is another fatal disease, in which the liver of grazing animals, especially sheep, is affected. This disease is caused by consuming feed contaminated by phomopsin, another mycotoxin produced by the fungus *Diaporthe toxica* (Kunz et al., 2021). The toxic effects and diseases caused by several mycotoxins are mentioned in Table 12.2.

TABLE 12.2
Some Important Mycotoxins and Their Adverse Effects on Human and Animal Health

Mycotoxins	Genus	Food Commodities	Toxic Effect and Ailments	Reference
Aflatoxins	*Aspergillus*	Cereals, dry fruits, oilseeds, spices, dairy products	Hepatotoxicity, genotoxicity, teratogenicity, mutagenicity, carcinogenicity, growth retardation, infertility, nutrients malabsorption, and damaged liver	Awuchi et al. (2022); Awuchi et al. (2021)
Ochratoxins	*Aspergillus*, *Penicillium*	Cereal, dried fruits, nuts, processed meat, spices, cheese, beverages	Carcinogenicity, hepatotoxicity, genotoxicity, immunotoxicity, neurotoxicity, and teratogenicity	Miri et al. (2024).
Fumonisins	*Fusarium, Aspergillus*	Cereals, asparagus; beer, soybeans, beans	Neural tube defect, liver and kidney failure, carcinogenicity, reproductive toxicity, neurotoxicity, immunotoxicity	Qu et al. (2022)
Patulins	*Penicillium*	Apples, grapes, pears, vegetables, cereals, cheese	Neurotoxicity; cytotoxicity; damaged liver; kidney failure; damaged gastrointestinal tract; liver, kidney, and endocrine disturbance; problems in immune and reproductive systems; group 3 carcinogen	Stoev et al. (2024)
Citrinins	*Aspergillus*, *Penicillium*, *Monascus*	Cereals	Cytotoxicity, mutagenicity, hepatotoxicity, genotoxicity, embryotoxicity, teratogenicity	Zargar and Wani (2023)
Trichothecenes	*Fusarium*	Cereals, peanuts	Cardiovascular system, hematopoietic system, and lung damage; growth retardation; reproductive system disturbance; fusariotoxicosis in pigs and poultry	Stoev et al. (2024)

Enniatins	*Fusarium*	Cereals, apples, rapeseeds, soybeans, eggs, potatoes, peanuts	Cytotoxicity, genotoxicity, cell cycle impairment, apoptosis, oxidative stress induction, mitochondrial membrane permeabilization changes, negative estrogenic effects	De Felice et al. (2023)
Phomopsins	*Diaporthe*	Lupin seeds, products, and pea seeds	Lupinosis (liver affected in sheep)	Kunz et al. (2021)
Sterigmatocystins	*Aspergillus*	Wheat, peanuts, corn, soybeans, and coffee beans	Carcinogenesis, cytotoxicity, oxidative stress increase, cell cycle–related gene expression abnormality, lipid metabolism disturbance, apoptosis	Zhou et al. (2023)
Ergot alkaloids	*Claviceps*, *Epichloe*	Cereals	Feed intake decreased and other malfunctions in pigs; decreased egg laying in poultry; hyperthermia, heavy dyspnea, tail loss, low milk yield, and even death of cattle	EFSA CONTAM (2020)
Moniliformin	*Fusarium*	Cereals	Keshan disease (heart ailment)	Ekwomadu et al. (2021)
Alternariol	*Alternaria*	Legumes, nuts, dried fruits, cereals, tuberous crops, vegetables, fruits, fruit-based food, oilseeds, spices	Cytotoxicity, mutagenicity, carcinogenicity, and endocrine disruptor effects; immunotoxicity, hematotoxicity, nephrotoxicity, vascular toxicity, and diarrhea	Saleh et al. (2024)

12.3.4 Regulations and Permissible Limits for Mycotoxins in Food

The European Union (EU) has accepted maximum permissible levels and published some strict regulations regarding the maintenance of some fatal mycotoxins in human food commodities or animal feedstuffs as less as possible. According to recently updated European regulations, the maximum levels of aflatoxin B1 and ochratoxin A in food should be maintained between the 0.1–12 µg/kg and 0.5–80 µg/kg range, respectively, depending upon the food item (Vila-Donat et al., 2024). The Union has considered 0.1 µg/kg of aflatoxin B1 content in processed cereal-based foods for children as maximum limit and mentioned that up to 4,000 µg/kg of fumonisin B1 and B2 can be present in unprocessed maize consumed by humans. While content of other toxins in maize, like ochratoxin A, aflatoxin B1, zearalenone, and deoxynivalenol, have a respective limitation up to 5 µg/kg, 2 µg/kg, 350 µg/kg, and 1,750 µg/kg, for wheat the accepted limits are 4 µg/kg, 2 µg/kg, 5 µg/kg, 1,250 µg/kg, and 100 µg/kg of aflatoxins, aflatoxin B1, ochratoxin A, deoxynivalenol, and zearalenone, respectively. In dairy products, 0.05 µg/kg amount of aflatoxin B1 is considered to be safe. In rice, the considered level of mycotoxins such as aflatoxin B1, ochratoxin B1, zearalenone, and deoxynivalenol is 5 µg/kg, 5 µg/kg, 100 µg/kg, and 1,250 µg/kg, respectively (Stoev et al., 2024). The EU, United States, and Chinese authorities have set the maximum limit up to 50 µg/L/kg for patulin fruit and apple juices, whereas the EU-accepted range of patulin level in food items is 10–50 µg/kg, depending upon the food. The European Food Safety Authority and Joint FAO/WHO Expert Committee on Food Additives accepted 0.25 µg/kg body weight as tolerable daily intake (TDI) for zearalenone. The Joint FAO/WHO Expert Committee on Food Additives mentioned 2 µg/kg body weight as the maximum tolerable intake per day for fumonisins, such as fumonisin B1, B2, and B3 (Awuchi et al., 2021).

The Standards Organisation of Nigeria (SON) has laid down some regulatory limits for aflatoxins in various food commodities, like raw groundnuts, maize seeds, sorghum, kuli-kuli (a type of fare prepared from groundnuts), sesame seeds to be 20 µg/kg, 10 µg/kg, 10 µg/kg, 4 µg/kg, and 4 µg/kg, respectively (Chilaka et al., 2022). The Food Safety and Standards Authority of India (FSSAI) has set some regulatory limits for some common mycotoxins that are frequently detected in food. The regulatory limits for aflatoxins in cereal, cereal-based products, nuts, pulses, betelnut, and oilseeds are 15 µg/kg each, and in dried figs and spices, a regulatory amount of 10 µg/kg is accepted. The regulatory limit of aflatoxin M1 in milk is 0.5 µg/kg; 20 µg/kg is specified as the regulatory limit for ochratoxin A in cereals like barley, wheat, and rye; and 50 µg/kg of patulin is accepted in apple juices and other ingredients of its juice. The presence of 1,000 µg/kg of vomitoxin in wheat has been accepted as its regulatory limit (Chatterjee et al., 2023).

12.4 TRADITIONAL DIAGNOSTIC METHODS FOR FOODBORNE FUNGAL INFECTIONS

12.4.1 Microscopy and Culture-Based Methods

Traditional diagnostic techniques for foodborne fungal infections primarily involve microscopy and culture-based methods. These methods rely on the growth of fungi in specific culture media, where visible colonies can be identified after incubation periods, which may extend up to a week for fungi (Mandal et al., 2011). Microscopy allows for the observation of morphological characteristics, such as hyphal structure and spore formation, which are essential for the identification of fungal species. Culture methods provide quantitative data on pathogen presence, enabling researchers to assess the level of contamination in food products (Zhao et al., 2014). Culture-based methods are often regarded as the gold standard due to their ability to yield definitive results regarding the presence of specific pathogens. For instance, culturing *E. coli* O157:H7 on selective media like Sorbitol MacConkey agar demonstrates high success rates in detecting this pathogen (Bennett et al., 2020). However, these methods require the meticulous preparation of culture media, inoculation of plates, and subsequent colony counting, which can be labor-intensive and time-consuming (Duncan et al., 2019).

12.4.2 Limitations and Challenges of Conventional Methods

Despite their historical significance, conventional methods face several limitations that can hinder effective diagnosis. One major challenge is the time required for results; traditional culture techniques can take several days to weeks to confirm the presence of pathogens (Zhao et al., 2014). This delay can be critical in food safety contexts, where rapid decision-making is necessary to prevent foodborne illnesses. Moreover, conventional methods may yield false negatives due to the presence of viable but non-culturable (VBNC) states of pathogens. VBNC pathogens can survive in a dormant state without growing in culture media, leading to underreporting of contamination levels (Lee et al., 2014). This issue is particularly concerning in food safety, as undetected pathogens pose a significant risk for outbreaks. Additionally, these methods can be hindered by interference from non-target organisms present in complex food matrices. The extraction of fungal DNA from such matrices can be challenging due to the presence of inhibitors that affect downstream analyses (Duncan et al., 2019). The specificity of culture-based methods may also be compromised by emerging serotypes or strains that do not grow well on standard media. In summary, while traditional diagnostic methods, such as microscopy and culture-based techniques, have been foundational in identifying foodborne fungal infections, their limitations necessitate the development and adoption of more rapid and sensitive diagnostic approaches.

12.4.3 Modern Diagnostic Tools for Foodborne Fungal Infections

Recent advancements in diagnostic techniques have significantly enhanced the ability to detect foodborne fungal infections. These modern tools include molecular techniques, rapid diagnostic assays and biosensors, and immunological methods, each contributing to improved sensitivity, specificity, and speed in identifying fungal pathogens in food products.

12.4.4 Molecular Techniques

Molecular techniques have revolutionized the detection of foodborne pathogens, particularly through the use of polymerase chain reaction (PCR), quantitative PCR (qPCR), and sequencing methods. PCR is a powerful tool that amplifies specific DNA sequences, allowing for the rapid detection of low levels of fungal pathogens in food samples (Kumar et al., 2021). This amplification process is critical because it enables the identification of pathogens even when they are present in minimal quantities, which is often the case in contaminated food products. The versatility of PCR has led to the development of various adaptations, including multiplex PCR and real-time PCR. *Multiplex PCR* allows for the simultaneous detection of multiple pathogens in a single reaction, significantly increasing efficiency (Zhao et al., 2014). *Real-time PCR* provides quantitative data on pathogen load, which is essential for assessing contamination levels and potential health risks. These molecular methods enhance both sensitivity and specificity compared to traditional culture-based techniques.

12.4.5 Rapid Diagnostic Assays and Biosensors

In addition to molecular techniques, rapid diagnostic assays and biosensors have emerged as critical tools in food safety. These methods are designed to provide quick results, significantly reducing the time required for pathogen detection compared to traditional culture methods (Zhang et al., 2022). For instance, biosensors utilize various detection principles—such as electrochemical and optical signals—to identify fungal pathogens effectively. The rapid nature of these assays is crucial in food industry settings, where timely decision-making can prevent widespread outbreaks. Biosensors can be tailored to specific pathogens and are capable of detecting low concentrations of fungi within complex food matrices. This capability makes them particularly valuable for monitoring food safety during processing and storage. The integration of biosensors into routine testing protocols can enhance overall food safety management by enabling continuous monitoring.

12.4.6 Immunological Techniques and Antibody-Based Assays

Immunological methods, such as enzyme-linked immunosorbent assays (ELISA), leverage antibody–antigen interactions to detect fungal toxins and pathogens in food products. These techniques offer good specificity due to the high affinity between antibodies and their target antigens (Smith and Jones, 2020). However, while immunological assays are effective for certain applications, they may suffer from lower sensitivity compared to molecular methods. This limitation can result in false negatives, particularly when pathogen levels are low or when the target antigen is not abundant. Despite these challenges, immunological techniques remain valuable tools in the detection of mycotoxins and other fungal metabolites in food products. They are often used in conjunction with molecular methods to provide a comprehensive approach to food safety diagnostics.

Modern diagnostic tools have significantly improved the ability to detect foodborne fungal infections. Molecular techniques like PCR offer rapid and sensitive detection capabilities, while rapid diagnostic assays and biosensors provide timely results essential for effective food safety management. Immunological techniques complement these methods by allowing for the detection of specific toxins. Together, these advancements enhance our capacity to monitor and control fungal contamination in food products.

12.4.7 Emerging Diagnostic Technologies for Foodborne Fungal Infections

The landscape of food safety diagnostics is rapidly evolving, particularly with the advent of emerging technologies that enhance the detection of foodborne fungal infections. These advancements include nano-based sensors, portable point-of-care devices, and the application of artificial intelligence (AI) and machine learning. Each of these technologies contributes to more efficient, sensitive, and rapid detection methods that are crucial for ensuring food safety.

12.4.8 Nano-Based Sensors

Nanotechnology is paving the way for innovative diagnostic tools that can detect fungal pathogens at very low concentrations. Nano-based sensors leverage the unique properties of nanomaterials to enhance sensitivity and specificity in pathogen detection (Lee et al., 2023). These sensors can operate on various principles, including electrochemical, optical, and mass-sensitive detection methods. Their small size allows for rapid response times and portability, making them ideal for on-site testing in diverse environments. The integration of nano-based sensors into point-of-care testing devices represents a significant advancement in food safety diagnostics. These devices enable the immediate analysis of food samples, allowing for quick decision-making regarding contamination and safety interventions. The ability to detect pathogens in real time is crucial in preventing foodborne outbreaks, particularly in settings such as food processing plants and restaurants, where immediate action may be necessary.

12.4.9 Portable and Point-of-Care Devices

The development of portable diagnostic devices has transformed the landscape of pathogen detection by allowing immediate testing in various settings. These devices reduce the need for extensive laboratory infrastructure, making them accessible for use in remote or resource-limited environments (Patel et al., 2023). Portable devices are designed to be user-friendly and can provide results within minutes to hours, significantly faster than traditional culture methods, which can take days. These point-of-care devices utilize various technologies, including biosensors and molecular assays, to facilitate the rapid detection of fungal pathogens. Their ability to deliver quick results is essential for effective food safety management, enabling stakeholders to make informed decisions about product recalls or other interventions to mitigate risks associated with contaminated food products.

12.4.10 Role of AI and Machine Learning in Fungal Pathogen Diagnostics

Artificial intelligence (AI) and machine learning are increasingly being applied to enhance diagnostic accuracy in food safety. These technologies can analyze complex datasets generated from various diagnostic modalities—including molecular techniques, biosensors, and traditional methods—to improve pathogen identification rates (Nguyen et al., 2024). By employing algorithms that learn from historical data, AI can identify patterns that may not be apparent through conventional analysis. Moreover, AI-driven tools can predict outbreaks based on environmental data and historical trends, allowing for proactive measures to be implemented before an outbreak occurs. This predictive capability is invaluable in public health contexts, where early intervention can prevent widespread illness. AI can also optimize testing protocols by streamlining laboratory workflows, reducing human error, and improving overall efficiency. The integration of AI into diagnostic processes represents a paradigm shift in how foodborne pathogens are detected and managed.

Advancements in diagnostic techniques for foodborne fungal infections reflect a significant shift toward more efficient, sensitive, and rapid detection methods. The integration of molecular biology with nanotechnology and AI is expected to further enhance food safety measures globally. As these technologies continue to develop, they hold the potential to transform food safety practices by enabling quicker responses to contamination threats and improving overall public health outcomes.

12.5 CONTROL AND PREVENTION STRATEGIES

12.5.1 Physical Methods

Physical methods employed for the detoxification of mycotoxins include both thermal as well as non-thermal techniques.

12.5.1.1 Non-Thermal Techniques and Thermal Techniques

They include, washing, sorting, separating, sieve cleaning, dehulling, density segregation, polishing, solvent extraction, adsorption of mycotoxins, etc. In addition, high hydrostatic pressure, cold plasma treatment, pulsed light and electric treatments, ultrasound methods, and irradiation treatments are some other commonly applied non-thermal methods to decontaminate mycotoxins in food (Sipos et al., 2021). Sorting is based on the fact that contaminated grains or particles have different organoleptic properties compared to healthy ones. Thus, they can be easily identified and removed from the whole mass. Since mycotoxins are soluble in water, organic solvents can be easily remediated physically by washing or by solvent extraction technique. The adsorptive removal of mycotoxins by employing both organic, inorganic, and mineral binders is another common method of decontamination. This surface phenomena involves interaction between the binders and target mycotoxins (Nahle et al., 2022). Ding et al. (2023) explained that the adsorptive removal of ochratoxin A is due to van der Waals forces and electrostatic interactions between the positively charged binders and the negatively charged ochratoxin A molecule.

The irradiation of food to combat mycotoxin contaminants involves the use of both ionizing and non-ionizing methods. Ionizing techniques include X-ray and γ-ray, and non-ionizing methods utilize ultraviolet and microwaves as toxin-decontaminating agents (Nguyen et al., 2024). NgolongNgea et al. (2020) highlighted the use of UV-C as a traditional physical treatment method to inactivate fungal pathogens and associated toxins. Ferreira et al. (2021) reported on how UV-C irradiation treatment for 1 hour at a 2.06 kJ/cm^2 dose can effectively cause photodegradation of fungal secondary metabolites in black and red rice. Aflatoxin (AFB1, AFB2, AFG1, and AFG2), deoxynivalenol, ochratoxin A, and zearalenone are noted to be degraded, whereas in brown rice, effectual detoxification was noted when dose and irradiation time of UV-C are increased to 6.18 kJ/cm^2 and 3 hours. AFB2, AFG2, deoxynivalenol, and zearalenone are degraded from brown rice. In the former case, no change in the properties of red and black rice grains is observed, but in case of higher-dose applications, their total phenolic content is reduced (Nahle et al., 2022; Nguyen et al., 2024). Heat treatments degrade heat-unstable mycotoxins, but several fungal secondary metabolites, like deoxynivalenol and zearalenone, are heat-resistant; thus, complete detoxification of these toxins is not possible by thermal treatment methods.

Extrusion, roasting, cooking, high-pressure cooking, nixtamalization, pasteurization, and instant catapult steam explosion are some common thermal techniques involved in the detoxification process. Since thermal techniques are always blamed for the change in taste, texture, and even nutrient content, their use is quite restricted to large-scale industries, where proper facilities are present to expose foodstuffs to heat treatments without having much effect on their natural properties (Stadler et al., 2020; Sipos et al., 2021; Ding et al., 2023).

12.5.2 Chemical Methods

The decontamination of fungal biosynthesized secondary metabolites by employing a plenty of chemicals is quite common nowadays. Chemical methods, like ammoniation, ozonation, peroxidation, etc., have been reported to neutralize or degrade mycotoxins present in food products (Ntungwe et al., 2024). However, the application of ammonization technique for decontaminating foodstuffs is not considered to be a safe mycotoxin-combatting measure among European countries. The application of bases, like NH_3, NaOH, KOH, $Ca(OH)_2$, and $NaHCO_3$; oxidizing agents, like H_2O_2, ClO_2, and O_3; and organic acids, like formic acid (HCOOH) and propionic acid (CH_3CH_2COOH), is all cited in literature for eliminating toxic mycotoxins. Currently, a mixture of glycerol ($C_3H_8O_3$) and calcium hydroxide, that is, $Ca(OH)_2$, was demonstrated to potentially detoxify fungal secondary mycotoxins at a very high rate (Conte et al., 2020; NgolongNgea et al., 2020; Ding et al., 2023; Santos et al., 2022).

12.5.2.1 Nanomaterials

Particles having size in the nanoscale, that is, 10^{-9}, are referred to as nanoparticles. Recently, metallic nanoparticles have evolved as effective mycotoxin-detoxifying agents; this is due to their antifungal activity. A vast literary evidence is present to justify the convincing antifungal effect of metallic nanoparticles (Hamouda et al., 2024; El-Abeid et al., 2024; Terea et al., 2024). Zhang et al. (2020) mentioned two different ways by which these nanoparticles can affect fungal strains and inhibit mycotoxin biosynthesis. The first method involves the encapsulation of antifungal compounds into the nanomaterials, which are released under desired conditions, and other strategies involve the direct penetration of nanosized particles into the cells of pathogenic fungi to inhibit it. Authors describe *metallic nanoparticles* as an efficacious photocatalyst that has convincing effect in mycotoxin decontamination.

A similar photocatalytic effect of metallic nanoparticles in controlling mycotoxins is also reported in the recent works of Lin et al. (2024) and Jing et al. (2024). Wu et al. (2020) reported the implementation of NaYF4:Yb, Tm@TiO2 nanoparticles at 10 μg/mL against DON mycotoxin in wheat. The outcomes confirmed 100% photocatalytic degradation of DON within 1 hour. Zhou et al. (2020) also claimed the successful synthesis and implementation of UCNP@TiO2 composite as a photocatalyst for the degradation of DON under the influence of near-infrared light, and 100% of DON was degraded as noted after a period of 120 minutes.

12.5.2.2 Ozonation

Ozonation is a simple and effective chemical strategy that involves the application of ozone (O_3) gas for the degradation of fungal secondary metabolites. Ozone acts as a strong oxidizing agent and thus generates free radicals, which can counter the mycotoxin functional groups by altering their molecular structures, and forms end products having lower molecular weight, with a reduced number of double bonds, as well as reduced toxicity levels. Since ozonation does not levy any harmful by-products after application against mycotoxin, this method of detoxification is considered highly significant (Sujayasree et al., 2022). Since its recognition by the Food and Drug Administration (FDA) as a safe antimicrobial agent, a huge research was conducted on its application to combat a wide range of microbial strains and toxins. Afsah-Hejri et al. (2020) highlighted aflatoxin B1 and aflatoxin G1 by employing ozone. They further added that treatment with ozone damages the hypertoxic site of the furan ring within aflatoxins. The double bond at C8–C9 of the furan ring of both aflatoxin B1 and aflatoxin G1 is attacked by ozone-generating ozonides, which later rearrange to another non-toxic end product, namely, molozonide. In toxic deoxynivalenol molecules, O_3 attacks the double bonds at C9–C10 position and also oxidizes the allylic carbon at the eighth position. The generation of N-oxide at the primary amine of the fumonisin B1 molecule was noted when O_3 interacts with this mycotoxin. This is due to the loss of two hydrogen atoms from the fumonisin B1 molecule. Wang et al. (2024) demonstrated the degradation of two mycotoxins, namely, alternariol and alternariol monomethyl ether, produced by *Alternaria* spp., by employing O_3 as the detoxification agent in aqueous solution and in orange juice. In orange juice, alternariol and alternariol monomethyl ether detoxification was noted to be 51.86% and 49.04%, respectively, after an exposure of ozone for 600 seconds, while in aqueous solution, the degradation percentages were quite high, namely, 57.95% and 40.24%, after 5 seconds of treatment. Even the effect of ozone in aqueous solution is so convincing that within 40 seconds of treatment, ~100% of these toxins were degraded.

12.5.3 Limitations of Physical and Chemical Decontamination Techniques

Physical detoxification techniques like heat treatments are reported to have severe effects on the organoleptic properties as well as the nutritional status of the food. The taste, texture, and nutrient content of detoxified food are changed. Also, there are several mycotoxins which are heat-resistant; thus, thermal techniques are insufficient to decontaminate those mycotoxins (Sipos et al., 2021). Santos et al. (2022)

mentioned physical methods as impractical techniques that are only restricted to large-scale industries having sophisticated equipment and also considered these methods as time-consuming and expensive, while chemical methods possess a huge threat toward both environment as well as humans since they can remain intact or as by-product in foodstuffs even after completion of the detoxification process. The organic or inorganic chemical methods utilized to detoxify the foodstuff may affect taste, texture, and nutritional value, and these chemicals may get accumulated into food items as such or as by-products, which directly affect the consumer as well, along with the mycotoxin. In addition, some chemicals may not show convincing detoxifying activity when exposed to increased levels of fungal secondary metabolites. The effectiveness of chemical detoxification methods even can't mask the fact that these techniques are highly expensive, change the natural properties of foodstuff, and require complex specifications to execute the process of mycotoxin detoxification. The most dangerous consequence of employing chemical systems is the development of strains into more toxic and antifungal treatment–resistant strains, which will be a pivotal threat toward food safety and the maintenance of its security. These limitations or negative outcomes of physical and chemical decontamination techniques shift the attraction of researchers toward biological alternatives (NgolongNgea et al., 2020; Nahle et al., 2022; Ntungwe et al., 2024).

12.5.4 Biological Methods: Biotechnological Approaches

The biological methods of degrading fungal secondary metabolites are getting a huge attention due to their efficiency in controlling foodborne fungal pathogens and detoxifying their associated mycotoxins into less severe compounds without changing the properties and nutritional attributes in both processed as well as non-processed food items.

12.5.4.1 Probiotics

The World Health Organization defines *probiotics* as "live microorganisms which when administered in adequate amounts confer a health benefit on the host." *Bacillus*, *Bifidobacterium*, *Lactobacillus*, and *Saccharomyces cerevisiae* are some commonly used probiotics (Liu et al., 2022). Several mechanisms are present through which these microbes can inhibit mycotoxin-producing fungal pathogens in food, such as inhibition of the mycotoxin biosynthesis pathway, biosorption, enzymatic degradation, etc. (Ding et al., 2023). Huang et al. (2018) employed three probiotics, namely, *Bacillus subtilis*, *Lactobacillus casein*, and *Candida utilis*, to detoxify aflatoxin B1 and zearalenone. *Bacillus subtilis* alone can degrade up to 38.38 and 42.18% of aflatoxin B1 and zearalenone, while the combined application of these three microbes in a 1:1:1 ratio can mean 44.9 and 45.49% decontamination of zearalenone and AFB1. The authors also combined the compounds from these three strains with enzymes of *Aspergillus oryzae* and assessed their effect on the detoxification process. The outcomes indicate comparatively higher degradation percentages, that is, 63.95% for AFB1 and 73.51% for zearalenone. Nahle et al. (2022) reviewed the degradation of mycotoxins by different species of *Actinobacteria*, *Bifidobacterium*, *Enterococcus*, and *Lactobacillus* and highlighted that most of the probiotic strains have bioremoval efficiency above 60% (with some exceptions) for common secondary fungal metabolites, like aflatoxins (AFB1 and AFM1), ochratoxin A, zearalenone, and patulin. As well, different genera of yeast probiotics (*Kluyveromyces, Kazachstania*, and *Saccharomyces*) also remove mycotoxins from 46 to 90% from different beverages, like milk, juices, etc. Similarly, Piotrowska et al. (2021) reviewed several previous works and enlisted 33 probiotic species belonging to 13 different genera, namely, *Alcaligenes*, *Bacillus*, *Brevibacterium*, *Cupriavidus*, *Devosia*, *Enterobacter*, *Escherichia*, *Lysinibacter*, *Lysinibacillus*, *Pediococcus*, *Pseudomonas*, *Rhodococcus*, and *Streptomyces*, that are successfully employed by researchers against plenty of toxic secondary fungal compounds belonging to aflatoxins, ochratoxins, ergot alkaloids, patulin, citrinin, trichothecenes, zearalenone, etc. Mischler et al. (2024) demonstrated above 90% degradation of zearalenone by employing *Bacillus licheniformis*, *B. megaterium*, and *Levilactobacillus brevis*. Teymourlouei et al. (2024) demonstrated the preparation of nanoadsorbents by entrapping nanozeolite-captured disrupted cell walls of *Saccharomyces cerevisiae* and *Candida albicans* within ca-alginate. The outcomes confirm comparatively better performance of immobilized, entrapped *C. albicans* cell wall adsorbents, with reduction efficiency of 89.49% against aflatoxin M1.

12.5.4.2 Enzymes

Enzymes are biocatalysts with high substrate specificity and thus can be used to degrade mycotoxins under moderate reaction conditions, like optimum pH, temperature, etc. Being biological, they do not have toxic effects on human health and the ecosystem; also, it is not yet known to alter food's natural properties. Oxidoreductases (like laccases, oxidases, peroxidases, and reductases) and lactonases are common enzymes that degrade aflatoxin B1. Similarly, laccases, monooxygenases, and peroxidases are oxidoreductases that have literary evidence to participate in zearalenone detoxification. Carboxypeptidases, amidohydrolases, lipases, and Nudix family hydrolases are reported to have ochratoxin A–decontaminating property. Lipase, aldolase, transferase, and oxidoreductase are some common patulin-degrading enzymes. Aminotransferases and carboxylesterases are two important fumonisin-degrading enzymes (Liu et al., 2024). Sun et al. (2024) Performed heterogeneous expression of laccases enzyme (PpLacl and AoLac2) extracted from *Pleurotus pulmonarius* and *Aspergillus oryzae* and expressed into probiotic yeast (*Pichia pastoris*) for decontaminating two mycotoxins aflatoxin B1, zearalenone and deoxynivalenol. The outcomes confirm that both enzymes can transform aflatoxin B1 and zearalenone into the compounds AFQ1 and 15-OH-ZEN, which are, respectively, 90 and 98% less toxic than the parent compounds. Also, these enzymes have been observed to cause more than 70% degradation of both aflatoxin B1 and zearalenone.

12.5.4.3 Biomolecules

The application of chitosan beads for the biosorption of foodborne fungal secreted metabolites is a modern approach (NgolongNgea et al., 2020). Abbasi Pirouz et al. (2020) claimed the successful bioremoval of a total of 11 fungal secondary metabolites, including four aflatoxins (AFB1, AFB2, AFG1, and AFG2), three trichothecenes

(T-2, HT-2 toxin, and deoxynivalenol), two fumonisins (FB1 and FB2), ochratoxin A and zearalenone from palm kernel cake by employing chitosan with more than ~50% removal. Gu et al. (2021) concluded that during thermal processing, the presence of fructose facilitates the detoxification of ochratoxin A. They also added that the addition of fructose leads to the generation of several non-toxic end products from ochratoxin A, like OTα-amide.

Ghobish et al. (2025) reported the preparation of a chitosan nanoadsorbent and employed these adsorbents alone as well as in combination with certain other biomolecules to facilitate the adsorptive removal of aflatoxin B1 from fish feed. The outcomes indicated that nano-chitosan composites, along with mannan oligosaccharide and bentonite in 1:1:1 ratio, achieved the highest adsorption, up to 74.77%, among other formulations. The results also confirmed that mild acidic conditions, that is, pH 6.1 and temperature 37°C, have facilitated the biosorption process.

12.5.5 Application of Natural Products

12.5.5.1 Plant Extracts

The utilization of plant extracts in inhibiting fungal pathogens and on the decontamination of associated mycotoxins in foodstuff has been claimed by many researchers. The remarkable antagonistic potential of these plant extracts is probably due to the presence of phytochemicals in them. The richness of pivotal phytocompounds, like alkaloids, flavonoids, tannins, phenols, saponins, and many others, makes them a potential candidate that can decontaminate foodstuffs polluted with noxious mycotoxin contaminants (Shams et al., 2024; Kalemba et al., 2024; Velazhahan et al., 2024). Some recent studies with remarkable outcomes are enlisted in Table 12.3.

12.5.5.2 Essential Oils

Another potential eco-friendly and safe mode of fungal secondary metabolite decontamination is the application of essential oils (Allagui et al., 2023). Essential oils are basically a complex mixture of a large number of different compounds, like phenols, terpenoids, flavonoids, etc., and extracted from a number of aromatic plants (Das et al., 2021). Essential oils are mostly surfactants and alter cell membrane permeability, making the interior of the fungal cell accessible to an extracellular substance that penetrates and induce cell lysis (Cai et al., 2023). Researchers claim the efficient antimicrobial properties of these oils and suggested that the bioactive compounds present in these oils as reason for their bactericidal and fungicidal efficiencies. Chlorophyll, carotenoids, essential oils, organic acids, aromatic acids, glucosinolates, glycosides, flavonoids, phytosterols, phenols/tannins, and terpenoids are some constituents directly associated with the biological properties of plants, and these substances were proved to have antimicrobial as well as many other health-benefiting effects, like anticarcinogenic, antioxidant, anti-inflammatory, antimutagenic, antigenotoxic, anthelmintic, and antiproliferative qualities, and thus either directly or indirectly cure diseases or inhibit disease-causing organisms (Shams et al., 2024). Zadravec et al. (2022) mentioned essential oils extracted from cloves, cinnamon, oregano, lemongrass, and palmarosa is proved to be effective against infections caused by *Fusarium*

TABLE 12.3
List of Recently Identified Plant Extracts That Combat Food Contaminating Mycotoxins

Sl. No.	Herbal Control Agent	Part	Mycotoxin Detoxified	Removal Efficiency	Reference
1	*Moringa oleifera* *Calotropis procera*	Leaf	Aflatoxin B1	100% 96%	Nisar et al. (2024)
2	*Heliotropium bacciferum* *Ocimum dhofarense* *Zataria multiflora*	Leaf	Aflatoxin B1	95% 93% 92%	Velazhahan et al. (2024)
3	*Syzygium aromaticum*	Flower buds	Gliotoxin	—	Talib et al. (2024)
4	*Curtisia dentata* *Aframomum melegueta* *Centella asiatica* *Nymphaea stellata*	Bark Seed Leaf Leaf and flower	Aflatoxin B1	71.13% 40.84% 37.86% 19.99% 18.54%	Kalemba et al. (2024)
5	*Hybanthus enneaspermus* *Eclipta prostrata* *Syzygium cumini* *Quercus infectoria* *Terminalia chebula* *Solanum trilobatum* *Phyllanthus niruri* *Myristica fragrans* *Cardiospermum halicacabum*	Dried powder	Aflatoxin B1	76.7% 75.8% 69% 69% 66% 63% 46.5% 46% 39.5%	Al-Owaisi et al. (2022)

verticillioides in maize seeds. Similarly, essentials oils from cinnamon, bay leaves, rosewood, and sassafras can inhibit the growth of *Aspergillus* spp., *Fusarium* spp., and *Penicillium* spp. Das et al. (2021) reviewed some previous works on essential oil–based nanoemulsion edible coating, particularly on fruits, and enlisted some essential oils and their contribution in protecting fruits after being coated with these oils. They mentioned that coating of essential oils on fruits can have convincing inhibitory effect on growth of fruit pathogenic fungi, inhibits biosynthesis of mycotoxins, and in some cases, enhancement of their shelf life was also noted. Thus, it can be concluded that the application of this essential oil-based nanoemulsion edible coating is a boon for the safety, security, and trade of fruits since it prevents post-harvest damage and thus supports the economy. Another previous study conducted, by Perczak et al. (2016), represent the effect of essential oils on zearalenone. A 20.02% and 24.7% reduction in zearalenone content was noted after 1 and 3 days' exposure to cinnamon leaf oil and cinnamon bark oil accounts for 90.93% reduction. Similarly, the application of lemon oil for some time can reduce this mycotoxin by 31.93% and 46.46%. In case of cedarwood, grapefruit, and eucalyptus, essential oil application reduces the amount of toxin by 25.36, 30.30, 32.74%, and 28.73, 33.59, 37.41%, after incubating 1 and 2 days, respectively, while zearalenone degradation by palmarosa oil is most convincing, achieving up to 99% reduction. Wan et al. (2020) reported that clove oil–based nanoemulsion application in barley crop has convincing effect on *Fusarium* spp. and induces subsequent decline in deoxynivalenol production. In Paris et al. (2020) the successful formation of *Cinnamomum zeylanicum* essential oil–encapsulated beads having convincing antagonistic activity toward *Alternaria alternata*, *Botrytis cinerea*, *Colletotrichum gloeosporioides,* and *Penicillium expansum*. Wang et al. (2019) reported above 97% inhibition of Alternaria alternata produced mycotoxins like alternariol, alternariol, monomethyl ether etc. Chaudhari et al. (2020) reported convincing inhibitory results of applying marjoram (*Origanum majorana*) essential oils on various storage molds belonging to the genera *Alternaria*, *Fusarium*, *Penicillium*, and *Aspergillus*. In addition, they have also been reported to have significant inhibitory effect on AFB1 mycotoxin. Ramsdam et al. (2021) demonstrated the complete inhibition of pathogenic *Aspergillus flavus* from maize grains by employing essential oils of different plants, namely, *Artemisia nilagirica*, *Curcuma longa*, *Gaultheria fragrantissima*, *Litsea cubeba*, and *Zingiber officinale* at concentrations less than 2 (μL/mL). Allagui et al. (2023) tested the efficiency of 30 essentials to combat fungal pathogens. The outcomes indicate that *Cinnamomum verrum* oils reflect complete inhibition of three pathogens, namely, *Alternaria alternata*, *Botrytis cinerea*, and *Penicillium italicum*, at a very low concentration of 0.4 mg/mL.

TABLE 12.4
Application of Essential Oils to Combat foodborne fungus and its associated mycotoxins.

Plant Species	Concentration	Pathogen/ Mycotoxin	Inhibition Efficiency (%)	Reference
Cinnamomum verrum	0.4 (mg/mL)	*Alternaria alternata*	100	Allagui et al. (2023)
		Botrytis cinerea	100	
		Penicillium italicum	100	
Foeniculum vulgar	0.4 (mg/mL)	*Botrytis cinerea*	94.1	
Cymbopogon nardus	0.4 (mg/mL)	*Botrytis cinerea*	99.1	
Syzygium aromaticum	0.4 (mg/mL)	*Alternaria alternata*	86.9	
		Penicillium italicum	87.4	
Pelargonium asperum	0.4 (mg/mL)	*Botrytis cinerea*	65.5	
Artemisia nilagirica	1.8 (μL/mL)	*Aspergillus flavus*	100	Ramsdam et al.
Curcuma longa	1.4 (μL/mL)	(in maize grains)	100	(2021)
Gaultheria	1.2 (μL/mL)		100	
fragrantissima	0.8 (μL/mL)		100	
Litsea cubeba	1.2 (μL/mL)		100	
Zingiber officinale				

(Continued)

TABLE 12.4 (*Continued*)
Application of Essential Oils to Combat foodborne fungus and its associated mycotoxins.

Plant Species	Concentration	Pathogen/ Mycotoxin	Inhibition Efficiency (%)	Reference
Cymbopogon citratus	1 (mg/mL)	*Stagonosporopsis cucurbitacearum*	100	Moumni et al. (2021)
Origanum majorana	1 (mg/mL)	*Alternaria alternata*	100	
Lanvandula hybrid	1 (mg/mL)	*Alternaria alternata*	90.5	
		Alternaria alternata	71.5	
Lavandula dentata	1 (mg/mL)	*Stagonosporopsis cucurbitacearum*	73.73	
Laurus nobilis	1 (mg/mL)	*Stagonosporopsis cucurbitacearum*	73.92	
Melaleuca alternifolia	1 (mg/mL)	*Stagonosporopsis cucurbitacearum*	65.29	
		Stagonosporopsis cucurbitacearum	73.73	
Zingiber officinale	2,000 (μg/mL)	FumonisinB1	98.35	Castro et al. (2020)
Cinnamomum zeylanicum	1,000 (μg/mL)	(*Fusarium*	91.04	
	250 (μg/mL)	*verticillioides*)	95.07	
Cymbopogon martinii	500 (μg/mL)		97.99	
Rosmarinus officinalis	2,000 (μg/mL)	*Aspergillus flavus*	93.2	da Silva Bomfim et al. (2020)
Cymbopogon martini	5 μg/mL	Zearalenone	99.29	Perczak et al. (2016)
Cinnamomum zeylanicum	0.5 μg/mL		90.93	
Artemisia herba-alba	5,000 (μL/mL)	*Aspergillus flavus*	100	Abu-Darwish et al. (2015)
	1,250 (μL/mL)	*Aspergillus niger*	100	
	2,500 (μL/mL)	*Aspergillus fumigatus*	100	
Lavandula multifida	0.64 (μL/mL)	*Aspergillus flavus*	100	Zuzarte et al. (2012)
Hedychium spp.	40,000 (μL/mL)	*Fusarium verticillioides*	100	Rajasekaran et al. (2012)
		Aspergillus flavus	100	
Ocimum gratissimum	0.6 (μL/mL)	*Fusarium nivale*	100	Prakash et al. (2011)
		Penicillium italicum	100	
		Aspergillus niger	85	
		Aspergillus flavus	91	

12.6 CHALLENGES AND FUTURE PERSPECTIVES

Over the last century, due to the rapid growth of the global human population, agricultural production has correspondingly increased. Microorganisms, especially filamentous fungi, yeasts, and bacteria (both Gram-positive and Gram-negative), full of protein and human-required nutrients, are receiving attention by consumers because of their health benefits (Kumar et al., 2008). Fungi pose a major threat to food security as the global population continues to expands. The effects of climate change further intensify the challenges in crop management and food storage, increasing the risk of new fungal strains that could lead to diseases and spoilage (Davies et al., 2021). By 2050, the global population is projected to reach 9.7 billion (United Nations, 2019), which may require a 119% increase in the global production of edible crops to meet demand (Berners-Lee et al., 2018). So there must be substantial reductions in food waste and spoilage. As a result, the food industry faces major challenges in addressing both current and future demands. In addition to concerns related to food transportation, storage infrastructure, water scarcity, and the impacts of climate change, a large amount of food is lost due to diseases affecting both crops and livestock (Davies et al., 2021). The widespread occurrence of infections caused by microbial contaminants of our food supply illustrates the intricate relationships between humans, plants, and pathogenic microorganisms on a global scale. Over time, the types of foodborne infections have changed dramatically; some old pathogens are now being under control or eliminated, while new ones are newly recognized. Despite progress in this area, foodborne diseases continue being a severe perfect global health hazard, affecting millions of people annually (Hassan et al., 2023).

Fungi and their by-products have played a major role in food science for many years. The growth and metabolic activity of fungi can have varied effects. On the positive side, fungi form beneficial relationships with plants, aiding in the production of valuable products, such as food. On the other hand, they are major contributors as undesirable effector, causing food spoilage, decay, and the formation of toxins.

Fungi are a major cause of crop losses during the later stages of food production, with an additional 10% of crops being lost post-harvest (Fisher et al., 2018). These spoilage fungi are especially responsible for the degradation of perishable foods, such as cheese, carbonated drinks, and condiments. Filamentous fungi, such as *Aspergillus niger*, can damage solid, fresh items, such as fruits, while spoilage yeasts can spoil liquid products, including preserves and drinks, even those designed to prevent the growth of spoilage microorganisms (Pitt and Hocking, 2009). The fungi contaminated livestock are at risk of fungal infections studied by Ahmad and Gholib (2016). Fungal spores, whether produced sexually or asexually, are dispersed widely through the air, making it easy for foods to become contaminated, colonized, and eventually spoiled. This contamination can occur at any stage—during processing, storage, or handling by manufacturers, retailers, or consumers—leading to food loss (Leyva et al., 2017). Certain fungi that infect food produce mycotoxins as by-products, which can have significant health risks, including the risk of cancer. Hence, monitoring and controlling mycotoxins in food items is a challenging process that requires careful management (Krisch et al., 2011).

Therefore, fungal contamination remains a major concern in the food industry, contributing to financial losses, inducing potential health risks to consumers and even causing death of consumers. The present research includes a comprehensive examination of the fungal species responsible for spoilage of foodstuffs (both raw and processed), highlighting their ecological traits, metabolic functions, and mode of action. Understanding the diversity and behavior of spoilage fungi is essential for developing control strategies that combat these food pathogens and ensure food safety without effecting their quality parameters, like taste, texture, etc. Ensuring food safety and combating foodborne ailments are the major priorities within the food industry (Ayofemi, 2020). Foodborne pathogens pose significant health risks and can lead to outbreaks. Globalization, changes in food production techniques, climate change, and antimicrobial resistance are some common factors that are associated with the contamination process. The findings reflect the requirement for developing effectual control and prevention measures throughout the food supply chain. In this regard, contamination during production, processing, transportation, and consumption is a mentionable challenge. These factors highlight the importance of implementing effective control and prevention strategies throughout the food supply chain. Key challenges include the risk of contamination at various stages, such as production, processing, transportation, and consumption (Hassan et al., 2023).

12.7 CASE STUDIES AND REAL-WORLD APPLICATIONS

12.7.1 Case Studies of Fungal Outbreaks and Their Impact on Public Health and the Food Industry

In 1960, about 1,00,000 turkeys died in the United Kingdom after having contaminated peanut feed. This incident attracted researchers to investigate the cause behind such outbreak. This led them to the discovery of a toxic compound produced by *Aspergillus flavus* as the reason behind those deaths. Thus, a new term, *mycotoxin*, was introduced, and further research on this topic established them as a global concern toward food safety and security (El-Sayed et al., 2022). But outbreaks of fatal diseases due to human exposure to fungal secondary metabolites had been documented much earlier. One such example is St. Anthony's fire, which is now referred to as ergotism caused by ergot alkaloids produced by *Claviceps* spp. The history of this disease is believed to date back to 600 BC in Assyria, where the soldiers of Julias Ceaser were reported to be infected by this disease. It was also described to occur in China in 1,100 BC. However, the proper literary evidence of this disease was reported by the French from the mid of the 9th century to the end of the 10th century, which was claimed to kill 20,000 to 50,000 people during this time. More outbreaks of ergotism were recorded from Germany during the 16th century. Another heartbreaking incident occurred in 1692 at Salem, Massachusetts, where 1,722,20,000 soldiers of the army of Peter 1 of Russia died after consuming bread made of ergot-contaminated flour. Several more cases of ergotism outbreak were reported from Russia, United Kingdom, and France between early and mid-20th century, affecting more than 30,000 people (Awuchi et al., 2021; Palanisamy, 2024). In 1974, 106 people from western India were affected and ultimately killed by infections caused due to the consumption of aflatoxin-contaminated maize. This is probably one of the earliest outbreaks of mycotoxicosis reported from India. Meanwhile, in 1982, the deaths of a huge population of chicks due to aflatoxin toxicity have been reported from the Chittoor district of Andhra Pradesh (Varshini and Gayathri, 2024). From October 1997 to March 1997, 3 outbreaks of gastrointestinal health issues were reported from schools in USA, and this incident was followed by 16 more outbreak events in schools from May 1998 to October 1998. Both incidents were linked to companies producing foodstuffs like burritos filled with pork meat, chicken, egg, bean, etc. Around 1,908 persons from 125 schools were affected by these incidents, and people, after consuming those burritos, experienced symptoms like abdominal cramps, headache, nausea, vomiting, etc. (Baines et al., 2018). On July 20, 2004, Kenya witnessed 317 cases of acute aflatoxicosis, which was caused by consuming maize contaminated with aflatoxins. Out of 317 cases, 125 patients died; thus, the disease represents a fatality rate of 39%, and 158 patients (~51%) affected by the disease were children below the 15-year age group, indicating the susceptibility of children toward this ailment (Omara et al., 2021; Musangi et al., 2024). Another African country, Tanzania, also experienced severe aflatoxicosis outbreak from 2016 to 2019. During the beginning of the outbreak, Aflatoxin-induced liver cancer led to the death of nearly 3 people per 1,00,000 people in Tanzania. Earlier, the toxin responsible for the outbreak was unknown, but later investigations and tests declared aflatoxin B1 as the culprit (Kinyenje et al., 2023). In 2023, lactating animals in a dairy farm in Lahore, Pakistan, were severely affected by aflatoxin B1, which reduced milk yield, increased percentages of abortion, and even led to the death of the herd (Penagos-Tabares et al., 2024). Another study revealed that mycotoxin deoxynivalenol facilitates replication, infection and damage caused by capripoxvirus in cattle leading to lumpy skin disease which is currently the most unprecedented disease in cattle (Kothapalli et al., 2024). On January 25, 2024, students of three elementary schools in the Gunungkidul district of Indonesia became sick after consuming snack sold near their schools. Vomiting, nausea, and stomach pain were some commonly observed symptoms in effected individuals. The results of laboratory tests confirmed the presence of microbes, including *Pseudomonas* sp., *Bacillus* cereus, *Staphylococcus* aureus, mold/yeast. The presence of mold/yeast in the food samples, like fresh egg macaroni, siomai, packaged seasoning powder, soy sauce, and chili sauce, clearly indicates the involvement of mycotoxin contaminants in poisoning (Handika et al., 2024).

12.7.2 Examples of the Successful Implementation of Control Measures in Food Production

One successful example of control measure implementation in food production is the MyToolBox Project (www.mytoolbox.eu). The MyToolBox project is a four-year project officially launched in March 2016 and funded by the European Commission, along with its 23 partners from 11 countries. This multi-actor partnership project has above 40% of its partners from industries including farmers and other partners include (academia, technology, small- and medium-sized enterprises, food industry, and policy stakeholders). This project has set several pre- and post-harvest objectives, which aimed to develop and synergize different management strategies along with the entire chain of food production to achieve a significant decline in food contamination and losses because of mycotoxins (Krska et al., 2016). One such example of pre-harvest management was cited by Focker et al. (2021), where four randomized blocks of winter wheat in England were inoculated with *Fusarium graminearum*, treated with two different fungicides, namely, Adepidyn (a newly developed succinate dehydrogenase inhibitor fungicide antagonistic to *Fusarium* spp.) and proline (standard fungicide) and left untouched. The experiments were performed to test the efficiency of this newly introduced fungicide to control *Fusarium* head blight disease and associated deoxynivalenol (DON) contamination of wheat crops. The outcomes confirm that Adepidyn has reduced deoxynivalenol levels in wheat kernels by 80%, which is higher than standard fungicide proline, with only 54% reduction in level of this particular mycotoxin, thus reflecting the relatively better performance of this new fungicide in promoting mycotoxin decontamination. This leads to the availability of 42,000, 129,000, and 219,000 extra tons of wheat for milling when 20, 50, and 80% or even more farmers use Adepidyn fungicide. Other projects run by the European Commission, like MycoGlobe (2004–2008), MycoRed (2012–2016), MycoKey (2016–2020), aimed to understand and formulate successful, sustainable combating strategies against mycotoxin contamination of food (Logrieco et al., 2021). Imade et al. (2021) highlighted that the application of Aflasafe (a fungal strain possessing biocontrol agent that combats mycotoxin-releasing fungi) in *Zea mays* and *Arachis hypogaea* led to the availability of Aflasafe products in some African countries, like Burkina Faso, Nigeria, Ghana, Kenya, Senegal, and The Gambia. This, in turn, indicates a strong possibility of these Aflasafe foods to be available in the remaining African nations in the coming future.

12.7.3 Lessons Learned from Past Outbreaks and Implications for Future Practices

The Food and Agriculture Organization mentioned that, at a global level, about 25% of cereal production is contaminated by fungal secondary metabolites. This led to an estimated loss of about 1 billion metric ton of both food and feed every year. Besides, these infections also have several negative implications on human health. Plenty of documented evidences are present in literature about the outbreak of fungal infections and their fatal effects on both humans and animals. As a result, several organizations at both world and country levels participated in research to understand the possible biosynthetic pathway and mode of entry of mycotoxins into the food. These organizations have studied the maximum and minimum amount of toxins that could affect human health and set some regulatory limits, total daily intake limits, weekly intake limits, etc. to made everyone aware of intake limitations and safety measures towards preparing or selecting a health food item. Small to large, every industry or entrepreneur engaged with food processing and packaging should follow the guidelines provided by them, and the amount of mycotoxin should not exceed the limits specified by them (Alpas et al., 2017). Important lessons that we learn from previous outbreaks of foodborne fungal toxic secondary metabolites and some important suggestions for the future are discussed in what follows:

1. Government should publish some strict guidelines regarding the levels of mycotoxins in food, methods of detection, best combating measures, along with a list of punishments for those who do not follow these rules.
2. There should be a strict intervention of government bodies, like food department, health officials, to check whether the guidelines published by

government bodies about maximum tolerable amount of mycotoxin were followed by the food industries or not. If any irregularity is observed, it should be reported as soon as possible, before it reaches market for trade.

3. Authorities should enormously research previous outbreak to understand the demographical features, susceptible target age groups, foods mostly contaminated, fungal strains, and toxins involved in poisoning process, etc.
4. Most importantly, public awareness should be increased on the health effects of consuming contaminated food. Government bodies like schools, colleges, universities, gram panchayats, non-government organizations, and other responsible institutions should voluntarily participate in this.
5. A large number of epidemiological studies should be conducted for better investigation and understanding of the outbreak.
6. Industries should trace the possible causes of contaminations and modify their existing processing systems into a more developed one that has more convincing effect on food commodities without affecting their taste, texture, nutritional attributes, etc.
7. More focus should be given on developing more efficient agricultural practices that could lead to the elimination of mycotoxins in raw foods. Good agricultural practices must be followed (Baines et al., 2018; Chilaka et al., 2022).

12.8 CONCLUSION

The safety and security of food commodities is one of the most important topics for the survival of humans. Food, in both raw and as well as processed forms, involves contamination by various biotic and abiotic stresses. One very serious abiotic factor includes fungal pathogens that highly affect cereals, millets, lentils, milk, cheese, meat, and various other processed, as well as unprocessed, foodstuffs. Thus, foodborne fungal pathogens possess huge risks toward the integrity of food products, making it a hot topic for research in modern time. These fungal pathogens contaminate a variety of food by secreting several secondary metabolites, and when these metabolic compounds enter a human or animal body, they induce various toxic effects on health. Sometimes diseases caused by these noxious metabolites can be fatal. Several evidences were discussed in this chapter regarding the outbreaks of these diseases and their effect on society. So there must be some techniques that can be utilized to combat these fungal strains, degrade their associate mycotoxins, and alleviate their effect on health. The detection of toxic fungal strains and metabolites have great importance as it allows the scientists and researchers to develop an efficient system to kill pathogens in food. Traditional techniques, such as microscopic examinations and culture-based methods, were employed to detect them. While other methods, like molecular methods, like PCR, qPCR, and sequencing methods, immunological methods like rapid diagnostic assays, antibody-based assays, biosensors, nano-based sensors, and portable care of devices also aided the detection of foodborne fungal pathogens. After decades of enormous research on developing potential identification techniques, some common and highly virulent fungal strains were investigated, and mycotoxins produced by them, such as aflatoxins (AFB1, AFB2, AFG1, AFG2, AFM1, AFM2), ochratoxins (OTA), and Citrinin, were reported to be released by members of the genera *Aspergillus* and *Penicillium*, trichothecenes (T-2, HT-2, DON, 3- acetyl- DON, 15-acetyl-DON, diacetoxyscirpenol, fusarenon X, neosolaniol, NX toxin, verrucarin), zearalenone, fumonisins, fusaric acid, fusarin C, were all reported from *Fusarium*, ergot alkaloids from *Claviceps*, phomopsin from *Diaporthe*, Alternaria toxins (Alternariol) from *Alternaria*, and many more toxins. Detection must be needed for a better understanding of the contamination process and also for the development of specific treatment techniques. Traditionally, physical and chemical treatment methods were employed to detoxify mycotoxins and for killing pathogens, but these methods have many negative outcomes, like non-eco-friendliness, expensiveness, lack of practicality, ineffectiveness, mutation of pathogens into more resistant as well as virulent one, and also exceptional changes in nutritional as well as organoleptic properties of food. This led to the development of more efficient and sustainable biological measures to combat these pathogens. The implementation of biological treatment measure by using probiotics, biomolecules, and enzymes is proved to be highly successful in treating these pathogenic strains with very minimal effect on properties of food. Besides, use of plant extracts and essential oils was reported to induce convincing detoxification of mycotoxins, block biosynthesis pathways and kills pathogens? As a whole, these techniques play an efficacious role in controlling foodborne fungal pathogens and thus facilitates the production of healthy, safe, and nutritious food items. The present chapter also highlights some previous reports of disease outbreaks in different countries that had affected the lives of many, and this helps in understanding how the mistakes done earlier help us understand the problems so that the gaps in the present detoxification process can be filled to achieve food safety, security, and sustainability of the globe.

REFERENCES

Abbasi Pirouz, A., J. Selamat, S. Zafar Iqbal, and N. Iskandar Putra Samsudin. "Efficient and Simultaneous Chitosan-mediated Removal of 11 Mycotoxins from Palm Kernel Cake." *Toxins* 12, no. 2 (2020): 115.

Abdel-Wahhab, M. A., and A. M. Kholif. "Mycotoxins in Animal Feeds and Prevention Strategies: A Review." *Asian Journal of Animal Sciences* 4, no. 3 (2010): 113–131.

Abu-Darwish, M. S., C. Cabral, M. J. Gonçalves, C. Cavaleiro, M. T. Cruz, T. Efferth, and L. Salgueiro. "Artemisia Herba-alba Essential Oil from Buseirah (South Jordan): Chemical Characterization and Assessment of Safe Antifungal and Anti-inflammatory Doses." *Journal of Ethnopharmacology* 174 (2015): 153–160.

Afsah-Hejri, L., P. Hajeb, and R. J. Ehsani. "Application of Ozone for Degradation of Mycotoxins in Food: A Review." *Comprehensive Reviews in Food Science and Food Safety* 19, no. 4 (2020): 1777–1808.

Ahmad, R. Z., and D. Gholib. "Important Mycotic Diseases in Animal Livestock Caused by Some Agricultural Fungi in Indonesia." *The International Journal of Tropical Veterinary and Biomedical Research* 1, no. 1 (2016): 38–44.

Al Khalaileh, N. I. "Prevalence of Ochratoxin a in Poultry Feed and Meat from Jordan." *Pakistan Journal of Biological Sciences* 21, no. 5 (2018): 239–244.

Allagui, M. B., M. Moumni, and G. Romanazzi. "Antifungal Activity of Thirty Essential Oils to Control Pathogenic Fungi of Postharvest Decay." *Antibiotics* 13, no. 1 (2023): 28.

Al-Owaisi, A., A. M. Al-Sadi, J. N. Al-Sabahi, S. P. Sathish Babu, M. M. A. Al-Harrasi, I. Hashil Al-Mahmooli, R. Abdel-Jalil, and R. Velazhahan. "In Vitro Detoxification of Aflatoxin B1 by Aqueous Extracts of Medicinal Herbs." *All Life* 15, no. 1 (2022): 314–324.

Alpas, H., F. Yeni, Y. Soyer, and J. Fletcher. "Recent Outbreaks of Human Pathogens on Plants (HPOPs) on Fresh Produce—Lessons Learned from the Practice." *Practical Tools for Plant and Food Biosecurity: Results from a European Network of Excellence* (2017): 77–96.

Andrade, J. C., S. Kumar, A. Kumar, L. Cernakova, and C. F. Rodrigues. "Application of Probiotics in Candidiasis Management." *Critical Reviews in Food Science and Nutrition* 62, no. 30 (2022): 8249–8264.

Awuchi, C. G., E. N. Ondari, S. Nwozo, G. A. Odongo, I. J. Eseoghene, H. Twinomuhwezi, C. U. Ogbonna, A. K. Upadhyay, A. O. Adeleye, and C. O. R. Okpala. "Mycotoxins' Toxicological Mechanisms Involving Humans, Livestock and Their Associated Health Concerns: A Review." *Toxins* 14, no. 3 (2022): 167.

Awuchi, C. G., E. N. Ondari, C. U. Ogbonna, A. K. Upadhyay, K. Baran, C. O. R. Okpala, M. Korzeniowska, and R. P. Guiné. "Mycotoxins Affecting Animals, Foods, Humans, and Plants: Types, Occurrence, Toxicities, Action Mechanisms, Prevention, and Detoxification Strategies—A Revisit." *Foods* 10, no. 6 (2021): 1279.

Ayofemi Olalekan Adeyeye, S. "Aflatoxigenic Fungi and Mycotoxins in Food: A Review." *Critical Reviews in Food Science and Nutrition* 60, no. 5 (2020): 709–721.

Baines, R. N., L. Manning, and J. M. Soon. "Mycotoxin Incidents Associated with Cereals: Lessons Learnt and Risk Reduction Strategies." *Quality Assurance and Safety of Crops & Foods* 10, no. 1 (2018): 1–16.

Beelman, R. B., M. D. Kalaras, and J. P. Richie Jr. "Micronutrients and Bioactive Compounds in Mushrooms: A Recipe for Healthy Aging?" *Nutrition Today* 54, no. 1 (2019): 16–22.

Bennett, J. W., and M. Klich. "Clinical Microbiology Reviews." *Mycotoxins* 16, no. 1 (2003): 497–516.

Bennett, J. W., M. Klich, and P. S. Dyer. "Fungal Pathogens: A Global Perspective on Their Impact on Food Safety." *Food Microbiology* 87 (2020): 103358.

Berners-Lee, M., C. Kennelly, R. Watson, and C. N. Hewitt. "Current Global Food Production Is Sufficient to Meet Human Nutritional Needs in 2050 Provided There Is Radical Societal Adaptation." *Elementa: Science of the Anthropocene* 6 (2018): 52.

Bhilwadikar, T., S. Pounraj, S. Manivannan, N. K. Rastogi, and P. S. Negi. "Decontamination of Microorganisms and Pesticides from Fresh Fruits and Vegetables: A Comprehensive Review from Common Household Processes to Modern Techniques." *Comprehensive Reviews in Food Science and Food Safety* 18, no. 4 (2019): 1003–1038.

Bintsis, T. "Foodborne Pathogens." *AIMS Microbiology* 3, no. 3 (2017): 529–563.

Borman, A. M., C. J. Linton, S. Miles, and E. M. Johnson. "Molecular Identification of Pathogenic Fungi." *Journal of Antimicrobial Chemotherapy* 61, no. 1 (2008): 7–12.

Botha, M. A., and A. Botes. "Mucor." In *Encyclopedia of Food Microbiology*, 2nd ed., edited by Carl A. Batt and Mary Lou Tortorello, 834–840. Cambridge: Academic Press, 2014. http://dx.doi.org/10.1016/B978-0-12-384730-0.00228-7.

Bousset, L., M. Ermel, B. Soglonou, and O. Husson. "A Method to Measure Redox Potential (Eh) and pH in Agar Media and Plants Shows That Fungal Growth Is Affected by and Affects pH and Eh." *Fungal Biology* 123, no. 2 (2019): 117–124.

Cai, T., L. Zhenye, G. Peng, G. Jialu, R. Wang, D. Guo, J. Yu, X. Lü, X. Xia, and C. Shi. "Antimicrobial and Antibiofilm Efficacy and Mechanism of Oregano Essential Oil Against *Shigella flexneri*." *Foodborne Pathogens and Disease* 20, no. 6 (2023): 209–221.

Casadevall, A., J. Heitman, and M. Buckley. *The Fungal Kingdom: Diverse and Essential Roles in Earth's Ecosystem*. Washington, DC: American Academy of Microbiology, 2008.

Castro, J. C., G. C. Pante, B. M. Centenaro, R. T. R. D. Almeida, E. J. Pilau, B. P. Dias Filho, S. A. G. Mossini, B. A. D. Abreu Filho, G. Matioli, and M. Machinski Jr. "Antifungal and

Antimycotoxigenic Effects of *Zingiber officinale*, *Cinnamomum zeylanicum*, and *Cymbopogon martinii* Essential Oils Against *Fusarium verticillioides*." *Food Additives & Contaminants: Part A* 37, no. 9 (2020): 1531–1541.

Cha, L., H. Feng, M. Wu, J. Xing, J. Li, and Q. Chen. "Effects of Extracellular Enzymes Secreted by Wild Edible Fungi Mycelia on the Surface Properties of Local Soil Colloids." *Environmental Technology* 44, no. 24 (2023): 3721–3730.

Chatterjee, S., A. Dhole, A. A. Krishnan, and K. Banerjee. "Mycotoxin Monitoring, Regulation and Analysis in India: A Success Story." *Foods* 12, no. 4 (2023): 705.

Chaudhari, A. K., V. K. Singh, S. Das, J. Prasad, A. K. Dwivedy, and N. K. Dubey. "Improvement of In Vitro and In Situ Antifungal, AFB1 Inhibitory and Antioxidant Activity of *Origanum majorana* L. Essential Oil through Nanoemulsion and Recommending as Novel Food Preservative." *Food and Chemical Toxicology* 143 (2020): 111536.

Chilaka, C. A., J. E. Obidiegwu, A. C. Chilaka, O. O. Atanda, and A. Mally. "Mycotoxin Regulatory Status in Africa: A Decade of Weak Institutional Efforts." *Toxins* 14, no. 7 (2022): 442.

Conte, G., M. Fontanelli, F. Galli, L. Cotrozzi, L. Pagni, and E. Pellegrini. "Mycotoxins in Feed and Food and the Role of Ozone in Their Detoxification and Degradation: An Update." *Toxins* 12, no. 8 (2020): 486.

Cousin, M. A., R. T. Riley, and J. J. Pestka. "Foodborne Mycotoxins: Chemistry, Biology, Ecology, and Toxicology." In *Foodborne Pathogens: Microbiology and Molecular Biology*, edited by P. M. Fratamico, A. K. Bhunia, and J. L. Smith, 163–226. Norfolk: Caister Academic Press, 2005.

D'Mello, J. P. F. *Mycotoxins in Plant Products*. Wallingford: CABI Publishing, 2001.

da Rocha, M. E. B., F. D. C. O. Freire, F. E. F. Maia, M. I. F. Guedes, and D. Rondina. "Mycotoxins and Their Effects on Human and Animal Health." *Food Control* 36, no. 1 (2014): 159–165.

da Silva Bomfim, N., C. Y. Kohiyama, L. P. Nakasugi, S. B. Nerilo, S. A. G. Mossini, J. C. Z. Romoli, J. M. Graton Mikcha, B. A. D. Abreu Filho, and M. Machinski Jr. "Antifungal and Antiaflatoxigenic Activity of Rosemary Essential Oil (*Rosmarinus officinalis* L.) Against *Aspergillus flavus*." *Food Additives & Contaminants: Part A* 37, no. 1 (2020): 153–161.

Das, S., A. Ghosh, and A. Mukherjee. "Nanoencapsulation-Based Edible Coating of Essential Oils as a Novel Green Strategy Against Fungal Spoilage, Mycotoxin Contamination, and Quality Deterioration of Stored Fruits: An Overview." *Frontiers in Microbiology* 12 (2021): 768414.

Davies, C. R., F. Wohlgemuth, T. Young, J. Violet, M. Dickinson, J. W. Sanders, C. Vallieres, and S. V. Avery. "Evolving Challenges and Strategies for Fungal Control in the Food Supply Chain." *Fungal Biology Reviews* 36 (2021): 15–26.

De Felice, B., L. J. Spicer, and F. Caloni. "Enniatin B1: Emerging Mycotoxin and Emerging Issues." *Toxins* 15, no. 6 (2023): 383.

Ding, L., M. Han, X. Wang, and Y. Guo. "Ochratoxin A: Overview of Prevention, Removal, and Detoxification Methods." *Toxins* 15, no. 9 (2023): 565.

Duncan, H., S. Smith, and R. Johnson. "Limitations of Traditional Microbiological Methods in Detecting Foodborne Fungi." *International Journal of Food Microbiology* 290 (2019): 1–8.

EFSA (European Food Safety Authority) and ECDC (European Centre for Disease Prevention and Control). "The European Union Summary Report on Trends and Sources of Zoonoses, Zoonotic Agents and Food-Borne Outbreaks in 2015." *EFSA Journal* 14 (2016): 4634–4865.

EFSA Panel on Contaminants in the Food Chain (EFSA CONTAM Panel), D. Schrenk, M. Bignami, L. Bodin, J. K. Chipman, J. del Mazo, B. Grasl-Kraupp, C. Hogstrand, L. Hoogenboom, J. C. Leblanc, and C. S. Nebbia. "Risk to Human Health Related to the Presence of Perfluoroalkyl Substances in Food." *EFSA Journal* 18, no. 9 (2020): e06223.

Ekwomadu, T. I., S. A. Akinola, and M. Mwanza. "Fusarium Mycotoxins, Their Metabolites (Free, Emerging, and Masked), Food Safety Concerns, and Health Impacts." *International Journal of Environmental Research and Public Health* 18, no. 22 (2021): 11741.

El-Abeid, S. E., M. A. Mosa, M. A. El-Tabakh, A. M. Saleh, M. A. El-Khateeb, and M. S. Haridy. "Antifungal Activity of Copper Oxide Nanoparticles Derived from Zizyphus Spina Leaf Extract Against Fusarium Root Rot Disease in Tomato Plants." *Journal of Nanobiotechnology* 22, no. 1 (2024): 28.

El-Sayed, R. A., A. B. Jebur, W. Kang, and F. M. El-Demerdash. "An Overview on the Major Mycotoxins in Food Products: Characteristics, Toxicity, and Analysis." *Journal of Future Foods* 2, no. 2 (2022): 91–102.

Endo, A. "Discovery and Development of Statins." *Natural Product Communications* 12, no. 8 (2017). https://doi.org/10.1177/1934578X1701200801.

Escott, C., J. M. Del Fresno, I. Loira, A. Morata, and J. A. Suárez-Lepe. "Zygosaccharomyces Rouxii: Control Strategies and Applications in Food and Winemaking." *Fermentation* 4, no. 3 (2018): 69.

FAO. "Assessing Quality and Safety of Animal Feeds." 2004. http://www.fao.org/3/y5159e/y5159e00.htm#Contents.

FAO, G. "Global Food Losses and Food Waste–Extent, Causes and Prevention." *SAVE FOOD: An Initiative on Food Loss and Waste Reduction* (2011): 9.

Ferreira, C. D., G. H. Lang, I. da Silva Lindemann, N. da Silva Timm, J. F. Hoffmann, V. Ziegler, and M. de Oliveira. "Postharvest UV-C Irradiation for Fungal Control and Reduction of Mycotoxins in Brown, Black, and Red Rice during Long-Term Storage." *Food Chemistry* 339 (2021): 127810.

Fisher, M. C., N. J. Hawkins, D. Sanglard, and S. J. Gurr. "Worldwide Emergence of Resistance to Antifungal Drugs Challenges Human Health and Food Security." *Science* 360, no. 6390 (2018): 739–742.

Focker, M., H. J. Van Der Fels-Klerx, N. Magan, S. G. Edwards, M. Grahovac, F. Bagi, D. Budakov, M. Suman, G. Schatzmayr, R. Krska, and M. De Nijs. "The Impact of Management Practices to Prevent and Control Mycotoxins in the European Food Supply Chain: MyToolBox Project Results." *World Mycotoxin Journal* 14, no. 2 (2021): 139–154.

Gallo, A., and G. Perrone. "Current Approaches for Advancement in Understanding the Molecular Mechanisms of Mycotoxin Biosynthesis." *International Journal of Molecular Sciences* 22, no. 15 (2021): 7878.

Ganesan, A. R., B. Balasubramanian, S. Park, R. Jha, I. Andretta, A. G. Bakare, and I. H. Kim. "Ochratoxin A: Carryover from Animal Feed into Livestock and the Mitigation Strategies." *Animal Nutrition* 7, no. 1 (2021): 56–63.

Gantenbein, S., E. Colucci, J. Kach, E. Trachsel, F. B. Coulter, P. A. Ruhs, K. Masania, and A. R. Studart. "Three-Dimensional Printing of Mycelium Hydrogels into Living Complex Materials." *Nature Materials* 22, no. 1 (2023): 128–134.

Ghobish, A. M., A. M. El-Nokrashy, A. E. Elshafey, and A. A. Tayel. "Investigating the Potential of Chitosan-Nanocomposites as a Bio-Based Adsorbent for Sustainable Aflatoxin Eradication from Fish Feed." *Egyptian Journal of Veterinary Sciences* 56, no. 8 (2025): 1719–1730.

Giacometti, J., A. B. Tomljanovic, and D. Josic. "Application of Proteomics and Metabolomics for Investigation of Food Toxins." *Food Research International* 54, no. 1 (2013): 1042–1051.

Goessens, T., T. Mouchtaris-Michailidis, K. Tesfamariam, N. N. Truong, F. Vertriest, Y. Bader, S. De Saeger, C. Lachat, and M. De Boevre. "Dietary Mycotoxin Exposure and Human Health Risks: A Protocol for a Systematic Review." *Environment International* (2024): 108456.

Gourama, H., and L. B. Bullerman. "Aspergillus Flavus and Aspergillus Parasiticus: Aflatoxigenic Fungi of Concern in Foods and Feeds: A Review." *Journal of Food Protection* 58, no. 12 (1995): 1395–1404.

Gu, K., D. Ryu, and H. J. Lee. "Ochratoxin A and Its Reaction Products Affected by Sugars During Heat Processing." *Food Chemistry* 348 (2021): 129038.

Hami, A., R. S. Rasool, N. A. Khan, S. Mansoor, M. A. Mir, N. Ahmed, and K. Z. Masoodi. "Morpho-Molecular Identification and First Report of *Fusarium equiseti* in Causing Chilli Wilt from Kashmir (Northern Himalayas)." *Scientific Reports* 11, no. 1 (2021): 3610.

Hamouda, R. A., M. S. Abdel-Hamid, N. Hagagy, and A. M. Nofal. "The Potent Effect of Selenium Nanoparticles: Insight into the Antifungal Activity and Preservation of Postharvest Strawberries from Gray Mold Diseases." *Journal of the Science of Food and Agriculture* 104, no. 11 (2024): 6756–6768.

Handika, D. O., M. I. Utama, D. R. Ananda, A. W. Maula, R. A. Ahmad, and S. H. Sukoco. "Foodborne Outbreak Investigation in Elementary School, Gunungkidul District, January 2024: A Cohort Retrospective Study Design." In *BIO Web of Conferences*, vol. 132, 03001. EDP Sciences, 2024.

Hassan, A., M. K. I. Khan, S. Fordos, A. Hasan, S. Khalid, M. Z. Naeem, and A. Usman. "Emerging Foodborne Pathogens: Challenges and Strategies for Ensuring Food Safety." *Biology and Life Sciences Forum* 31, no. 1 (December 2023): 32.

Hassanain, N. A., M. A. Hassanain, W. M. Ahmed, R. M. Shaapan, A. M. Barakat, and H. A. El-Fadaly. "Public Health Importance of Foodborne Pathogens." *World Journal of Medical Sciences* 9, no. 4 (2013): 208–222.

Hawksworth, D. L. "The Magnitude of Fungal Diversity: The 1.5 Million Species Estimate Revisited." *Mycological Research* 105, no. 12 (2001): 1422–1432.

Heredia, N., and S. García. "Animals as Sources of Food-Borne Pathogens: A Review." *Animal Nutrition* 4, no. 3 (2018): 250–255.

Hou, X., L. Liu, D. Xu, D. Lai, and L. Zhou. "Involvement of LaeA and Velvet Proteins in Regulating the Production of Mycotoxins and Other Fungal Secondary Metabolites." *Journal of Fungi* 10, no. 8 (2024): 561.

Houbraken, J. A., J. C. Frisvad, and R. A. Samson. "Taxonomy of *Penicillium citrinum* and Related Species." *Fungal Diversity* 44 (2010): 117–133.

Huang, W., J. Chang, P. Wang, C. Liu, Q. Yin, Q. Zhu, F. Lu, and T. Gao. "Effect of the Combined Compound Probiotics with Mycotoxin–Degradation Enzyme on Detoxifying Aflatoxin B1 and Zearalenone." *The Journal of Toxicological Sciences* 43, no. 6 (2018): 377–385.

Hulvova, H., P. Galuszka, J. Frebortova, and I. Frebort. "Parasitic Fungus *Claviceps* as a Source for Biotechnological Production of Ergot Alkaloids." *Biotechnology Advances* 31, no. 1 (2013): 79–89.

Hutt, P. B., and P. B. I. Hutt. "A History of Government Regulation of Adulteration and Misbranding of Food." *Food Drug Cosmetic Law Journal* 39 (1984): 2–73.

Imade, F., E. M. Ankwasa, H. Geng, S. Ullah, T. Ahmad, G. Wang, C. Zhang, O. Dada, F. Xing, Y. Zheng, and Y. Liu. "Updates on Food and Feed Mycotoxin Contamination and Safety in Africa with Special Reference to Nigeria." *Mycology* 12, no. 4 (2021): 245–260.

Ince, B., I. Uludag, B. Demirbakan, C. Ozyurt, B. Ozcan, and M. K. Sezginturk. "Lateral Flow Assays for Food Analyses: Food Contaminants, Allergens, Toxins, and beyond." *TrAC Trends in Analytical Chemistry* 169 (2023): 117418.

International Agency for Research on Cancer. *IARC Monographs on the Evaluation of Carcinogenic Risks to Humans*, 1–136. World Health Organization, 2024.

Jabeen, N., and K. Shahzad. "Studies on Identification of the Foodborne Fungal Pathogens and Their Pathogenic Effects on Human Health." *Acta Scientific Agriculture* 3 (2019): 186–191.

Jackson, L. S., T. Beacham-Bowden, S. E. Keller, C. Adhikari, K. T. Taylor, S. J. Chirtel, and R. I. Merker. "Apple Quality, Storage, and Washing Treatments Affect Patulin Levels in Apple Cider." *Journal of Food Protection* 66, no. 4 (2003): 618–624.

Jha, N. S. "Common Adulterants and Contaminants." In *Rapid Detection of Food Adulterants and Contaminants*, 25–61. 2016. https://doi.org/10.1016/B978-0-12-420084-5.00002-0.

Jing, G., Y. Wang, M. Wu, W. Liu, S. Xiong, J. Yu, W. Li, W. Liu, and Y. Jiang. "Photocatalytic Degradation and Pathway from Mycotoxins in Food: A Review." *Food Reviews International* 40, no. 1 (2024): 276–292.

Kalemba, M. R. K., R. Makhuvele, and P. B. Njobeh. "Phytochemical Screening, Antioxidant Activity of Selected Methanolic Plant Extracts and Their Detoxification Capabilities Against AFB1 Toxicity." *Heliyon* 10, no. 2 (2024).

Khan, R. "Mycotoxins in Food: Occurrence, Health Implications, and Control Strategies—A Comprehensive Review." *Toxicon* (2024): 108038.

Kinyenje, E., R. Kishimba, M. Mohamed, A. Mwafulango, E. Eliakimu, and G. Kwesigabo. "Aflatoxicosis Outbreak and Its Associated Factors in Kiteto, Chemba, and Kondoa Districts, Tanzania." *PLOS Global Public Health* 3, no. 8 (2023): e0002191.

Kolawole, O., J. Meneely, A. Petchkongkaew, and C. Elliott. "A Review of Mycotoxin Biosynthetic Pathways: Associated Genes and Their Expressions Under the Influence of Climatic Factors." *Fungal Biology Reviews* 37 (2021): 8–26.

Kothapalli, P. K., S. Nagaraj, S. M. Pubbineedi, M. N. B. Nair, P. Natesan, and G. B. M. Reddy. "Mycotoxin Deoxynivalenol (DON) Increases Replication of the Lumpy Skin Disease Virus in MDBK Cells." *Indian Journal of Animal Health* (2024) https://doi.org10.36062/ijah.2024.16323.

Krisch, J., R. Tserennadmid, and C. Vagvolgyi. "Essential Oils Against Yeasts and Moulds Causing Food Spoilage." In *Science Against Microbial Pathogens: Communicating Current Research and Technological Advances*, Badajoz, Spain, 2011.

Krska, R., M. De Nijs, O. McNerney, M. Pichler, J. Gilbert, S. Edwards, M. Suman, N. Magan, V. Rossi, H. J. Van der Fels-Klerx, and F. Bagi. "Safe Food and Feed Through an Integrated Toolbox for Mycotoxin Management: The MyToolBox Approach." *World Mycotoxin Journal* 9, no. 4 (2016): 487–495.

Kumar, V., M. S. Basu, and T. P. Rajendran. "Mycotoxin Research and Mycoflora in Some Commercially Important Agricultural Commodities." *Crop Protection* 27, no. 6 (2008): 891–905.

Kumar, V., R. Singh, and A. Sharma. "Advances in Molecular Diagnostics of Mycotoxins: A Review." *Food Control* 121 (2021): 107588.

Kunz, B. M., A. Voß, J. Dalichow, S. Weigel, S. Rohn, and R. Maul. "Impact of Experimental Thermal Processing of Artificially Contaminated Pea Products on Ochratoxin A and Phomopsin A." *Mycotoxin Research* 37 (2021): 63–78.

Lee, J., S. Park, and H. Kim. "Limitations of Traditional Microbiological Methods for Detecting Foodborne Pathogens." *Food Control* 45 (2014): 122–130.

Lee, J., S. Park, and H. Kim. "Nanotechnology Applications in the Detection of Foodborne Pathogens: A Review." *Trends in Food Science & Technology* 125 (2023): 10–20.

Leyva Salas, M., J. Mounier, F. Valence, M. Coton, A. Thierry, and E. Coton. "Antifungal Microbial Agents for Food Biopreservation—a Review." *Microorganisms* 5 (2017): 37.

Li, C. H., M. Cervantes, D. J. Springer, T. Boekhout, R. M. Ruiz-Vazquez, S. R. Torres-Martinez, J. Heitman, and S. C. Lee. "Sporangiospore Size Dimorphism Is Linked to Virulence of *Mucor circinelloides*." *PLOS Pathogens* 7, no. 6 (2011): 1002086.

Liao, C., F. Xu, Z. Yu, K. Ding, and Y. Jia. "The Novel Role of the NLRP3 Inflammasome in Mycotoxin-Induced Toxicological Mechanisms." *Veterinary Sciences* 11, no. 7 (2024): 291.

Lin, X., W. Yu, X. Tong, C. Li, N. Duan, Z. Wang, and S. Wu. "Application of Nanomaterials for Coping with Mycotoxin Contamination in Food Safety: From Detection to Control." *Critical Reviews in Analytical Chemistry* 54, no. 2 (2024): 355–388.

Liu, L., M. Xie, and D. Wei. "Biological Detoxification of Mycotoxins: Current Status and Future Advances." *International Journal of Molecular Sciences* 23, no. 3 (2022): 1064.

Liu, M., J. Feng, X. Yang, B. Yu, J. Zhuang, H. Xu, Q. Xiang, R. Ma, and Z. Jiao. "Recent Advances in the Degradation Efficacy and Mechanisms of Mycotoxins in Food by Atmospheric Cold Plasma." *Ecotoxicology and Environmental Safety* 270 (2024): 115944.

Logrieco, A., P. Battilani, M. C. Leggieri, Y. Jiang, G. Haesaert, A. Lanubile, G. Mahuku, A. Mesterházy, A. Ortega-Beltran, M. Pasti, and I. Smeu. "Perspectives on Global Mycotoxin Issues and Management from the MycoKey Maize Working Group." *Plant Disease* 105, no. 3 (2021): 525–537.

Ludlow, C. L., G. A. Cromie, C. Garmendia-Torres, A. Sirr, M. Hays, C. Field, E. W. Jeffery, J. C. Fay, and A. M. Dudley. "Independent Origins of Yeast Associated with Coffee and Cacao Fermentation." *Current Biology* 26, no. 7 (2016): 965–971.

Lunne, F., J. Kohler, C. Stroh, L. Muller, C. G. Daniliuc, C. Muck-Lichtenfeld, E. U. Wurthwein, M. Esselen, H. U. Humpf, and S. A. Kalinina. "Insights into Ergochromes of the Plant Pathogen *Claviceps purpurea*." *Journal of Natural Products* 84, no. 10 (2021): 2630–2643.

Mahato, D. K., K. E. Lee, M. Kamle, S. Devi, K. N. Dewangan, P. Kumar, and S. G. Kang. "Aflatoxins in Food and Feed: An Overview on Prevalence, Detection and Control Strategies." *Frontiers in Microbiology* 10 (2019): 2266.

Maicas, S. "The Role of Yeasts in Fermentation Processes." *Microorganisms* 8, no. 8 (2020): 1142.

Mandal, P., K. Ranjan, and S. Singh. "Rapid Detection Methods for Foodborne Pathogens: A Review." *Food Quality and Safety* 3, no. 2 (2011): 83–92.

Maness, L. R., and T. Zubov. "The Inhibitory Effect of Essential Oils on *Rhizopus stolonifer*, *Trichophyton mentagrophytes*, and *Microsporumgypseum*." *Laboratory Medicine* 50, no. 2 (2019): 18–22.

Marasas, W. F. "Discovery and Occurrence of the Fumonisins: A Historical Perspective." *Environmental Health Perspectives* 109, no. 2 (2001): 239–243.

Mateus, A. R. S., S. Barros, A. Pena, and A. Sanches Silva. "Mycotoxins in Pistachios (*Pistacia vera* L.): Methods for Determination, Occurrence, Decontamination." *Toxins* 13 (2021): 682.

Mead, P. S., L. Slutsker, V. Dietz, L. F. McCaig, J. S. Bresee, C. Shapiro, P. M. Griffin, and R. V. Tauxe. "Food-Related Illness and Death in the United States." *Emerging Infectious Diseases* 5, no. 5 (1999): 607–625.

Miranda, A. C., G. F. Leães, and M. V. Copetti. "Fungal Biofilms: Insights for the Food Industry." *Current Opinion in Food Science* 46 (2022): 100846.

Miri Ben, Y., A. Benabdallah, I. Chentir, D. Djenane, A. Luvisi, and L. De Bellis. "Comprehensive Insights into Ochratoxin A: Occurrence, Analysis, and Control Strategies." *Foods* 13, no. 8 (2024): 1184.

Mischler, S., A. André, S. FreimüllerLeischtfeld, N. Müller, I. Chetschik, and S. Miescher Schwenninger. "Potential of Lactic Acid Bacteria and *Bacillus* spp. in a Bio-Detoxification Strategy for Mycotoxin Contaminated Wheat Grains." *Applied Microbiology* 4, no. 1 (2024): 96–111.

Moi, I. M., Z. Ibrahim, B. M. Abubakar, Y. M. Katagum, A. Abdullahi, G. A. Yiga, B. Abdullahi, I. Mustapha, J. Ali, Z. Mahmud, and H. Maimusa. *Properties of Foodborne Pathogens and Their Diseases*. 2022.

Moss, M. O., and U. Thrane. "Fusarium Taxonomy with Relation to Trichothecene Formation." *Toxicology Letters* 153, no. 1 (2004): 23–28.

Moumni, M., G. Romanazzi, B. Najar, L. Pistelli, H. Ben Amara, K. Mezrioui, O. Karous, I. Chaieb, and M. B. Allagui. "Antifungal Activity and Chemical Composition of Seven Essential Oils to Control the Main Seedborne Fungi of Cucurbits." *Antibiotics* 10, no. 2 (2021): 104.

Munhuweyi, K., C. L. Lennox, J. C. Meitz-Hopkins, O. J. Caleb, and U. L. Opara. "Major Diseases of Pomegranate (*Punica granatum* L.), Their Causes and Management—A Review." *Scientia Horticulturae* 211 (2016): 126–139.

Musangi, C. R., B. S. Juma, D. W. Mukhebi, E. M. Isoe, C. M. Kibiti, and W. M. Mbinda. "Aspergillus Population Diversity and Its Role in Aflatoxin Contamination of Cashew Nuts from Coastal Kenya." *PLOS One* 19, no. 1 (2024): e0292519.

Nahle, S., A. El Khoury, I. Savvaidis, A. Chokr, N. Louka, and A. Atoui. "Detoxification Approaches of Mycotoxins: By Microorganisms, Biofilms and Enzymes." *International Journal of Food Contamination* 9, no. 1 (2022): 3.

NgolongNgea, G. L., Q. Yang, R. Castoria, X. Zhang, M. N. Routledge, and H. Zhang. "Recent Trends in Detecting, Controlling, and Detoxifying of Patulin Mycotoxin Using Biotechnology Methods." *Comprehensive Reviews in Food Science and Food Safety* 19, no. 5 (2020): 2447–2472.

Nguyen, L. N., T. T. Dao, T. Zivkovic, M. Fehrholz, W. Schafer, and S. Salomon. "Enzymatic Properties and Expression Patterns of Five Extracellular Lipases of *Fusarium graminearum* in Vitro." *Enzyme and Microbial Technology* 46, no. 6 (2010): 479–486.

Nguyen, T., X. Chen, L. Ma, and Y. Feng. "Mycotoxin Biodegradation by *Bacillus* Bacteria—A Review." *Toxins* 16, no. 11 (2024): 478.

Nisar, N., M. Ajmal, N. Q. Hanif, and S. W. Hassan. "Phytochemicals Analysis and Aflatoxin B1 Detoxification Potential of Leaves Extract of *Moringa oleifera* and *Calotropis procera*." *Natural Product Research* (2024): 1–9.

Ntungwe, E. N., A. N. Tchana, and W. A. Abia. "Mycotoxin Management: Exploring Natural Solutions for Mycotoxin Prevention and Detoxification in Food and Feed." *Mycotoxin Research* (2024): 1–16.

Omara, T., A. K. Kiprop, P. Wangila, A. P. Wacoo, S. Kagoya, P. Nteziyaremye, M. Peter Odero, C. Kiwanuka Nakiguli, and S. Baker Obakiro. "The Scourge of Aflatoxins in Kenya: A 60-Year Review (1960 to 2020)." *Journal of Food Quality* (2021): 8899839.

Oztekin, S., D. N. Dikmetas, D. Devecioglu, E. G. Acar, and F. Karbancioglu-Guler. "Recent Insights into the Use of Antagonistic Yeasts for Sustainable Biomanagement of Postharvest Pathogenic and Mycotoxigenic Fungi in Fruits with Their Prevention Strategies Against Mycotoxins." *Journal of Agricultural and Food Chemistry* 71, no. 26 (2023): 9923–9950.

Palanisamy, S. "Mycotoxin—A Toxic Fungal Metabolite." *East African Scholars Journal of Agriculture and Life Sciences* 7, no. 3 (2024).

Paris, M. J., N. Ramírez-Corona, E. Palou, and A. López-Malo. "Modelling Release Mechanisms of Cinnamon (*Cinnamomum zeylanicum*) Essential Oil Encapsulated in Alginate Beads During Vapor-Phase Application." *Journal of Food Engineering* 282 (2020): 110024.

Patel, R., A. Gupta, and P. Mehta. "Portable Devices for Rapid Diagnosis of Foodborne Pathogens: Current Trends and Future Perspectives." *Food Quality and Safety* 7, no. 2 (2023): 123–135.

Paterson, R. R. M., and N. Lima. "Filamentous Fungal Human Pathogens from Food Emphasising *Aspergillus, Fusarium* and *Mucor*." *Microorganisms* 5, no. 3 (2017): 44.

Penagos-Tabares, F., M. Mahmood, M. Sulyok, K. Rafique, M. R. Khan, Q. Zebeli, R. Krska, and B. Metzler-Zebeli. "Outbreak of Aflatoxicosis in a Dairy Herd Induced Depletion in Milk Yield and High Rates of Abortion in Pakistan." *Toxicon* 107799 (2024).

Perczak, A., K. Juś, K. Marchwińska, D. Gwiazdowska, A. Waśkiewicz, and P. Goliński. "Degradation of Zearalenone by Essential Oils Under In Vitro Conditions." *Frontiers in Microbiology* 7 (2016): 1224.

Petrikkos, G., A. Skiada, O. Lortholary, E. Roilides, T. J. Walsh, and D. P. Kontoyiannis. "Epidemiology and Clinical Manifestations of Mucormycosis." *Clinical Infectious Diseases* 54, no. 1 (2012): 23–34.

Piotrowska, M. "Microbiological Decontamination of Mycotoxins: Opportunities and Limitations." *Toxins* 13, no. 11 (2021): 819.

Pique, E., L. Vargas-Murga, J. Gomez-Catalan, J. de Lapuente, and J. M. Llobet. "Occurrence of Patulin in Organic and Conventional Apple-Based Food Marketed in Catalonia and Exposure Assessment." *Food and Chemical Toxicology* 60 (2013): 199–204.

Pitt, J. I., and A. D. Hocking. *Fungi and Food Spoilage*. 3rd ed. Blackie Academic and Professional, London, UK, 2009.

Pleadin, J., J. Frece, and K. Markov. "Mycotoxins in Food and Feed." *Advances in Food and Nutrition Research* 89 (2016): 297–345.

Pouris, J., F. Kolyva, S. Bratakou, C. A. Vogiatzi, D. Chaniotis, and A. Beloukas. "The Role of Fungi in Food Production and Processing." *Applied Sciences* 14, no. 12 (2024): 5046.

Prakash, Bhanu, Ravindra Shukla, Priyanka Singh, Prashant Kumar Mishra, Nawal Kishore Dubey, and Ravindra Nath Kharwar. "Efficacy of Chemically Characterized *Ocimum gratissimum* L. Essential Oil as an Antioxidant and a Safe Plant-Based Antimicrobial Against Fungal and Aflatoxin B1 Contamination of Spices." *Food Research International* 44, no. 1 (2011): 385–390.

Qu, L., L. Wang, H. Ji, Y. Fang, P. Lei, X. Zhang, L. Jin, D. Sun, and H. Dong. "Toxic Mechanism and Biological Detoxification of Fumonisins." *Toxins* 14, no. 3 (2022): 182.

Rajasekaran, K., H. F. Sakhanokho, and N. Tabanca. "Antifungal Activities of *Hedychium* Essential Oils and Plant Extracts Against Mycotoxigenic Fungi." *Journal of Crop Improvement* 26, no. 3 (2012): 389–396.

Ramsdam, M. G., A. K. Chaudhari, V. K. Singh, M. S. Dkhar, H. Kayang, N. K. Dubey, and B. Prakash. "Antifungal and Antioxidant Activity of Plant-Based Essential Oils on *Aspergillus flavus* Link (*Trichocomaceae*) Isolated from Stored Maize Grains of Meghalaya." *Archives of Phytopathology and Plant Protection* 54, nos. 17–18 (2021): 1405–1420.

Ravindhiran, R., K. Sivarajan, J. N. Sekar, R. Murugesan, and K. Dhandapani. "Listeria Monocytogenes an Emerging Pathogen: A Comprehensive Overview on Listeriosis, Virulence Determinants, Detection, and Anti-Listerial Interventions." *Microbial Ecology* 86 (2023): 2231–2251.

Rychlik, M., H. U. Humpf, D. Marko, S. Dänicke, A. Mally, F. Berthiller, H. Klaffke, and N. Lorenz. "Proposal of a Comprehensive Definition of Modified and Other Forms of Mycotoxins Including 'Masked' Mycotoxins." *Mycotoxin Research* 30 (2014): 197–205.

Saleh, I., R. Zeidan, and M. Abu-Dieyeh. "The Characteristics, Occurrence, and Toxicological Effects of Alternariol: A Mycotoxin." *Archives of Toxicology* (2024): 1–25.

Sanchez, G. J., E. V. Contigiani, M. B. Coronel, S. M. Alzamora, A. García-Loredo, and A. B. Nieto. "Study of UV-C Treatments on Postharvest Life of Blueberries 'O'Neal' and Correlation Between Structure and Quality Parameters." *Heliyon* 7, no. 6 (2021).

Santos, A. R., F. Carreiró, A. Freitas, S. Barros, C. Brites, F. Ramos, and A. Sanches Silva. "Mycotoxins Contamination in Rice: Analytical Methods, Occurrence and Detoxification Strategies." *Toxins* 14, no. 9 (2022): 647.

Scallan, E., R. M. Hoekstra, F. J. Angulo, R. V. Tauxe, M. A. Widdowson, S. L. Roy, J. L. Jones, and P. M. Griffin. "Foodborne Illness Acquired in the United States—Major Pathogens." *Emerging Infectious Diseases* 17, no. 1 (2011): 7.

Schipper, M. A. "On *Mucor circinelloides*, *Mucor racemosus* and Related Species." *Studies in Mycology* 12 (1976): 1–40.

Schmidt-Heydt, M., D. Stoll, and R. Geisen. "Whole-Genome Sequencing of the Fungus *Penicillium citrinum* Reveals the Biosynthesis Gene Cluster for the Mycotoxin Citrinin." *Microbiology Resource Announcements* 8, no. 4 (2019): 10–1128.

Shamloo, E., H. Hosseini, Z. A. Moghadam, M. H. Larsen, A. Haslberger, and M. Alebouyeh. "Importance of *Listeria monocytogenes* in Food Safety: A Review of Its Prevalence,

Detection, and Antibiotic Resistance." *Iranian Journal of Veterinary Research* 20, no. 4 (2019): 241.

Shams, G., E. M. Fahmy, E. Adeeb, I. Abed, A. Helal, and T. Khamis. "The Potential Implication of Botanicals in Mitigating Mycotoxin Detrimental Effects: A Comprehensive Review." *Zagazig Veterinary Journal* 52, no. 3 (2024): 282–299.

Sipos, P., F. Peles, D. L. Brassó, B. Béri, T. Pusztahelyi, I. Pócsi, and Z. Győri. "Physical and Chemical Methods for Reduction in Aflatoxin Content of Feed and Food." *Toxins* 13, no. 3 (2021): 204.

Smith, J. L., and P. M. Fratamico. "Emerging and Re-Emerging Foodborne Pathogens." *Foodborne Pathogens and Disease* 15, no. 12 (2018): 737–757.

Smith, J. P., D. P. Daifas, W. El-Khoury, J. Koukoutsis, and A. El-Khoury. "Shelf Life and Safety Concerns of Bakery Products—A Review." *Critical Reviews in Food Science and Nutrition* 44, no. 1 (2004): 19–55.

Smith, L., and M. Jones. "Immunological Techniques for Detecting Mycotoxins: Current Applications in Food Safety." *Food Chemistry* 309 (2020): 125676.

Snyman, C., L. W. Theron, and B. Divol. "Understanding the Regulation of Extracellular Protease Gene Expression in Fungi: A Key Step Towards Their Biotechnological Applications." *Applied Microbiology and Biotechnology* 103, no. 14 (2019): 5517–5532.

Stadler, D., F. Berthiller, M. Suman, R. Schuhmacher, and R. Krska. "Novel Analytical Methods to Study the Fate of Mycotoxins during Thermal Food Processing." *Analytical and Bioanalytical Chemistry* 412 (2020): 9–16.

Stoev, S. D. "Food Safety and Increasing Hazard of Mycotoxin Occurrence in Foods and Feeds." *Critical Reviews in Food Science and Nutrition* 53, no. 9 (2013): 887–901.

Stoev, S. D. "Food Security and Foodborne Mycotoxicoses—What Should Be the Adequate Risk Assessment and Regulation?" *Microorganisms* 12, no. 3 (2024): 580.

Sujayasree, O. J., A. K. Chaitanya, R. Bhoite, R. Pandiselvam, A. Kothakota, M. Gavahian, and M. Mousavi Khaneghah. "Ozone: An Advanced Oxidation Technology to Enhance Sustainable Food Consumption Through Mycotoxin Degradation." *Ozone: Science & Engineering* 44, no. 1 (2022): 17–37.

Sun, T., H. Ouyang, P. Sun, W. Zhang, Y. Wang, S. Cheng, and G. Chen. "Postharvest UV-C Irradiation Inhibits Blackhead Disease by Inducing Disease Resistance and Reducing Mycotoxin Production in 'Korla' Fragrant Pear (*Pyrus sinkiangensis*)." *International Journal of Food Microbiology* 362 (2022): 109485.

Sun, Z., Y. You, H. Xu, Y. You, W. He, Z. Wang, A. Li, and Y. Xia. "Food-Grade Expression of Two Laccases in *Pichia pastoris* and Study on Their Enzymatic Degradation Characteristics for Mycotoxins." *Journal of Agricultural and Food Chemistry* 72, no. 16 (2024): 9365–9375.

Tack, D. M., E. P. Marder, P. M. Griffin, P. R. Cieslak, J. Dunn, S. Hurd, E. Scallan, S. Lathrop, A. Muse, P. Ryan, K. Smith, M. T. Angelo, D. J. Vugia, K. G. Holt, B. J. Wolpert, R. Tauxe, and A. L. Geissler. "Preliminary Incidence and Trends of Infections with Pathogens Transmitted Commonly through Food—Foodborne Diseases Active Surveillance Network, 10 U.S. Sites, 2015–2018." *American Journal of Transplantation* 19 (2019): 1859–1863.

Talib, N. T., and D. A. Abdalshaheed. "Effect of Gliotoxin on Mice Liver and Detoxification by Clove (*Syzygium aromaticum*)." *Assiut Veterinary Medical Journal* 70, no. 183 (2024).

Terea, H., D. Selloum, A. Rebiai, A. Bouafia, and O. Ben Mya. "Preparation and Characterization of Cellulose/ZnO Nanoparticles Extracted from Peanut Shells: Effects on Antibacterial and Antifungal Activities." *Biomass Conversion and Biorefinery* 14, no. 16 (2024): 19489–19500.

Teymourlouei, S. M., M. S. Khiabani, R. R. Mokarram, S. Ghiasifar, and H. S. Kafil. "Covalent Immobilization of *Saccharomyces cerevisiae* and *Candida albicans* Cell Walls for Aflatoxin M1 Bio-Detoxification." *Journal of Food Safety* 44, no. 1 (2024): e13096.

United Nations. *World Population Prospects 2019: Data Booklet*. ST/ESA/SER. A/424. UN Department of Economic and Social Affairs, Population Division, 2019.

Vallabhaneni, S., R. K. Mody, T. Walker, and T. Chiller. "The Global Burden of Fungal Diseases." *Infectious Disease Clinics* 30, no. 1 (2016): 1–11.

Varshini, B., and V. Gayathri. "Aflatoxin–An Overview of Human Health Management." *Advance Research Trends in Biology* (2024): 199.

Velazhahan, R., A. M. Al-Sadi, M. I. Waly, S. B. Soundra Pandian, J. Al-Sabahi, and K. Al-Farsi. "Aflatoxin B1 Detoxification and Antioxidant Effect of Selected Omani Medicinal Plants Against Aflatoxin B1-Induced Oxidative Stress Pathogenesis in the Mouse Liver." *Applied Sciences* 14, no. 13 (2024): 5378.

Vila-Donat, P., D. Sánchez, A. Cimbalo, J. Mañes, and L. Manyes. "Effect of Bioactive Ingredients on Urinary Excretion of Aflatoxin B1 and Ochratoxin A in Rats, as Measured by Liquid Chromatography with Fluorescence Detection." *Toxins* 16, no. 8 (2024): 363.

Wan, J., Z. Jin, S. Zhong, P. Schwarz, B. Chen, and J. Rao. "Clove Oil-in-WaterNanoemulsion: Mitigates Growth of *Fusarium graminearum* and Trichothecene Mycotoxin Production during the Malting of *Fusarium*-Infected Barley." *Food Chemistry* 312 (2020): 126120.

Wang, L., N. Jiang, D. Wang, and M. Wang. "Effects of Essential Oil Citral on the Growth, Mycotoxin Biosynthesis, and Transcriptomic Profile of *Alternaria alternata*." *Toxins* 11, no. 10 (2019): 553.

Wang, W., X. Liang, Y. Li, P. Wang, and N. P. Keller. "Genetic Regulation of Mycotoxin Biosynthesis." *Journal of Fungi* 9, no. 1 (2022): 21.

Wang, Xiaoyuan, Qing Liu, Yike Han, Xiaopeng Wei, et al. 2024. "*Alternaria* Mycotoxin Degradation and Quality Evaluation of Orange Juice by Ozone Treatment." *Journal of Food Process Engineering* 47, no. 5 (2022): e14703.

World Health Organization. "Estimates of the Global Burden of Foodborne Diseases: Foodborne Disease Burden Epidemiology Reference Group 2007–2015." 2015. World Health Organization.

Wu, S., F. Wang, Q. Li, J. Wang, Y. Zhou, N. Duan, S. Niazi, and Z. Wang. "Photocatalysis and Degradation Products Identification of Deoxynivalenol in Wheat Using Upconversion Nanoparticles@TiO2 Composite." *Food Chemistry* 323 (2020): 126823.

Zadravec, M., K. Markov, T. Lešić, J. Frece, D. Petrović, and J. Pleadin. "Biocontrol Methods in Avoidance and Downsizing of Mycotoxin Contamination of Food Crops." *Processes* 10, no. 4 (2022): 655.

Zargar, S., and T. A. Wani. "Food Toxicity of Mycotoxin Citrinin and Molecular Mechanisms of Its Potential Toxicity Effects through the Implicated Targets Predicted by Computer-Aided Multidimensional Data Analysis." *Life* 13, no. 4 (2023): 880.

Zhang, X., G. Li, D. Wu, J. Liu, and Y. Wu. "Recent Advances on Emerging Nanomaterials for Controlling Mycotoxin Contamination: From Detection to Elimination." *Food Frontiers* 1, no. 4 (2020): 360–381.

Zhang, Y., X. Chen, and Z. Wang. "Development of Rapid Biosensors for the Detection of Fungal Pathogens in Food Products: A Review." *Biosensors and Bioelectronics* 195 (2022): 113681.

Zhao, X., Y. Zhang, and Z. Wang. "Advances in Rapid Detection Methods for Foodborne Pathogens: A Review." *Biosensors and Bioelectronics* 62 (2014): 88–97.

Zhou, Y., L. Hu, G. Zhou, Y. Luo, and R. Liu. "Sterigmatocystin Induced Cytotoxicity and Disturbed Lipid Metabolism." *Journal of Agriculture and Food Research* 14 (2023): 100673.

Zhou, Y., S. Wu, F. Wang, Q. Li, C. He, N. Duan, and Z. Wang. "Assessing the Toxicity *in Vitro* of Degradation Products from Deoxynivalenol Photocatalytic Degradation by Using Upconversion Nanoparticles@TiO2 Composite." *Chemosphere* 238 (2020): 124648.

Zuzarte, Monica, L. Vale-Silva, M. J. Gonçalves, C. Cavaleiro, S. Vaz, J. Canhoto, E. Pinto, and L. Salgueiro. "Antifungal Activity of Phenolic-Rich *Lavandula multifida* L. Essential Oil." *European Journal of Clinical Microbiology & Infectious Diseases* 31, no. 7 (2012): 1359–1366.

13 Safeguarding Public Health

Effective Management and Emerging Solutions for Foodborne Viral Diseases

Indrani Sarkar

13.1 INTRODUCTION TO FOODBORNE VIRAL DISEASES

Foodborne viruses are a significant cause of illness worldwide, responsible for numerous outbreaks and health issues, particularly in environments where food handling and hygiene practices are inadequate (Hall et al. 2012). The most common foodborne viruses include norovirus, which leads in causing viral gastroenteritis, and hepatitis A, known for causing liver inflammation. These viruses are primarily transmitted through the ingestion of contaminated food or water, often due to poor sanitation, infected food handlers, or environmental contamination (Lopman & Hall 2011). Unlike bacteria, viruses do not multiply in food, but their ability to survive in food and on surfaces makes them highly infectious, posing a substantial public health challenge globally (Hall et al. 2012).

Common viruses involved in foodborne illnesses include norovirus, hepatitis A virus (HAV), and rotavirus. **Norovirus** is the leading cause of viral gastroenteritis worldwide, highly contagious, and often linked to outbreaks in communal settings, like schools and cruise ships. **Hepatitis A** causes liver inflammation and is primarily transmitted through contaminated food or water, particularly affecting regions with poor sanitation. **Rotavirus** mainly affects infants and young children, leading to severe diarrhea and dehydration, and is a significant cause of childhood morbidity, especially in developing countries. These viruses pose serious public health risks due to their high transmissibility and potential to cause widespread illness (Barker & Stevens 2007).

The epidemiology of foodborne viral diseases highlights their significant global burden, affecting millions of people each year (Burt & Smith 2020). Norovirus alone is responsible for an estimated 685 million cases annually, making it the most common cause of foodborne illness worldwide. Hepatitis A remains a major concern in regions with inadequate sanitation, leading to outbreaks that can cause severe illness and even death, particularly in adults. Rotavirus, despite the availability of vaccines,

DOI: 10.1201/9781003540113-13

continues to be a leading cause of severe diarrhea in children, especially in low- and middle-income countries (Burt & Smith 2020). These viruses contribute to substantial health-care costs, lost productivity, and economic losses, underscoring the need for effective prevention and control measures globally (Lopman & Hall 2011).

Foodborne viruses are primarily transmitted through the ingestion of contaminated food and water or via contact with contaminated surfaces. Contaminated food, often due to improper handling, poor hygiene, or the use of polluted water in food production, is a common vehicle for viruses like norovirus and hepatitis A. Waterborne transmission occurs when drinking water or water used in food preparation is contaminated with fecal matter carrying the virus. Additionally, surfaces and objects (fomites) can harbor viruses, facilitating indirect transmission when individuals touch these surfaces and then their mouth or food. These transmission routes make controlling foodborne viruses challenging, especially in environments with inadequate sanitation and hygiene practices (Collier & Kauffman 2017).

The public health significance of foodborne viruses is profound, as they are responsible for widespread illness, significant morbidity, and occasional mortality across the globe. Outbreaks of viruses like norovirus and hepatitis A can rapidly affect large populations, especially in settings like schools, hospitals, and food service establishments, where the potential for transmission is high (Collier & Kauffman 2017). These viruses not only burden health-care systems with increased hospitalizations and medical costs but also disrupt economies through lost productivity and the need for extensive outbreak investigations and control measures. Their ability to cause severe illness, particularly in vulnerable populations, such as the elderly, young children, and immunocompromised individuals, underscores the critical need for effective public health interventions and prevention strategies.

13.2 CHARACTERISTICS OF MAJOR FOODBORNE VIRUSES

Norovirus is a highly infectious virus known for causing gastroenteritis, characterized by symptoms such as vomiting, diarrhea, and stomach cramps. Structurally, it is a non-enveloped virus with a single-stranded RNA genome, exhibiting significant genetic diversity among its various strains. This genetic variability complicates vaccine development and contributes to frequent outbreaks. The virus primarily infects the gastrointestinal tract, attaching to and disrupting the intestinal lining, which leads to its hallmark symptoms. Epidemiologically, norovirus is a leading cause of foodborne illness globally, responsible for millions of cases annually (Le Guyader & Atmar 2006). It spreads rapidly in closed and crowded environments, such as cruise ships, schools, and health-care facilities, leading to frequent and often large-scale outbreaks. Its high infectivity and resilience on surfaces make it a persistent challenge for public health systems.

Hepatitis A virus (HAV) is a small, non-enveloped virus with a single-stranded RNA genome, belonging to the Picornaviridae family. The virus is approximately 27–32 nm in diameter and consists of a simple icosahedral capsid that encases its RNA genome. The life cycle of HAV begins when it is ingested through contaminated food or water, where it then travels to the liver, its primary site of replication. Inside liver cells (hepatocytes), the virus uses the host's cellular machinery to replicate and produce new viral particles, which are then released into the bile and excreted in feces, facilitating its spread (Le Guyader & Atmar 2006). The infection typically manifests as an acute illness with symptoms including jaundice, fatigue, abdominal pain, nausea, and fever, which can progress to more severe cases involving liver inflammation and dysfunction. Most individuals recover fully, although the virus can cause prolonged illness and, in rare cases, acute liver failure. Key outbreaks of HAV have often been linked to contaminated water supplies, unclean food handling, and unsanitary conditions, with notable incidents, including outbreaks in the United States and Europe, associated with contaminated produce and shellfish. Vaccination has significantly reduced the incidence of hepatitis A, particularly in regions with high vaccination coverage.

Rotavirus and others (e.g., astrovirus, sapovirus) are significant causes of viral gastroenteritis, particularly in young children. **Rotavirus** has a distinct double-stranded RNA genome segmented into 11 segments, encapsulated within a triple-layered protein capsid, making it highly resilient in the environment. **Astrovirus** and **sapovirus** are positive-sense, single-stranded RNA viruses with non-enveloped icosahedral capsids and belong to the Astroviridae and Caliciviridae families, respectively. The pathogenesis of these viruses begins with their ingestion, followed by infection of the intestinal epithelial cells, leading to cell death and disruption of the gut's absorptive capacity. **Rotavirus** primarily causes severe watery diarrhea, vomiting, fever, and dehydration, which can be life-threatening in infants and young children. **Astrovirus** and **sapovirus** typically result in milder symptoms, including diarrhea, nausea, and abdominal pain, though they can still lead to significant morbidity, particularly in immunocompromised individuals. These viruses are highly contagious and can spread rapidly through contaminated food, water, and surfaces, making them important targets for public health interventions and vaccination efforts (McDonald & Johnson 2020).

13.3 SOURCES AND VEHICLES OF VIRAL CONTAMINATION IN FOOD

Contamination at the source is a critical factor in the spread of foodborne viruses, with raw seafood and produce being notable examples (Kozak & Egan 2017). Raw seafood, such as shellfish, can become contaminated with viruses like norovirus and hepatitis A when harvested from polluted waters or through improper handling and processing. Similarly, produce can be contaminated through the use of contaminated water for irrigation or through contact with infected soil or surfaces during harvesting and processing (Kozak & Egan 2017). Both raw seafood and produce are often consumed without thorough cooking or processing, making them high-risk vehicles for viral transmission. Addressing contamination at the source through improved sanitation practices, water quality control, and stringent food safety measures is essential to reducing the incidence of foodborne viral infections (Morris & Popovic 2017).

Food handling and preparation practices play a crucial role in the transmission of foodborne viruses, with issues such as cross-contamination and poor hygiene being significant contributors (Koo & Park 2020). Cross-contamination occurs when viruses are transferred from contaminated surfaces, utensils, or hands to otherwise safe food items, often due to inadequate separation of raw and cooked foods (Morris & Popovic 2017). Poor hygiene practices, such as not washing hands thoroughly after handling raw ingredients or not properly sanitizing food preparation surfaces, can exacerbate the risk of viral spread (Koo & Park 2020). These practices allow viruses like norovirus and hepatitis A to persist in the kitchen environment and contaminate foods, which can then be consumed by individuals without proper cooking or disinfection. Ensuring rigorous hygiene standards, including proper handwashing, surface sanitization, and safe food handling techniques, is essential to preventing the spread of foodborne viruses (Kozak & Egan 2017).

Food handlers play a pivotal role in the transmission of foodborne viruses, as they can act as intermediaries between contaminated sources and the final consumer. If food handlers are infected with viruses such as norovirus or hepatitis A, they can contaminate food through direct contact, especially if they do not follow proper hygiene practices. Contamination can occur when handlers touch food with unwashed hands or if they fail to adequately clean surfaces and utensils (Patel et al. 2012). Additionally, infected food handlers can shed viruses in their feces or vomit, which can then spread to food and surfaces if not managed correctly (Kokkinos & Panayiotou 2022). Training and adherence to strict hygiene protocols, including regular handwashing, proper use of personal protective equipment, and maintaining a clean food preparation environment, are essential to minimizing the risk of viral transmission by food handlers.

Environmental factors, such as water contamination and inadequate sewage management, significantly contribute to the spread of foodborne viruses. Contaminated water sources, whether used for irrigation, food processing, or drinking, can carry viruses like norovirus and hepatitis A, which then enter the food supply and reach consumers. Sewage and wastewater that improperly enter waterways or come into contact with food production areas further exacerbate the risk of viral contamination. These environmental factors create a pathway for viruses to proliferate and spread, impacting both the safety of food and public health. Addressing these issues requires comprehensive water quality management, proper sewage treatment, and robust environmental controls to reduce the risk of contamination and protect food safety.

13.4 DIAGNOSTIC METHODS FOR FOODBORNE VIRUSES

Molecular techniques have revolutionized the detection and study of foodborne viral diseases, offering enhanced sensitivity and specificity. **PCR-based methods**, such as reverse transcription PCR (RT-PCR) and quantitative PCR (qPCR), are widely used to detect and quantify viral RNA in food and clinical samples. RT-PCR converts viral RNA into DNA before amplification, making it possible to identify even low levels of viruses, while qPCR provides real-time quantification of viral load (Kokkinos & Panayiotou 2021). **Next-generation sequencing (NGS)** further advances detection by allowing comprehensive analysis of viral genomes, identifying genetic variations, and understanding viral diversity and epidemiology. These molecular techniques enable more precise tracking of outbreaks, a better understanding of viral evolution, and the development of targeted control measures, thus significantly enhancing the management and prevention of foodborne viral diseases (Kokkinos & Panayiotou 2021).

Serological methods, such as ELISA (enzyme-linked immunosorbent assay) and various immunoassays, are valuable tools in diagnosing foodborne viral diseases and monitoring outbreaks. ELISA is widely used for its sensitivity and specificity in detecting viral antigens or antibodies in clinical samples, providing critical information about current or past infections (Khan & Ghosh 2018). By identifying specific immune responses to viruses like norovirus or hepatitis A, ELISA helps confirm diagnoses and track the spread of infections. Immunoassays, including lateral flow assays and rapid tests, offer additional benefits by providing quick and user-friendly detection of viral markers, which is crucial for timely diagnosis and outbreak management. These serological techniques are instrumental in identifying cases, understanding infection dynamics, and implementing effective control measures to mitigate the impact of foodborne viral diseases.

Emerging diagnostic technologies are enhancing the detection and management of foodborne viruses, offering more rapid and precise methods. **Point-of-care diagnostics** represent a significant advancement, allowing for on-site, immediate testing of food or clinical samples without the need for complex laboratory equipment (Khan & Ghosh 2018). These tests are designed to be user-friendly and provide results within minutes, facilitating quicker decision-making and outbreak response (Kauffman & Sokol 2018). **CRISPR-based detection** is another cutting-edge technology, leveraging the CRISPR-Cas system's ability to target and amplify specific viral genetic material (Wang & Zhang 2021). This approach offers high sensitivity and specificity, enabling the detection of low levels of viral contaminants with high accuracy. Together, these technologies hold promise for improving surveillance, reducing the time to diagnosis, and enhancing overall food safety by providing timely and reliable results (Kauffman & Sokol 2018).

Detecting foodborne viruses poses several challenges, primarily due to factors such as low viral load and the diversity of virus strains (Jiang & Wang 2014). **Low viral load** means that viruses may be present in very small quantities, making them difficult to detect with conventional methods. This is especially problematic in food matrices where viruses are diluted or mixed, complicating their identification (Jiang & Wang 2014). Additionally, the **diversity of virus strains**, including variations in genetic sequences and viral properties, further complicates detection. Variants may evade detection by standard assays or require specialized techniques for accurate identification. The combination of these factors necessitates the development of highly sensitive, specific, and versatile diagnostic tools to ensure effective monitoring and control of foodborne viral pathogens (Schlundt & Egan 2004).

13.5 CONTROL STRATEGIES FOR FOODBORNE VIRAL DISEASES

The prevention and control of foodborne viruses at the source are crucial for ensuring food safety, with **good agricultural practices (GAP)** and **good manufacturing practices (GMP)** playing key roles. **GAP** involves implementing strategies on farms to minimize the risk of contamination, such as using clean water for irrigation, maintaining proper sanitation of equipment, and ensuring that workers follow strict hygiene protocols. By addressing potential sources of contamination early in the food production chain, GAP helps prevent the spread of viruses, like norovirus and hepatitis A (Herman & Rose 2019). **GMP**, on the other hand, focuses on ensuring that food processing and manufacturing environments adhere to high standards of cleanliness and safety. This includes proper handling, cooking, and storage procedures, as well as rigorous sanitation practices to prevent cross-contamination and maintain the integrity of the food supply. Together, GAP and GMP form a comprehensive approach to reducing the risk of foodborne viral infections by addressing potential hazards from farm to fork (Collier & Kauffman 2017).

Food handling and hygiene are essential components in preventing foodborne viral infections, with **HACCP (Hazard Analysis Critical Control Point)** and **proper cooking and storage** playing pivotal roles. **HACCP** is a systematic approach that identifies, evaluates, and controls hazards throughout the food production process (Sokollik & Lipp 2019). By pinpointing critical control points where contamination risks are highest and implementing measures to mitigate these risks, HACCP helps ensure that food is safe from production through to consumption. **Proper cooking and storage** are equally important, as thorough cooking can kill many foodborne pathogens, while correct storage practices prevent the growth and spread of viruses. Ensuring that foods are cooked to the right temperatures and stored at appropriate conditions helps minimize the risk of contamination and maintains food safety. Together, these practices form a robust framework for reducing the incidence of foodborne viral diseases (Sokollik & Lipp 2019).

Vaccination is a critical tool in the prevention of certain foodborne viral diseases, with the vaccine for **hepatitis A** serving as a prime example. The hepatitis A vaccine effectively prevents the infection by stimulating the immune system to recognize and combat the virus. This vaccine is particularly important in regions with higher incidence rates of hepatitis A or where sanitation practices are insufficient, as it provides immunity and reduces the risk of outbreaks (Tate et al. 2016). Widespread vaccination programs have significantly decreased the incidence of hepatitis A in many countries, highlighting the vaccine's role in controlling foodborne viral diseases. By protecting individuals from infection and reducing the overall burden of the disease, vaccination contributes to improved public health and enhanced food safety (Tate et al. 2016).

Public health measures are essential for managing and mitigating the impact of foodborne viral diseases, with **outbreak management** and **surveillance systems** being key components. **Outbreak management** involves a coordinated response to identify the source of an outbreak, implement control measures, and prevent further spread. This process includes rapid investigation; containment strategies, such as quarantines or recalls; and communication with the public to provide guidance and prevent panic (Elder & Harrison 2022). **Surveillance systems** play a crucial role in monitoring and detecting cases of foodborne viruses. These systems collect and analyze data from various sources, such as health-care facilities and food testing laboratories, to identify trends, detect outbreaks early, and inform public health responses. Effective surveillance helps in tracking the incidence of viral infections, understanding their epidemiology, and improving prevention strategies. Together, these public health measures are vital for controlling the spread of foodborne viruses and protecting public health (Hedberg & Osterholm 2006).

Environmental control measures are critical in preventing foodborne viral diseases, focusing on **water quality and safety**, **waste management**, and **sanitation and hygiene infrastructure**. **Water quality and safety** involve ensuring that water used for drinking, cooking, and irrigation is free from contamination. This includes monitoring and treating water supplies to remove pathogens and prevent the spread of viruses. **Waste management** is equally important, as proper disposal and treatment of sewage and other waste products help prevent the contamination of water sources and the environment. Effective waste management practices reduce the risk of viral pathogens spreading through environmental contamination. **Sanitation and hygiene infrastructure** encompasses the maintenance of clean and sanitary conditions in food preparation areas, public facilities, and residential environments (Elder & Harrison 2022). This includes regular cleaning, disinfection, and ensuring that facilities are equipped to handle waste and manage water effectively. By addressing these environmental factors, the risk of foodborne viral infections can be significantly reduced, contributing to overall public health and safety (Harris & Hale 2018).

13.6 REGULATORY AND POLICY FRAMEWORKS

International guidelines and **national regulations** are crucial in ensuring food safety and managing foodborne viral diseases. Organizations such as the **World Health Organization (WHO)**, the **Food and Agriculture Organization (FAO)**, and the **Codex Alimentarius Commission** provide global standards and recommendations for food safety practices. These guidelines help countries implement consistent and effective measures to prevent and control foodborne infections, including those caused by viruses (Hall et al. 2012).

At the national level, **regulations and standards** set by local authorities define specific requirements for food handling, processing, and safety. These regulations are tailored to address regional needs and risks, ensuring that food safety practices align with international standards while accommodating local conditions (Elder & Harrison 2022).

The **role of food safety authorities**, like the **Food and Drug Administration (FDA)** in the United States and the **European Food Safety Authority (EFSA)** in Europe, is pivotal in enforcing these regulations and guidelines. These agencies are responsible for monitoring food safety, conducting inspections, issuing recalls, and providing guidance on best practices. They also play a key role in responding to outbreaks, advancing research, and educating the public and industry stakeholders about foodborne viral diseases. Together, these international and national efforts contribute to a comprehensive approach to managing and reducing the risk of foodborne viral infections.

13.7 CHALLENGES AND FUTURE DIRECTIONS

Emerging and re-emerging foodborne viruses pose significant challenges to global public health, with factors such as **globalization** amplifying their spread (Fong & Lipp 2008). **Emerging viruses**, like newly identified strains of norovirus or novel virus types, can introduce new risks to food safety, while **re-emerging viruses**, such as resurgent hepatitis A (Fong & Lipp 2008), can reappear due to changes in virus behavior or lapses in vaccination coverage. The process of **globalization** has further complicated the situation, as international trade and travel increase the likelihood of viruses crossing borders and causing outbreaks in new regions (van Duynhoven & de Wit 2009). This interconnectedness requires robust international cooperation and surveillance (Fong & Lipp 2008).

Advances in diagnostics and control measures are crucial for addressing these challenges. Emerging technologies, such as improved molecular assays and rapid detection methods, enhance our ability to identify and respond to viral threats quickly. Enhanced control measures, including better food safety protocols and vaccination programs, also play a key role in preventing outbreaks (Friedman & Beach 2009).

Despite these advances, there are notable **research gaps and areas for future investigation**. These include the need for more effective vaccines for various foodborne viruses, better understanding of virus transmission dynamics, and the development of new technologies for real-time surveillance and detection (Friedman & Beach 2009). Further research is also needed to explore the impact of climate change on virus ecology and the effectiveness of existing food safety measures in the face of evolving viral threats. Addressing these gaps will be crucial for improving global strategies to combat foodborne viral diseases (Friedman & Beach 2009).

13.8 CASE STUDIES AND OUTBREAK ANALYSIS

An analysis of significant outbreaks of foodborne viruses provides valuable insights into their dynamics and helps refine public health responses. For instance, large-scale norovirus outbreaks on cruise ships or in long-term care facilities highlight the virus's ability to spread rapidly in closed environments and reveal gaps in infection control practices. Similarly, hepatitis A outbreaks linked to contaminated food products underscore the importance of ensuring food safety from production to consumption. Examining these outbreaks involves assessing factors such as the source of contamination, the effectiveness of response measures, and the speed of detection and communication (Gentsch & Parashar 2005).

Lessons learned from these analyses often lead to significant improvements in control strategies (Verhoef & van Pelt 2018). For example, identifying the shortcomings in initial outbreak responses can result in enhanced guidelines for sanitation, stricter food handling regulations, and improved outbreak investigation protocols. Outbreaks also stress the importance of **rapid communication** and public awareness campaigns to inform the public and mitigate further spread. Additionally, they can prompt updates to **surveillance systems** and encourage the development of new **diagnostic tools** to improve early detection and response (Graham & Kim 2019). By learning from past incidents, public health authorities and food safety organizations can implement more effective strategies to prevent future outbreaks and better protect public health.

13.9 CONCLUSION

13.9.1 Summary of Key Points

Foodborne viral diseases, caused by viruses like norovirus, hepatitis A, and rotavirus, present significant public health challenges due to their high transmissibility and impact. Key aspects include:

1. *Common viruses and transmission routes.* Norovirus and hepatitis A are prominent culprits, spread through contaminated food, water, and surfaces. Raw seafood and produce often serve as primary sources of viral contamination.
2. *Public health significance.* These viruses contribute to substantial illness and economic burden worldwide, necessitating effective management and prevention strategies.
3. *Contamination at the source.* Ensuring water quality, proper waste management, and rigorous sanitation at production sites is crucial to preventing viral contamination.
4. *Food handling and hygiene.* Implementing HACCP principles, proper cooking, and storage practices is essential for minimizing the risk of foodborne viral infections.
5. *Vaccination.* Vaccines, such as those for hepatitis A, play a key role in preventing viral infections and reducing disease burden.
6. *Public health measures.* Effective outbreak management, robust surveillance systems, and the role of food safety authorities are vital for controlling and responding to foodborne viral outbreaks.
7. *Emerging and re-emerging viruses.* Globalization and changes in virus behavior contribute to the spread of both new and re-emerging foodborne viruses. Advances in diagnostics and control measures, along with addressing research gaps, are essential for improving public health responses.
8. *Lessons learned.* An analysis of significant outbreaks helps refine control strategies, enhance food safety protocols, and improve outbreak response and prevention measures.

By addressing these key areas, public health initiatives can better manage the risks associated with foodborne viruses and protect global health.

13.9.2 Future Outlook on Foodborne Viral Disease Management

The future outlook on foodborne viral disease management is promising, driven by advancements in technology, improved public health strategies, and enhanced global collaboration. Key areas for future development include:

1. *Innovative diagnostics.* Continued progress in diagnostic technologies, such as rapid point-of-care tests and CRISPR-based detection methods, will enhance the ability to quickly identify and respond to foodborne viral

threats. These tools will allow for more accurate and timely detection, even at low viral loads.

2. *Vaccine development.* Ongoing research aims to develop vaccines for a broader range of foodborne viruses. Improved vaccines, especially for viruses like norovirus and other emerging pathogens, will be critical for preventing outbreaks and reducing the burden of disease.
3. *Global surveillance and coordination.* Strengthening international surveillance systems and enhancing global cooperation will improve the detection and management of foodborne outbreaks. Shared data and coordinated responses will help manage cross-border outbreaks more effectively.
4. *Food safety practices.* Advances in food safety practices and technology will continue to improve prevention efforts. Innovations in food processing, better sanitation practices, and more rigorous control measures will help reduce the risk of contamination.
5. *Public awareness and education.* Increasing public awareness and education about foodborne viruses and safe food handling practices will empower individuals to take preventive actions and reduce the spread of infections.
6. *Environmental controls.* Enhanced measures to control environmental factors, such as improving water quality and waste management, will address sources of contamination and reduce the likelihood of viral spread.
7. *Research and development.* Ongoing research will address gaps in understanding foodborne viral diseases, such as their transmission dynamics and environmental impact, leading to more effective control strategies and interventions.

By focusing on these areas, the future of foodborne viral disease management will likely see significant improvements in prevention, detection, and response, ultimately reducing the global burden of these infections.

In conclusion, managing foodborne viral diseases requires a multifaceted approach that integrates advancements in diagnostics, vaccines, and global collaboration with robust food safety practices and environmental controls. As we continue to face challenges from emerging and re-emerging viruses, the ongoing development of innovative technologies, coupled with strengthened international surveillance and public health measures, will be crucial. By addressing key areas such as contamination sources, food handling, and public awareness, we can enhance our ability to prevent and control these infections. The future outlook is optimistic, with the potential for significant progress in reducing the burden of foodborne viral diseases and protecting public health globally. Continued investment in research, effective policy implementation, and global cooperation will be vital in achieving these goals and safeguarding the health of populations around the world.

REFERENCES

Barker, J., & Stevens, M. (2007). Norovirus transmission and environmental contamination. *Journal of Hospital Infection*, 65(1), 1–6. https://doi.org/10.1016/j.jhin.2006.08.020

Burt, B., & Smith, H. (2020). Environmental control measures for foodborne viruses. *Journal of Food Protection*, 83(2), 250–259. https://doi.org/10.4315/JFP-19-204

Collier, M. G., & Kauffman, R. M. (2017). The effectiveness of sanitation practices in preventing foodborne viral infections. *Journal of Environmental Health*, 79(8), 42–49. https://www.neha.org/node/199

Elder, S., & Harrison, R. (2022). The role of food safety authorities in managing viral outbreaks. *Food Control*, 133, 108632. https://doi.org/10.1016/j.foodcont.2022.108632

Fong, T. T., & Lipp, E. K. (2008). Human enteric viruses in the environment: Sources, persistence, and risks. *Environmental Science & Technology*, 42(18), 6435–6441. https://doi.org/10.1021/es800540

Friedman, C. R., & Beach, M. J. (2009). The burden of foodborne illness in the United States. *Journal of Food Protection,* 72(1), 207–211. https://doi.org/10.4315/0362-028X–72.1.207

Gentsch, J. R., & Parashar, U. D. (2005). The epidemiology of rotavirus. *Journal of Infectious Diseases*, 192(1), 13–19. https://doi.org/10.1086/430535

Graham, D. Y., & Kim, J. (2019). Foodborne viral pathogens: A review. *Journal of Clinical Gastroenterology*, 53(5), 405–413. https://doi.org/10.1097/MCG.0000000000001077

Hall, A. J., Curns, A. T., McDonald, L. C., & Parashar, U. D. (2012). Norovirus disease in the United States. *Emerging Infectious Diseases*, 18(1), 80–87. https://doi.org/10.3201/eid1801.110295

Harris, J. P., & Hale, C. (2018). The role of environmental factors in foodborne viral outbreaks. *Environmental Health Perspectives,* 126(3), 037009. https://doi.org/10.1289/EHP2865

Hedberg, C. W., & Osterholm, M. T. (2006). Foodborne disease outbreaks: A comprehensive review. *Foodborne Pathogens and Disease,* 3(4), 1–17. https://doi.org/10.1089/fpd.2006.3.1

Herman, J. A., & Rose, J. (2019). Improving food safety through effective public health measures. *Food Safety and Quality Assurance*, 7(4), 239–246. https://doi.org/10.1111/fsq.12124

Jiang, X., & Wang, M. (2014). Hepatitis A virus: Epidemiology, pathogenesis, and prevention. *Vaccines*, 2(1), 145–158. https://doi.org/10.3390/vaccines2010145

Kauffman, R. M., & Sokol, A. (2018). New frontiers in foodborne viral disease prevention. *Current Opinion in Food Science*, 22, 16–22. https://doi.org/10.1016/j.cofs.2018.03.003

Khan, M. S., & Ghosh, S. (2018). Advances in the diagnosis and treatment of foodborne viruses. *Clinical Microbiology Reviews*, 31(2), e00009–17. https://doi.org/10.1128/CMR.00009-17

Kokkinos, P., & Panayiotou, C. (2021). CRISPR-based methods for detection of foodborne pathogens. *Journal of Food Safety*, 41(4), e12877. https://doi.org/10.1111/jfs.12877

Kokkinos, P., & Panayiotou, C. (2022). Emerging trends in foodborne viral diseases. *Food and Environmental Virology*, 14(1), 1–11. https://doi.org/10.1007/s12560-021-09523-4

Koo, H., & Park, J. (2020). The global impact of foodborne diseases: A systematic review. *Journal of Food Protection*, 83(12), 2020–2036. https://doi.org/10.4315/JFP-20-158

Kozak, M., & Egan, P. (2017). Advances in foodborne virus detection technologies. *Food Analytical Methods*, 10(5), 1436–1446. https://doi.org/10.1007/s11483-017-9537-0

Le Guyader, F. S., & Atmar, R. L. (2006). Foodborne viruses: Detection and control measures. *International Journal of Food Microbiology*, 112(3), 229–245. https://doi.org/10.1016/j.ijfoodmicro.2006.06.002

Lopman, B. A., & Hall, A. J. (2011). Norovirus: The most common cause of acute gastroenteritis in the United States. *Clinical Infectious Diseases*, 52(8), 1018–1025. https://doi.org/10.1093/cid/cir035

McDonald, S. M., & Johnson, S. (2020). Genomic studies of foodborne viruses: Insights into evolution and control. *Journal of Virology*, 94(8), e02135–19. https://doi.org/10.1128/JVI.02135-19

Morris, J. G., & Popovic, T. (2017). Hepatitis A vaccination and global prevention strategies. *Vaccine*, 35(24), 3111–3116. https://doi.org/10.1016/j.vaccine.2017.03.065

Patel, M. M., Pitzer, V. E., Zheng, Y., & Cortese, M. M. (2012). Global morbidity and mortality of norovirus and rotavirus in children less than 5 years of age. *Journal of Infectious Diseases*, 206(1), 14–23. https://doi.org/10.1093/infdis/jis344

Schlundt, J., & Egan, P. (2004). Global food safety: Challenges and strategies. *World Health Organization, WHO Technical Report Series*, 927, 1–147. https://www.who.int/foodsafety/publications/techreport/en/

Sokollik, C., & Lipp, E. K. (2019). Advances in point-of-care diagnostics for foodborne viruses. *Frontiers in Microbiology*, 10, 1352. https://doi.org/10.3389/fmicb.2019.01352

Tate, J. E., Burton, A. H., & Boschi-Pinto, C. (2016). Rotavirus surveillance—Worldwide, 2001–2018. *Morbidity and Mortality Weekly Report*, 65(50), 1415–1421. https://doi.org/10.15585/mmwr.mm6550a3

van Duynhoven, Y., & de Wit, M. (2009). Global burden of foodborne diseases: A review. *PLoS Medicine,* 6(5), e1000053. https://doi.org/10.1371/journal.pmed.100005

Verhoef, L., & van Pelt, W. (2018). The impact of foodborne viruses on public health and safety. *Public Health Reports*, 133(3), 322–328. https://doi.org/10.1177/003335491813300309

Wang, H., & Zhang, Y. (2021). Innovations in foodborne virus diagnostics: A review. *Journal of Applied Microbiology*, 130(6), 2092–2105. https://doi.org/10.1111/jam.15116

14 Foodborne Parasitic Diseases Diagnostic and Control Measures

Bamidele Tolulope Odumosu, Chukwuemeka Amarachi Esther, Romiluyi Bosede, and Victor Agbo Olofu

14.1 INTRODUCTION

Foodborne diseases are a global concern, and they pose a significant public health risk (Larkin, 2004). These diseases are usually caused by the presence of unwanted organisms/pathogens, such as bacteria, viruses, fungi, and parasites, that are present in food or water and can be transmitted through consumption (Bintsis, 2017). The direct consequences of consuming food or water already contaminated with pathogens are sickness and manifestation of symptoms, while the indirect consequences include significant economic losses, decrease in working population, and increased hospital stay and cost. In this chapter, the focus is on foodborne parasites, which include protozoa, helminths, and trematodes. Foodborne parasites are organisms that can cause illness when ingested through contaminated food or water (Eng et al., 2015). They infect all sorts of living organisms, including fish, shellfish, fresh fruits, vegetables, and are prevalent in the flesh and internal organs of animals commonly used for meat. Foodborne diseases associated with parasites can manifest in different ways with a range of clinical symptoms and are typically severe or life-threatening situations. Foodborne parasitic infection manifestations include gastrointestinal disorders, such as diarrhea, abdominal pain, nausea, vomiting, or organ-specific infections, depending on the specific parasites involved (Chávez-Ruvalcaba et al., 2021). Foodborne biological agents pose a significant threat to public health, with illnesses and deaths caused by these parasites being highly consequential. The global burden and impact of foodborne parasitic infections are significant, with an estimated millions of populations of affected individuals worldwide annually (Torgerson et al., 2015). The global burden of this disease is more pronounced in developing countries in Africa and Asia, where there is limited access to clean and safe water, good sanitation, and proper safety measures (WHO, 2014). Foodborne parasites are complex organisms that pose greater challenges in detection and control compared to bacteria. For effective risk assessment, food safety personnel must be aware of both the prevalence and the significance of these medically important foodborne organisms. The WHO's assessment of foodborne pathogens, conducted by the Foodborne

DOI: 10.1201/9781003540113-14

Disease Burden Epidemiology Reference Group, acknowledged the lack of knowledge regarding the burden of FBPs. The report lists the disability-adjusted life years (DALYs) caused by 31 foodborne pathogens, including parasites (Gabriël et al., 2019), estimating a total of 7,195,014 DALYs, 90,391,678 illnesses, and 51,468 deaths for the year 2010. The emergence and re-emergence of parasitic disease associated with food and water is a continuous public health concern due to the effects of globalization of food supply, changes in consumers preferences, and climate change. To ensure these problems are addressed, effective diagnostic and control measures are crucial to mitigate the impact of these infections and protect public health. In this chapter, a comprehensive overview of the foodborne parasite will be discussed, and the specific diagnostic and control measures highlighting the current state and future considerations.

14.2 TYPES OF FOODBORNE PARASITES

Foodborne parasites can be categorized into three main types: protozoa, helminths, and ectoparasites. Here is an overview of each type:

1. Protozoa
2. Helminths
3. Trematodes

14.2.1 Protozoa

Protozoa are multicellular organisms which are diverse in nature. They occupy a variety of ecology niches. There are over 30 phyla of protozoa; however, only few enteric ones belonging to *Apicomplexa*, *Metamonada*, *Amoebozoa*, and *Microspora* are responsible for foodborne human diseases and are important causes of water and foodborne diseases (Nichols, 2000). Among the listed phyla, notable foodborne pathogens include *Toxoplasma gondii* belonging to Apicomplexa, *Giardia lamblia* (Metamonada), and *Blastocystis hominis* (Stramenopila). They are important because of their widespread distribution in nature and easy contamination of food and water (Dawson, 2005). The diseases associated with them are also severe and can also be transmitted via food and water. For instance, *Toxoplasma*, which is responsible for toxoplasmosis, is responsible for miscarriages in women and birth defects, while *Giardia* causes giardiasis, resulting in diarrhea and malnutrition in children. Their peculiarity is not in their life cycle, because humans may not necessarily be their host—for example, Toxoplasma—however, the consumption of uncooked or undercooked food substances, such as meat, is considered a major risk in the case of toxoplasmosis (Patton et al., 2002).

14.2.2 Helminths

Helminths are parasitic worms that are typically visible to the naked eye. They are classified into three phyla comprising of Platyhelminthes, Nematoda, and Acanthocephala; however, foodborne parasites among these phyla are the flukes (class

Trematoda) and the worms (class Cestoda), belonging to the Platyhelminthes and Nematodes, respectively. They exhibit a wide range of life histories, including direct fecal–oral transmission (e.g., ascaris roundworms), development through free-living stages (e.g., hookworm larvae in the environment), or reliance on invertebrate vectors (e.g., schistosome snail vectors). Nematodes or roundworms are major intestinal worms and filarial worms that cause lymphatic filariasis (LF), while platyhelminths or trematodes include *schistosomes* and *cestodes* (tapeworms), altogether causing various kinds of foodborne-related diseases. Humans and helminths' coexisting relationship dates back to the earliest human records (Cox, 2002).

14.2.3 Trematodes

Trematodes are an important class of parasitic flatworms noted for their significant roles in foodborne diseases worldwide. There are over 70 species of trematodes from different genera and families that can cause foodborne infections. Examples include *Brachylaimidae*, *Diplostomidae*, *Echinostomatidae*, *Fasciolidae*, *Gymnophallidae*, *Heterophyidae*, *Lecithodendriidae*, *Microphallidae*, and *Strigeidae* (Toledo et al., 2006). They are also known to cause intestinal infections in humans; however, the most prevalent is the liver fluke (*Fasciola hepatica*) and *Clonorchis sinensis* (Chinese fluke), which are normally found in water, undercooked food, or raw meat, fish, or vegetables. The impact of trematodes in human health is significant, although their symptoms range from mild to severe, depending on the specific pathogen involved. All foodborne trematodes belong to the subclass Digenea. Digenean trematodes have sexual reproduction in their definitive hosts (e.g., humans or mammals) and asexual reproduction in their intermediate host. Their life cycles are complex and involve one or more intermediate hosts (usually snail), and their life cycle plays an important role in their infectivity in man (Farrar et al., 2023).

14.3 TRANSMISSION DYNAMICS OF FOODBORNE PARASITES

14.3.1 Contaminated Water

Pathogens can be transmitted through a variety of pathways, each characterized by distinct mechanisms that facilitate the spread of infection. Understanding these pathways is crucial for developing effective strategies to prevent and control the spread of diseases. Here, direct and indirect transmission are the primary modes of transmission.

14.3.1.1 Direction Transmission

The direct consumption of contaminated food or water is a significant transmission pathway for numerous pathogens, leading to both foodborne and waterborne diseases. Direct transmission of foodborne parasites is common, and it usually occurs through food ingestion or water contaminated by these pathogens (Appelbee et al., 2005). Due to the abundance of the pathogens and reproduction, during their life cycle, these parasites, eggs, and larva are often shed into the environment via water

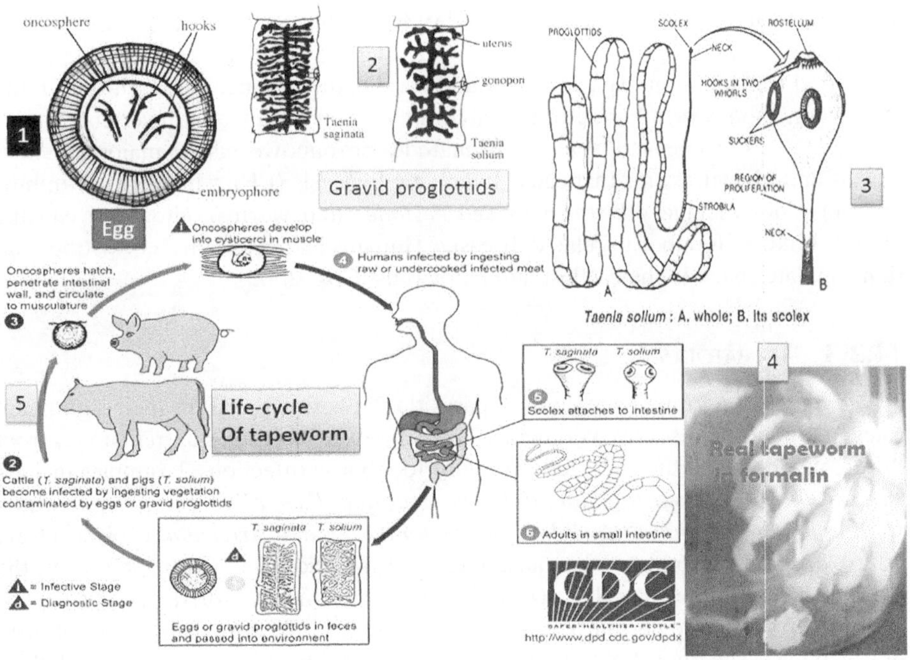

FIGURE 14.1 Life cycle of a tapeworm.

Source: Centers for Disease Control and Prevention.

and food. Direct transmission can occur via raw or undercooked meat, seafood, and unwashed fruits and vegetables, serving as primary culprits (Figure 14.1).

Consuming raw or undercooked meat or fish usually increases the risk of infection, as these may harbor the parasites or their eggs—for example, *Taenia solium* (Dróżdż et al., 2003) (Figure 14.1). Aside from food, contaminated water sources are vital for the direct transmission of these parasites to humans. Polluted rivers and lakes and wells situated close to septic tanks and sewers all play a significant role in the direct transmission of foodborne parasites.

Indirect transmission is the second most common route of infection, which cannot be overemphasized because it also plays a moderate role in the spread of foodborne pathogens.

14.3.1.2 Indirect Transmission

This can easily occur via contamination with food handling equipment, kitchen utensils, and cutleries, serving as carriage of these parasites into the human system (Kagan et al., 2002). Cross-contamination from infected animals, such as livestock, pets, and infected humans, further increases the risk. Carriage of parasitic larvae by vectors such as mosquitoes and cockroaches contributes significantly to the transmission of foodborne parasites. The fecal–oral route is also an indirect form of transmission, whereby the parasite eggs or larvae are shed in human feces, and it contaminates the soil or water (Idowu and Rowland, 2006). This egg may later be consumed via contaminated water or food. Last but not the least, the environment continues to serve as a reservoir for all living things and a center for the spread of numerous infectious pathogens. Numerous parasite species can spread easily through contaminated soil and water, which are popular reservoirs. These two locations are frequently subjected to a variety of contaminants, including sewage contamination, agricultural runoff, and anthropogenic sources that are high in parasite materials. Each of them is essential to maintaining the parasites' environmental persistence. Poor sanitation and hygiene, a wet environment, and inappropriate waste disposal are other human-related problems. The environment itself can serve as a parasite reservoir, with contaminated soil and water offering ideal conditions for parasite spread. Parasites persist in the environment as a result of sewage contamination, agricultural runoff, and soil tainted by feces. The transmission cycle is completed by human-related variables, such as eating raw or undercooked food, not washing your hands, and using improper cooking and freezing techniques.

14.4 FOODBORNE PARASITES OF MEDICAL IMPORTANCE

Here we describe common foodborne parasites that have been linked to various medical diagnoses.

1. *Taenia solium.* Often referred to as pork tapeworms, these parasites are among the largest ribbon-like worms that people may contract, reaching a maximum length of 10 m when fully developed. Larval cysts found in undercooked pork can cause infection when consumed. After hatching in the stomach, these cysts quickly mature into adult worms that live in the gut and feed on the nutrients the host consumes.
2. *Echinococcus granulosus.* This is a small tapeworm, measuring only 3–7 mm in length, that causes a severe disease known as cystic echinococcosis (CE). This parasite is typically spread directly by the ingestion of undercooked meat or by inadvertently consuming an egg from dairy products. Its life cycle typically involves carnivores, such as dogs, and livestock, like sheep. The eggs are extremely durable and can stay infectious for months, even under freezing conditions (Barosi and Umhang, 2024). *Echinococcus granulosus*, while transmitted through similar routes as *Taenia solium*, causes a different condition (cystic echinococcosis) and has a distinct life cycle
3. *Echinococcus multilocularis.* This tapeworm has an uneven geographical distribution, yet it is present in both North American and European countries. Its prevalence is gradually rising. Its life cycle generally includes foxes and small rodents, but it can also occur in domestic dogs and cats. In humans, it causes alveolar echinococcosis, a disease that forms cysts in internal organs. These cysts can reproduce and spread like tumors, potentially becoming fatal if untreated (Wen et al., 2019). This infection poses a risk to hunters who handle infected fox carcasses and people who forage for berries and mushrooms contaminated by fox feces.

4. *Toxoplasma gondii*. It is a single-celled parasitic protozoan that can infect nearly all warm-blooded mammals, with its life cycle commonly occurring between cats and rodents. This parasite is prevalent in most countries and is one of the most widespread protozoan parasites impacting humans. Human infection rates vary significantly, ranging from 10 to 80% across different regions. Typically, the parasite remains dormant in the host's tissues for their entire lifetime, with most infected individuals showing no symptoms and being unaware of their infection (Sullivan and Jeffers, 2012).
5. *Cryptosporidium* spp. These protozoan parasites are mainly transmitted through water or food contaminated with infected water (Widerström et al., 2014). Several outbreaks have been linked to unpasteurized cider and milk, as well as contaminated shellfish. The parasite is prevalent worldwide, including in the UK, often causing infections due to fecal contamination of water supplies by infected livestock. In healthy individuals, the disease results in severe watery diarrhea, which usually resolves on its own. It is advised to thoroughly wash fresh produce, even "ready-washed" lettuces.
6. *Trichinella spiralis*. Also known as the intracellular "pork roundworm," *Trichinella spiralis* is the causative agent of trichinellids, a muscle infection (Ottstein et al., 2009). This infection occurs after consuming raw or undercooked pork or pork products, such as smoked sausages. Other sources can include game meats, like wild boar and even walrus. Infected meat contains cysts housing tiny larvae, which are invisible to the naked eye. Once digested, these larvae develop into adult worms that mate and produce thousands of new larvae. These larvae then migrate to muscle tissues and encyst, lying dormant until the current host is consumed by another predator.
7. *Opisthorchiidae*. This family of flatworms, or flukes, is primarily found in Southeast Asia, with some species also present in Europe and Russia. Infection occurs through consuming raw or undercooked freshwater fish infected with larvae from snails. Within the fish, these larvae develop into another form, and when ingested by a mammal (such as a human), they mature into adult worms that inhabit the bile duct and gallbladder. The adult worms produce eggs excreted in feces, which then hatch and infect new snails upon reaching a freshwater source.
8. *Ascaris* spp. These are the largest human intestinal roundworms, growing up to 35 cm in length, and are the most common human parasites, affecting around 25% of the global population. Upon ingestion, the eggs hatch into larvae in the intestine and begin an extraordinary journey: They travel through the bloodstream to the lungs, ascend the airways to the throat, and are then swallowed back into the stomach. Finally, they return to the intestine, where they mature into adult worms.
9. *Trypanosoma cruzi*. This protozoan parasite causes Chagas disease, a condition known for its slow progression. The parasite infects various cells and organs in the body, including the heart, over many years. In the early stages, symptoms are often absent or mild. However, as the disease advances, it can lead to serious and sometimes fatal cardiac or intestinal problems.

14.5 LIFE CYCLE OF FOODBORNE PARASITES

As is found in all parasites, the life cycle of a typical foodborne parasite is a complex type which involves multiple hosts and environmental stages. The final or definitive hosts during the life cycle are the hosts in which the parasites reach maturity and reproduce. For many foodborne parasites, humans or other mammals serve as the final hosts, while the intermediate hosts harbor the parasite during its larval or asexual stages. The intermediate host stage is usually the infective stage, where they are ingested or consumed and matured in the definitive host (Ortiz et al., 2024). For foodborne parasites, there can be one or more intermediate hosts involved in the life cycle. We would look at some of the notable life cycles as well as their definitive and intermediate host in the following section.

1. *Taenia* spp. *(tapeworms).* Intermediate host: Animals such as cattle or pigs ingest the parasite's eggs, which hatch into larvae and form cysts in their muscles. Definitive host: When humans eat undercooked or raw meat that contains these cysts, the larvae develop into adult tapeworms within the intestines, where they can reproduce. Their eggs are expelled into the environment through feces of an infected host, then ingested by an intermediate host. The parasite eggs can survive for an extended period in the environment before transitioning into the next stage within the intermediate host. The larvae (cysticerci) is formed in the intermediate host, such as pigs or cattle, and embedded themselves in the muscle of the host in the form of cysts. Adult worms are usually released in the final host after consuming undercooked or raw mat containing these cysticerci. The larvae mature into adult tapeworms, which can grow several meters long and produce more egg to continue the growth cycle (Gabriel et al., 2019).
2. *Trichinella* spp. These have a unique life cycle, where the same host can act as both the intermediate and the definitive host. Here is how it works: The intermediate hosts are usually animals, such as pigs, wild boars, and other carnivores or omnivorous species. Intermediate hosts: These are usually animals, such as pigs, wild boars, bears, and other carnivores or omnivores. The larvae encyst within the muscle tissues of these animals and remain dormant until the meat is consumed by another host. The adult worms are formed when humans consume undercooked or raw meat containing these encysted larvae, wherein the larvae are released in the intestines. Here, they mature into adult worms, which can reproduce and continue their life cycle. New larvae are formed after the adult worms reproduce in the intestines; the newly formed larvae migrate through the bloodstream to the muscle tissues. They then encyst in the muscle fibers, ready to be transmitted to the next host through the consumption of infected meat (Western College of Veterinary Medicine, 2021; McKerr et al., 2022).
3. *Toxoplasma gondii.* This is a protozoan parasite with a complex life cycle involving both intermediate and definitive hosts. A wide range of warm-blooded animals, including birds and mammals (such as rodents and humans), serve as intermediate hosts, while the family Felidae (cat families)

are the only known definitive hosts. In these hosts, the parasite undergoes sexual reproduction, and the oocysts are shed in the feces of the infected cats to the environment. The rapid multiplication stage occurs inside the intermediate host, where the oocyst forms tachyzoites after ingestion of the oocyst via undercooked meat.

4. *Trematodes (flukes).* Freshwater snails are commonly the first intermediate hosts for trematodes. In this stage, their eggs hatch into larvae (miracidia) that infect the snails. In the second intermediate host, the larvae develop into another form (cercariae) that leaves the snail and infects a second intermediate host, often a fish or another aquatic organism. At the last stage of the final host, the parasite is consumed via secondary means, either through ruminant, fish, or other aquatic organism, allowing the parasite to mature and reproduce in their intestines or other organs. (See Figure 14.2.)
5. *Nematodes (roundworms) and cestodes (tapeworms).* Some nematodes reside in insects or other small animals, which serve as their intermediate hosts for reproduction. Their definitive host, humans, are usually infected by ingesting the larvae through contaminated food or water. The larvae then mature into adult worms in the human intestines. These complex life cycles ensure the parasites' survival and transmission across different

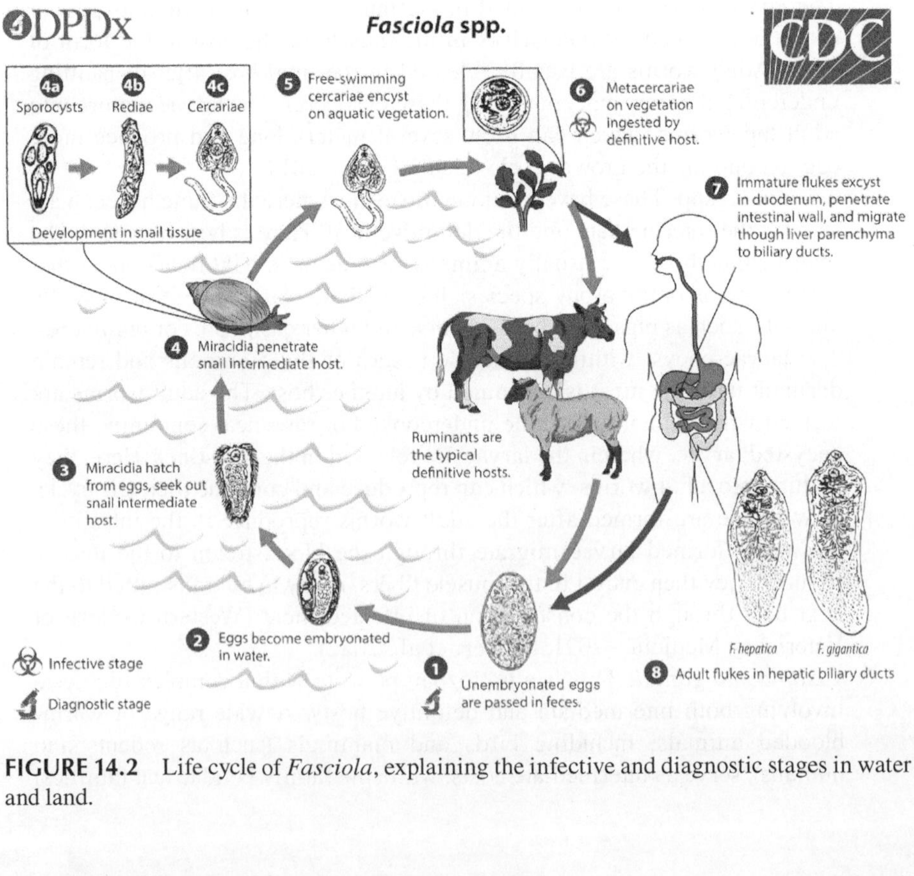

FIGURE 14.2 Life cycle of *Fasciola*, explaining the infective and diagnostic stages in water and land.

environments and hosts. Understanding these cycles is crucial for developing effective control and prevention strategies.

14.6 DIAGNOSTIC TECHNIQUES FOR FOODBORNE PARASITES

An accurate and timely diagnosis of foodborne parasites is crucial for the effective treatment and prevention of further transmission. The diagnosis involves a combination of techniques for the identification of specific parasites and at each stage of infection (Figure 14.3). We shall look at some of these approaches in the following section.

14.6.1 Conventional Methods

The conventional method of diagnosis includes microscopy examination of stool samples to check for visible parasite eggs, cyst, and larvae using specific staining methods which allow for their visibility under the microscope. In biomedical laboratory, during microscopy, fecal samples are usually collected to investigate gastrointestinal parasites (protozoa or helminth), by detecting live mobile trophozoites (protozoan), cysts (also known as protozoa inactive state), or helminth eggs in stool sample. Other microscopic examinations include the following:

1. *Wet mount.* In this procedure, each fecal specimen is prepared in two batches of unstained and temporary stained preparation (Figure 14.4). In the unstained, also known as saline wet mount, the fecal sample is placed on a clean dry microscope slide and emulsified (made softer and freer) using one or two drops of saline and the application of temporary stain (iodine)

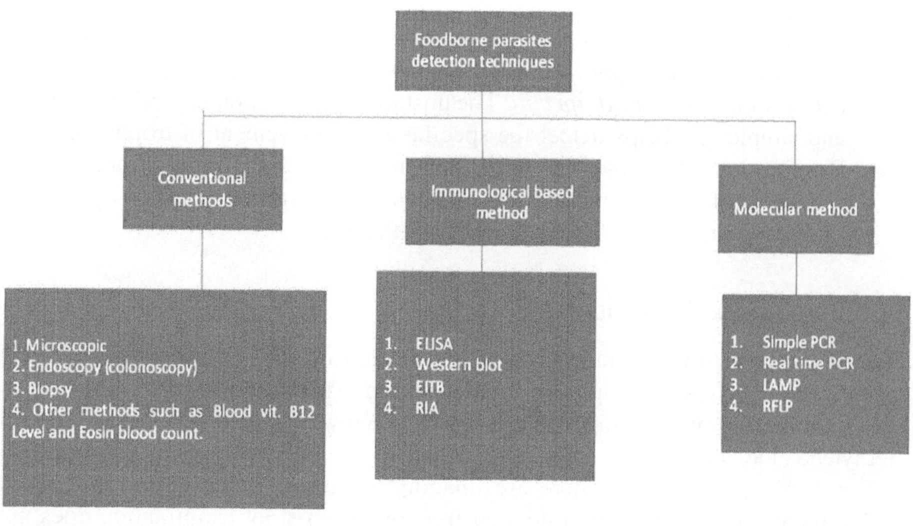

FIGURE 14.3 Diagnostic techniques for foodborne parasites.

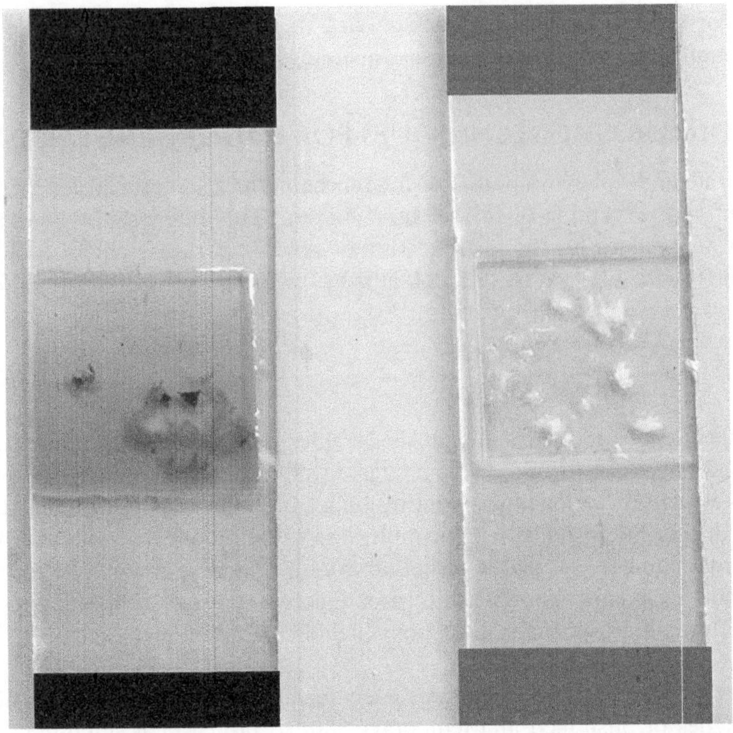

FIGURE 14.4 *Left to right:* Preparation of stool sample slide using iodine wet mount for the identification of trophozoite cysts of protozoa, ova/larvae of helminths, and saline wet mount for identification of protozoa cysts.

before being examined under the microscope. This procedure helps detect the active motile trophozoites of *Entamoeba histolytica*, *Giardia lamblia*, and as well as *Balantidium coli*. The unstained preparation procedure is fast and simple and helps detect the specific active movement of trophozoites. Despite this advantage, the setback of analyzing the internal structure is usually not factual; it is difficult to conclusively diagnose based on cysts or trophozoite hence decrease sensitivity of the test.

14.6.2 Staining Procedures

There have been several staining methods to improve the temporary stained wet mount, which include the improved Ziehl–Neelsen (ZN) with the preparation of 10% H_2SO_4, auramine-phenol stain, cold method of Kinyoun's acid-fast stain, Safranin methylene blue, and Giemsa stains.

For diagnostic methods, there are imaging techniques which are also applied in the diagnosis of foodborne infection in cases when stool examination does not provide adequate information, or if the results of the preceding diagnostics are negative and the symptoms persist, then imaging techniques can be applied. This include the following:

1. *Endoscopy (colonoscopy).* The endoscopy procedure is used to detect parasites when stool examination does not reveal the cause of diarrhea or infection (Hosoe et al., 2014). It involves the use of a flexible thin tube with a light and camera which is often inserted through the anus to examine the lining of the gastrointestinal tract for abnormalities, such as inflammations, ulcers, or growth.
2. *Biopsy.* This involves the collection of tissue samples for the detection of parasites or their eggs, for example, *Clonorchis sinesis* (trematode) and *Anisakis simplex* (nematode). Biopsy could be endoscopic or surgical biopsy, depending on the nature of the infection. A pathologist takes a sample of the body tissue and examines it under the microscope for further evidence of intestinal damage or complications. This process is usually carried out when other diagnostic procedures are inconclusive or whenever the symptom persists.

14.6.3 Other Diagnostic Procedures

These include investigating hematological parameters and nutritional status. *Blood vitamin B12 level:* Some parasitic infections, particularly those affecting the small intestine, can interfere with the adsorption of vitamin B12; hence, it is necessary to carry out this investigation. A low vitamin B12 level might suggest the presence of a parasitic infection in the gastrointestinal tract (Sobczyńska-Malefora et al., 2021). *For instance, Eosin blood count:* An elevated eosinophil count in the blood can be indicative of parasitic infections, especially those involving tissue invasion (e.g., helminths, certain protozoa). The absence of eosinophilia (normal eosin blood count is between 0 and 0.4×10^1) may be due to possible parasitic causes. For example, aside from localized subcutaneous swelling, eosinophilia can be used to diagnose Gnathostoma, a parasitic nematode.

14.6.4 Immunological Analysis Methods

This method involves antibody–antigen reaction, in which a specific antigen binds to a specific antibody, which may be monoclonal or polyclonal antibody. However, to achieve more selectivity, sensitivity, reproducibility, and reliability in the detection of pathogens, monoclonal antibodies are more appropriate. The degree of sensitivity and specificity observed using this method is dependent on the position of epitope of the antibody that binds to the specific antigen. The advantage of this procedure is that it is rapid and reliable, sensitive enough to detect pathogens in low numbers in food, time-efficient, labor-saving, and it reduces human error. However, the drawbacks include high cost, the need for additional enrichment steps, reaction with other pathogen antigen possibly giving a false positive result, and an inability to identify the pathogen's cell that are damaged.

1. *Enzyme-linked immunosorbent assay (ELISA).* There are test kits available for use, for example, Clonorchis Ab (IgG) micro-ELISA (Genedia Cs/Pw Ab ELISA Green Cross MS). This test is more sensitive and specific in the detection of foodborne parasites.
2. *Western blot.* This type of serological testing makes use of IgG Western blot to diagnose foodborne parasites, such as *Paragonimus* spp. (Trematode). An improved Western blot has been proven to be more sensitive and specific for human paragonimiasis.
3. *Enzyme-linked immunoelectrotransfer blot (EITB).* This is an advanced method which makes use of soluble *T. solium* metacestode antigen for the serodiagnosis of taeniasis. Hence, the procedure is highly sensitive and specific.
4. *Radioimmunoassay (RIA).* Here, the immunoassay is labelled with an isotope and used to diagnose *Anisakis simplex.* However, there are a lot of drawbacks, which include expensive equipment, short shelf life of isotope, complexity of test, problem with data interpretation and availability of isotope, hence the use of ELISA.

14.6.5 Molecular Methods

These methods are used because of their high sensitivity and specificity.

1. *Polymerase chain reaction.* It is a process that allows for selective amplification from complex genome. The procedure makes use of enzyme to amplify DNA in vitro using a small quantity of sample, hence making the method specific for detecting pathogens in stool. Despite the high sensitivity and specificity of PCR-based methods in detecting parasites, the main drawbacks are the long processing time and the fact that it does not provide quantitative data.
2. *Real-time PCR.* This method addresses the drawback in the polymerase chain reaction method. It maintains and ensures the PCR method is specific and sensitive, in addition to being able to immediately detect the specific target of interest. Although this method is expensive, it has high performance and produces rapid results. It is essential in detecting *T. gondii.*
3. *Loop-mediated isothermal radiation (LAMP).* This method allows DNA to be amplified under isothermal conditions, which ensures the process is sensitive, specific, and rapid. The procedure saves time, and it is economical, as an inexpensive incubator can be improvised with water bath or block heater. It is used by parasitologists to detect *Taenia*, *Toxoplasma gondii*, and *Opisthorchis viverrini*, among others.
4. *Restriction fragment length polymorphism (RFLP).* This is a common molecular method for detecting species, including genotypes of parasites,

such as *Toxoplasma gondii*. In this procedure, PCR products are digested using restriction enzymes or endonucleases.

14.7 CONTROL OF FOODBORNE PARASITES

Having established the fact that foodborne parasites represent a significant public health concern that affects millions of people worldwide, it is important to device a control measure for these parasites to ensure food safety and public health. These measures encompass a range of strategies, including proper food handling practices, rigorous hygiene standards, effective sanitation protocols, and comprehensive public education campaigns (WHO, 2020a). Additionally, regulatory frameworks and surveillance systems play vital roles in monitoring and controlling the prevalence of parasites.

We shall look at various control measures in different categories and sectors which have direct impact on the food consumption chain or may directly impact the food supply chain. Here are some control measures in different practices and sectors.

14.7.1 Agriculture

14.7.1.1 Agricultural Practices

Effective control of agricultural practices has direct influence on food safety, which is paramount in ensuring the delivery of safe and wholesome food to consumers. Food safety begins with safety practice at the point of source, which aims to reduce the number of contaminants before getting to the end consumers. The use of fertilizers should not only be regulated but must also be applied only when necessary to avoid contamination of crops and soil. The use of organic fertilizers, if not properly handled and stored, may result in the growth of parasites, which can shed their eggs in the fertilizers, especially those that are made from manure. There should be proper timing for the application of the fertilizers to farms during the early planting season, especially to reduce the chances of survival of the pathogens before the crops are harvested. Additionally, the sanitation of farm equipment is necessary to ensure certain tools are free from parasite contamination. This is important because by cleaning farm equipment before and after use, the chances of transferring parasites or their eggs to growing crops during use are reduced. Sanitation also applies to post-harvesting, during which parasite eggs can be harvested along with the crops; hence, sanitation is essential during agricultural practice.

14.7.1.2 Safe Use of Irrigation Water

Irrigation water is inevitable in places with less rain or during dry season; hence, the use of irrigation in farming is common. However, to control the proliferation of parasites in agricultural products, safe irrigation practices must be practiced. Some safe irrigation practices include:

1. *Source selection*. The source of the water for irrigation is very important; hence, the selection of clean, uncontaminated water as source of irrigation

is important in this aspect. Surface water sources, like rivers and streams, are more prone to contamination by parasites because they are exposed to various forms of contamination and other anthropogenic waste, so groundwater or treated water should be preferred when possible and affordable.

2. *Timing of irrigation.* Irrigation water should be applied during the time when the crops are less likely to come into direct contact with the water, such as through drip irrigation systems, which apply water directly to the soil and not to the foliage. Overhead irrigation should be avoided when harvesttime is close, to reduce the risk of contaminating the edible parts of plants and minimize the spread of waterborne parasites that can contaminate the crops directly.

14.7.1.3 Animal Health Management

Animal health management is pivotal in controlling food parasites, which can greatly affect food safety. Sick animals are usually found spreading all kinds of infectious pathogens via their waste, for example, feces; hence, proper veterinary care must be ensured for farm animals to prevent the spread of infectious pathogens. For instance, regular deworming can reduce the process of shedding parasite eggs in their excreta to the environment or water surrounding the farm, while adequate vaccination is needed to prevent infections that can pave the way for secondary parasitic infection. Examples of these types of infection include *Anaplasma* (bacterial), which can lead to babesiosis (parasitic) in cattle, and *Salmonella* (bacterial), which can lead to coccidiosis (parasitic) in poultry. The improvement and maintenance of farm hygiene reduce the chance of parasitic infection or contamination in the farm which may lead to foodborne disease. Therefore, regular monitoring and surveillance of animal health can help detect signs of infections and prevent further spread among farm animals, which often spread to the food chain. It is important that farmers are educated on these simple practices to improve and manage animal health in animal husbandry. These practices, if in regular adherence, can reduce the incidence of foodborne parasites in the food chain.

14.7.2 Food Processing and Treatments

Food processing is a key factor in nutrition; however, improper food processing and treatments may lead to contamination by foodborne pathogens. There are different types of food processing methods and treatment which can be used in controlling foodborne parasites. Here we will look at some of them briefly.

14.7.2.1 Thermal Processing

Thermal processing includes the use of heat to preserve food, and this may include direct cooking and pasteurization of certain food samples. In the case of direct cooking, the application of heat for the preparation of food allows parasites and their larva present in food substance to die due to the continuous high temperature of around 80°C and above, which kills all living organisms, while pasteurization involves heating liquid food such as dairy products or juice to a specific temperature for a set time to kill all pathogens present without compromising food quality or taste. Although some bacterial spores can resist heat treatment, most parasites die in the presence of heat. The recommended temperature for cooking different food can be obtained from nutritionists to prevent total loss of food nutrients.

14.7.2.2 Freezing

Freezing or refrigerating food at sufficiently low temperatures (e.g., −20°C or lower) can kill certain parasites, such as *Trichinella* in pork. This method is particularly useful for raw or undercooked foods like sushi, where freezing is used to control parasites like *Anisakis*. Food items such as fish are recommended to be frozen at −30°C for at least 15 hours in a commercial blast freezer, or at −20°C for at least 7 days in a domestic freezer, to kill *Anisakis*, while some other parasitic worms require longer or shorter periods at freezing temperatures. The same condition applies to the freezing of meat; however, the duration should be up to 6 to 20 days to destroy certain worms, for example, trichinae found in pork meat.

14.7.2.3 High-Pressure Processing (HPP)

HPP is a non-thermal method that uses extreme pressure to inactivate parasites in food samples. This method of food processing is highly recommended for food products or samples that cannot undergo heat treatment, such as certain seafood and ready-to-eat meals. The process involves uniform pressure up to 87,000 psi at a temperature range of 4 to 25°C. The advantage of this method is that it helps preserve the food quality and extend its shelf life without any significant changes in food quality and its properties.

14.7.2.4 Irradiation

The method of food irradiation uses ionizing energy to kill parasites and other pathogens and preserve the food from spoilage. The process involves exposing the food to controlled levels of ionizing radiation, such as gamma rays, electron beams, and X-rays. This method is effective for a variety of foods, including meats, spices, and fresh produce, and helps extend shelf life by eliminating the risk of parasite contamination.

14.7.2.5 Salting, Curing, and Drying

These traditional methods are believed to be able to reduce water activity in food, making the environment inhospitable for parasites. However, there are concerns that there may be incomplete parasite inactivation, especially those at the larva stage, whereas over-salting can affect food quality and safety. A combination of these methods suggests effective preservation and processing.

14.7.3 Life Cycle Targets

In controlling parasites, another viable approach is to study the life cycles of parasites to find the weak links where the cycles may be broken, because incomplete life cycles will hinder their proliferation. The disruption of parasite life cycles is crucial if their presence in food must be properly controlled, and this can be achieved in a number of ways, although some may not be practicable due to their peculiarities. The goal of this method is to totally eradicate parasite presence; hence, it must target either the parasite, the egg, or the larva so that the cycle can be broken. Biological control of the parasite's life cycle is a good strategy which negatively affects their growth in the environment and reduces the chances of foodborne infection and is sustainable. The biological control of parasites targeting their life cycle involves the use of natural compounds possessing anti-parasitic activities, such as essential oils (from lavender trees), garlic (*Allium sativa*), and lemon balm, which has specific adverse effect on parasites and their insect hosts and is capable of disrupting their life cycles. Parasites in the environment can be controlled if their intermediary hosts are dislodged from the environment. Many other parasites live in one or more animals during the different stages of their life cycle, and so control of these animals could potentially reduce parasite numbers. There are anthelminthic drugs that can be administered to these animals to eliminate adult worms that are producing eggs and thereby reduce the incidence of cysticercosis. Parasites often persist in the environment as cysts or eggs, which are resistant to desiccation and other stresses. *Fasciola* cysts, for example, can survive in running water for as long as 122 days, and in pasture for up to a year, and *Giardia* cysts can survive for months in cold water. Preventing the cyst from being shed to the environment can cut short their dissemination into the environment. Vaccination of farm animals or livestock can prevent them from being infected by these parasites in the environment, thereby reducing their infection rate. Other biological ways include the use of chemicals for pre-harvest and post-harvest control, which is also effective but has certain limitations, especially concerning the effect of chemicals on the food itself.

However, it is impossible to completely control populations of wild animals, and it is not practical to attempt to eliminate some intermediate hosts, such as snails and small crustaceans, from every water source. Hence, sustainable ways, as described earlier, may only serve as a form of reduction of the proliferation of the parasite, leading to the reduction of foodborne diseases associated with parasite infection.

14.7.4 Regulations and Guidelines

Regulations and guidelines for controlling food parasites are critical in ensuring food safety and public health. These regulations are established by international bodies, national governments, and industry organizations to minimize the risk of foodborne parasitic infections. Here is an overview of how these regulations and guidelines work:

14.7.4.1 International Standards and Guidelines

- *Codex Alimentarius.* The *Codex Alimentarius*, established by the Food and Agriculture Organization (FAO) and the World Health Organization (WHO), sets international food standards, guidelines, and codes of practice. It includes provisions for controlling foodborne parasites in various food products, such as meat, fish, and produce. These guidelines help harmonize food safety practices across countries and ensure that imported and exported food meets safety standards.

- *World Health Organization (WHO).* The WHO provides global recommendations and guidelines for controlling foodborne parasites. This includes identifying high-risk parasites, offering guidance on monitoring and controlling these parasites in food, and promoting best practices for food safety. The WHO guidelines are revised from time to time to ensure that pathogens recently reported or existing are not left out.

14.7.4.2 National Regulations

1. **United States**
   - *Food and Drug Administration (FDA).* The FDA sets regulations for food safety, including guidelines for controlling parasites in food. This includes standards for fish and fishery products to control parasites such as tapeworms. The FDA's Food Safety Modernization Act (FSMA) also requires food facilities to implement preventive controls to address foodborne parasites.
   - *United States Department of Agriculture (USDA).* The USDA regulates meat, poultry, and egg products in the United States. It enforces standards to prevent parasitic contamination in these products, including inspection protocols for parasites like *Trichinella* in pork and *Toxoplasma* in meat.
2. **European Union (EU)**
   - *European Food Safety Authority (EFSA).* EFSA provides scientific advice and sets food safety standards in the EU, including controls for foodborne parasites. The EU has specific regulations for food hygiene (Regulation (EC) No. 852/2004) that require food businesses to identify and control parasites through HACCP (Hazard Analysis Critical Control Point) systems.
3. **China**
   - *China National Food Safety Standards.* These standards, overseen by the National Health Commission, include specific requirements for controlling foodborne parasites in meat, seafood, and fresh produce. These standards often align with international practices to ensure food safety for both domestic consumption and export.
4. **India**
   - *Food Safety and Standards Authority of India (FSSAI).* This statutory body oversees and regulates the manufacture, storage, and distribution of food in India. They also set regulations and guidelines to prevent the proliferation of foodborne pathogens in alignment with global practices. The FSSAI operates under the authority of the Ministry of Health and Family Welfare, Government of India.
5. **Nigeria**
   - *National Agency for Food and Drug Administration and Control (NAFDAC).* NAFDAC is a government agency in Nigeria that sets guidelines and standards to control foodborne parasites, particularly focusing on imported food products, local ingredients, and local food processing practices. NAFDAC ensures manufacturers obey strict rules and

regulations that align with international guidelines for the production of foods and drug administration in Nigeria.

14.7.4.3 Industry-Specific Guidelines

- *Seafood industry.* Parasite control plans: In industries like seafood, specific guidelines are established to control parasites such as *Anisakis* in fish. This includes recommendations for freezing, cooking, and other processes to kill parasites before the seafood reaches consumers.
- *Meat processing.* Good manufacturing practices (GMPs): Meat processing facilities often follow GMPs that include specific measures for controlling parasites like *Trichinella* in pork. This may involve mandatory freezing, cooking, or testing protocols.
- *Produce industry.* Safe agricultural practices: Guidelines for the produce industry include safe agricultural practices to prevent contamination of fruits and vegetables with parasites like *Cyclospora* and *Cryptosporidium*. This may involve safe water use, proper fertilization, and worker hygiene.

14.7.4.4 Monitoring and Enforcement

- *Inspection and auditing.* Regulatory agencies conduct regular inspections and audits of food processing facilities, farms, and import/export operations to ensure compliance with parasite control standards. Non-compliance can result in penalties, recalls, or closure of facilities.
- *Certification programs.* Some countries and industries have certification programs that verify adherence to food safety standards, including parasite control. Certified facilities are regularly monitored to ensure they maintain high standards of food safety.

14.7.4.5 Consumer Education and Awareness

Effective public health campaigns on food handling can be organized by health societies, governments, and international agencies to educate consumers about the risks of foodborne parasites and how to minimize them. Awareness includes advice on method cooking (to prevent undercooking). Improper cooking or undercooking, as discussed previously, often leads to contamination and foodborne infection by parasites. Therefore, awareness will increase consumers' cooking habits, and it will go a long way in controlling foodborne infection. Awareness of food handling procedures is also paramount in the process of controlling the proliferation of parasites in food. Proper food handling, such as preservation and labelling of food items, is important. For instance, labels may indicate a specific temperature a food item should be cooked in order to be sure of total elimination of pathogens and persisting cysts or an instruction indicating that a food sample has been previously frozen to eliminate parasites.

14.8 PUBLIC HEALTH INTERVENTIONS

Public health interventions for foodborne pathogens are crucial in reducing the burden, incidence, and impact of foodborne diseases associated with parasites. These interventions include surveillance, outbreak investigation, and control measures.

TABLE 14.1
General Control Measures for Specific Foodborne Parasites

Parasites	Control Measures	References
Toxoplasma gondii	1. Meat should be cooked to safe temperatures (≥63°C/145°F) to kill the parasites. 2. Handwashing after handling raw meat to avoid self-infection and transmission of parasites. 3. Pasteurization of milk and milk products. Unpasteurized milk and products should be avoided. 4. Control stray cat populations, and responsible pet ownership should be disseminated to the populace. 5. Proper sanitation practice should be adhered to when gardening.	1. Jones and Dubey (2010) 2. CDC (2020b), *Toxoplasmosis*
Giardia intestinalis	1. Proper boiling of water from potentially contaminated sources to kill parasites. 2. Thorough handwashing with soap and water to prevent parasite infestation. 3. Fruits and vegetables should be washed thoroughly before consumption. 4. Practice good hygiene in food preparation areas.	CDC (2020a), *Giardiasis*
Entamoeba histolytica	1. Ensure safe drinking water (boil or filter). 2. General sanitation and hygiene practices should be strictly adhered to. 3. Food should be cooked thoroughly, especially meat, to avoid infestation. 4. Food from unsanitary sources and vendors should be avoided.	WHO (2017), *Entamoeba histolytica*
Ascaris lumbricoides	1. General sanitation and proper waste disposal should be promoted. 2. The public should be educated on the importance of handwashing, especially after using the bathroom and toilet. 3. Ensure access to clean water and proper sanitation facilities.	WHO (2022), soil-transmitted helminth infections
Cyclospora cayetanensis	1. Fresh produce should be washed thoroughly with clean water or saline water (especially berries and leafy greens). 2. Use safe water for irrigation of crops. 3. Educate food handlers about hygiene practices.	CDC (2021), *Cyclospora* infection
Trichinella spiralis	1. Educate about proper food handling and cooking practices. 2. Pork and wild games should be thoroughly cooked at appropriate temperatures (≥63°C/145°F) to kill parasites. 3. Freeze pork for at least 3 weeks at −15°C (5°F) to kill larvae.	WHO (2020b), *Trichinellosis*

The investigation and management of foodborne disease outbreaks are complex, requiring expertise in various fields, including clinical medicine, epidemiology, laboratory science, food microbiology and chemistry, food safety and control, as well as risk communication and risk management. Effective outbreak resolution demands this multidisciplinary approach (White et al., 2023). Many outbreaks are not adequately investigated, due to the lack of required skills or the expectation that a single field investigator should master all these areas without proper training (Hoffman et al., 2005). The approach at field level is to be an instrument for proper investigation by conducting initial epidemiological, environmental, and laboratory investigations and implementing appropriate control measures by alerting investigators to seek assistance in a more complex situation. These guidelines provide a good template for decision-makers at different levels for coordination and identifying resources and in establishing an environment conducive to the effective management of foodborne disease outbreaks. According to the World Health Organization, these guidelines are designed for public health practitioners, food and health inspectors, district and national medical officers, laboratory personnel, and others involved in investigating and controlling foodborne disease outbreaks. They can be adapted to address the specific needs of individual countries and local contexts.

Surveillance systems play a crucial role in detecting and monitoring foodborne diseases, allowing for timely responses to outbreaks. These systems collect data from various sources, including state and local health departments, to track the occurrence of foodborne illnesses, as stated by the Centers for Disease Control and Prevention (CDC, 2024). Outbreak investigations are essential in identifying the source and mode of transmission, which helps in implementing effective control measures (WHO, 2008).

Control measures involve a combination of strategies, such as improving food safety practices, enhancing hygiene standards, and enforcing regulations. Public health authorities often conduct educational campaigns to raise awareness about safe food handling and preparation practices among consumers and food industry workers. Additionally, regulatory agencies set and enforce standards for food production, processing, and distribution to minimize contamination risks. Technological advancements, such as whole genome sequencing, have enhanced the ability to identify and track foodborne pathogens, leading to more precise and effective interventions (CDC, 2024). Collaboration between public health agencies, food industry stakeholders, and international organizations is also vital in addressing the global nature of foodborne diseases.

Moreover, the World Health Organization (WHO) has developed a handbook to help countries measure their foodborne disease burden and identify food safety system needs and data gaps, thereby strengthening national infrastructure and better protecting public health. Advances in detection technologies, such as matrix–assisted laser desorption/ionization time-of-flight (MALDI-TOF) mass spectrometry and Fourier transform infrared spectroscopy, have also improved the identification and typing of foodborne pathogens (WHO, 2008). These efforts, combined with a "one health" approach that considers the interconnectedness of humans, animals, and environmental health, are crucial for effective foodborne disease management.

Overall, these interventions aim to protect public health by preventing foodborne illnesses, reducing their severity, and minimizing their socioeconomic impact (WHO, 2008).

14.9 CONCLUSION

Foodborne parasites are a public health challenge, and the various diseases associated with them also pose significant health challenges globally because it affects millions of people annually, with high cost of treatment and economic cost. The complexity of these diseases demands a dynamic and holistic approach toward diagnosing, controlling, and preventing the proliferation of these pathogens. A reduction in the dissemination of these pathogens requires joint collaborative efforts of individuals, researchers, and government agencies in creating awareness among farmers and consumers, conducting more research into the ways of preventing their continuous multiplication and state-of-the-art facilities for proper diagnostic testing and investigation of foodborne parasites.

REFERENCES

Appelbee, A., Thompson, R., Olson, M. 2005. Giardia and Cryptosporidium in mammalian wildlifeñcurrent status and future needs. *Trends Parasitol.* 21: 370–376.

Barosi, R., Umhang, G. 2024. Presence of *Echinococcus* eggs in the environment and food: A review of current data and future prospects. *Parasitology* 21: 1–16.

Bintsis, T. 2017. Foodborne pathogens. *AIMS Microbiol.* 3 (3): 529–563.

CDC. 2024. *Surveillance System (FDOSS).* https://www.cdc.gov/fdoss/index.html (accessed August 11, 2024).

Centers for Disease Control and Prevention (CDC). (2020a). *Giardiasis.* https://www.cdc.gov/parasites/giardia/index.html (accessed October 15, 2024).

Centers for Disease Control and Prevention (CDC). (2020b). *Toxoplasmosis.* https://www.cdc.gov/parasites/toxoplasmosis/index.html (accessed September 22, 2024).

Centers for Disease Control and Prevention (CDC). (2021). *Cyclospora infection.* https://www.cdc.gov/parasites/cyclosporiasis/index.html (accessed September 18, 2024).

Chávez-Ruvalcaba, F., Chávez-Ruvalcaba, M. I., Moran Santibañez, K., Muñoz-Carrillo, J. L., León Coria, A., Reyna Martínez, R. 2021. Foodborne parasitic diseases in the neotropics–a review. *Helminthologia* 58 (2): 119–133.

Cox, F. E. 2002. History of human parasitology. *Clin Microbiol Rev.* 15 (4): 595–612.

Dawson, D. 2005. Foodborne protozoan parasites. *Int J Food Microbiol.* 103 (2): 207–227.

Dróżdż, J., Demiaszkiewicz, A. W., Lachowicz, J. 2003. Expansion of the Asiatic parasite *Ashworthius sidemi* (Nematoda, Trichostrongylidae) in wild ruminants in Polish territory. *Parasitol Res.* 89 (2): 94–97.

Eng, S. K., Pusparajah, P., Mutalib, N.-S.A., Ser, H., Chan, K., Lee, L.-H. 2015. *Salmonella*: A review on pathogenesis, epidemiology and antibiotic resistance. *Front Life Sci.* 8: 284–293.

Farrar, J., Hotez, P. J., Junghanss, T., Kang, G., Lalloo, D., White, N. J., Garcia, P. J., eds. 2023. *Manson's Tropical Infectious Diseases.* Elsevier Health Sciences.

Gabriël, S., Mwape, K. E., Phiri, I. K., Devleesschauwer, B., Dorny, P. 2019. Taenia solium control in Zambia: The potholed road to success. *Parasite Epidemiol Control* 4: e00082.

Hoffman, R. E., Greenblatt, J., Matyas, B. T. 2005. Capacity of state and territorial health agencies to prevent foodborne illness. *Emerg Infect Dis.* 11 (1): 11–16.

Hosoe, N., Ogata, H., Hibi, T. 2014. Endoscopic imaging of parasites in the human digestive tract. *Parasitol Int*. 63 (1): 216–220.

Idowu, O. A., Rowland, S. A. 2006. Oral fecal parasites and personal hygiene of food handlers in Abeokuta, Nigeria. *Afr Health Sci*. 6 (3): 160–164.

Jones, J. L., Dubey, J. P. 2010. Toxoplasmosis. In R. Friedman (Ed.), *Foodborne Diseases* (pp. 148–167). Elsevier.

Kagan, L. J., Aiello, A. E., Larson, E. 2002. The role of the home environment in the transmission of infectious diseases. *J Commun Health*. 27(4): 247–267

Larkin, M. 2004. Foodborne illnesses—a global concern. *Lancet Infect Dis*. 4 (4): 250–251.

McKerr, C., Chalmers, R. M., Elwin, K., et al. 2022. Cross-sectional household transmission study of Cryptosporidium shows that *C. hominis* infections are a key risk factor for spread. *BMC Infect Dis*. 22: 114.

Nichols, G. 2000. Food-borne protozoa. *British Med Bull*. 56: 209–235.

Ortiz, J. B., Uys, M., Seguino, A., et al. 2024. Foodborne helminthiasis. *Curr Clin Microbiol Rep*. 11 (3): 153–165.

Ottstein, B., Pozio, E., Nöckler, K. 2009. Epidemiology, diagnosis, treatment, and control of trichinellosis. *Clin Microbiol Rev*. 22 (1): 127–145.

Patton, S., Faulkner, C., Anderson, A., Smedley, K., Bush, E. 2002. *Toxoplasma gondii infection in sows and market-weight pigs in the United States and its potential impact on consumer demand for pork*. National Pork Board Report. http://www. pork. org/PorkScience/Documents/00-130.

Sobczyńska-Malefora, A., Delvin, E., McCaddon, A., Ahmadi, K. R., Harrington, D. J. 2021. Vitamin B12 status in health and disease: A critical review. Diagnosis of deficiency and insufficiency–clinical and laboratory pitfalls. *Crit Rev Clin Lab Sci*. 58 (6): 399–429.

Sullivan, W. J. Jr., Jeffers, V. 2012. Mechanisms of Toxoplasma gondii persistence and latency. *FEMS Microbiol Rev*. 36 (3): 717–733.

Toledo, R., Esteban, J. G., Fried, B. 2006. Immunology and pathology of intestinal trematodes in their definitive hosts. *Adv Parasitol*. 63: 285–365.

Torgerson, P. R., Devleesschauwer, B., Praet, N., et al. 2015. World Health Organization estimates of the global and regional disease burden of 11 foodborne parasitic diseases, 2010: A data synthesis. *PLoS Med*. 12 (12): e1001920.

Wen, H., Vuitton, L., Tuxun, T., Li, J., Vuitton, D. A., Zhang, W., McManus, D. P. 2019. Echinococcosis: Advances in the 21st century. *Clin Microbiol Rev*. 13: 32(2)

Western College of Veterinary Medicine. 2021. *Trichinella species in people and animals*. https://wcvm.usask.ca/learnaboutparasites/parasites/trichinella-people-animals.php.

White, A. E., Garman, K. N., Hedberg, C., Pennell-Huth, P., Smith, K. E., Sillence, E., Baseman, J., Scallan, W. E. 2023. Integrated food safety centers of excellence working group. improving foodborne disease surveillance and outbreak detection and response using peer networks-the integrated food safety centers of excellence. *J Public Health Manag Pract*. 29 (3): 287–296.

Widerström, M., Schönning, C., Lilja, M., Lebbad, M., Ljung, T., Allestam, G., Lindh, J. 2014. Large outbreak of *Cryptosporidium hominis* infection transmitted through the public water supply, Sweden. *Emerging Infectious Diseases*, 20 (4): 581–589.

World Health Organization (WHO). 2008. *Foodborne disease outbreaks: Guidelines for investigation and control*. World Health Organization. https://iris.who.int/bitstream/handle/10665/43771/9789241547222_eng.pdf.

World Health Organization (WHO). 2014. *WHO initiative to estimate the global burden of foodborne diseases: Fourth formal meeting of the Foodborne Disease Burden Epidemiology Reference Group (FERG): Sharing new results, making future plans, and preparing ground for the countries*.

World Health Organization (WHO). 2017. *Entamoeba histolytica*. https://www.who.int/news-room/fact-sheets/detail/amoebiasis.

World Health Organization (WHO). 2020a. *Foodborne parasites*. http://www.who.int/news-room/fact-sheets/detail/food-safety.

World Health Organization (WHO). 2020b. *Trichinellosis*. https://www.who.int/news-room/fact-sheets/detail/trichinellosis.

World Health Organization (WHO). 2022. *Soil-transmitted helminth infections*. https://www.who.int/news-room/fact-sheets/detail/soil-transmitted-helminth-infection.

15 Antimicrobial Resistance in Poultry Pathogens

Implications for Food Safety and the Need for Enhanced Quality Control Standards

Gourav Biswas, Neeraja Mano, Adon Babu, and Tony Grace

15.1 INTRODUCTION

Antimicrobials are a critical group of drugs for the treatment of human infections. Still, most of their use is global, and about 73% of applications are dedicated to animals raised for food production (Van Boeckel et al. 2017). In animal agriculture, antimicrobials act as an adjunct to veterinary treatments, contribute as surrogates for growth promoters, and help maintain high hygiene standards in some regional settings (O'Neill 2016). Even though the application of these antimicrobials is crucial in the manufacturing of quality food products, the laws regulating the usage of these drugs are lax in most places. The unchecked usage of antimicrobials in the food industry has largely contributed to the development of antimicrobial resistance in the bacterial population worldwide (Aparicio-Blanco et al. 2024)

Antimicrobial resistance has been recognized as a medical concern of global proportions that poses risks to public health. An approximate 1.27 million deaths are attributed annually to antimicrobial resistance in 2019, which is estimated to increase up to a staggering number of 10 million deaths annually by 2050 (Thompson 2022).

Meat production for developed countries has levelled over the years since 2000 (Van Boeckel et al. 2019). The meat production in developing economies like Africa, Asia, and South America, however, has been rising at a rate of 68%, 64%, and 40%, respectively. Asia accounts for most of the production of pigs at 56%, as well as chickens at 54% of world stocks (Van Boeckel et al. 2019). The increased demand of the meat industry in the global food market has significantly increased the usage of antimicrobials in animal agriculture, which has resulted in foodborne pathogens gaining resistance to those antimicrobials (Van Boeckel et al. 2015, 2019).

Modern animal husbandry practice often includes systematic applications of antimicrobials to fulfil production requirements to reduce losses (Silbergeld et al. 2008). However, the high usage of these drugs has increased the emergence of AMR in

DOI: 10.1201/9781003540113-15

populations of livestock (Munk et al. 2018), which is comparable to the case observed in human populations (Murray et al. 2022). Livestock is one of the primary routes through which enteric AMR genes are introduced and spread through the food chain and then dispersed throughout the environment (Munk et al. 2018). The increased usage of antimicrobials in animal agriculture sectors in Europe has been directly related to the rise of antimicrobial resistance (AMR) in livestock (Munk et al. 2018)

Antibiotics are mostly used in the rearing of chickens to enhance growth and treat infectious diseases; their administration without prescription is rated among the biggest contributors to the emergence of antibiotic-resistant bacteria (Adams et al. 2023). The improper use of antibiotics in veterinary medicine has led to an increasing number of antibiotic-resistant bacterial strains in recent times. In 2016, the World Organization for Animal Health (OIE) stated that 87.13% of antibiotics were applied in geographical regions like America, Asia, and East Oceania, where tetracyclines and penicillins emerged as the most commonly used antibiotics in these regions (Saraiva et al. 2021).

Therefore, this has precipitated an emerging recognition of AMR's interrelated risks to human health, animal health, and the environment, and thus the necessity to use a "one health" approach to tackle AMR in food systems.

15.2 IMPLICATIONS OF AMR IN PUBLIC HEALTH AND AMR IN POULTRY

Even though AMR is not a new occurrence, recent studies in the field point to the fast-mounting rate at which it is spreading in the community. This has put significant burden on the global health care (Tillotson and Zinner 2017). The rampant usage of antimicrobials across different sectors has exerted selective pressure on the bacteria, which evolved and fixed resistance-related genes that can be transmitted among species (Samreen et al. 2021). An increased resistance to antibiotics, including third-generation carbapenems, polymyxins, and cephalosporins, is reported worldwide (Pfeifer et al. 2010; Rizzo et al. 2019). The emergence of strains resistant to broad-spectrum antibiotics, like carbapenem and colistin, as well as last-resort antibiotics, like vancomycin, paints a critical picture for the health-care system (Samreen et al. 2021). Approximately 2 million people are infected with antibiotic-resistant strains of bacteria in the European Union alone, which leads to roughly 25,000 deaths annually ("Antimicrobial Consumption—Annual Epidemiological Report for 2018" 2024).

The antimicrobial-resistant strains of bacteria can transmit resistance genes to humans via the food chain (Rousham et al. 2018). The AMR genes can enter the food chain through animal products colonized by resistant bacteria, which can be attributed to the usage of antimicrobials in animal agriculture (Igbinosa et al. 2016). Many common infections are resistant to first-line treatment, the causative bacteria of which are also found in livestock (Parascandola and Weed 2001).

Poultry, mainly chicken, is a cheap source of good-quality protein and is in rising demand, mainly in low- and middle-income countries (LMICs) (Van Boeckel et al. 2015; Ritchie and Roser 2019). Countries like India and China have also witnessed a high growth in poultry production as the demand for chicken increases exponentially (Van Boeckel et al. 2019). The number of antimicrobials exhibiting more than 50% resistance (P50) in the poultry industry of low- and middle-income countries (LMICs) has increased by 2.73-fold (Laxminarayan et al. 2015). Global regulations for antimicrobial usage in animal farming have varied over time. The European Union banned the use of antimicrobials in animal feed in 2006 (Munk et al. 2018), while the FDA in the USA had sanctioned the use of antimicrobials in animal feed without veterinary prescription from 1951 (Murray et al. 2022). Despite guidelines restricting antibiotic growth promoters (AGPs), countries like Brazil, India, and South Africa continue their use in poultry production (Munk et al. 2018; Sagar et al. 2023). India, however, took a significant step by banning the unregulated use of antimicrobials in animal farming on July 19, 2019, as part of its commitment to the "one health" approach advocated by the WHO. Despite regulatory efforts, antimicrobial resistance (AMR) in livestock, including poultry, remains a growing challenge. Europe, for example, has reported increasing AMR in swine and poultry, highlighting the ongoing need for stricter controls and alternative farming practices (Sánchez-Vargas et al. 2011; Chlebicz and Śliżewska 2018).

The vertical nature of poultry production leads to the introduction and transfer of poultry-associated foodborne pathogens during multiple processing phases (Ricke 2021). Antimicrobials like tetracycline, colistin, ciprofloxacin, tylosin, neomycin, amoxicillin, trimethoprim, sulfonamides, doxycycline, erythromycin, and tiamulin are commonly utilized to combat these foodborne pathogens (Chowdhury et al. 2021). The World Organization for Animal Health (OIE) reported in 2016 that 87.13% of antimicrobials were used in regions such as the Americas, Asia, and East Oceania, with tetracyclines and penicillins being the most frequently applied (Saraiva et al. 2021).

15.3 ANTIMICROBIAL RESISTANCE (AMR) IN THE CONTEXT OF *SALMONELLA* SP. AND THE MECHANISMS UNDERLYING AMR IN *SALMONELLA*

Salmonella is an enteropathogenic bacterium from the family of *Enterobacteriaceae* which contributes largely to human gastroenteritis workload and could be hosted by mammals, amphibians, or reptiles (Sánchez-Vargas et al. 2011). Yearly, infections due to *Salmonella* are reported to cause almost 93 million instances of gastroenteritis, leading to approximately 155,000 deaths all around the world (Lamichhane et al. 2024). Between February and July 2022, about 884 cases were documented due to *Salmonella* in 48 states in the United States, mainly related to poultry and poultry-related products (Lamichhane et al. 2024). Since it is classified as a category B pathogen, *Salmonella* is associated with moderate disease outbreak rates and relatively low mortality rates. It can infect avian species of all ages; however, young chickens and turkeys are highly susceptible to infection within the first two weeks of life. The infected birds have a bad physical state, with ruffled feathers, lethargy, anorexia, huddling behavior, diarrhea, pasty vents, reduced egg-laying, and on postmortem inspection, signs of organ swelling and hemorrhaging in the liver and spleen (Olnood et al. 2015; Wibisono et al. 2020). In both humans and animals, antibiotics are the main form of therapy (Chen et al. 2013).

In patients presenting with an overt clinical illness, multiple antibiotics are frequently employed (Galán 2021). Chloramphenicol and trimethoprim/sulfamethoxazole were some of the first antibiotics used in the treatment of salmonellosis (Stoycheva and Murdjeva 2006). Presently, drugs of choice to treat *Salmonella* infections in immunocompromised individuals include third-generation quinolones, among which are fluoroquinolones, like ciprofloxacin and ofloxacin (Jibril et al. 2021). However, because resistance continues to increase against fluoroquinolones, other treatment options now also include cephalosporins, such as ceftriaxone, and macrolides, such as azithromycin (Hailu et al. 2021; Dhengare et al. 2021). Besides antibiotics, vaccines also play a very important role in the prevention and management of *Salmonella* infections; however, the available vaccines for typhoid fever only induce moderate and transient immunity in human populations (Milligan et al. 2018; Panhuis et al. 2013).

The mechanisms of antimicrobial resistance of *Salmonella* represent an important area of study in the continuous battle of bacterial pathogens and antibiotics. Being a highly well-known pathogen due to its high adaptability, *Salmonella* continues developing sophisticated ways of resistance. Therefore, this represents great challenges in successful therapeutic management. It is significant for understanding the persistence of the pathogen to reduce its impact on public health. In the next discussion, we present the subtle mechanism *Salmonella* used to evade the action of antibacterial agents:

Enzymatic degradation and modification. Salmonella secretes a variety of enzymes that degrade the effectiveness of antibiotics through several kinds of chemical modifications. For example, an aminoglycoside-modifying enzyme, such as acetyltransferase AAC(6')-Ib-cr, facilitates resistance by bringing about acetylation, phosphorylation, or adenylation modification of antibiotic structures. Enzymic modifications cause the activity of antibiotics so that they may not perform their action on the targets (Ramirez and Tolmasky 2010).

Structural alterations of the bacterial barriers. By altering cell wall and membrane structures, *Salmonella* creates a permeation barrier that becomes impermeable to antibiotics, so this structure integrity improvement allows antibiotics to reach the targets in intracellular environments to lower the effectivity (Akinyemi and Ajoseh 2017; Abdi et al. 2019; Saha et al. 2022).

Genomic mutations of the target sites. Mutations within core genetic loci allow *Salmonella* to affect the conformation of antibiotic target sites such that susceptibility to antibiotics is reduced. For example, mutations in the gene rpoB confer resistance by changing RNA polymerase structure to enable it to resume RNA synthesis when the antibiotic is present. This pliability within the genome then represents the nature by which *Salmonella* grants resistance (Reygaert 2018).

Horizontal gene transfer and genetic mobilization. Horizontal acquisition of resistance genes by plasmids, transposons, and integrons is key to the rapid adaptation of *Salmonella*. Genetic exchange enables the adaptation of *Salmonella* in acquiring new resistant traits, which propels the growth of AMR among different bacterial species and species' barriers (McMillan et al. 2020).

Interference of antibiotic target sites. Salmonella may protect sensitive target sites from the inhibition imposed by antibiotics. For instance, the plasmid-mediated quinolone resistance protein, Qnr, acts by competing with quinolones for the DNA gyrase and topoisomerase binding sites; thus, the replication and survival of the bacteria in the antibiotic is assured (Jacoby et al. 2015).

Low-level genetic adaptations. Resistance can be due to subtle yet important mutations, such as point mutations. These mutations, as shown by mutations in the rpoB gene that confer resistance to rifampicin, precisely adjust target site structures, thereby ensuring bacterial survival without compromising essential cellular functions (Patel et al. 2023).

Reduced membrane permeability. The changes in LPS structures reduce the net negative charge of the outer membrane of *Salmonella*, lowering the binding affinity of cationic antimicrobial peptides, such as polymyxins. Such changes significantly impair the association of drugs with and the destabilization of bacterial membranes (Martynowycz et al. 2019).

Efflux pump systems. One of the effective mechanisms of resistance used by *Salmonella* is efflux pumps. These transport systems are termed *active effluxes*, where antibiotics are pumped out from the intracellular environment and most times are associated with MDR. Often, efflux pump genes are located within mobile genetic elements that maximize the movement of resistance traits (Gaurav et al. 2023).

The multiple mechanisms by which *Salmonella* avoids antimicrobial activities illustrate fitness adaptability. Such a complex situation demands new antimicrobial agents, along with novel strategies against the misuse of antibiotics in health-care settings, as well as agricultural applications. With greater knowledge of these mechanisms, there is an open avenue for new strategies that will counter the continued evolution of *Salmonella* and its resistance mechanisms (Castro-Vargas et al. 2020; Cosby et al. 2015; Mashe et al. 2023; Osman et al. 2021).

Further, significant genetic diversity found within and between different serotypes of *Salmonella* adds complexity to the treatment of this pathogen (Giannelli et al. 2017; Benedetto et al. 2018; Maddux et al. 2017; Mancini et al. 2021). This has created an urgent need to develop other control strategies for *Salmonella* infections not only in animals but also in humans. Possible interventions may include probiotics, prebiotics, bacteriophages, antimicrobial peptides, essential oils, and better vaccines.

15.4 ANTIMICROBIAL RESISTANCE IN THE CONTEXT OF *E. COLI*

The Gram-negative, facultatively anaerobic, rod-shaped bacterium of the genus *Escherichia Escherichia coli* is usually a non-pathogenic gut commensal (Tenaillon et al. 2010; Jenkins 2000). However, some strains have virulence factors that can cause diseases including diarrhea, hemorrhagic colitis, urinary tract infections, and even meningitis (Croxen and Finlay 2010; Jerse et al. 1992; Pouillot et al. 2012). Most *Escherichia coli* strains are generally not pathogenic; nevertheless, there are ongoing fears that pathogenic strains may be transmitted from animals to man and that these sometimes lead to illness in chickens produced for commercial purposes (Ievy et al. 2022). The results of studies conducted in the United States indicate that *E. coli* strains isolated from livestock have significantly higher levels of antibiotic resistance than human clinical isolates (Tadesse et al. 2012). An inadequate hygiene involved in the slaughter process acts to increase the contamination of meat with *E. coli.*, thus posing serious concerns to food safety (Laury et al. 2009). Other researchers have reported that *E. coli* strains from contaminated meat and meat products exhibit a high rate of resistance to most used antibiotics (Mølbak 2004). Overuse of antibiotics on livestock is among the major factors that promote antibiotic resistance (Moreno et al. 2008; Okeke et al. 1999), and antibiotic resistance acts as an opportunistic precursor that enhances this resistance through mechanisms like horizontal gene transfer and genetic changes (Laxminarayan and Brown 2001; Hughes and Andersson 2015). MDR *E. coli* strains commonly possess multiple antibiotic resistance genes (Nikaido 2009). This, thus, has led to severe illness and death in populations; therefore, an urgent approach needs to be taken on this public health issue (Been et al. 2014).

15.5 CONCLUSION AND FUTURE REMARKS

The crisis of the emergence of antimicrobial resistance to key antibiotics, including ceftriaxone, norfloxacin, and vancomycin, brings to the attention that there is a potential for horizontal transfer of antibiotic-resistant *Salmonella*, *E. coli* from poultry to humans. Combining this with resistance to macrolides, tetracycline, and ciprofloxacin underscores the need for enhanced surveillance with policy interventions as part of the poultry production system to limit the growth of the resistance. Such results point to a heightened urgent call to control antibiotic resistance in chicken populations as important to safeguard human health and have these critical antibiotics remain effective in clinical use. Additionally, further study on the genetic factors underlying antibiotic resistance in *Salmonella* strains from these chicken groups is needed to help in establishing specific intervention measures and ensuring the continued efficacy of antibiotics both in veterinary and human medicine.

In the future, adopting a holistic "one health" approach must be espoused to appreciate the interrelationship between human, animal, and environmental health. Future studies should bridge the knowledge gap between the use of antibiotics in poultry, antibiotic resistance in bacteria, and the potential implications of such an event for human health. Research into alternative approaches for the promotion of health and the growth of poultry products without antibiotic use is emerging. Probiotics, prebiotics, and phage therapy should bring solutions to the possible replacement of antibiotics and ensure sustainable poultry production.

REFERENCES

"Antimicrobial Consumption—Annual Epidemiological Report for 2018." 2024. Accessed December 5. https://www.ecdc.europa.eu/en/publications-data/surveillance-antimicrobial-consumption-europe-2018.

Abdi, Milad, Shiva Mirkalantari, and Nour Amirmozafari. 2019. "Bacterial Resistance to Antimicrobial Peptides." *Journal of Peptide Science: An Official Publication of the European Peptide Society* 25 (11): e3210-NA. John Wiley and Sons Ltd. https://doi.org/10.1002/psc.3210.

Adams, James R.G., Jai Mehat, Roberto La Ragione, and Shahriar Behboudi. 2023. "Preventing Bacterial Disease in Poultry in the Post-Antibiotic Era: A Case for Innate Immunity Modulation as an Alternative to Antibiotic Use." *Frontiers in Immunology*. https://doi.org/10.3389/fimmu.2023.1205869.

Akinyemi, Kabiru Olusegun and Samuel Oluwasegun Ajoseh. 2017. "Factors Contributing to the Emergence and Spread of Antibiotics Resistance in Salmonella Species." In *Current Topics in Salmonella and Salmonellosis*. https://doi.org/10.5772/67701.

Aparicio-Blanco, Juan, Nikhar Vishwakarma, Claus Michael Lehr, Clive A. Prestidge, Nicky Thomas, Richard J. Roberts, Chelsea R. Thorn, and Ana Melero. 2024. "Antibiotic Resistance and Tolerance: What Can Drug Delivery Do Against This Global Threat?" *Drug Delivery and Translational Research* 14 (6): 1725. Springer. https://doi.org/10.1007/S13346-023-01513-6.

Been, Mark de, Val F. Lanza, María de Toro, Jelle Scharringa, Wietske Dohmen, Yu Du, Juan Hu, et al. 2014. "Dissemination of Cephalosporin Resistance Genes between Escherichia Coli Strains from Farm Animals and Humans by Specific Plasmid Lineages." *PLoS Genetics* 10 (12). https://doi.org/10.1371/journal.pgen.1004776.

Benedetto, Gianluigi De, Laura Salvini, Stefano Gotta, Paola Cescutti, and Francesca Micoli. 2018. "Investigation on Sugar-Protein Connectivity in Salmonella O-Antigen Glycoconjugate Vaccines." *Bioconjugate Chemistry* 29 (5). https://doi.org/10.1021/acs.bioconjchem.8b00178.

Castro-Vargas, Rafael Enrique, María Paula Herrera-Sánchez, Roy Rodríguez-Hernández, and Iang Schroniltgen Rondón-Barragán. 2020. "Antibiotic Resistance in Salmonella Spp. Isolated from Poultry: A Global Overview." *Veterinary World* 13 (10). https://doi.org/10.14202/vetworld.2020.2070–2084.

Chen, Hung Ming, Yue Wang, Lin Hui Su, and Cheng Hsun Chiu. 2013. "Nontyphoid Salmonella Infection: Microbiology, Clinical Features, and Antimicrobial Therapy." *Pediatrics and Neonatology*. https://doi.org/10.1016/j.pedneo.2013.01.010.

Chlebicz, Agnieszka and Katarzyna Śliżewska. 2018. "Campylobacteriosis, Salmonellosis, Yersiniosis, and Listeriosis as Zoonotic Foodborne Diseases: A Review." *International Journal of Environmental Research and Public Health*. https://doi.org/10.3390/ijerph15050863.

Chowdhury, Sukanta, Sumon Ghosh, Mohammad Abdul Aleem, Shahana Parveen, Md Ariful Islam, Md Mahbubur Rashid, Zubair Akhtar, and Fahmida Chowdhury. 2021. "Antibiotic Usage and Resistance in Food Animal Production: What Have We Learned from Bangladesh?" *Antibiotics (Basel, Switzerland)* 10 (9). https://doi.org/10.3390/ANTIBIOTICS10091032.

Cosby, Douglas E., Nelson A. Cox, Mark Harrison, Jeanna L. Wilson, R. Jeff Buhr, and Paula J. Fedorka-Cray. 2015. "Salmonella and Antimicrobial Resistance in Broilers: A Review." *Journal of Applied Poultry Research* 24 (3): 408–26. Elsevier BV. https://doi.org/10.3382/japr/pfv038.

Croxen, Matthew A. and B. Brett Finlay. 2010. "Molecular Mechanisms of Escherichia Coli Pathogenicity." *Nature Reviews Microbiology*. https://doi.org/10.1038/nrmicro2265.

Dhengare, Archana, Savita Pohekar, Prerana Sakharwade, Sheetal Sakharkar, Samrudhi Gujhar, Vaishali Tembare, and RuchirraAnkar. 2021. "Case Report of Typhoid Fever (Salmonella Typhi)." *Journal of Pharmaceutical Research International*. https://doi.org/10.9734/jpri/2021/v33i58a34089.

Galán, Jorge E. 2021. "Salmonella Typhimurium and Inflammation: A Pathogen-Centric Affair." *Nature Reviews Microbiology*. https://doi.org/10.1038/s41579-021-00561-4.

Gaurav, A., Perwez Bakht, Mahak Saini, Shivam Pandey, Ranjana Pathania. 2023. "Role of Bacterial Efflux Pumps in Antibiotic Resistance, Virulence, and Strategies to Discover

Novel Efflux Pump Inhibitors." *Microbiology (Reading, England)* 169 (5): NA-NA. Society for General Microbiology. https://doi.org/10.1099/mic.0.001333.

Giannelli, C., E. Cappelletti, R. Di Benedetto, F. Pippi, M. Arcuri, V. Di Cioccio, L. B. Martin, A. Saul, and F. Micoli. 2017. "Determination of Free Polysaccharide in Vi Glycoconjugate Vaccine against Typhoid Fever." *Journal of Pharmaceutical and Biomedical Analysis* 139. https://doi.org/10.1016/j.jpba.2017.02.042.

Hailu, Woinshet, Yosra A. Helmy, Geoffrey Carney-Knisely, Michael Kauffman, Dean Fraga, and Gireesh Rajashekara. 2021. "Prevalence and Antimicrobial Resistance Profiles of Foodborne Pathogens Isolated from Dairy Cattle and Poultry Manure Amended Farms in Northeastern Ohio, the United States." *Antibiotics* 10 (12). https://doi.org/10.3390/antibiotics10121450.

Hughes, Diarmaid and Dan I. Andersson. 2015. "Evolutionary Consequences of Drug Resistance: Shared Principles across Diverse Targets and Organisms." *Nature Reviews Genetics*. https://doi.org/10.1038/nrg3922.

Ievy, Samina, M. Nazmul Hoque, Md Saiful Islam, Md Abdus Sobur, Fatimah Muhammad Ballah, M. Shaminur Rahman, Md Bahanur Rahman, Jayedul Hassan, Mohammad Ferdousur Rahman Khan, and Md Tanvir Rahman. 2022. "Genomic Characteristics, Virulence, and Antimicrobial Resistance in Avian Pathogenic Escherichia Coli MTR_BAU02 Strain Isolated from Layer Farm in Bangladesh." *Journal of Global Antimicrobial Resistance* 30. https://doi.org/10.1016/j.jgar.2022.06.001.

Igbinosa, Etinosa O., Abeni Beshiru, Lucy U. Akporehe, Faith E. Oviasogie, and Owen O. Igbinosa. 2016. "Prevalence of Methicillin-Resistant Staphylococcus Aureus and Other Staphylococcus Species in Raw Meat Samples Intended for Human Consumption in Benin City, Nigeria: Implications for Public Health." *International Journal of Environmental Research and Public Health* 13 (10). Int J Environ Res Public Health. https://doi.org/10.3390/IJERPH13100949.

Jacoby, George A., Marian A. Corcoran, and David C. Hooper. 2015. "Protective Effect of Qnr on Agents Other than Quinolones That Target DNA Gyrase." *Antimicrobial Agents and Chemotherapy* 59 (11). https://doi.org/10.1128/AAC.01292-15.

Jenkins, R. 2000. "Bacteria in Biology, Biotechnology and Medicine (5th Edn) Paul Singleton (Ed.): Wiley, Chichester, 1999, pp. Xiv+489. Price Â£22.50 Paperback, ISBN 0-471-98880-4." *Biochemistry and Molecular Biology Education* 28 (4). https://doi.org/10.1016/s1470-8175(00)00007–2.

Jerse, A.E., Jun Yu, Ben D. Tall, James B. Kaper. 1992. "A Genetic Locus of Enteropathogenic Escherichia Coli Necessary for the Production of Attaching and Effacing Lesions on Tissue Culture Cells." *Journal of Pediatric Gastroenterology and Nutrition* 15 (2). https://doi.org/10.1097/00005176-199208000-00024.

Jibril, Abdurrahman Hassan, Iruka N. Okeke, Anders Dalsgaard, and John Elmerdahl Olsen. 2021. "Association between Antimicrobial Usage and Resistance in Salmonella from Poultry Farms in Nigeria." *BMC Veterinary Research* 17 (1). https://doi.org/10.1186/s12917-021-02938-2.

Lamichhane, Bibek, Asmaa M.M. Mawad, Mohamed Saleh, William G. Kelley, Patrick J. Harrington, Cayenne W. Lovestad, Jessica Amezcua, et al. 2024. "Salmonellosis: An Overview of Epidemiology, Pathogenesis, and Innovative Approaches to Mitigate the Antimicrobial Resistant Infections." *Antibiotics*. https://doi.org/10.3390/antibiotics13010076.

Laury, Angela, Alejandro Echeverry, and Mindy Brashears. 2009. "Fate of Escherichia Coli O157:H7 in Meat." In *Safety of Meat and Processed Meat*. https://doi.org/10.1007/978-0-387-89026-5_2.

Laxminarayan, R., T. Van Boeckel, and A. Teillant. 2015. *The Economic Costs of Withdrawing Antimicrobial Growth Promoters from the Livestock Sector*, Vol. 78. OECD Food, Agriculture and Fisheries Papers. https://doi.org/10.1787/5js64kst5wvl-en.

Laxminarayan, Ramanan, and Gardner M. Brown. 2001. "Economics of Antibiotic Resistance: A Theory of Optimal Use." *Journal of Environmental Economics and Management* 42 (2). https://doi.org/10.1006/jeem.2000.1156.

Maddux, Jacob T., Zachary R. Stromberg, Roy Curtiss, and MelhaMellata. 2017. "Evaluation of Recombinant Attenuated Salmonella Vaccine Strains for Broad Protection against Extraintestinal Pathogenic Escherichia Coli." *Frontiers in Immunology* 8 (OCT). https://doi.org/10.3389/fimmu.2017.01280.

Mancini, Francesca, Francesca Micoli, Francesca Necchi, Mariagrazia Pizza, Francesco Berlanda Scorza, and Omar Rossi. 2021. "GMMA-Based Vaccines: The Known and The Unknown." *Frontiers in Immunology*. https://doi.org/10.3389/fimmu.2021.715393.

Martynowycz, Michael W., Amy Rice, Konstantin Andreev, Thatyane M. Nobre, Ivan Kuzmenko, Jeff Wereszczynski, and David Gidalevitz. 2019. "Salmonella Membrane Structural Remodeling Increases Resistance to Antimicrobial Peptide LL–37." *ACS Infectious Diseases* 5 (7). https://doi.org/10.1021/acsinfecdis.9b00066.

Mashe, Tapfumanei, Gaetan Thilliez, Blessmore V. Chaibva, Pimlapas Leekitcharoenphon, Matt Bawn, Moses Nyanzunda, Valerie Robertson, et al. 2023. "Highly Drug Resistant Clone of Salmonella Kentucky ST198 in Clinical Infections and Poultry in Zimbabwe." *NPJ Antimicrobials and Resistance* 1 (1). https://doi.org/10.1038/s44259-023-00003-6.

McMillan, Elizabeth A., Charlene R. Jackson, and Jonathan G. Frye. 2020. "Transferable Plasmids of Salmonella Enterica Associated With Antibiotic Resistance Genes." *Frontiers in Microbiology*. https://doi.org/10.3389/fmicb.2020.562181.

Milligan, Rachael, Mical Paul, Marty Richardson, and Ami Neuberger. 2018. "Vaccines for Preventing Typhoid Fever." *Cochrane Database of Systematic Reviews*. https://doi.org/10.1002/14651858.CD001261.pub4.

Mølbak, K. 2004. "Spread of Resistant Bacteria and Resistance Genes from Animals to Humans—The Public Health Consequences." *Journal of Veterinary Medicine Series B: Infectious Diseases and Veterinary Public Health* 51. https://doi.org/10.1111/j.1439-0450.2004.00788.x.

Moreno, Andrea, Helia Bello, Drago Guggiana, Mariana Domínguez, and Gerardo González. 2008. "Extended-Spectrum β Lactamases Belonging to CTX-M Group Produced by Escherichia Coli Strains Isolated from Companion Animals Treated with Enrofloxacin." *Veterinary Microbiology* 129 (1–2). https://doi.org/10.1016/j.vetmic.2007.11.011.

Munk, Patrick, Berith Elkær Knudsen, Oksana Lukjacenko, Ana Sofia Ribeiro Duarte, Liese Van Gompel, Roosmarijn E.C. Luiken, Lidwien A.M. Smit, et al. 2018. "Abundance and Diversity of the Faecal Resistome in Slaughter Pigs and Broilers in Nine European Countries." *Nature Microbiology* 3 (8). https://doi.org/10.1038/s41564-018-0192-9.

Murray, Christopher J.L., Kevin Shunji Ikuta, Fablina Sharara, Lucien Swetschinski, Gisela Robles Aguilar, Authia Gray, Chieh Han, et al. 2022. "Global Burden of Bacterial Antimicrobial Resistance in 2019: A Systematic Analysis." *The Lancet* 399 (10325). https://doi.org/10.1016/S0140-6736(21)02724–0.

Nikaido, Hiroshi. 2009. "Multidrug Resistance in Bacteria." *Annual Review of Biochemistry* 78 (1): 119–46. Annual Reviews. https://doi.org/10.1146/annurev.biochem.78.082907.145923.

O'neill, J.I.M. 2014. "Antimicrobial Resistance: Tackling a Crisis for the Health and Wealth of Nations." *Review on Antimicrobial Resistance*, 20, 1–16.

Okeke, Iruka N., Adebayo Lamikanra, and Robert Edelman. 1999. "Socioeconomic and Behavioral Factors Leading to Acquired Bacterial Resistance to Antibiotics in Developing Countries." *Emerging Infectious Diseases*. https://doi.org/10.3201/eid0501.990103.

Olnood, Chen G., Sleman S.M. Beski, Mingan Choct, and Paul A. Iji. 2015. "Use of Lactobacillus Johnsonii in Broilers Challenged with Salmonella Sofia." *Animal Nutrition* 1 (3). https://doi.org/10.1016/j.aninu.2015.07.001.

Osman, Abdinasir Yusuf, Sharifo Ali Elmi, David Simons, Linzy Elton, Najmul Haider, Mohd Azam Khan, Iekhsan Othman, Alimuddin Zumla, David McCoy, and Richard Kock.

2021. "Antimicrobial Resistance Patterns and Risk Factors Associated with Salmonella Spp. Isolates from Poultry Farms in the East Coast of Peninsular Malaysia: A Cross-Sectional Study." *Pathogens* 10 (9). https://doi.org/10.3390/pathogens10091160.

Panhuis, Willem G. van, John Grefenstette, Su Yon Jung, Nian Shong Chok, Anne Cross, Heather Eng, Bruce Y. Lee, et al. 2013. "Contagious Diseases in the United States from 1888 to the Present." *New England Journal of Medicine* 369 (22). https://doi.org/10.1056/nejmms1215400.

Parascandola, M., and D.L. Weed. 2001. "Causation in Epidemiology." *Journal of Epidemiology and Community Health* 55 (12): 905. https://doi.org/10.1136/JECH.55.12.905.

Patel, Yesha, Vijay Soni, Kyu Y. Rhee, and John D. Helmann. 2023. "Mutations in RpoB That Confer Rifampicin Resistance Can Alter Levels of Peptidoglycan Precursors and Affect β-Lactam Susceptibility." *MBio* 14 (2). https://doi.org/10.1128/mbio.03168-22.

Pfeifer, Yvonne, Angela Cullik, and Wolfgang Witte. 2010. "Resistance to Cephalosporins and Carbapenems in Gram-Negative Bacterial Pathogens." *International Journal of Medical Microbiology: IJMM* 300 (6): 371–79. Int J Med Microbiol. https://doi.org/10.1016/J.IJMM.2010.04.005.

Pouillot, Flavie, Maryline Chomton, Hélène Blois, Celine Courroux, Julien Noelig, Philippe Bidet, Edouard Bingen, and Stéphane Bonacorsi. 2012. "Efficacy of Bacteriophage Therapy in Experimental Sepsis and Meningitis Caused by a Clone O25b: H4-ST131 Escherichia Coli Strain Producing CTX-M–15." *Antimicrobial Agents and Chemotherapy* 56 (7). https://doi.org/10.1128/AAC.06330-11.

Ramirez, Maria S. and Marcelo E. Tolmasky. 2010. "Aminoglycoside Modifying Enzymes." *Drug Resistance Updates* 13 (6). https://doi.org/10.1016/j.drup.2010.08.003.

Reygaert, Wanda C. 2018. "An Overview of the Antimicrobial Resistance Mechanisms of Bacteria." *AIMS Microbiology* 4 (3): 482–501. American Institute of Mathematical Sciences (AIMS). https://doi.org/10.3934/microbiol.2018.3.482.

Ricke, Steven C. 2021. "Strategies to Improve Poultry Food Safety, a Landscape Review." *Annual Review of Animal Biosciences* 9 (February): 379–400. Annu Rev Anim Biosci. https://doi.org/10.1146/ANNUREV-ANIMAL-061220-023200.

Ritchie, Hannah, Pablo Rosado, and Max Roser. 2019. "Meat and Dairy Production." Published online at OurWorldinData.org. Retrieved from: 'https://ourworldindata.org/meat-production' [Online Resource]

Rizzo, Kyle, Sam Horwich-Scholefield, and Erin Epson. 2019. "Carbapenem and Cephalosporin Resistance among Enterobacteriaceae in Healthcare-Associated Infections, California, USA1." *Emerging Infectious Diseases* 25 (7): 1389–96. Emerg Infect Dis. https://doi.org/10.3201/EID2507.181938.

Rousham, Emily K., Leanne Unicomb, and Mohammad Aminul Islam. 2018. "Human, Animal and Environmental Contributors to Antibiotic Resistance in Low-Resource Settings: Integrating Behavioural, Epidemiological and One Health Approaches." *Proceedings. Biological Sciences* 285 (1876). Proc Biol Sci. https://doi.org/10.1098/RSPB.2018.0332.

Sagar, P., Ajmal Aseem, Santosh Kumar Banjara, and Shobi Veleri. 2023. "The Role of Food Chain in Antimicrobial Resistance Spread and One Health Approach to Reduce Risks." *International Journal of Food Microbiology* 391–93: 110148. Elsevier BV. https://doi.org/10.1016/j.ijfoodmicro.2023.110148.

Saha, Sukumar, Elder Pupo, Afshin Zariri and Peter van der Ley. 2022. "Lipid A Heterogeneity and Its Role in the Host Interactions with Pathogenic and Commensal Bacteria." *MicroLife* 3: uqac011-NA. Oxford University Press (OUP). https://doi.org/10.1093/femsml/uqac011.

Samreen, Iqbal Ahmad, Hesham A. Malak, and Hussein H. Abulreesh. 2021. "Environmental Antimicrobial Resistance and Its Drivers: A Potential Threat to Public Health." *Journal of Global Antimicrobial Resistance* 27 (December): 101–11. Elsevier. https://doi.org/10.1016/J.JGAR.2021.08.001.

Sánchez-Vargas, Flor M., Maisam A. Abu-El-Haija, and Oscar G. Gómez-Duarte. 2011. "Salmonella Infections: An Update on Epidemiology, Management, and Prevention." *Travel Medicine and Infectious Disease*. https://doi.org/10.1016/j.tmaid.2011.11.001.

Saraiva, Mauro M.S., Kelvin Lim, Daniel Farias Marinho do Monte, Patrícia Emília Naves Givisiez, Lucas Bocchini Rodrigues Alves, Oliveiro C. Freitas Neto, Samuel Kariuki, Angelo Berchieri, Celso José Bruno de Oliveira, and Wondwossen A. Gebreyes. 2021. "Antimicrobial Resistance in the Globalized Food Chain: A One Health Perspective Applied to the Poultry Industry." *Brazilian Journal of Microbiology: [Publication of the Brazilian Society for Microbiology]* 53 (1): 465–86. Springer Science and Business Media LLC. https://doi.org/10.1007/s42770-021-00635-8.

Silbergeld, Ellen K., Jay P. Graham, and Lance B. Price. 2008. "Industrial Food Animal Production, Antimicrobial Resistance, and Human Health." *Annual Review of Public Health* 29 (1): 151–69. Annual Reviews. https://doi.org/10.1146/annurev.publhealth.29.020907.090904.

Stoycheva, Mariana V., and Marianna A. Murdjeva. "Antimicrobial Therapy of Salmonelloses—Current State and Perspectives." *Folia medica* 48, no. 1 (2006): 5–10.

Tadesse, Daniel A., Shaohua Zhao, Emily Tong, Sherry Ayers, Aparna Singh, Mary J. Bartholomew, Patrick F. McDermott. 2012. "Antimicrobial Drug Resistance in Escherichia Coli from Humans and Food Animals, United States, 1950–2002." *Emerging Infectious Diseases* 18 (5): 741–49. Centers for Disease Control and Prevention (CDC). https://doi.org/10.3201/eid1805.111153.

Tenaillon, Olivier, David Skurnik, Bertrand Picard, and Erick Denamur. 2010. "The Population Genetics of Commensal Escherichia Coli." *Nature Reviews Microbiology*. https://doi.org/10.1038/nrmicro2298.

Thompson, Tosin. 2022. "The Staggering Death Toll of Drug-Resistant Bacteria." *Nature*. https://doi.org/10.1038/d41586-022-00228-x.

Tillotson, Glenn S. and Stephen H. Zinner. 2017. "Burden of Antimicrobial Resistance in an Era of Decreasing Susceptibility." *Expert Review of Anti-Infective Therapy* 15 (7): 663–76. Expert Rev Anti Infect Ther. https://doi.org/10.1080/14787210.2017.1337508.

Van Boeckel, Thomas P., Charles Brower, Marius Gilbert, Bryan Bt Grenfell, Simon A. Levin, Timothy P. Robinson, Aude Teillant, and Ramanan Laxminarayan. 2015. "Global Trends in Antimicrobial Use in Food Animals." *Proceedings of the National Academy of Sciences of the United States of America* 112 (18): 5649–54. Proceedings of the National Academy of Sciences. https://doi.org/10.1073/pnas.1503141112.

Van Boeckel, Thomas P., Emma E. Glennon, Dora Chen, Marius Gilbert, Timothy P. Robinson, Bryan T. Grenfell, Simon A. Levin, Sebastian Bonhoeffer, and Ramanan Laxminarayan. 2017. "Reducing Antimicrobial Use in Food Animals." *Science (New York, N.Y.)* 357 (6358): 1350–52. American Association for the Advancement of Science (AAAS). https://doi.org/10.1126/science.aao1495.

Van Boeckel, Thomas P., João Pires, Reshma Silvester, Cheng Zhao, Julia Song, Nicola G. Criscuolo, Marius Gilbert, Sebastian Bonhoeffer, and Ramanan Laxminarayan. 2019. "Global Trends in Antimicrobial Resistance in Animals in Low- and Middle-Income Countries." *Science (New York, N.Y.)* 365 (6459): NA-NA. American Association for the Advancement of Science (AAAS). https://doi.org/10.1126/science.aaw1944.

Wibisono, Freshindy Marissa, Freshinta Jellia Wibisono, Mustofa Helmi Effendi, Hani Plumeriastuti, Akvyan Rafi Hidayatullah, Erwan Budi Hartadi, and Eka Dian Sofiana. 2020. "A Review of Salmonellosis on Poultry Farms: Public Health Importance." *Systematic Reviews in Pharmacy*. https://doi.org/10.31838/srp.2020.9.69.

16 A Closer Look at Food Testing Agencies, Acts, Regulations, and Patents

Ashim Jyoti Pathak, Karpumpuli Mili, Pankaj Chetia, and Minakshi Puzari

16.1 INTRODUCTION

Food safety and quality are jeopardized because of innumerable pathogens that cause various foodborne diseases, algal toxins that cause acute diseases, and fungal toxins that may have a chronic upshot that may be acutely toxic (Borchers et al., 2010). It is crucial to recognize that food pollution can be chemical and biological contamination and that each is as harmful to human health. Most microbial species that cause biological contamination are pathogenic, including viruses, bacteria, fungi, and protozoa (Garvey, 2019). Food is necessary for survival, so its safety is a fundamental human right. Billions of people are at the prospect of unsafe food, several million are becoming sick, and hundreds of thousands perish every year. In the past, eating tainted food due to deliberate or inadvertent personal behavior and the government's failure to guarantee food safety and quality had resulted in human tragedies and economic catastrophes (Fung et al., 2018). The terms "quality" and "food safety" are frequently used in the media and interchangeably. *Food safety* has emerged as a characteristic feature of food quality, and government agencies are pressuring the feed and food sectors to uphold extensive quality management systems to improve food safety, revamp the food inspection system, and improve consumer information to restore consumer confidence in foods (Röhr et al., 2005). There will be an immediate need to increase the effectiveness and efficiency of food chains due to projections that the world's population will exceed 9 billion people by the year 2050. To become fully regenerative, it requires 70% more food, food production systems, and food chains (King et al., 2017). Every year, millions of people are hospitalized and die due to foodborne illnesses, which makes food safety a vast global problem. 70% of foodborne illnesses are associated with catering or food service businesses, and hospital settings have the highest risk of fatal outbreaks. Hospital patients are particularly vulnerable to foodborne illnesses; therefore, cleanliness is crucial. Nutritious and customized diets are essential for the patients' speedy recovery; as a result, the prepared foods must be secure, healthful, high-quality, and supplied at appropriate times (Bou-Mitri et al., 2018). As these challenges rise yearly, researchers are trying to reduce contamination and increase food shelf life and nutritional value.

DOI: 10.1201/9781003540113-16

Some frequently used methods for controlling contamination are thermal control, refrigeration, and chemicals, which adversely affect the food's sensory, physical, and biochemical nature (Rahman et al., 2022). Food scientists and technologists routinely study food's chemical composition and physical characteristics as part of their product development, quality management, or research activities. To meet the ever-growing requirement for food concerns, food safety and quality have escalated the importance of analysis for determining the composition and for censorious product analysis. To make a consolidated decision, scientists must successfully perform all three major steps in the analysis: (1) selecting and preparing samples, (2) performing the assay, and (3) calculating and expounding on the results (Nielsen, 2024).

Throughout the globe, regulations have been sanctified by provincial, state, and federal authorities to protect consumers against health hazards and the skullduggery associated with consuming and selling foods. Based on the development of the country involved, these vary from comparatively primitive controls over basic aspects of food adulteration and sanitation to sophisticated and detailed legislation framed to provide wide protection against different hazards, even the ones resulting from food processing techniques and production (Chapman & Morrison, 1966). Developing countries face the problem of providing sufficient food for their increasing population. Policies and programs are actively promoted to enhance food production and agriculture. Thus, there is a need to take expedient and appropriate steps to ensure quality and food safety for export and even domestic consumption. An effective and implementable national food safety policy is needed to guarantee that the food supplied to the customers is nutritious, good-quality, adequate, and wholesome (Nyor, 2014).

16.2 AN OUTLINE OF FOOD SAFETY AND QUALITY

Food quality and safety are the two important components of the food industry, each serving a distinct but interrelated purpose (Kubde et al., 2016). Within the last decade, food safety has become a crucial component of food quality. Food quality and safety are major media topics these days. Food safety has become a defining feature of food quality in recent years. The food and feed industries are under pressure from consumers to develop extensive quality management systems to promote food safety and to enhance consumer information, to restore consumer confidence in food and reorganize food inspection systems (Röhr et al., 2005). In the past, nutritional components, including established risk factors for specific chronic diseases and supplements containing nutrients, were not included in the definition of *food safety*. Due to growing concerns around genetically engineered foods, functional foods, high-nutrient additive levels, and nutritional supplements are now being considered (Burlingame & Pineiro, 2007). The quality concept in the food industry prioritizes three major factors: gratification of consumer's expectations and perceptions, adherence to the product's intended purpose, and perceptions. In general, the clientele focuses on the first two, but their expectations vary and include many varying factors, such as price, appearance, taste, nutritious content, production system, and fair trade, among others (Savov & Kouzmanov, 2009). In the agri-food sector, food safety has received focus. It is the underpinning of all the leads considered on various activity levels ranging from farm to enterprises unabridged on national, regional, and international levels (Panghal et al., 2018). The problem of food-related diseases and linked illnesses is a common anxiety all across the globe. HACCP (Hazard Analysis Critical Control Point) and food safety/quality management are engaged in tackling the issue. It is a food safety strategy that employs organized protective methods to protect consumers and foods from chemical, physical, and biological contaminants. Mainly implemented in production processes as well as post-production processes to make sure that no contamination is present, HACCP and food safety cannot be separated because the right application of HACCP is required to ensure food safety (Awuchi, 2023). It is considered one of the best food safety management systems worldwide. NASA developed it to guarantee the safety of food used in space programs, and this has been adopted as an international standard across the food industry for the production, storage, transportation, marketing, and packaging of food products (Radu et al., 2023).

16.3 FOOD TESTING AGENCIES

Food testing agencies play an important role in guaranteeing the security and caliber of food items. Various tests are conducted to detect contaminants, verify nutritional content, and ensure docility with regulatory standards. International standard-setting bodies and governments provide a legislative scaffold to promulgate food safety and ensure legal trade practices. Food safety authorities enforce regulations to protect customers against contaminated, unsafe, fraudulently presented, or harmful foods. It is implemented by prohibiting the sale of food that does not adhere to these requirements and preventing the introduction of hazards into the food supply (Bietlot & Kolakowski, 2012).

16.3.1 Role Performed by Food Testing Agencies

Food quality and safety inspections are extensively conducted in the food industry for many reasons: to appraise management systems, procure certifications to certain food quality and safety standards, confirm legal compliance, and so on. Food safety scandals have heightened consumer interest in food safety and quality, which has prompted the people and commercial food industries to create a range of food quality and safety standards (Kotsanopoulos & Arvanitoyannis, 2017). Food safety is a fundamental component of all national public health systems and the vitality of many national health programs. FSL (Food Safety Law) ensures the quality and safety of food by reviewing foods and food products for the presence of adulterants and contaminants and checking product quality and nutritive value (Figure 16.1) (Attrey, 2017). The Food Standards Agency (FSA), a self-dependent government department that safeguards public health and consumers' food interests, for example, places a high premium on ensuring that food is safe for consumption (Turner et al., 2021).

16.3.2 Global Scenario of Food Testing Agencies

Food safety and quality (FSQ) has been given utmost importance because it is one of the most important issues of the 21st century, with factors like increasing population and unstable global economy enhancing the deprecation of the same. Irrespective

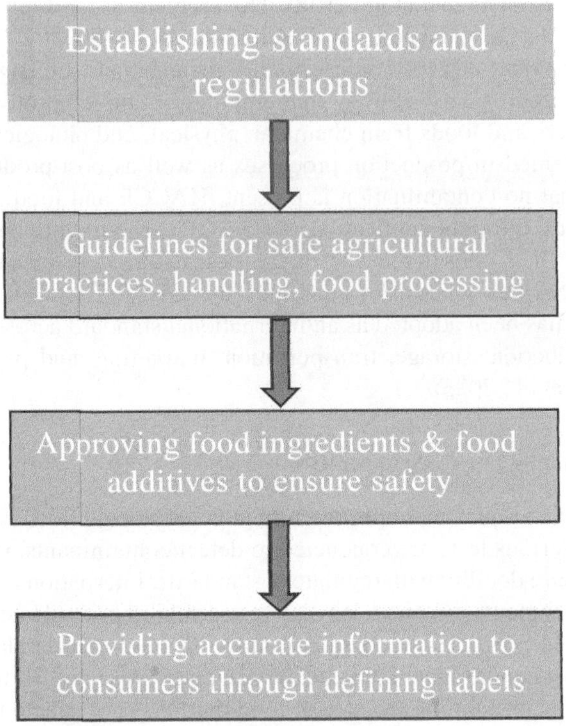

FIGURE 16.1 Some of the important functions performed by food testing agencies.

of the extraneous efforts by governing bodies, many nations have not achieved the requisite standards of FSQ (Patil et al., 2023). Some food testing agencies actively maintain global food safety, and requirements are- NFSI, CDC, FDA, EFSA, etc. These organizations work collaboratively and independently to make sure that food products are safe, meet regulatory standards, and are correctly labelled for consumers worldwide (Table 16.1).

National Food Safety Initiative. NFSI was started in 1997 to decrease the onset and risk of foodborne diseases. It ended in 2001, but at the Centers for Disease Control and Prevention (CDC), the erstwhile NFSI funding and activities have been established as an ongoing food safety program (Hoffman et al., 2005).

Centers for Disease Control and Prevention. In 1999, the CDC released a series of yearly state-based maps that demonstrated the rapid fluctuations in the prevalence of obesity, marking the identification of obesity as a national health issue and an epidemic. The CDC provided the environment and intervention-specific guidelines needed to implement these tactics (Dietz, 2015).

In 2016, the CDC published its second series of Prevention Status Reports, which reviewed the state of public health practices and policies, with an emphasis on ten key public health issues: overindulgence in alcohol, car accidents, food safety, nutrition, physical exercise, obesity, overdosing on prescription drugs, health

TABLE 16.1
Some Important Global Food Safety Agencies

SL. No.	Food Safety Agencies	Functions	Reference
1	National Food Safety Initiative (NFSI)	Reduce the onset and dangers of foodborne diseases	(Hoffman et al., 2005)
2	Centers for Disease Control and Prevention (CDC)	Focus on public health policy	(Lipcsei & Kambhampati, 2016)
3	China National Center for Food Safety Risk Assessment (CFSA)	Technical guidance for national food safety surveillance	(Pei et al., 2015)
4	Food Standards Australia New Zealand (FSANZ)	Reduction in the risk of manufactured, processed, and prepared foods	(Baines et al., 2003)
5	International Food Safety Authorities Network (INFOSAN)	Prevention of the international spread of tainted food and stopping outbreaks of foodborne illnesses	(Savelli & Mateus, 2020)
6	Food Standards Agency (FSA)	Protection of public health and customers' interest	(Hurt, 2002)
7	Codex Alimentarius Commission (CAC)	Defense of its interest in food safety matters	(Dawson, 1995)
8	European Food Safety Authority (EFSA)	Testing of dietary incidence to mycotoxins for people's health and that of associated animals	(Eskola et al., 2018)
9	HACCP (Hazard Analysis Critical Control Point)	Tackling foodborne diseases and associated problems	(Awuchi, 2023)

care–associated infections (HAI), heart attacks, strokes, HIV, teen pregnancy, and tobacco use (Lipcsei & Kambhampati, 2016).

Food and Drug Administration. It authorizes medications and oversees their security once they are sold. The FDA is consolidating its post-marketing surveillance through new resources and staff to meet the upcoming challenges of the altering world of therapeutics. The success of the FDA in decreasing drug-related adverse effects relies a large chunk on the negative reports it gets straight from health-care professionals and indirectly from manufacturers (Ahmad, 2003).

Canadian Food Inspection Agency. The increased communication between scientists, risk managers, risk assessors, and consumers to face food safety challenges is an intricate and interdependent food system. CFIA needs to recognize that risk communication is a vital component of the quality risk management process for which the problems of uncertainty and subjectivity may arise (Charlebois & Summan, 2015). CFIA is improving its risk-based approach to inspection by developing a framework that targets access to the different risks connected with food establishments. The new risk assessment method considers food safety threats and may be used to calculate the amount of surveillance needed to control risks effectively. It enables the agency to be more unambiguous and prognosticate in its actions with industry partners (Zanabria et al., 2018).

European Food Safety Authority. EFSA assesses dietary exposure to mycotoxins for public health and the well-being of companion animals and farms at the request of the European Commission, and it has evaluated the safety and effectiveness of feed additives designed to reduce mycotoxin contamination in feeds during the European authorization process for these additives (Eskola et al., 2018). EFSA's main mandate is to analyze new monomers and additives for plastics to support their authorization (roughly ten substances per year)—a small job compared to the field (Grob, 2019).

Food Standards Agency. FSA was established by a parliamentary act in 2000 to safeguard the people's health and the interests of consumers concerning food (Hurt, 2002).

Codex Alimentarius Commission. CAC mainly demonstrates how the European community represents itself regarding food safety issues in the Codex Alimentarius Commission. This mainly aims to demonstrate how the European community has strengthened its influence as a policymaker in the Codex Commission. Defending certain values and the necessity to take into account elements other than science when setting food standards, requiring labels, and creating traceability for food generated from biotechnology are among the responsibilities of the European members of the Codex Commission (Poli, 2004). It was founded in 1962 to carry out the FAO/WHO Joint Food Standards Program. The CAC aims to ensure fair procedures in the food sector and to safeguard consumers' health. It comprises 189 members, including 188 member countries and the European Union (UN) (Dawson, 1995).

International Food Safety Authorities Network (INFOSAN). The World Health Organization and the Food and Agriculture Organization of the United States collaborated to develop the International Food Safety Authorities Network (INFOSAN) in 2004. The primary goals of this worldwide network are to stop the movement of tainted food across international borders, stop food-related disease outbreaks, and consolidate food hygiene systems worldwide to reduce the incidence of food-associated diseases. Members of INFOSAN have not evaluated its worth methodically or rigorously, nor has it ever been regarded as an active community of practice (Savelli & Mateus, 2020).

Food Standards Australia New Zealand (FSANZ). Australian and New Zealand consumers understand that food security is a major international issue. It affects their status as premier food-exporting nations and the health and well-being of the Australian population. To elevate food and trade safety, these countries work together through FSANZ and other cooperative agreements (Ghosh, 2014). They aim to develop standards for food handling practices, premises, transport vehicles, and equipment in food businesses that are outcome-focused rather than suggestive. The intention is to reduce the risk of manufactured, processed, and prepared foods for carrying microorganisms, other contaminants, or foreign materials that can cause foodborne illness or food to become rather unsuitable (Baines et al., 2003).

China National Center for Food Safety Risk Assessment (CFSA). In recent years, China has been hit by food safety issues, and these issues have varied from microbiological issues to melamine in milk powders, Sudan dyes in duck eggs, clenbuterol in pork, and plasticizer in beverages. These have caused safety problems; people's health has been affected, significantly impairing food trade and international reputation (Liu et al., 2013). CFSA provides technical guidance for national food safety surveillance, quality control, and personal training. Moreover, it collects, scrutinizes, and manages surveillance data, recognizes food hygiene problems, and puts forward proposals on time (Pei et al., 2015).

16.3.3 A Local (India) Account on Food Testing Agencies

In India, food testing agencies play an important role in ensuring food safety and quality. Some of the prominent ones include the following.

16.3.3.1 Food Safety and Standards Authority of India (FSSAI)

FSSAI was created to rein over food regulatory compliance in India, as per the Food Safety and Standard Act 2006. It has done a lot of groundwork for properly executing the new food safety regime. Moreover, many challenges need to be faced to strengthen the regulatory framework for robust enforcement of the FSS Act of 2006 (Dhara et al., 2021).

16.3.3.2 Bureau of Indian Standards (BIS)

It establishes requirements for many goods, including food. Moreover, it certifies products so that they meet quality standards. It is the standardization body of India. It formulates standards in demand with the national priorities for different sectors. Recently, it created a committed department for standardization in the Ayush systems (Dua et al., 2023).

16.3.3.3 Central Food Technological Research Institute (CFTRI)

It is an NABL-accredited and ISO-certified institution of CSIR, Government of India. It is involved in creative research in food science and nutrition and is the country's only food and research institute. The research programs are cross-functional and incorporate value addition to agricultural resources, nutraceuticals and health foods, food testing and analysis, fruit and vegetable processing, food processing, etc., aside from basic research in the biochemistry of lipids, carbohydrates, and proteins, molecular nutrition, bioactive, spices, and flavors (Rajasekharan, 2015).

16.4 CORROBORATING SAFE FOOD: A LOOK AT THE FOOD SAFETY ACTS

According to the Food and Agricultural Organization (FAO) and the World Health Organization (WHO), food safety is all feasible risks—acute and chronic—that may jeopardize a consumer's health when they consume food. Any food can become hazardous in the increasingly intricate global food market system if it is not manufactured or handled with appropriate hygiene procedures. Recent food accident cases show that once a contamination epidemic happens, it is challenging to fully sanitize products and challenging to identify the source of the issue within the global food chain system (Henson & Heasman, 1998). A major issue with food safety in one region of the world may have far-reaching effects on many others. Food safety risks can be mitigated, or food contamination can be prevented in advance by using food regulations in the food production system to monitor foodborne hazards. Corroborating food safety involves checking and verifying that protocols and guidelines are being adhered to regularly. The execution of this process is necessary to ensure the safety of food products for consumption and regulatory compliance for food businesses. Here is a thorough look at the process used to confirm food safety.

16.4.1 Regulatory Compliance

Inspections and audits. One way to make sure a food manufacturer or individual is abiding by particular guidelines, requirements, or rules is through food safety audits and inspections (Powell et al., 2013). An *audit* is a "systematic and independent evaluation to ascertain if quality and safety results and planned arrangements adhere to objectives and whether these arrangements are implemented suitably and effectively to achieve the required goals." Food safety agencies routinely inspect food manufacturing, processing, and handling facilities to ensure that food safety regulations are being followed. Unannounced inspections ensure that establishments follow safety procedures even when they are not prepared for an audit (Kotsanopoulos & Arvanitoyannis, 2017).

Certification and accreditation. Many countries promote certification and accreditation to reduce food safety hazards and boost public confidence in the food industry. Facilities often obtain certifications from recognized bodies, such as the Safe Quality Food (SQF) program or Hazard Analysis Critical Control Point (HACCP), which third-party organizations audit (Guo et al., 2019). The certification mark on the food label is important to prove food quality. Facilities may also be accredited by national or international organizations, confirming that they meet specific food safety standards.

16.4.2 Testing and Verification

16.4.2.1 Microbiological Testing

Pathogen testing. Microbiological hazards in the production environment are common in slaughterhouses and facilities where food is produced or processed. The ongoing microbiological hazards in the food processing environment can cause contamination of processed goods, which poses serious health risks to consumers and financial losses for producers (Collins et al., 2022). This is a major concern for the general public's health and for the food and feed business. Microorganisms that can lead to foodborne diseases, including bacteria, viruses, parasites, and fungi, frequently undetectable to human sight, can infect food at any stage of manufacturing, processing, distribution, or preparation. Regular testing of food products for pathogens like *Salmonella*, *E. coli*, and *Listeria* helps ensure food is free from harmful microorganisms. It is commonly known that certain microbial risks might endure in food processing environments (FoPE). Facilities that produce cheese (Fox et al., 2011), produce salmon or crab meat, and process meat and meat products have been reported to harbor persistent strains of *Listeria monocytogenes* (Elson et al.,

2019). Since *Salmonella* can survive for a long time in dry production conditions and low-moisture products, its persistence in high-lipid and low-moisture matrices like chocolate, halva made of sesame, tahini, or peanut butter, and their production environments provide additional challenges. There have also been reports of *Salmonella* persistence in pig and poultry slaughterhouses (Zeng et al., 2021).

Environmental testing. Cross-contamination can be avoided by testing food facilities' surfaces, equipment, and surroundings for contaminants. Every link in the food chain has the potential to become contaminated by microorganisms. Processing environments have to be regarded as a major source of contamination because of production-related pollution or inefficient cleaning and disinfection practices. One strategy to ensure food safety is implementing environmental monitoring programs (EMPs). These programs are designed to verify cleaning and sanitation processes, as well as other pathogen control measures, to prevent environmental contamination of the final product (De Oliveira Mota et al., 2021). The atmosphere of and tools from food handling facilities are frequently linked to *Listeria monocytogenes* outbreaks and product recalls, primarily due to inadequate cleanliness (Zoellner et al., 2018). Ready-to-eat (RTE) deli meat caused an outbreak in Canada in 2008 due to slicer contamination (Simmons & Wiedmann, 2018). According to a 1998 French foodborne illness outbreak study, equipment contamination was linked to 40% of the contaminated cases (De Oliveira Mota et al., 2021).

16.4.2.2 Chemical Testing

Residue analysis. Food items are verified to satisfy safety standards through testing for contaminants, such as heavy metals, pesticide residues, and other impurities. Numerous chemicals that are classified as organochlorine pesticides include lindane; 1,1,1-trichloro-2,2-bis(p-chlorophenyl); ethane (dichloro-diphenyl dichloroethene) (DDT); cyclodienes, like dieldrin, chlordane, and heptachlor; and hexachlorobenzene, among others. These chemicals can be harmful to both humans and animals (Gladen & Rogan, 1991). Nutritional consumption is considered the main method of human exposure to DDT and its metabolite, DDE. Humans store DDT longer than any other animal tested, including monkeys, and it is primarily kept in adipose tissue. Both, however, have been demonstrated to induce cancer in animals, and the IARC has categorized them as likely (category 2B) carcinogens for humans. Strong evidence has been found relating DDE to testicular germ cell tumors and DDT to liver cancer in human cohorts exposed to relatively high levels of exposure (Borchers et al., 2010). It has been determined that heavy metals like lead (Pb), mercury (Hg), and cadmium (Cd) are extremely hazardous to human health. Drinking water tainted by lead pipes or brass fittings may have a lead content of up to 8%. However, lead is still present in the environment and seems to be mostly found in food, especially in soils used to cultivate crops. Meat and milk can be significantly contaminated by local Pb contamination of pastures, and fish typically have high Pb contents as well (Llobet et al., 2003).

Most mercury emissions originate from natural sources, with the balance coming from human activities, such as mining, metal processing, cement manufacturing, and coal burning. Methyl mercury is created when inorganic mercury (Hg) enters aquatic habitats by sediment bacteria. Methyl mercury is then bioaccumulated and biomagnified in the aquatic food chain. Methyl mercury is absorbed and disseminated throughout the body within 30 to 40 hours after consuming fish. According to Borchers et al. (2010), it mostly builds up in the liver and kidney, but it can also be discovered in the brain in amounts of 10%.

Additive testing. Thousands of structurally varied chemical compounds, primarily from naturally occurring sources, are found in the human diet. In addition, chemicals like colorants, minerals, and flavor-imparting agents are added on purpose (Mepham, 2011). Additional chemicals that cause chemical alterations and introduce molecules not typically found in organic agricultural goods may find their way into the food during food processing and preparation. In addition, chemicals are added to produce specific technical effects, like color and consistency, such as emulsification, flavoring, sweetening, and other physical effects, in addition to preservation (Pressman et al., 2017). So it is necessary to make sure that food additives are safe and conform to regulatory limits.

16.4.3 Consumer Feedback and Complaints

Customer feedback about food safety concerns can be used to pinpoint possible issues and areas needing change. Resolving consumer concerns regarding food safety in a timely and serious manner aids in problem identification and resolution (Djekic et al., 2023).

Guidelines are required to comply with international standards for safe food because food handlers and marketing officers lack the knowledge and training necessary to do so. According to the internationally recognized Codex Alimentarius Commission standards and codes of practice, the local government must adopt and implement national food safety standards (Jairath & Purohit, 2013). Several laws and rules about food safety are in place to guarantee that the food supply is fit for human consumption. Here are some key acts and regulations associated with food safety, primarily focusing on the United States, but also touching on international standards.

16.4.3.1 United States

1. *Federal Food, Drug, and Cosmetic Act (FDCA) (1938).* This foundational law gave the Food and Drug Administration (FDA) authority to oversee the safety of food, drugs, and cosmetics. It includes provisions related to food labelling, adulteration, and misbranding. Without trying to cover every aspect of regulation, it lays forth a core set of goals: products that are wholesome, safe, effective, and properly labelled. The Food and Drug Administration aims to carry out these goals using the most effective and efficient controls possible (Hutt, 1973).
2. *Food Additives Amendment (1958).* Food additives are now subject to regulatory scrutiny through the Food Additive Amendment (1958) to the FDCA, which also gave the FDA the power to demand data from manufacturers, proving that a chemical is reasonably free of harm before allowing it into the food supply. Since then, the FDA has established guidelines for chemistry and toxicity studies required to evaluate the safety of compounds classified as generally recognized as safe (GRAS) and food additives. It is crucial

to understand that the safety evaluation standards applied to food additives and substances undergoing GRAS review are fundamentally the same; the key distinctions lie in the timeline for market entry and the origin of the expert reviewers assessing the publicly available safety data (Pressman et al., 2017).

3. *Nutrition Labelling and Education Act (NLEA) (1990).* This act mandates nutrition labelling on most packaged foods and requires that labels be truthful and not misleading (Asirvatham et al., 2017). Food labelling was optional unless the product was fortified with proteins, minerals, or vitamins, as per the 1973 regulations of the Food and Drug Administration (FDA). The following important step requires all processed food products to show uniform nutrition data, and it was the NLEA that was passed into law in 1990. Additionally, it gave the FDA permission to establish guidelines for health and nutrient content claims (Silverglade, 1996).
4. *Food Safety Modernization Act (FSMA) (2011).* The Food and Drug Administration's (FDA) regulations governing the safety of imported and domestically produced foods have been updated significantly and broadly by the US Food Safety Modernization Act (FSMA). The US Congress gives the FDA more authority under FSMA, and it is directed to create laws that will force the food sector to concentrate on the prevention of foodborne diseases (DeWaal & Plunkett, 2013).
5. *Meat Inspection Act (1906).* This Act, along with the Poultry Products Inspection Act (1957), established requirements for the inspection of meat and poultry products to ensure they are sanitary and correctly labelled (MILLER, 1956).
6. *Bioterrorism Act (2002).* The Bioterrorism Act of 2002 is formally called the "Public Health Security and Bioterrorism Preparedness and Response Act of 2002." It was enacted in the United States in response to the 2001 anthrax attacks and the growing threat of bioterrorism following the September 11 attacks. The Act aimed to enhance the country's capability to respond to and prepare for bioterrorism and related public health emergencies (Martin, 2004). Part of this Act includes provisions for the safety of food supply and the requirement for food facilities to register with the FDA. It also includes measures for tracking and tracing food products (Dias et al., 2010; Martin, 2004).

16.4.3.2 International Standards

1. *General Food Law Regulation (EC) 178/2002 (European Union).* This regulation created the European Food Safety Authority (EFSA), laid out the broad guidelines and criteria of food law, and outlined the processes for handling food safety–related matters within the EU. The European "vertical" and "horizontal" regulations and directives specify the statutory data that must be included on food labels. These 'horizontal' regulations specifically apply to food products that are meant to be supplied to mass caterers (restaurants, hospitals, canteens, etc.), as well as those that are to be provided as such to the final customer (retail stage). These include

fundamental standards for quality and safety, ingredient composition, shelf life, and storage conditions.

Concurrently, precise regulations were developed that applied "vertically" to certain products, including establishing food names and content, registering origin or distinctiveness, grading quality, assisting farmers, and stabilizing markets (Cheftel, 2005).

2. *Food Safety and Standards Act (2006) (India).* The Food and Safety Standard Act (FSSA) 2006 established the Food Safety and Standards Authority of India (FSSAI), which has the jurisdiction to oversee food regulation compliance in India (Ajay, 2018; Dhara et al., 2021). FSSAI unifies the food-related laws under FSSA; controls food production, processing, distribution, storage, and importation; establishes guidelines for food items; limits the use of additive foods, toxins, antibiotics, pesticides, and other substances; raises consumer awareness; and guarantees access to nutritious, safe food for human consumption, along with related issues or matters (Dhara et al., 2021).
3. *Food Safety Law (China).* China's Food Safety Law is comprehensive and includes provisions for food safety supervision, inspection, and penalties for violations. On February 28, 2009, China's National People's Congress Standing Committee passed the Food Safety Law (FSL) of the People's Republic (PR) of China, which was enacted on June 1, 2009 (Jia & Jukes, 2013)—the fundamental law that guarantees food safety and quality and safeguards consumers' health. It was significantly revised on April 24, 2015, and enacted on October 1, 2015, to enhance food safety oversight (Geng et al., 2015).

16.4.4 Food Safety Regulations: Promoting Public Health and Consumer Confidence

Food safety regulations are crucial in promoting public health and consumer confidence. These regulations aim to minimize the risk of food-associated illnesses, contamination, and fraud by establishing standards and procedures for handling, processing, and distributing food. Consumer alertness of food safety and nutrition is key to leading healthy lives and preventing disease. Numerous foodborne diseases have been linked to improper food management by consumers. People must be prepared to alter their eating and storage habits to lower the chance of contracting a foodborne illness. A high correlation exists between consumer understanding of appropriate food handling techniques and changes in such behaviors (Losasso et al., 2012). Building customer trust in food safety is one of the first concerns for food safety agencies (De Jonge et al., 2010). The security and nutritional content of food products have a direct impact on people's quality of life. Nevertheless, growing and widely reported outbreaks of foodborne illnesses harm consumers' trust in the nutritional content, safety, and caliber of the food products they buy.

As a result, customers are asking for more information about the ingredients and country of origin of the food goods they buy to eat. This has increased the significance of tracking a food product from the point of origin to the grocery store throughout the supply chain. An increasingly popular tool for surveillance of food supplies and enhancing the standards and safety of food items through the supply chain is traceability (Ayokanmbi & Oluwoye, 2020). Regarding the European Union, *traceability* is defined as "the capacity to monitor any food, feed, animal raised for food, or material intended for human consumption throughout every phase of its creation, processing, and delivery" (Arvanitoyannis, 2016). It is critical to track changes in consumer confidence over time to evaluate the effectiveness of food security organizations' initiatives to enhance the risk analysis procedure and shield the public from food hazards (De Jonge et al., 2010). Here is a breakdown of how food safety regulations contribute to these goals: Food goods must be labelled accurately and informatively concerning ingredients, allergies, and nutrition. This aids customers in making wise decisions. Best-by and expiration dates are required on labels to assist consumers in maintaining food safety at home. Additional guarantees of food safety and quality come from certifications, like those from the Safe Quality Food (SQF) program or the Hazard Analysis Critical Control Point (HACCP) program (Guo et al., 2019). Respect for international standards, such as those set forth by *Codex Alimentarius*, guarantees that food safety procedures meet international norms, promote trade internationally, and build consumer confidence (Masson-Matthee, 2007). Since people cannot evaluate food safety before consuming it, mass media can significantly impact fostering or eroding consumer confidence in food safety. Therefore, consumers heavily rely on information from others when assessing and evaluating food safety. Studies on media coverage of food safety issues have mostly focused on particular food incidents and products (De Jonge et al., 2010). Governments and agencies seek to safeguard public health, avert foodborne illnesses, and foster consumer confidence in the food supply by putting these policies into effect and upholding them.

16.5 IMPORTANCE OF PATENTING IN FOOD SAFETY AND PROTECTION OF COMPANIES' INTELLECTUAL PROPERTY RIGHTS (IPR)

Food safety patenting and intellectual property rights (IPR) protection are critical to innovation, safeguarding unique technologies, and maintaining customer confidence (Tansey, 2002). They play a crucial role in helping businesses maintain long-term growth in the food sector, avoid counterfeiting, and gain competitive advantages. Without strong patent protections, companies would be more vulnerable to losing their investment returns and jeopardizing progress in food safety. The term "intellectual property" describes the class of intangible rights that shield the highly valued creation of human intellect for economic purposes (Yamin, 2003). Through IPR, authors and inventors can prevent the unauthorized use or duplication of their creations. Strong IPRs are justified globally by offering the right reward, incentive, and security to encourage investment in research and development (R&D), resulting in discoveries that improve people's lives (Tansey, 2002). The recent geographical spread of IPRs is facilitated chiefly by the World Trade Organization (WTO) through Trade-Related Aspects of Intellectual Property Rights (TRIPs). IPRs, such as patents and plant breeders' rights, often aim to encourage the private sector to create seeds with improved yields or unique qualities to enhance food security and agrobiodiversity management (Cullet, 2004.) IPRs have grown in importance in various sectors during the last several decades. This includes, for example, agricultural biotechnology, where IPRs offer a fundamental incentive for the growth of the private sector in this field. Because of the close link between agriculture and food security, the extension of IPRs to this sector is particularly important. IPRs in agriculture give rise to particular concerns about farmers' control over their knowledge and resources. Small and subsistence farmers often lose out to agribusinesses in the long run from IPRs because of the royalties farmers have to pay to obtain protected seeds and the associated restrictions on storing, replanting, and selling seed that has been conserved (Yamin, 2003). Theoretically, it seems imperative that plant breeders maintain control over their varieties for them to continuously develop, enhance, and modify their stock to meet shifting demands and environmental circumstances. IPR holders will have the same level of influence on farmers' ability to preserve and replant seeds as they have in nations like the United States, where contractual requirements frequently strengthen IPR protection (Hamilton, 2001). However, applying genetic use restriction technology would provide a special challenge since it would allow patent holders to make sure that farmers fully respect their rights. Patents safeguard businesses' intellectual property rights and advances in food safety. They guarantee legal protection, stimulate R&D investment, give a competitive advantage, and advance industry-wide improvements in food safety standards.

16.6 CONCLUSION

The intricate relationship between food testing agencies, acts, regulations, and patents is critical to make sure food products' safety, quality, and innovation. Food testing agencies, like the FDA and EFSA, are essential in evaluating food safety, protecting public health, and enforcing regulations that govern food production and distribution. Acts and regulations, like the Food Safety Modernization Act (FSMA) and the European Union's General Food Law, set the legal framework that governs food safety, emphasizing preventive controls and responsibility across the entire supply chain. Patents encourage innovation by protecting novel food technologies, processes, and ingredients while raising complex legal and ethical questions about intellectual property, access to essential food products, and public welfare. Balancing innovation with regulatory compliance ensures that advancements in the food sector benefit both consumers and industry stakeholders.

REFERENCES

Ahmad, S. R. (2003). Adverse drug event monitoring at the food and drug administration. *Journal of General Internal Medicine*, *18*(1), 57–60.

Ajay, V. (2018). *Food Safety Standards and its Growing Role in Recent Times in India* (SSRN Scholarly Paper 3130684).

Arvanitoyannis, I. S. (2016). *Authenticity of Foods of Animal Origin* (0 ed.). CRC Press.

Asirvatham, J., McNamara, P. E., & Baylis, K. (2017). Informational campaign effects of the Nutrition Labeling and Education Act (NLEA) of 1990 on diet. *Cogent Social Sciences*, *3*(1), 1327684.

Attrey, D. P. (2017). Chapter 12—Role of public health food safety laboratories in detection of adulterants/contaminants. In R. K. Gupta, Dudeja, & Singh Minhas (Eds.), *Food Safety in the 21st Century* (pp. 161–175). Academic Press.

Awuchi, C. G. (2023). HACCP, quality, and food safety management in food and agricultural systems. *Cogent Food & Agriculture*, *9*(1), 2176280.

Ayokanmbi, F. M., & Oluwoye, J. (2020). Improving consumer confidence in food safety and nutritional quality. *Journal of Multidisciplinary Engineering Science and Technology*, *7*(9).

Baines, J., Crerar, S., & Johnson, T. (2003). Evaluating the effectiveness of the joint Australia New Zealand. *Food Standards Code. Food Control*, *14*(6), 439–446.

Bietlot, H. P., & Kolakowski, B. (2012). Risk assessment and risk management at the Canadian Food Inspection Agency (CFIA): A perspective on the monitoring of foods for chemical residues. *Drug Testing and Analysis*, *4*(S1), 50–58.

Borchers, A., Teuber, S. S., Keen, C. L., & Gershwin, M. E. (2010). Food safety. *Clinical Reviews in Allergy & Immunology*, *39*(2), 95–141.

Bou-Mitri, C., Mahmoud, D., El Gerges, N., & Jaoude, M. A. (2018). Food safety knowledge, attitudes and practices of food handlers in lebanese hospitals: A cross-sectional study. *Food Control*, *94*, 78–84.

Burlingame, B., & Pineiro, M. (2007). The essential balance: Risks and benefits in food safety and quality. *Journal of Food Composition and Analysis*, *20*(3), 139–146.

Chapman, R. A., & Morrison, A. B. (1966). Regulatory agencies and food safety. *Canadian Medical Association Journal*, *94*(12), 609–613.

Charlebois, S., & Summan, A. (2015). A risk communication model for food regulatory agencies in modern society. *Trends in Food Science & Technology*, *45*(1), 153–165.

Cheftel, J. C. (2005). Food and nutrition labelling in the European Union. *Food Chemistry*, *93*(3), 531–550.

Collins, Á. B., Floyd, S., Gordon, S. V., & More, S. J. (2022). Prevalence of *Mycobacterium bovis* in milk on dairy cattle farms: An international systematic literature review and meta-analysis. *Tuberculosis*, *132*, 102166.

Cullet, P. (2004). *Food security and intellectual property rights in developing countries*. RIBios (Réseau interdisciplinaire biosécurité): Institut universitaire d'études du développement (IUED).

Dawson, R. J. (1995). The role of the Codex Alimentarius Commission in setting food standards and the SPS agreement implementation. *Food Control*, *6*(5), 261–265.

De Jonge, J., Van Trijp, H., Renes, R. J., & Frewer, L. J. (2010). Consumer confidence in the safety of food and newspaper coverage of food safety issues: A longitudinal perspective. *Risk Analysis*, *30*(1), 125–142.

De Oliveira Mota, J., Boué, G., Prévost, H., Maillet, A., Jaffres, E., Maignien, T., Arnich, N., Sanaa, M., & Federighi, M. (2021). Environmental monitoring program to support food microbiological safety and quality in food industries: A scoping review of the research and guidelines. *Food Control*, *130*, 108283.

DeWaal, C. S., & Plunkett, D. W. (2013). The food safety modernization act. *Food Protection Trends*, *328*.

Dhara, D., Biswas, S., Das, S. K., & Biswas, O. (2021). Status of food safety and food security in India in the perspective of FSSAI. *Indian Journal of Animal Health*, *60*(2-Spl). https://doi.org/10.36062/ijah.2021.spl.01821

Dias, M. B., Reyes-Gonzalez, L., Veloso, F. M., & Casman, E. A. (2010). Effects of the USA PATRIOT Act and the 2002 Bioterrorism Preparedness Act on select agent research in the United States. *Proceedings of the National Academy of Sciences*, *107*(21), 9556–9561. https://doi.org/10.1073/pnas.0915002107

Dietz, W. H. (2015). The response of the US Centers for Disease Control and Prevention to the obesity epidemic. *Annual Review of Public Health*, *36*, 575–596. https://doi.org/10.1146/annurev-publhealth-031914-122415

Djekic, I., Dimitrijevic, B., & Smigic, N. (2023). Consumer complaints associated with food quality. *Journal of Consumer Protection and Food Safety*, *18*(1), 57–69.

Dua, P. K., Naik, R., & Jain, S. (2023). Standardization in AYUSH systems: Bureau of Indian Standards (BIS) and International Organization for Standardization (ISO) perspectives. *International Journal of Ayurveda Research*, *4*(1), 5.

Elson, R., Awofisayo-Okuyelu, A., Greener, T., Swift, C., Painset, A., Amar, C. F. L., Newton, A., Aird, H., Swindlehurst, M., Elviss, N., Foster, K., Dallman, T. J., Ruggles, R., & Grant, K. (2019). Utility of whole genome sequencing to describe the persistence and evolution of *Listeria monocytogenes* strains within crabmeat processing environments linked to two outbreaks of listeriosis. *Journal of Food Protection*, *82*(1), 30–38.

Eskola, M., Altieri, A., & Galobart, J. J. W. M. J. (2018). Overview of the activities of the European food safety authority on mycotoxins in food and feed. *World Mycotoxin Journal*, *11*(2), 277–290.

Fox, E., Hunt, K., O'Brien, M., & Jordan, K. (2011). *Listeria monocytogenes* in Irish Farmhouse cheese processing environments. *International Journal of Food Microbiology*, *145*, S39–S45.

Fung, F., Wang, H.-S., & Menon, S. (2018). Food safety in the 21st century. *Biomedical Journal*, *41*(2), 88–95.

Garvey, M. (2019). Food pollution: A comprehensive review of chemical and biological sources of food contamination and impact on human health. *Nutrire*, *44*(1), 1.

Geng, S., Liu, X., & Beachy, R. (2015). New Food Safety Law of China and the special issue on food safety in China. *Journal of Integrative Agriculture*, *14*(11), 2136–2141.

Ghosh, D. (2014). Food safety regulations in Australia and New Zealand Food Standards. *Journal of the Science of Food and Agriculture*, *94*(10), 1970–1973.

Gladen, B. C., & Rogan, W. J. (1991). Effects of perinatal polychlorinated biphenyls and dichlorodiphenyl dichloroethene on later development. *The Journal of Pediatrics*, *119*(1 Pt 1), 58–63.

Grob, K. (2019). The role of the European Food Safety Authority (EFSA) in a better European regulation of food contact materials—some proposals. *Food Additives & Contaminants: Part A*, *36*(12), 1895–1902.

Guo, Z., Bai, L., & Gong, S. (2019). Government regulations and voluntary certifications in food safety in China: A review. *Trends in Food Science & Technology*, *90*, 160–165.

Hamilton, N. D. (2001). Legal issues shaping society's acceptance of biotechnology and genetically modified organisms. *Drake Journal of Agricultural Law*, *6*, 81.

Henson, S., & Heasman, M. (1998). Food safety regulation and the firm: Understanding the compliance process. *Food Policy*, *23*(1), 9–23.

Hoffman, R. E., Greenblatt, J., Matyas, B. T., Sharp, D. J., Esteban, E., Hodge, K., & Liang, A. (2005). Capacity of state and territorial health agencies to prevent foodborne illness. *Emerging Infectious Diseases*, *11*(1), 11–16. https://doi.org/10.3201/eid1101.040334

Hurt, E. (2002). Nutrition labelling: European Union and United Kingdom perspectives. *Asia Pacific Journal of Clinical Nutrition*, *11*(2), S77–S79.

Hutt, P. B. (1973). Philosophy of regulation under the federal food, drug and cosmetic act. *Food, Drug, Cosmetic Law Journal*, *28*(3), 177–188.

Jairath, M. S., & Purohit, P. (2013). Food safety regulatory compliance in India: A challenge to enhance agri-businesses. *Indian Journal of Agricultural Economics*, *68*(3), 431–448.

Jia, C., & Jukes, D. (2013). The national food safety control system of China—a systematic review. *Food Control*, *32*(1), 236–245.

Journals (www.scholarconsult.com), S. (n.d.). *The Role of Regulatory Agencies in Food Quality Control in Nigeria*. Retrieved August 28, 2024, from https://www.academia.edu/6016428/The_Role_of_Regulatory_Agencies_in_Food_Quality_Control_in_Nigeria

King, T., Cole, M., Farber, J. M., Eisenbrand, G., Zabaras, D., Fox, E. M., & Hill, J. P. (2017). Food safety for food security: Relationship between global megatrends and developments in food safety. *Trends in Food Science & Technology*, *68*, 160–175.

Kotsanopoulos, K. V., & Arvanitoyannis, I. S. (2017). The role of auditing, food safety, and food quality standards in the food industry: A review. *Comprehensive Reviews in Food Science and Food Safety*, *16*(5), 760–775.

Kubde, S., Pattankar, J., & Kokiwar, P. (2016). Knowledge and food hygiene practices among food handlers in food establishments. *International Journal of Community Medicine and Public Health*, 251–256.

Lipcsei, L., & Kambhampati, A. (2016). Improving food safety through prevention: CDC's food safety prevention status report. *Journal of Environmental Health*, *79*(2), 46–48.

Liu, S., Xie, Z., Zhang, W., Cao, X., & Pei, X. (2013). Risk assessment in Chinese food safety. *Food Control*, *30*(1), 162–167.

Llobet, J. M., Falcó, G., Casas, C., Teixidó, A., & Domingo, J. L. (2003). Concentrations of arsenic, cadmium, mercury, and lead in common foods and estimated daily intake by children, adolescents, adults, and seniors of Catalonia, Spain. *Journal of Agricultural and Food Chemistry*, *51*(3), 838–842.

Losasso, C., Cibin, V., Cappa, V., Roccato, A., Vanzo, A., Andrighetto, I., & Ricci, A. (2012). Food safety and nutrition: Improving consumer behaviour. *Food Control*, *26*(2), 252–258.

Martin, W. (2004). Legal and public policy responses of states to bioterrorism. *American Journal of Public Health*, *94*(7), 1093–1096.

Masson-Matthee, M. D. (2007). *The Codex Alimentarius Commission and Its Standards*. T.M.C. Asser Press.

Mepham, B. (2011). Food additives: An ethical evaluation. *British Medical Bulletin*, *99*(1), 7–23.

Miller, A. R. (1956). The federal meat inspection act. *Food, Drug, Cosmetic Law Journal*, *11*(11), 565–571.

Nielsen, S. S. (2024). Introduction to food analysis. In B. P. Ismail & S. S. Nielsen (Eds.), *Nielsen's Food Analysis* (pp. 3–14). Springer International Publishing.

Nyor, J. T. (2014). The role of regulatory agencies in food quality control in Nigeria. *SCSR J. Agribus*, *1*, 01–05.

Panghal, A., Chhikara, N., Sindhu, N., & Jaglan, S. (2018). Role of Food Safety Management Systems in safe food production: A review. *Journal of Food Safety*, *38*(4), e12464.

Patil, S. M., Ramu, R., Shirahatti, P. S., Prasad, M. N. N., Kumari, V. B. C., Sujay, S., & Tejaswini, M. (2023). Global food safety and quality: A worldwide scenario. In *Food Microbial and Molecular Biology*. Apple Academic Press.

Pei, X., Li, N., Guo, Y., Liu, X., Yan, L., Li, Y., Yang, S., Hu, J., Zhu, J., & Yang, D. (2015). Microbiological food safety surveillance in China. *International Journal of Environmental Research and Public Health*, *12*(9), Article 9.

Poli, S. (2004). The European Community and the adoption of international food standards within the Codex Alimentarius Commission. *European Law Journal*, *10*(5), 613–630.

Powell, D. A., Erdozain, S., Dodd, C., Costa, R., Morley, K., & Chapman, B. J. (2013). Audits and inspections are never enough: A critique to enhance food safety. *Food Control*, *30*(2), 686–691.

Pressman, P., Clemens, R., Hayes, W., & Reddy, C. (2017). Food additive safety: A review of toxicologic and regulatory issues. *Toxicology Research and Application*, *1*, 2397847317723572.

Radu, E., Dima, A., Dobrota, E. M., Badea, A.-M., Madsen, D. Ø., Dobrin, C., & Stanciu, S. (2023). Global trends and research hotspots on HACCP and modern quality management systems in the food industry. *Heliyon*, *9*(7), e18232.

Rahman, M., Hasan, M. S., Islam, R., Rana, R., Sayem, A. S. M., Sad, M. A. A., Matin, A., Raposo, A., Zandonadi, R. P., Han, H., Ariza-Montes, A., Vega-Muñoz, A., & Sunny, A. R. (2022). Plasma-activated water for food safety and quality: A review of recent developments. *International Journal of Environmental Research and Public Health*, *19*(11), Article 11.

Rajasekharan, R. (2015). Central food technological research institute, Mysuru. *Current Science, 108*(8), 1438–1440.

Röhr, A., Lüddecke, K., Drusch, S., Müller, M. J., & Alvensleben, R. v. (2005). Food quality and safety—consumer perception and public health concern. *Food Control, 16*(8), 649–655.

Savelli, C. J., & Mateus, C. (2020). Looking inside the international food safety authorities network community website. *Journal of Food Protection, 83*(11), 1889–1899.

Savov, A. V., & Kouzmanov, G. B. (2009). Food quality and safety standards at a glance. *Biotechnology & Biotechnological Equipment, 23*(4), 1462–1468.

Silverglade, B. A. (1996). The nutrition labeling and education act—progress to date and challenges for the future. *Journal of Public Policy & Marketing, 15*(1), 148–150.

Simmons, C. K., & Wiedmann, M. (2018). Identification and classification of sampling sites for pathogen environmental monitoring programs for *Listeria monocytogenes*: Results from an expert elicitation. *Food Microbiology, 75*, 2–17.

Tansey, G. (2002). Patenting our food future: Intellectual property rights and the global food system. *Social Policy & Administration, 36*(6), 575–592.

Turner, P. J., Andoh-Kesson, E., Baker, S., Baracaia, A., Barfield, A., Barnett, J., Brunas, K., Chan, C.-H., Cochrane, S., Cowan, K., Feeney, M., Flanagan, S., Fox, A. T., George, L., Gowland, M. H., Heeley, C., Kimber, I., Knibb, R., Langford, K., . . . O'Brien, J. (2021). Identifying key priorities for research to protect the consumer with food hypersensitivity: A UK Food Standards Agency priority setting exercise. *Clinical & Experimental Allergy, 51*(10), 1322–1330.

Yamin, F. (2003). *Intellectual property rights, biotechnology and food security*. Institute of Development Studies.

Zanabria, R., Racicot, M., Cormier, M., Arsenault, J., Ferrouillet, C., Letellier, A., Tiwari, A., Mackay, A., Griffiths, M., Holley, R., Gill, T., Charlebois, S., & Quessy, S. (2018). Selection of risk factors to be included in the Canadian Food Inspection Agency risk assessment inspection model for food establishments. *Food Microbiology, 75*, 72–81.

Zeng, H., De Reu, K., Gabriël, S., Mattheus, W., De Zutter, L., & Rasschaert, G. (2021). Salmonella prevalence and persistence in industrialized poultry slaughterhouses. *Poultry Science, 100*(4), 100991.

Zoellner, C., Ceres, K., Ghezzi-Kopel, K., Wiedmann, M., & Ivanek, R. (2018). Design elements of *Listeria* environmental monitoring programs in food processing facilities: A scoping review of research and guidance materials. *Comprehensive Reviews in Food Science and Food Safety, 17*(5), 1156–1171.

17 Microbial Contribution to Fermented Foods and Quality Enhancement

Indrani Gogoi, Shyamalima Saikia, and Pankaj Chetia

17.1 INTRODUCTION

Microorganisms are crucial for human health, bioenergy, and food production owing to their remarkable stability, resilience, and adaptability. Microorganisms are pivotal in converting substrates into fermented products with diverse metabolic pathways. Microbial diversity is essential for producing unique flavors, aromas, and textures of fermented foods, as it enhances their sensory appeal and shelf life. Diverse microbial cultures and different plant- and animal-based formulations have been used to produce different ferments around the world for many years. Fermented foods and beverages are considered to be one of the oldest processed foods consumed by humans. Historically, fermentation techniques had been integrated into the cultural heritage of numerous civilizations, which originated in the Middle East and were swiftly disseminated to Western regions (Hugenholtz et al., 2013). Some other evidences suggest the gathering and consumption of fermented mushrooms in China approximately 6,000 years ago (Wang, 1989). Similarly, the extensive uses of fermented vegetable products, bread, and beverages have been traced back to a pan-continental origin, suggesting that these techniques were developed across vast geographical areas and highlighting their importance in the global culinary landscape. However, fermented dairy products (yoghurt, sweet and savory fermented milk) emerged independently in Europe, the Middle East, and the Indian subcontinents (Campbell-platt, 1994). Earlier, fermented foods were highly valued because of their shelf life and organoleptic properties. However, people were not aware of the fermentation process and the microbial involvement during that period, and the principles of fermenting food products were conveyed verbally to subsequent generations and utilized in practice. Subsequently, studies on the nutritional and functional values of such products facilitated by enzymatic transformation have offered a deeper understanding of the process and helped attain more interest among researchers. With the advent of genetic and metabolic engineering and synthetic biology, researchers can improve microbial productivity for various biotechnological processes. However, high-throughput screening and sequencing studies have opened up an avenue for determining the microbial diversity of fermented foods.

DOI: 10.1201/9781003540113-17

Fermentation is a traditional biotechnology that evolved from natural methods of enhancing vitamins, essential amino acids, and proteins and improving visual appeal, taste, and aroma in the presence of microorganisms (Campbell-Platt, 1994). The fermented food microbiome comprises all microorganisms that inhabit the food throughout the whole life of the product, from the raw ingredients to the final consumed states (Louw et al., 2023). Microorganisms involved in this process play an active role by showing changes in the chemical and physical characteristics of the foods. The oxygen requirement may vary when converting suitable substrates into a desirable product. Therefore, "fermentation" should not be confined exclusively to anaerobic dissimilation, as commonly understood in conventional definitions (Maxon, 1955). The fermentation process harnesses the activity of diverse microbial communities, enhancing the preservation and safety of foods and significantly improving their nutritional properties and sensory qualities (Selhub et al., 2014). For example, lactic acid bacteria, often found in various fermented foods, play a crucial role in enhancing both flavor and texture. Simultaneously, they also inhibit the growth of contaminants by producing antimicrobial compounds using allelopathic activity to help prolong shelf life and promote food safety (Bertuzzi et al., 2018; Xiang et al., 2020). The varied microbial communities in fermented foods stem from intricate ecological and metabolic interactions shaped by numerous factors, including substrate, temperature, pH, and fermentation duration. These interactions regulate the fermentation process and determine the qualities of end products. Consequently, these characteristics can differ significantly among fermented foods due to the unique contributions of specific microorganisms (Paul Ross et al., 2002).

17.2 MICROBIAL CONSORTIA OF FERMENTED FOODS AND BEVFRAGES

17.2.1 Diversity of Microbial Communities in Fermented Foods and Beverages

For millennia, humans have leveraged the transformative ability of microbial fermentation to improve, preserve, and diversify their culinary traditional practices, leading to an extensive array of fermented food products that delight the senses and provide numerous health benefits (Selhub et al., 2014). This diversity arises from the definite microbial communities associated with fermentation processes, which vary significantly across the types and other functional properties of the food (Leeuwendaal et al., 2022). They serve a multifaceted role in fermentation within any naturally occurring ecosystem. While lactic acid bacteria (LAB) are primarily recognized for dairy product fermentations, their significance in non-dairy product fermentation was explored by analyzing their contributions to fermented fish products and vegetable yoghurts during the early 1990s. In rice wine and beer fermentation, they also play essential roles. Among lactic acid bacteria, the species of *Lactobacillus*, *Enterococcus*, *Leuconostoc*, *Pediococcus*, *Weissella*, etc. are commonly associated with fermented foods and beverages. *Leuconostoc mesenteroides* have displayed their role in fermenting steamed rice bread, salted fish products, and cereals from South Asian countries (Rhee et al., 2011). *L. mesenteroides* exhibits a faster growth rate in plant materials compared to other LAB, regardless of temperatures and salt concentrations, and also promotes the proliferation of other LAB by generating CO_2 and short-chain fatty acids (SCFA), which alter the surrounding environment (Kumar et al., 2013; Rhee et al., 2011). Another LAB, *Lactobacillus plantarum* AS1, having probiotic characteristics, isolated from South Indian fermented food kallappam showed great adhesiveness on HT-29 cell lines and prevented colonization of pathogenic *Vibrio paramaelyticus* (Kumar et al., 2013). A recent study demonstrated that fermenting blueberry juice with a combination of LAB strains, including *Levilactobacillus brevis*, *Lactiplantibacillus plantarum*, *Pediococcus acidilactici*, *Lacticaseibacillus casei*, and *Lacticaseibacillus rhamnosus*, improved the quality of fruit harvest and extended the shelf life (Lugo-zarate et al., 2024). In industries, commercial LAB strains are used as starter cultures in processes that improve the quality of products by introducing beneficial changes and enhancing fermentation efficiency. Studies have confirmed that *Lactobacillus acidophilus* Bauer strain L0938 is effective in meat fermentation, where it enhances the shelf life of dry-cured meat by improving color stability and reducing oxidation (Stadnik et al., 2022). Some important modes for improving the quality of fermented foods through microorganisms are illustrated in Figure 17.1 and detailed in Table 17.1.

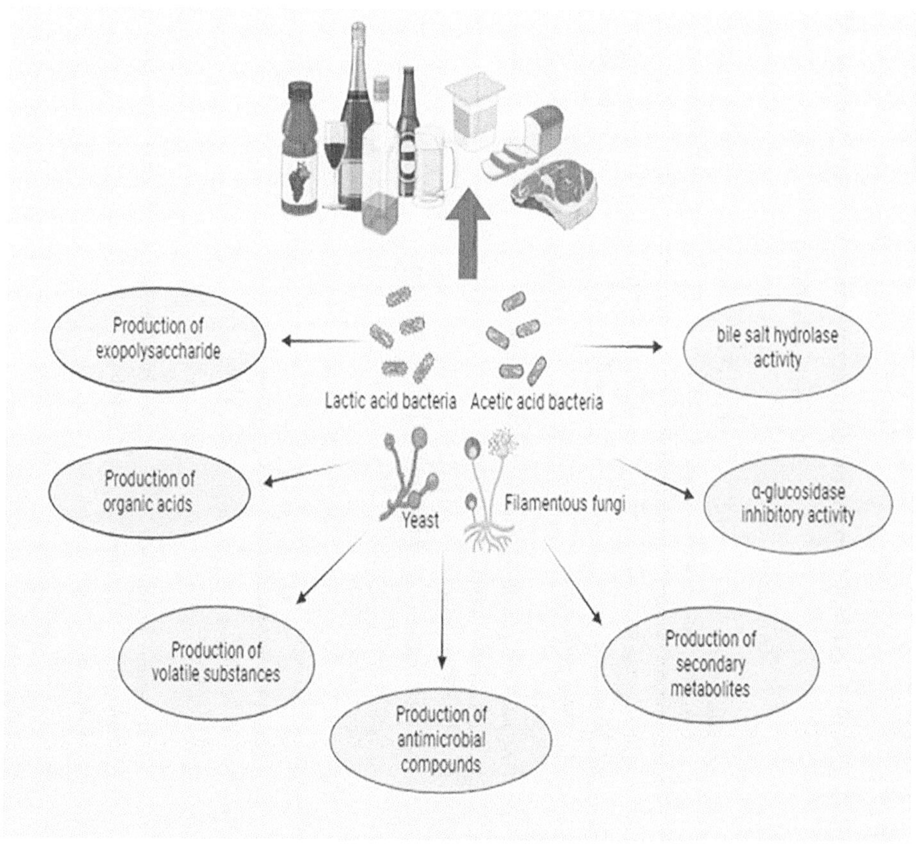

FIGURE 17.1 Different modes of nutritional and functional quality enhancement of microorganisms.

TABLE 17.1
Role of Different Microbial Communities in Fermenting Various Foods and Beverages

Community	Organisms	Fermented Foods/ Beverages	Category	Microbial Role in Fermentation	References
LAB	*Lactobacillus kefiranofaciens*, *L. kefiri*, *Leuconostoc mesenteroides*, *Lactococcus lactis*, *Acetobacter* sp.	Kefir (France, Ireland, UK)	Acidic beverages; obtained from fermented milk	Act as starter culture; improve nutritional properties	(Wang et al., 2022; Yilmaz et al., 2022)
	Lactobacillus plantarum, *L. fermentum*, *Corynebacterium* spp., *Acetobacter* spp.	Ogi (Western Africa, mainly Nigeria)	Fermented millet or maize or sorghum product	Potential starter culture	(Oyedeji et al., 2013)
	Lactobacillus plantarum, *Lactobacillus paraplantarum*, *Pediococcus stilesii*, *P. pentosaceus*	Meju (Korea)	Fermented soybean block	Reduction of an organic and volatile fatty amine, trimethylamine (TMA)	(Park et al., 2020)
	Lactobacillus sakei, *L. alimentarius*, *L plantarum*, *Lactococcus lactis*, *Weissella hellenica*	Harbin (Northeast China)	Naturally dry fermented sausage	Enhance flavor and produce volatile compounds, such as (E,E)-2,4-decadienal, nonanal, hexanal, octanal, 1-often-3-ol, and benzaldehyde.	(Hu et al., 2022)
	Species of *Lactobacillus*, *Micrococcus*, and *Staphylococcus*	Suanyu (Northeast China)	Fermented fish product	Contributed to the production of lactic, malic, acetic, and tartaric acids during fermentation	
	Lactobacillus fermentum, *L. rhamnosus*, *L. helveticus*, *L. acidipiscis*	Nakham (Meghalaya, India)	Fermented fish product	Production of exopolysaccharide, known for having probiotic attributes, anti-bacterial activity	(Das et al., 2020)

AAB	*Komagataeibacter rhaeticus*, *Acetobacter must*, and *Gluconobacter potus*	Kombucha (New Zealand, East and Central Asia)	Fermented tea beverage	Potential probiotic strains having antioxidant, antimicrobial, and bile salt hydrolase activity	(Wang et al., 2024)
	Acetobacter indonesiensis, *A. cerevisiae*, *A. orientalis*, *A. tropicalis*, *A. labarum*, *A. pasteurianus*, and *A. syzygii*	Mulberry fruit vinegar (Korea)	Traditional fermented vinegar prepared from *Cudrania tricuspidata*	High acetic acid producers; α-glucosidase inhibitory activity	(Yim et al., 2015)
	Acetobacter lambici, *A. orientalis*, *Gluconobacter cerevisiae*	Lambic beer (Belgium)	Alcoholic acidic beer with barley malt	Responsible for acidic flavor	(De Roos and De Vuyst, 2018)
Yeast	*Pichia anomala*, *Candida glabrata*, and *Saccharomycopsis fibuligera*	Kodo ko jaanr	Alcoholic fermented beverage produced from finger millet	Act as probiotics	(Ilango and Antony, 2021)
	Saccharomyces cerevisiae, *S. fibuligera*, *Candida tropicalis*, *Pichia kudriavzevii*	Chhang and Jau chhang (Himachal Pradesh, India)	Mild alcoholic sweet beverages made from barley	Act as an amylolytic starter	(Thakur et al., 2015)
	S. cerevisiae, *S. bayanus*, *S. fibuligera*, *S. capsularis*, *Pichia anomala*, *P. burtonii*,	Jhand, Rakshi, Huaun thon (Nepal)	Undistilled alcoholic beverages made from rice, maize, wheat, millet, etc.	Serve as predominant microbes present in starter murcha	(Anal, 2019)
	Endomycopsis and *Saccharomyces*	Sato (Thailand)	Traditional beverage prepared from glutinous rice	Act as a starter culture	
	Schizosaccharomyce pombe, *Saccharomycodes ludwiggi*, *Kloeckera apiculate*, *Saccharomyces cerevisiae*, *Zygosaccharomyces bailii*, *Brettanomyces bruxellensis*, *B. lambicus*, *B. custersii*, *Candida* spp., and *Pichia* spp.	Kombucha (East and Central Asia)	Fermented tea	Convert sucrose into fructose and glucose and finally produce ethanol	

In the microbial niche involved in alcoholic fermentation, yeast, *Saccharomyces cerevisiae*, is predominant, and its capability to convert sugar to ethanol and CO_2 under anaerobic and aerobic conditions forms the fundamental principle of this fermentation process (Dashko et al., 2014). It can generally produce beverages with an alcoholic content of 6–14% from various sources of sugars, including grapes, barley, etc., that give distinct flavor and aroma. For example, wine and cider are produced from grapes and apple juice, whereas beer is made from liquid wort-malt from germinated barley. Palm juice, agave juice, beetroot, sugarcane, potato, and honey are important sugar sources of indigenous alcoholic fermentation in different regions worldwide. Both wild and cultivated yeast strains are used in alcoholic fermentation (Leeuwendaal et al., 2022). Alcoholic beverages also harbor a variety of other microorganisms aside from *S. cerevisiae*, particularly LAB and non-*Saccharomyces* yeasts (*Hanseniaspora*, *Lachancea*, *Kluyveromyces*, *Metschnikowia*, *Pichia*, *Schizosaccharomyces*, *Starmerella*, and *Torulaspora*), which are prominent during spontaneous fermentation (Binati et al., 2020; Torres et al., 2021; Tufariello et al., 2021). Many alcoholic beverages are produced by amylolytic starters in the presence of Ascomycete strains *Aspergillus oryzae* and *A. sojae* (Tamang et al., 2016).

Acetic acid fermentation is an essential process mediated by microbes, facilitated by alcohol dehydrogenase and aldehyde dehydrogenase. Acetic acid bacteria (AAB) oxidize ethanol into acetic acid, making them valuable for the industrial production of vinegar, vitamin C, and cellulose. The oxidative metabolism of ABA, such as *Escherichia coli*, *Acetobacter pasteurianus*, *Gluconobacter* spp., etc., was involved in traditional rice vinegar production (Nanda et al., 2001). Members of the Acetobacteraceae family have been identified in various spontaneous fermentations, such as the cocoa bean fermentation, the production of sour beers, and other mildly acidic beverages, like kombucha, milk kefir, and water kefir. While AAB are often present in fermented foods and beverages, their occurrence and physiological roles still need to be understood due to challenges in cultivation and isolation. AABs remain viable in a low-oxygen environment but cannot be cultivated, which is a primary issue with laboratory cultivation (De Roos and De Vuyst, 2018). However, high-throughput sequencing, advanced identification techniques like MALDI-TOF MS, metagenomics studies, and culture-independent methods have made the cultivation and identification of these bacteria considerably more reliable and straightforward (De Roos and De Vuyst, 2018). An inter-delta/PCR, PCR-RFLP, ERIC-PCR, and 16s sequencing of Chinese traditional vinegar, "Shanxi aged vinegar," revealed the dominance of *Acetobacter pasteurianus* in the acetic acid fermentation stage, while other species of *Acetobacter*, viz., *A. senegalensis*, *A. indonesiensis*, *A. molarum*, *A. orientalis*, and *Gluconobacter oxydans*, were detected at the initial stage of acetic acid fermentation (Wu et al., 2012). Like LAB, AAB also interacts with a wide range of microbes to produce fermented foods with unique tastes and diversity of metabolites. In previous studies, it was reported that the inoculation of AAB with LAB and yeast decreases the pH of the culture and inhibits *Enterobacter* and *Klebsiella* contamination in the dough (Li et al., 2021).

17.2.2 Traditional Fermented Foods as a Reservoir of Microbial Community

The use of microorganisms in producing fermented foods from locally available plant and animal sources has been a practice dating back to prehistoric times. They are selected by adapting to the substrate and modifying the fermentation conditions. Filamentous fungi (*Aspergillus*, *Actinomucor*, *Mucor*, *Neurospora*, *Penicillium*, *Rhizopus*, *Monascus*, etc.), yeasts (*Candida*, *Geotrichum*, *Hansenula*, *Kluyveromyces*, *Pichia*, *Saccharomyces*, *Zygosaccharomyces*, etc.), and bacteria (LAB and AAB) are the three major types of microorganisms linked to traditional fermented foods (Tamang, 1998). In many East and Southeast Asian fermented foods, filamentous fungi are commonly present, while bacteria or mixed cultures of bacteria and yeast are more widely used to prepare fermented foods in Africa, America, and Europe (Geisen, 1993; Tamang, 1998). In the preparation of famous traditional fermented foods of Japan, miso, natto, and tempeh, which are famous for their longevity, also use mixed cultures of *Lactobacillus* spp., *Aspergillus oryzae*, *S. cerevisiae*, and *B. subtilis* (Negrete-romero et al., 2021). Owing to the diverse agro-climatic conditions and dietary traditions of various ethnic tribes, all three groups of microorganisms are involved in producing fermented foods and beverages in India (Tamang et al., 2020). Moreover, the decrease in pH during acidic fermentation hinders the proliferation of other microorganisms, establishing the basis for biopreservation, a phenomenon commonly observed in several fermented foods. During fermentation of traditional fermented olive greens, *Lactobacillus plantarum* produces an antifungal compound which inhibits the mycelial growth of *Candida pelliculosa* in pH ranges from 3.8 to 7.0, showing effectiveness in biopreservation (Abouloifa et al., 2021). Phenyllactic acid obtained from LAB such as *Lactobacillus*, *Leuconostoc*, and *Weissella* through phenylalanine metabolism has broad-spectrum inhibitory activity against yeast, fungi, bacteria, and molds (Mishra et al., 2022). Besides pH, bacterial count is a vital indicator of fermentation. A recent study revealed that in the popular fermented food "kimchi," LAB count increased quickly throughout fermentation and ended up at $\geq 1 \times 10^8$ CFU/mL. However, the highly acidic atmosphere and low temperature made it difficult for the bacteria to adapt, which altered the makeup of the microbial population (Song et al., 2021). Traditional fermentation enriches food substrates with proteins, amino acids, and vitamins, thereby boosting the nutritional value of the raw material. The Firmicutes present are renowned for their ability to produce exopolysaccharides, such as glucan, galactan, and dextran. Subsequent studies have shown that several strains of Firmicutes capable of producing exopolysaccharides play a critical role in enhancing the nutritional and functional properties of fermented foods (Kavitake et al., 2022). Furthermore, the temporal and spatial stability and resilience of microbial composition led to successful fermentation, resulting in high-quality foods. In Jiuqu (Chinese liquor), fermentation, moisture, and temperature are crucial elements, as lowering the fermentation temperature decreases the abundance of thermotolerant LAB, potentially affecting the generation of flavor compounds (Wang et al., 2022). Some of the well-known traditional fermented foods and beverages are listed in Table 17.1, along with their microbial compositions.

17.3 MICROBIAL INTERACTION WITH FERMENTED FOOD PRODUCTS

The intricate world of microbial interactions within plant-based fermented food products is a fascinating realm that has garnered attention in the scientific community. Fermentation processes often involve the concerted efforts of multiple microorganisms, each contributing unique metabolic capabilities and properties to the final product (Smid and Lacroix, 2013). During fermentation, the plant materials undergo specific modifications, leading to changes in their physical properties. These alterations can subsequently impact the nutritional and organoleptic properties of the fermented food products. Most fermentation processes in nature and industry are powered by simple or complex microbial populations (Smid and Lacroix, 2013). During fermentation, microbial interactions can occur by various mechanisms, which may be direct or indirect. Direct mechanism involves physical interaction or the action of signaling molecules, while indirect mechanism may involve other microbial strains that influence environmental factors to alter physiological parameters (Sieuwerts et al., 2008). These interactions can be either positive (commensalism, mutualism, cooperation) or negative (amensalism and antagonism), significantly affecting the organoleptic characteristics of the final products. Positive interactions typically involve the production of metabolites and secondary metabolites during fermentation, and advantages are derived from compounds released during yeast autolysis. Conversely, negative interaction arises from competition for available nutrients.

Years back, research revealed that interaction between *Lactobacillus bulgaricus* and *Streptococcus thermophilus* served as a crucial element in the manufacture of yoghurt and Italian and Swiss cheeses, where *L. bulgaricus* stimulates the acid production of *S. thermophilus* (Moon and Reinbold, 1976). Though LAB is ubiquitous in fermentation, the co-cultures of LAB are somewhat overlooked due to the paucity of studies on their interactions in the literature. However, studies are available on the concurrent existence of LAB with yeasts or, to a lesser extent, with AAB in different fermented foods. Similarly, the AAB (*Acetobacter lambic* and *Gluconobacter cerevisiae*) enhanced the acidic flavor of lambic beers with lactic acid–producing bacteria (De Roos and De Vuyst, 2018). Cultivating propionic acid bacteria (PAB) alongside LAB demonstrates a form of commensalism, as the lactic acid produced by LAB serves as an ideal carbon source for PAB growth. Co-culturing *Lactobacillus plantarum* SM39 (LAB) and *Propionibacterium freudenreichii* DF13 (PAB) can produce high levels of folate and vitamin B12, comparable to those from genetically engineered strains (Smid and Lacroix, 2013).

Other types of positive interaction found in fermentation are bidirectional, including mutualism and cooperation. The successful growth of microbes during fermentation implies an excellent example of positive interaction among them. For instance, in alcohol production, *Aspergillus* was the predominant strain in the initial stage, and *Bacillus*, *Cunninghamia lanceolata*, and *Listeria monocytogenes* functioned at the later stage. By fortifying certain molds, the succession can encourage the growth of advantageous staphylococci and prevent the spread of pathogens during the cheese-making process. The brewing process of Huangjiu, a Chinese rice wine, involves the crucial involvement of bacteria, molds, and yeasts to preserve the functional attributes of the product. While *S. cerevisiae* and *Aspergillus* are involved in the fermentation of alcohol and the acquisition of taste, the enzymes released by molds aid in the saccharification and liquefication processes (Wu et al., 2015; Yang et al., 2022). The synergistic interactions between LAB and yeast have also been observed in fermenting cereals and non-cereal-based traditional West African products. The coexistence of microbes does not work only on the substrates but also with each other to cause division of labor between them. The microbes exchange metabolites and molecular signaling in the presence of the substrates, which greatly differ according to the environment and compositions of the raw materials.

The production of antimicrobial compounds, probiotics, and antioxidants during fermentation can be viewed as counterproductive interactions. The metabolic versatility of bacteria is a boon for food preservation and biosafety. LAB strains derived from fermented vegetable juices, yak's yoghurt, and various other sources demonstrate notable antimicrobial activity against enteropathogens. LAB produces organic acids that penetrate the cytoplasm of bacterial pathogens. This process gradually lowers intracellular pH by accumulating ionized organic acids, ultimately leading to cell lysis (Zhang et al., 2011). Organic acids produced during fermentation, such as acetic and propionic acid, exhibit antagonistic properties that inhibit the growth of pathogenic bacteria, yeasts, and filamentous fungi. Another bacterial by-product that exerts broad-spectrum antagonistic activity against bacteria sharing the same niche is bacteriocin. Bacteriocins can be grouped as either heat-stable lantibiotics with low molecular weight (class I) or heat- and pH-stable non-lantibiotics with molecular weight <10 kDa (class II) (Smaoui et al., 2024). The mode of action of bacteriocin varies with its type. However, many studies have reported that class I bacteriocins like nisin, subtilin, and lactocin, as well as class II bacteriocins, viz., plantaricin, enterocin, pentocin, pediocin, etc., are produced by the LAB community of fermented foods (Sharma et al., 2022).

17.4 CHANGES IN MICROBIOLOGICAL AND ENZYMATIC PROPERTIES ASSOCIATED WITH FERMENTATION

Fermentation is considered a sustainable approach for maximizing available food resources to address the global food crisis. In 2017, fermented food comprising different species of microorganisms was listed as a superfood that provides nutritional advantages and increases product quality. The microbes used for fermentation supplemented in fermented foods are referred to as microbial factories, which participate in a series of biochemical reactions and ultimately lead to the formation of the desired product. For example, they induce hydrolytic reactions that release nutrients and bioactive molecules, such as amino acids and peptides, from the raw material. One example is the production of lactose-free or lactose-reduced products from fermented milk by the bacteria containing beta-galactosidases, breaking down lactose during fermentation (Xiang et al., 2019). LAB is used to manufacture fermented dairy products, such as yoghurt, shrikhand, kefir, koumiss, and cheese, which increases the content of galactose, lactic acid, free amino acid, vitamin B complex, and fatty acids compared to non-fermented milk (Figure 17.2). Similarly, rice fermented with

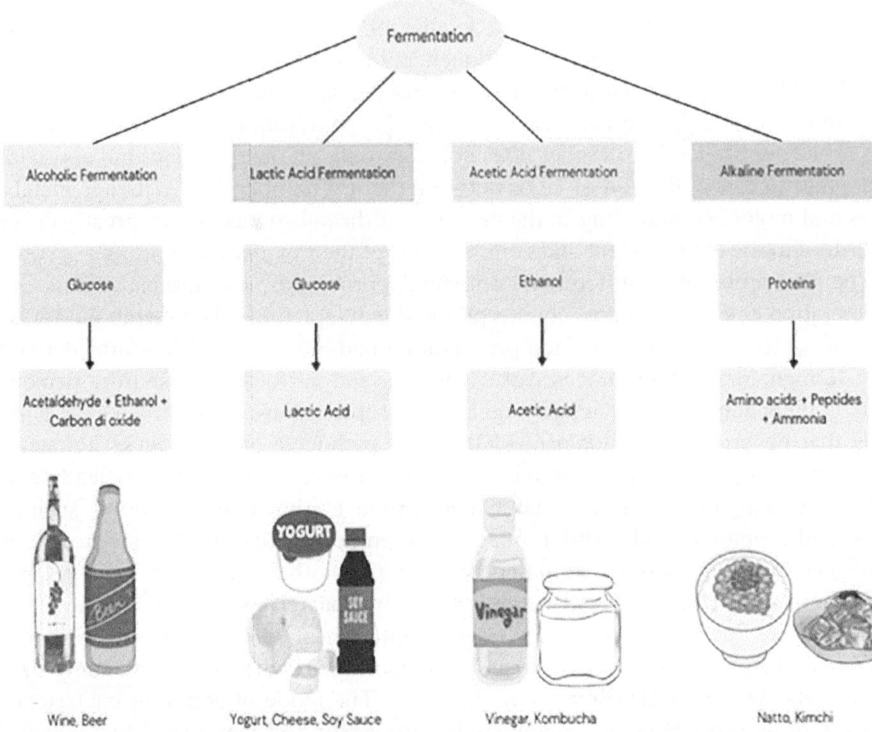

FIGURE 17.2 Significant types of fermentations, biosynthetic pathways, and products.

S. bouldardii was found to contain good levels of sterols, unsaturated fatty acids, phenolics, and vitamins, as well as other health-promoting properties, such as those against Alzheimer's disease, cardiovascular disease, diabetes, and cancer.

17.5 USE OF BENEFICIAL MICROBES IN UPGRADING THE NUTRITIONAL QUALITY OF FERMENTED FOOD

During fermentation, the metabolic activities of these complex microbial communities initiate various biochemical reactions, resulting in the production of multiple bioactive compounds, which impart new functional properties that positively impact on human health (Sharma et al., 2020). The microorganisms play a crucial role in preventing the formation of harmful chemical substances during fermentation (Skowron et al., 2022). The nutritional improvement in fermented foods is achieved through three primary mechanisms: catabolic and anabolic processes, the breakdown of complex compounds, and the synthesis of vitamins and other essential growth factors (Skowron et al., 2022). Important elements affecting the properties of fermented meals include the makeup of the substrates utilized and the particular bacteria that cause fermentation. Moreover, the period of the fermentation has a great role to play on how well the final product turns out (Knez et al., 2023).

17.5.1 Functional and Nutritional Properties of Fermented Food

Fermentation-based microorganisms contribute several functional properties, including antioxidant, antimicrobial, and probiotic effects, as well as fibrinolytic activity, the production of peptides and polyglutamic acid, and the breakdown of antinutritive compounds (Tamang et al., 2016). During fermentation, bacteria produce active peptides, vitamins, minerals, and enzymes, such as peptidases and proteinases, while also removing non-nutritive substances. Notably, specific peptides, including conjugated linoleic acids (CLA), bacteriocins, and sphingolipids, offer significant health benefits, such as reducing blood pressure, exhibiting potent antibacterial properties, and showing promising anticancer activities (Şanlier et al., 2019). Additionally, some bioactive peptides demonstrate antioxidant, antiallergenic, antimicrobial, anti-inflammatory, antidiabetic, and opioid-antagonistic properties (Şanlier et al., 2019).

Fermentation also enhances the bioavailability and absorption of amino acids, their derivatives, polyphenols, and minerals. Short-chain fatty acids (SCFAs), organic acids, and exopolysaccharides play a key role in stimulating the host's immune system by reducing inflammation, inhibiting pathogens, and maintaining the integrity of the gut barrier (Mukherjee et al., 2023). Moreover, enzymes generated during fermentation help break down biological macromolecules into simpler forms, thereby improving digestibility, facilitating the uptake of food particles, and boosting nutrient bioavailability (Tamang et al., 2016). For example, the digestibility of fermented proteins is enhanced by reducing inhibitors of digestive enzymes, such as trypsin and chymotrypsin, and by producing microbial proteases that partially degrade and release proteins from the food matrix (Mukherjee et al., 2023).

The enzymes produced by microbes responsible for improving the color and flavor of fermented meat products are proteases, catalases, and esterases, which contribute to the formation of small active molecules, like amino acids, esters, peptides, and other volatile SCFAs which work in synergy with endogenous enzymes (Tamang et al., 2016). This process significantly enhances digestive function; increases the content of essential amino acids, vitamins, and bifidobacterium; and thereby improves the nutritional and health benefits (Wang et al., 2021). In several Asian and West African countries, *Bacillus subtilis* and *B. amyloliquefaciens* are also utilized for fermenting soybean and locust beans (Kimura and Yokoyama, 2019). Akin and Ozcan (2017) investigated the functional characteristics of fermented milk products enriched with plant proteins. They found that adding plant proteins increases the total protein and free amino acid levels compared to non-fermented milk products (Akin and Ozcan, 2017).

17.5.2 Lactic Acid Fermentation and Nutrient Quality of Fermented Food

Lactic acid fermentation is a commonly employed method for preserving food, utilizing the ability of LAB to generate various volatile bioactive compounds during the process (Gallo et al., 2020). LAB are extensively applied in the fermentation of fruits, vegetables, meat, fish, and dairy products, as well as in improving the flavor of sausages, wine, and bread (Szutowska, 2020). Additionally, they inhibit microbe-induced spoilage, extending the shelf life of food products (Tamang et al., 2016).

Sangija et al. (2022) assessed the impact of lactic acid fermentation on the nutritional quality of *Solanum villosum* and *S. scabrum* using *Lactobacillus plantarum* and *Leuconostoc mesenteroides.* All fermented products remained stable after six months of storage at both ambient (37°C) and refrigerated (4°C) temperatures (Sangija et al., 2022). Similarly, another study investigated the effect of lactic acid fermentation on spirulina (*Arthrospira platensis*) in a vegetable-based soybean drink using the probiotic bacterium *Lactiplantibacillus plantarum* (Niccolai et al., 2020).

17.5.3 Alcoholic Fermentation and Nutrient Quality of Fermented Food

Alcoholic fermentation is a well-known process for fermenting and preserving sugar-rich products, such as fruits, rice, and vegetables. This type of fermentation is primarily facilitated by various yeasts, along with some fungi and bacteria, with *S. cerevisiae* being the most commonly utilized microorganism for this purpose (Lopes et al., 2018). Non-saccharomyces yeasts are also good ethanologens and are known for adding a unique aroma to beverages rather than ethanol yield (Walker and Walker, 2018). The nutritional composition of *bhaati jaanr*, a traditional rice-based beverage, revealed that it contains high protein, crude fiber, and mineral content (Tamang et al., 2016).

17.5.4 Acetic Acid Fermentation and Nutrient Quality of Fermented Food

ABA are renowned for their ability to oxidize sugars into alcohol, subsequently producing organic acids as end products (De Roos and De Vuyst, 2018). These bacteria also contribute to the sour taste in other fermented beverages, including wine, cider, beer, and soft drinks. Furthermore, ABA produces gluconic acid, l-sorbose, and bacterial cellulose, which have potential applications in the food and biomedical industries (Gomes et al., 2018). Among their most significant contributions to the food industry is the production of vinegar, including industrial vinegar, rice vinegar, wine vinegar, and traditional balsamic vinegar (De Roos and De Vuyst, 2018). The characteristic flavor of cocoa is developed through the fermentation of cocoa beans, where yeast and lactic acid bacteria dominate the early stages of fermentation due to the low pH and high sugar content of the pulp. As the temperature rises and alcohol accumulates, ABA becomes predominant (Neffe-Skocińska et al., 2023).

17.5.5 Probiotics and Antimicrobial Compounds in Enhancing Food Quality

Probiotics are living bacteria that offer the host health advantages when ingested in appropriate amounts (Tamang et al., 2016). These probiotic properties can significantly combat harmful microflora, enhancing the safety and extending the shelf life of fermented foods. The most commonly utilized probiotic bacteria are part of the LAB group, particularly *Lactobacillus* and *Enterococcus*, as well as the genus

Bifidobacterium. However, in recent years, yeasts and other microorganisms have also been developed as potential probiotics (Ouwehand and Röytiö, 2015).

LAB are generally considered safe (GRAS) microorganisms known for producing various antimicrobial compounds, including bacteriocins, diacetyl, organic acids, and other metabolites (Reis et al., 2012). In a study by Ricci et al. (2021), the antimicrobial properties of fermented vegetables were examined using *Lacticaseibacillus* strains. The findings revealed that all fermented extracts exhibited antimicrobial activity both *in vitro* and in food samples, surpassing the effectiveness of commercial preservatives when tested against spoilage microorganisms and foodborne pathogens (Ricci et al., 2021). These promising results suggest a potential for developing innovative natural preservatives that could enhance the safety and shelf life of various food products. Moreover, propionic acid, produced by PAB, is widely used in the food industry as a fungistatic agent for preventing fungal growth in dairy and bakery items (Song et al., 2021). Additionally, it is used to inhibit the growth of bacteria and yeasts in syrups, sauces, and certain fresh fruits (Sharma et al., 2022).

17.6 ADVANCEMENT OF IMPROVING FOOD AND BEVERAGE QUALITY

The production of metabolites such as acetic acid, lactic acid, ethanol, CO_2, and antimicrobial peptides during fermentation is influenced by the used substrate, species, and number of microorganisms (Xiang et al., 2020). For example, *E. coli* has limited fermentation properties compared to LAB, which results in lower-end products (Tamang et al., 2016). During the industrial production of food, alcoholic beverages, and pharmaceuticals, the growth and activity of fermenting organisms are enhanced by providing required materials, such as nitrogen, carbon, trace elements, salts, and vitamins (Siddiqui et al., 2023). The temperature, oxygen, and pH concentrations are also adjusted to ensure the highest possible yield of the desired products. Additionally, the entire procedure is meticulously watched to make sure that no toxic substances, hazardous chemicals, or diseases are introduced to organisms that could have an impact on the result (Siddiqui et al., 2023).

Emerging fermentation-based technologies have significantly transformed the food and beverage industry. Xu et al. (2020) reported an improved nutritional profile of carrot juice by using *L. gasseri* as a probiotic, with a 27% reduction in sugar content. Acidogenic fermentation plays a crucial role in producing carboxylic acids and alcohol by increasing redox potential, which enhances metabolic pathways. This process improves substrate utilization rates and product yields by facilitating electron flow and reducing activation energy (Siddiqui et al., 2023). This technique, known as electro-fermentation, holds great potential for advancing fermentation technology by incorporating an external electricity supply. Ultrasonication is also used in large-scale wine production, enhancing the aroma, color, and phenolic profile of the wine (Zhang et al., 2021; Siddiqui et al., 2023). A previous study also assessed the effect of ultrasound treatment on volatile compounds and the sensory profile of red wine and observed a significant increase in free volatile compounds, including C6 alcohols, terpenes, as well as an increase in electrical conductivity, aroma, and chromatic characteristics (Zhang et al., 2021)

17.7 CHALLENGES AND FUTURE PERSPECTIVES

The future of fermentation and food safety is poised for significant advancements that harmonize traditional practices with modern technologies. Research in fermentation technology is evolving with the integration of recent advances that enhance the functional and nutritional properties as well as the safety of fermented foods. Moreover, addressing food safety risks and promoting sustainable food systems will become increasingly essential. New fermentation methods have been developed to produce functional foods that provide nutrition and confer health benefits. However, many ethnic fermented foods still require investigation into their microbial compositions and nutritional advantages. There is significant potential to uncover various functional properties of the microbes linked to these fermented foods, which could serve as a valuable resource for providing probiotic and antimicrobial-producing organisms. Future perspectives in food safety will require robust risk assessment and management methodologies. Detecting foodborne pathogens and contaminants at deficient levels is necessary to ensure higher safety standards. Critical approaches, such as mass spectrometry, next-generation sequencing, high-throughput screening, and omics, are impactful in ensuring food safety. Thus, incorporating microbiology, food science, nutrition, and engineering insights will provide a more holistic view of safety implications in fermentation practices.

17.8 CONFLICT OF INTEREST

The authors declare no conflict of interest.

REFERENCES

Abouloifa, H., Gaamouche, S., Rokni, Y., Hasnaoui, I., Bellaouchi, R., Ghabbour, N., Karboune, S., Brasca, M., D'Hallewin, G., Ben Salah, R., Saalaoui, E., and Asehraou, A. 2021. Antifungal activity of probiotic *Lactobacillus* strains isolated from natural fermented green olives and their application as food bio-preservative. *Biological Control*, 152:104450.

Akin, Z., and Ozcan, T. 2017. Functional properties of fermented milk produced with plant. *LWT—Food Science and Technology*'86:25–30.

Anal, A. K. 2019. Quality ingredients and safety concerns for traditional fermented foods and beverages from Asia: A review. *Fermentation* 5(1).

Bertuzzi, A. S., Walsh, A. M., Sheehan, J. J., Cotter, P. D., Crispie, F., Mcsweeney, P. L. H., and Kilcawley, K. N. 2018. Omics-based insights into flavor development and microbial succession within surface-ripened cheese. *MSystems* 3(1):10–1128.

Binati, R. L., Junior, W. J. F. L., Luzzini, G., Slaghenau, D., Ugliano, M., and Torriani, S. 2020. Contribution of non- *Saccharomyces* yeasts to wine volatile and sensory diversity: A study on *Lachancea thermotolerans*, *Metschnikowia* spp. and *Starmerella bacillaris* strains isolated in Italy. *International Journal of Food Microbiology* 318.

Campbell-Platt, G. 1994. Fermented foods—a world perspective. *Food Research International* 27:253–257.

Cruz-Cansino, G. A., Beatriz, P., and Cardoso-Ugarte, G. A. 2021. Nutritional contributions and health associations of traditional fermented foods. *Fermentation* 7(4):1–23.

Das, S., Mishra, B. K., and Hati, S. 2020. Techno-functional characterization of indigenous Lactobacillus isolates from the traditional fermented foods of Meghalaya, India. *Current Research in Food Science* 3:9–18.

Dashko, S., Zhou, N., Compagno, C., and Pi, J. 2014. Why, when, and how did yeast evolve alcoholic fermentation?*FEMS Yeast Research* 14:826–832.

De Roos, J., and De Vuyst, L. 2018. Acetic acid bacteria in fermented foods and beverages. *Current Opinion in Biotechnology* 49:115–119.

Gallo, M., Passannanti, F., Colucci, R., Nigro, F., Schiattarella, P., Zappulla, S., Budelli, A., and Nigro, R. 2020. Lactic fermentation of cereals aqueous mixture of oat and rice fl ours with and without glucose addition. *Heliyon* 6:e04920.

Geisen, R. 1993. Fungal starter cultures for fermented foods: Molecular aspects. *Trends in Food Science & Technology* 4(8):251–256.

Gomes, R. J., de Fatima Borges, M., de Freitas Rosa, M., Castro-Gómez, R. J. H., and Spinosa, W. A. 2018. Acetic acid bacteria in the food industry: Systematics, characteristics and applications. *Food Technology and Biotechnology* 56(2):139.

Hu, Y., Zhang, L., Wen, R., Chen, Q., and Kong, B. 2022. Role of lactic acid bacteria in flavor development in traditional Chinese fermented foods: A review. *Critical Reviews in Food Science and Nutrition* 62(10):2741–2755.

Hugenholtz, F., Adair, J., and Kleerebezem, M. 2013. Modulation of the microbial fermentation in the gut by fermentable carbohydrates. *Bioactive Carbohydrates and Dietary Fibre* 2(2):133–142.

Ilango, S., and Antony, U. 2021. Probiotic microorganisms from non-dairy traditional fermented foods. *Trends in Food Science & Technology* 118:617–638.

Kavitake, D., Suryavanshi, M. V, and Kandasamy, S. 2022. Bacterial diversity of traditional fermented food, Idli by high thorough-put sequencing. *Journal of Food Science and Technology* 59(10):3918–3927.

Kimura, K., and Yokoyama, S. 2019. Trends in the application of *Bacillus* in fermented foods. *Current Opinion in Biotechnology* 56:36–42.

Knez, E., Kadac-Czapska, K., and Grembecka, M. 2023. Fermented vegetables and legumes vs. lifestyle diseases: Microbiota and more. *Life* 13(4):044.

Kumar, R. S., Kanmani, P., Yuvaraj, N., Paari, K. A., Pattukumar, V., Arul, V., Kumar, R. S., Kanmani, P., Yuvaraj, N., Paari, K. A., Pattukumar, V., and Arul, V. 2013. Traditional Indian fermented foods: A rich source of lactic acid bacteria. *International Journal of Food Sciences and Nutrition* 64(4):415–428.

Leeuwendaal, N. K., Stanton, C., O'Toole, P. W., and Beresford, T. P. 2022. Fermented foods, health and the gut microbiome. *Nutrients* 14(7):1–26.

Li, Z., Zhou, M., Cui, M., Wang, Y., and Li, H. 2021. Improvement of whole wheat dough fermentation for steamed bread making using selected phytate-degrading Wickerhamomyces anomalus P4. *Journal of Cereal Science* 100:103261.

Lopes, R. P., Mota, M. J., Pinto, C. A., Sousa, S., Lopes, J. A., Gomes, A. M., Delgadillo, I., and Saraiva, J. A. 2018. Physicochemical and microbial changes in yogurts produced under different pressure and temperature conditions. *LWT—Food Science and Technology* 99:423–430.

Louw, N. L., Lele, K., Ye, R., Edwards, C. B., and Wolfe, B. E. 2023. Microbiome assembly in fermented foods. *Annual Review of Microbiology* 381–402.

Lugo-Zarate, L., Delgado-Olivares, L., Cruz-Cansino, N. S., Guillermo, L., Shantal, N., and Estrada-luna, D. 2024. Blackberry juice fermented with two consortia of lactic acid bacteria and isolated whey: Physicochemical and antioxidant properties during storage. *International Journal of Molecular Sciences* 25(16):8882.

Maxon, W. D. 1955. Continuous fermentation: A discussion of its principles and applications. *Applied Microbiology* 3(2):110–22.

Mishra, S., Aravind, S. M., Charpe, P., Ajlouni, S., Ranadheera, C. S., and Chakkaravarthi, S. 2022. Traditional rice-based fermented products: Insight into their probiotic diversity and probable health benefits. *Food Bioscience* 50:102082.

Moon, N. J., Reinbold, G. W. 1976. Commensalism and competition in mixed cultures of *Lactobacillus bulgaricus* and *Streptococcus thermophilus*. *Journal of Food Protection* 39(5):337–341.

Mukherjee, P., Raj, N., and Sivaprakasam, S. 2023. Harnessing valorization potential of whey permeate for D - lactic acid production using lactic acid bacteria Generally Recognized as Safe National Centre for Biotechnology Information. *Biomass Conversion and Biorefinery* 13(17):15639–15658.

Nanda, K., Taniguchi, M., Ujike, S., and Ishihara, N. 2001. Characterization of acetic acid bacteria in traditional acetic acid fermentation of rice vinegar (komesu) and unpolished rice vinegar (kurosu) produced in Japan. *Applied and Environmental Microbiology* 67(2):986–990.

Neffe-Skocińska, K., Karbowiak, M., Kruk, M., Kołożyn-Krajewska, D., and Zielińska, D. 2023. Polyphenol and antioxidant properties of food obtained by the activity of acetic acid bacteria (AAB)—a systematic review. *Journal of Functional Foods* 107.

Niccolai, A., Bažec, K., Rodolfi, L., Biondi, N., Zlatić, E., Jamnik, P., and Tredici, M. R. 2020. Lactic acid fermentation of *Arthrospira platensis* (Spirulina) in a vegetal soybean drink for developing new functional lactose-free beverages. *Frontiers in Microbiology* 11:560684.

Ouwehand, A. C., and Röytiö, H. 2015. Probiotic fermented foods and health promotion. In *Advances in Fermented Foods and Beverages*. Elsevier Ltd.

Oyedeji, O., Ogunbanwo, S. T., and Onilude, A. A. 2013. Predominant lactic acid bacteria involved in the traditional fermentation of fufu and ogi, two nigerian fermented food products. *Food and Nutrition Sciences* 04(11):40–46.

Park, S. K., Jo, D. M., Yu, D., Khan, F., Lee, Y. B., and Kim, Y. M. 2020. Reduction of trimethylamine off-odor by lactic acid bacteria isolated from korean traditional fermented food and their in situ application. *Journal of Microbiology and Biotechnology* 30(10):1510–1515.

Paul Ross, R., Morgan, S., and Hill, C. 2002. Preservation and fermentation: Past, present and future. *International Journal of Food Microbiology* 79(1–2):3–16.

Rcis, J. A., Paula, A. T., Casarotti, S. N., and Penna, A. L. B. 2012. Lactic acid bacteria antimicrobial compounds: Characteristics and applications. *Food Engineering Reviews* 4:124–140.

Rhee, S. J., Lee, J., and Lee, C. 2011. Importance of lactic acid bacteria in Asian fermented foods. In *Microbial Cell Factories*, Vol. 10(Suppl 1), pp. 1–13. BioMed Central.

Ricci, A., Bertani, G., Maoloni, A., Bernini, V., Levante, A., Neviani, E., and Lazzi, C. 2021. Antimicrobial activity of fermented vegetable byproduct extracts for food applications. *Foods* 10(5):1092.

Sangija, F., Martin, H., and Matemu, A. 2022. Effect of lactic acid fermentation on the nutritional quality and consumer acceptability of African nightshade. *Food Science & Nutrition* 3128–3142.

Şanlier, N., Gökcen, B. B., and Sezgin, A. C. 2019. Health benefits of fermented foods. *Critical Reviews in Food Science and Nutrition* 59(3):506–527.

Selhub, E. M., Logan, A. C., and Bested, A. C. 2014. Fermented foods, microbiota, and mental health: Ancient practice meets nutritional psychiatry. *Journal of Physiological Anthropology* 33:1–12.

Sharma, B. R., Halami, P. M., and Tamang, J. P. 2022. Novel pathways in bacteriocin synthesis by lactic acid bacteria with special reference to ethnic fermented foods. *Food Science and Biotechnology* 31(1):1–16.

Sharma, R., Garg, P., Kumar, P., Bhatia, S. K., and Kulshrestha, S. 2020. Microbial fermentation and its role in quality improvement of fermented foods. *Fermentation* 3(1):1–20.

Siddiqui, S. A., Erol, Z., Rugji, J., Taşçı, F., Kahraman, H. A., Toppi, V., Musa, L., Di Giacinto, G., Bahmid, N. A., Mehdizadeh, M., and Castro-Muñoz, R. 2023. An overview of fermentation in the food industry—looking back from a new perspective. *Bioresources and Bioprocessing* 10(1).

Sieuwerts, S., De Bok, F. A. M., Hugenholtz, J., and Van Hylckama Vlieg, J. E. T. 2008. Unraveling microbial interactions in food fermentations: From classical to genomics approaches. *Applied and Environmental Microbiology* 74(16):4997–5007.

Skowron, K., Budzyńska, A., Grudlewska-Buda, K., Wiktorczyk-Kapischke, N., Andrzejewska, M., Wałecka-Zacharska, E., and Gospodarek-Komkowska, E. 2022. Two faces of fermented foods—the benefits and threats of its consumption. *Frontiers in Microbiology* 13:845166.

Smaoui, S., Echegaray, N., Kumar, M., Chaari, M., D'Amore, T., Shariati, M. A., Rebezov, M., and Lorenzo, J. M. 2024. Beyond conventional meat preservation: Saddling the control of bacteriocin and lactic acid bacteria for clean label and functional meat products. *Applied Biochemistry and Biotechnology* 196(6):3604–3635.

Smid, E. J., Lacroix, C. 2013. Microbe—microbe interactions in mixed culture food fermentations. *Current Opinion in Biotechnology* 24(2):148–154.

Song, Z., Hu, Y., Chen, X., Li, G., Zhong, Q., He, X., and Xu, W. 2021. Correlation between bacterial community succession and propionic acid during gray sufu fermentation. *Food Chemistry* 353:129447.

Stadnik, J., Kęska, P., Gazda, P., Siłka, Ł., and Kołożyn-Krajewska, D. 2022. Influence of LAB fermentation on the color stability and oxidative changes in dry-cured meat. *Applied Sciences* 12(22):11736.

Szutowska, J. 2020. Functional properties of lactic acid bacteria in fermented fruit and vegetable juices: A systematic literature review. *European Food Research and Technology* 246(3):357–372.

Tamang, J P. 1998. Role of microorganisms in traditional fermented foods. *Indian Food Industry*17(3):162–166.

Tamang, Jyoti P, Watanabe, K., and Holzapfel, W. H. 2016. Diversity of microorganisms in global fermented foods and beverages. *Frontiers in Microbiology* 7:377.

Tamang, Jyoti Prakash, Cotter, P. D., Endo, A., Han, N. S., Kort, R., Liu, S. Q., Mayo, B., Westerik, N., and Hutkins, R. 2020. Fermented foods in a global age: East meets West. *Comprehensive Reviews in Food Science and Food Safety* 19(1):184–217.

Thakur, N., Savitri, Saris, P. E. J., and Bhalla, T. C. 2015. Microorganisms associated with amylolytic starters and traditional fermented alcoholic beverages of North Western Himalayas in India. *Food Bioscience* 11:92–96.

Torres, R., Braulio, G., Zarzoso, E., Reguant, C., and Bordons, A. 2021. Microbial interactions in alcoholic beverages. *International Microbiology* 25(1):1–5.

Tufariello, M., Rizzuti, A., Palombi, L., Ragone, R., Capozzi, V., Gallo, V., Mastrorilli, P., and Grieco, F. 2021. Non-targeted metabolomic approach as a tool to evaluate the chemical profile of sparkling wines fermented with autochthonous yeast strains. *Food Control* 126:108099.

Walker, G. M., and Walker, R. S. K. 2018. Enhancing yeast alcoholic fermentations. In *Advances in Applied Microbiology*, Vol. 105. Elsevier Ltd.

Wang, B., Rutherfurd-Markwick, K., Liu, N., Zhang, X.-X., and Mutukumira, A. N. 2024. Probiotic potential of acetic acid bacteria isolated from kombucha in New Zealand in vitro. *The Microbe* 4:100130.

Wang, H., Huang, Y., and Huang, Y. 2021. Microbiome diversity and evolution in stacking fermentation during different rounds of Jiang-flavoured Baijiu brewing. *LWT* 143:111119.

Wang, H. H. (1989). Fermented rice products. In *Rice: Volume I. Production/Volume II. Utilization*, pp. 614–642. Boston, MA: Springer US.

Wang, X., Wang, B., Sun, Z., Tan, W., Zheng, J., and Zhu, W. 2022. Effects of modernized fermentation on the microbial community succession and ethyl lactate metabolism in Chinese baijiu fermentation. *Food Research International* 159:111566.

Wu, C., Liu, R., Huang, W., Rayas-Duarte, P., Wang, F., and Yao, Y. 2012. Effect of sourdough fermentation on the quality of Chinese Northern-style steamed breads. *Journal of Cereal Science* 56(2):127–133.

Wu, Y., Ma, H., Zheng, M., and Wang, K. 2015. Bioresource technology lactic acid production from acidogenic fermentation of fruit and vegetable wastes. *Bioresource Technology* 191:53–58.

Xiang, H., Sun-Waterhouse, D., Waterhouse, G. I., Cui, C., and Ruan, Z. 2019. Fermentation-enabled wellness foods: A fresh perspective. *Food Science and Human Wellness* 8(3):203–243.

Xiang, W., Zhang, N., Lu, Y., Zhao, Q., Xu, Q., Rao, Y., and Liu, L. 2020. Effect of *Weissel-lacibaria* co-inoculation on the quality of Sichuan Pickle fermented by Lactobacillus plantarum. *LWT—Food Science and Technology* 121:108975.

Xu, Y., Hlaing, M. M., Glagovskaia, O., Augustin, M. A., and Terefe, N. S. 2020. Fermentation by probiotic lactobacillus gasseri strains enhances the carotenoid and fibre contents of carrot juice. *Foods* 9(12).

Yang, Q., Yao, H., Liu, S., and Mao, J. 2022. Interaction and application of molds and yeasts in chinese fermented foods. *Frontiers in Microbiology* 12:1–12.

Yilmaz, B., Elibol, E., Shangpliang, H. N. J., Ozogul, F., and Tamang, J. P. 2022. Microbial communities in home-made and commercial kefir and their hypoglycemic properties. *Fermentation* 1–10.

Yim, E. J., Jo, S. W., Lee, E. S., Park, H. S., Ryu, M. S., Uhm, T. B., Kim, H. Y., and Cho, S. H. 2015. Fermentation characteristics of mulberry (*Cudrania tricuspidata*) fruit vinegar produced by acetic acid bacteria isolated from traditional fermented foods. *Food Science and Preservation* 22(1):108–118.

Zhang, P., Ma, W., Meng, Y., Zhang, Y., Jin, G., and Fang, Z. 2021. Wine phenolic profile altered by yeast: Mechanisms and influences. *Comprehensive Reviews in Food Science and Food Safety* 20(4):3579–3619.

Zhang, Y., Zhang, L., Du, M., Yi, H., Guo, C., Tuo, Y., Han, X., Li, J., Zhang, L., and Yang, L. 2011. Antimicrobial activity against *Shigella sonnei* and probiotic properties of wild lactobacilli from fermented food. *Microbiological Research* 167(1):27–31.

18 Food Protein Production through Recombinant Microbial Technology

Xiao Zheng and Jianguo Zhang

18.1 INTRODUCTION

As an important nutrient for human beings, food protein also is the main undertaker of human life activities and is obtained from food traditionally. According to "Dietary Nutrient Intake of Chinese Residents, 2022," the protein intake of adult men and that of adult women were suggested as 65 g/d and 55 g/d, respectively. For specific people, including athletes, the elderly, and pregnant women, protein intake plays the additional roles of preventing physical issues, such as edema, fatigue, and malnutrition. And the optimal protein intake of specific people was considered as a higher level than the recommended amount for the general population. The demand for food protein will increase by 30–50% to 265 million tons, considering that global population is expected to increase to 10 billion by 2050 (He et al. 2024; Leger et al. 2021; Linder 2023). At present, the majority of global food protein production relies on large-scale agricultural crops and aquaculture production technology, which fluctuates due to weather. In addition, the aquaculture industry also has serious issues of environmental pollution, large land occupation area, and water resource supply. The fluctuation in quality and quantity of protein influences the total protein supply and, thus, human health simultaneously. According to World Resources Institute, one calorie of chicken, and even lesser amount of another meat, can be produced from nine calories of soy, wheat, and corn. Therefore, traditional food protein production techniques cannot meet humanity's future protein needs because of its inefficiency. Currently, global food protein production is about 200 million tons; hence, Good Food Institute (GFI) has pointed out that the Earth will not be able to feed the world's nearly 10 billion people if no alternative protein is supplied by 2050 (Falcon, Naylor, and Shankar 2022). Therefore, it is urgent to innovate and develop large-scale, low-cost, and sustainable food protein production technologies.

Microorganisms have the advantage of a high protein content, an essential amino acid required by the human body, and can be used as a new alternative protein source for the food industry. Microbial protein manufacturing based on renewable biomass raw materials brings protein yield as high as several kilograms per cubic meter per hour, which is several orders of protein yield of traditional food protein production technologies. According to World Resources Institute, 1 million tons of fungal protein produced from renewable raw materials by fermentation can save 5 million cows

DOI: 10.1201/9781003540113-18

and 1 billion chicken, which would alleviate the severe shortage of land and water resources greatly. In addition, microbial protein manufacturing also has the characteristics of short production cycle, high manufacturing efficiency, efficient energy utilization, no disturbance to the environment, and low carbon dioxide emission (Durkin et al. 2022; Banks et al. 2022). For decades, the fermentation industry has relied mainly on a few mature microbial species, and microbial food protein only accounts for 10–15% of total food protein using submerged fermentation (Areniello et al. 2023a). With the development of modern recombinant microbial technology, more proportion of food protein from microbial species is estimated to cover the food protein demand, which provides efficient protein for a sustainable society.

18.2 MICROBIAL FOOD PROTEINS

Microbial food protein has the advantages of high rates of digestion and absorption for humans, rich dietary fiber and micronutrients content, and maintaining skeletal muscle mass with low risk of obesity and disease (Liu, Aimutis, and Drake 2024). As shown in Figure 18.1, microbial protein can be divided into microalgae protein, bacterial protein, yeast protein, and filamentous fungal protein according to different microbial species, such as *Bacillus* and *Corynebacterium glutamate* as bacteria, *Komagataella phaffii* and *Saccharomyces cerevisiae* as yeast, and *Aspergillus niger* as fungi (Alves et al. 2023). The general procedure of microbial food protein includes fermentation and centrifugation, nucleic acid removal, centrifugation, enzymatic hydrolysis, extraction, purification, separation, sterilization, and drying.

18.2.1 Microalgae Proteins

Microalgae contains chlorophyll A and can carry out photosynthesis, which immobilizes carbon dioxide mainly through the CBB cycle (Figure 18.2). There are about

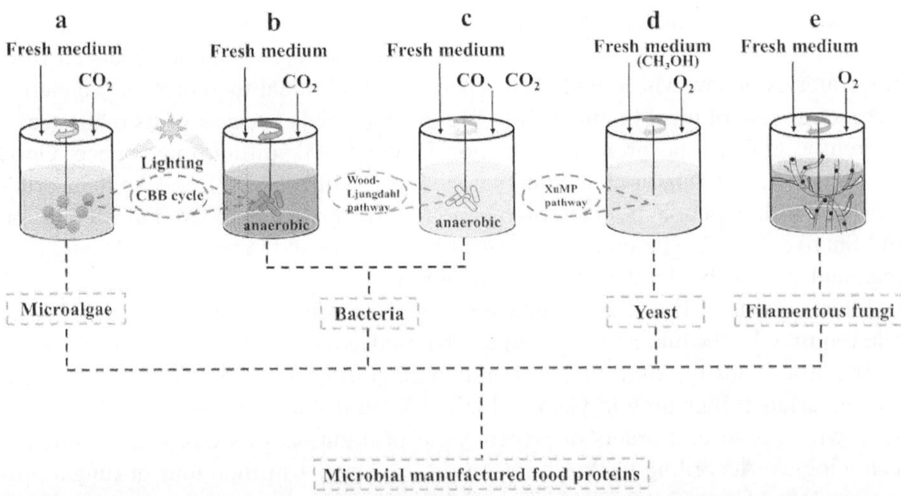

FIGURE 18.1 Schematic diagram of food protein production by different microorganisms.

200 microalgae species in the world, including eukaryotic microalgae of the *Chlorophyceae*, *Bacillariophyceae*, *Chrysophyceae*, and *Rhodophyta*, and also prokaryotic microalgae of the *Cyanophyceae* and *Chlorella*. Microalgal protein is popular in the food market, such as in baked goods, beverages, ice cream, and infant formula foods (Ijaola et al. 2024; Geada et al. 2021), which is recognized as rich in protein, amino acids, polyunsaturated fatty acids, polysaccharides, vitamins and minerals, and other nutrients (Mosibo, Ferrentino, and Udenigwe 2024; Zhou et al. 2022; Karabulut, Purkiewicz, and Goksen 2024). However, the nutrient content in microalgae is also affected by growth stage and culture conditions. For example, the protein content of microalgae increased to 60–70% at the condition of sufficient nitrogen source, while the lipid content of microalgae shifted up to 85–90% at the condition of insufficient nitrogen source (Demarco et al. 2022; Song et al. 2024). Generally, the protein content of industrialized *Spirulina* can reach 60–70%, so microalgae protein is a reliable source of food protein. Additionally, the content of essential amino acids of microalgae protein is higher than that of wheat, corn, and other agricultural crop proteins, which is comparable to that of meat and soybeans. In addition, microalgae have the advantages of high biomass production, simple culture condition, and high photosynthesis efficiency. The annual yield of microalgae can reach 225 t dry biomass per hectare, which is over ten times higher than that of agricultural crops (Manuel et al., 2018).

18.2.2 Bacterial Proteins

To accumulate protein, bacteria producing protein is divided into methanol bacteria, methane-oxidizing bacteria, hydrogen-oxidizing bacteria, chemoautotrophic

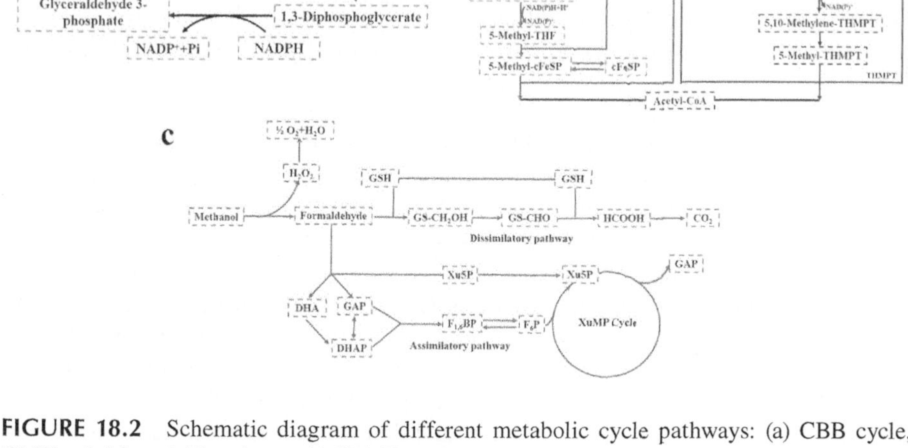

FIGURE 18.2 Schematic diagram of different metabolic cycle pathways: (a) CBB cycle, (b) Wood–Ljungdahl pathway, (c) XuMP cycle.

bacteria, and so on (Woern and Grossmann 2023; Shahzad et al. 2024; Li et al. 2023b; Jain et al. 2023). Hydrocarbons (such as natural gas) and methanol are generally used as substrates for methanol bacteria, methane-oxidizing bacteria, and hydrogen-oxidizing bacteria for high valued-added protein production because of the high cost of methanol via different metabolic pathways. Chemoautotrophs use carbon monoxide or carbon dioxide as carbon source through the Wood–Ljungdahl pathway to produce high value-added products (Kerckhof et al. 2021; Khoshnevisan et al. 2022; Wang et al. 2021) (Figure 18.2b). The protein content of bacteria generally accounts for more than 75% of its dry weight, which is 2.5 times that of meat and 1.7 times that of soybeans. And the protein production efficiency of bacteria is much higher than that of plants and animals. The energy consumption of bacterial protein is only 1/5 of that of meat (beef, pork, poultry), which can significantly reduce the cost of food protein, as well as water consumption. Chemoautotrophic bacteria can use greenhouse gases to produce proteins, which is an environmentally friendly way. Considering the high nucleic acid content of bacterial protein after extraction procedure, it must be further processed to become edible protein raw materials for humans. *Bacillus*, *Nocardia*, *Microcus*, and *Pseudomonas* are the main bacteria for food protein production in an industrial scale (Zhang et al. 2021; Chen et al. 2024; Areniello et al. 2023b; Teng et al. 2024).

18.2.3 Yeast Protein

Yeast is a traditional producer of food protein, with the advantages of easy cultivation and high growth rate. Methylotrophic yeast is very impressive for food protein production, such as *Komagataella phaffii*, *Ogataea polymorpha*, *Candida boidinii* (Tsuda and Nonaka 2024). Methylotrophic yeast uses methanol as the sole carbon source through the xylulose monophosphate pathway (Eskandari ct al. 2024) (Figure 18.2c). Along the metabolic pathway, methanol is first oxidized to formaldehyde by alcohol oxidase, and then formaldehyde reacts with xylulose5-phophate, resulting in glyceraldehyde 3-phosphate and dihydroxyacetone phosphate entering the glycolysis pathway to participate in cell metabolism. Methylotrophic yeast can accumulate up to 80% protein. However, the low methanol utilization rate of methylotrophic yeast delays the cell growth rate because of formaldehyde toxicity. Researchers have used methylotrophic yeast to produce recombinant proteins, considering its relatively high cost.

For non-methylotrophic yeast, *Saccharomyces cerevisiae* has been closely related to human life for wine brewing and bread fermentation for many years. As a model eukaryote, *S. cerevisiae* is currently the most fully investigated without toxins and is generally recognized as safe by the US FDA. Over 80% biomass of the *S. cerevisiae* cell is protein, fat, dietary fiber, etc. *S. cerevisiae* protein is widely used in dairy products, protein drinks, vegetarian meat, baked food, flavorings, protein powder, energy bars, and protein bars because of its 81% digestion ratio, nine essential amino acids contributing 47% of total amino acids, no allergenic components, favorable flavor of high glutamic acid content, color enhancement, and water retention (Jach et al. 2022; Ma et al. 2023; Zhao et al. 2024). Compared with agricultural protein, yeast protein is very suitable for meat products and dairy products processing because of its foam stability, high emulsification, and low viscosity. Yeast protein releases about 1/20 carbon dioxide of animal protein, which is close to agricultural protein; about 1/200 water of animal protein and 1/10 of plant protein; and about 1/100 land of animal protein and 1/10 of plant protein (Choi, Geonjin, and Jak 2022).

18.2.4 Filamentous Fungal Protein

Filamentous fungal protein (Lubeck and Lubeck 2022; Amara and El-Baky 2023; Derbyshire 2022), such as *Fusarium*, *Fusarium yellowstone*, *Aspergillus oryzae*, and so on (Yang et al. 2024), also has the advantages of being free from environmental restrictions, fast growth rate, land-saving resources, and waste utilization (Wang et al. 2023; Li et al. 2023a). Fungal protein accounts for 50% of its dry cell weight (Zeng et al. 2023). The essential amino acids profile of filamentous fungal protein is similar to that of eggs, which can be used for infants and teenagers for development (Awasthi et al. 2023), and can to produce alternative meat for its meat-like texture/structure. With *F. filamentum* as the first edible fungal protein, its safety, production process, ribonucleic acid removal technology, and nutritional composition (protein) have been developed in detail in the 1980s. *F. filamentum* protein has nine essential amino acids, no cholesterol, low saturated fatty acid and is rich in dietary fiber. However, 8–25% of the nucleic acids of fungal cell proteins dry weight limit its application because of uric acid (Achor, Richardson, and Draper 1981). So fungal protein is expected to minimize its RNA content for food protein production.

18.3 RECOMBINANT MICROBIAL TECHNOLOGY

With the development of recombinant genetic technology, recombinant protein expression by microbial cell factory stands at the center of its applications. Food protein production through recombinant microbial technology shows bright potential as a new technology of genetic engineering and well-established microbial horses.

18.3.1 Gene Editing Technology

Gene editing allows modification of a specific gene in a genome through efficient and precise insertion, deletion, or replacement to alter the genetic information and phenotype of microbial cells (Figure 18.3a). Several technologies have been developed sequentially through the years to improve the efficiency and precision of gene editing. Firstly, homologous recombination is used to introduce a heterologous gene or inactivate and repair a specific gene or repair a defective gene by exchanging the homologous arms of heterologous gene with the target gene (Carroll 2014). Then, zinc finger nuclease technology (ZFNs), transcription activator-like effector nuclease technology (TALEN), clustered regularly spaced short palindromic repeats–associated nuclease technology (CRISPR-Cas), etc. (Figure 18.3b, c, d) are used to increase gene editing efficiency and decrease off-target ratio (Kim and Kim 2014; Chaudhary, Pratap, and Sharma 2016; Gupta and Musunuru 2014). ZFN technology still has issues of high cost, off-target phenomenon, and low affinity. TALEN is relatively simple to design for use with a relative high specific affinity to DNA

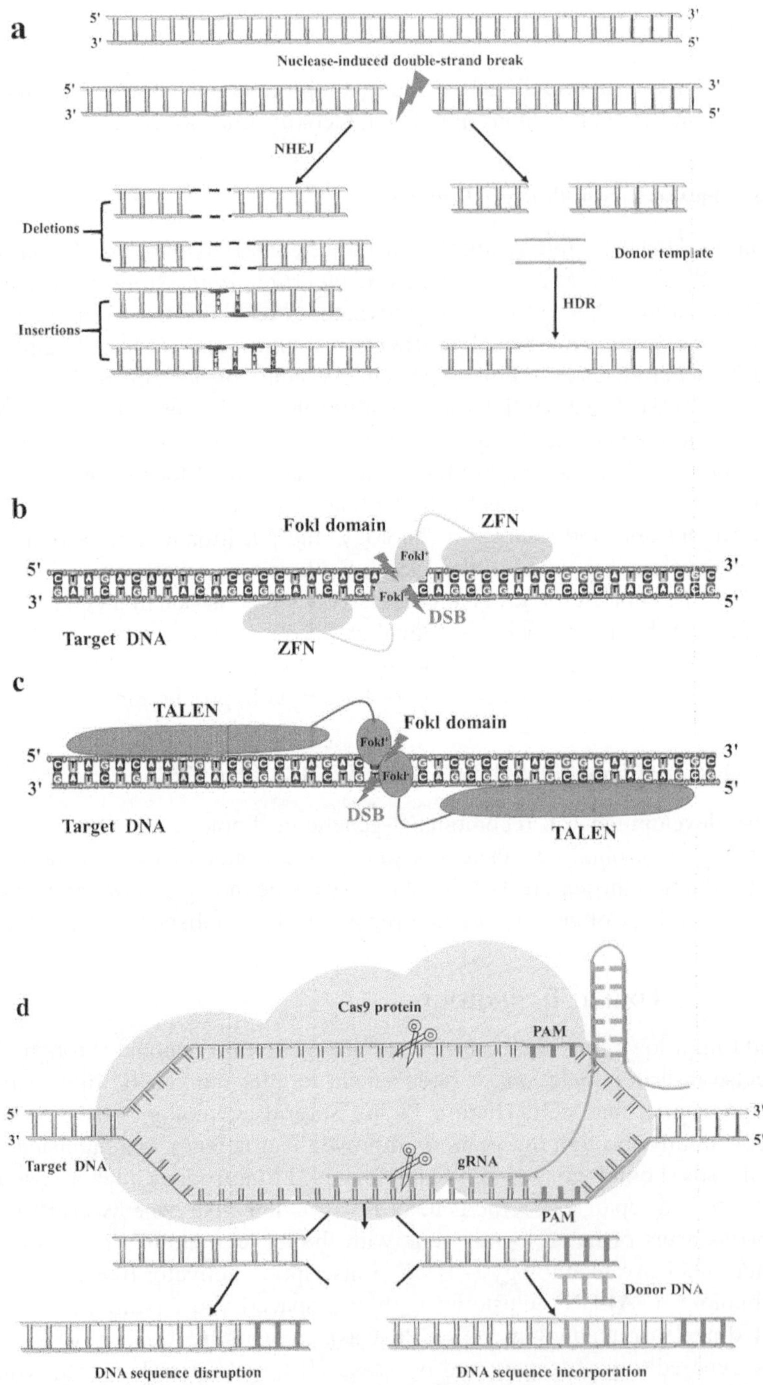

FIGURE 18.3 Schematic diagram of different gene editing methods: (a) Gene insertion, deletion, and replacement; (b) zinc finger nuclease technology; (c) transcription activator-like effector nuclease technology; (d) clustered regularly spaced short palindromic repeats–associated nuclease technology.

sequence, although with the problems of long construction cycle and target process. CRISPR-Cas technology is a cutting-edge, most efficient, and specific gene editing technology demonstrating capabilities of multiple sites editing simultaneously and the lowest off-target ratio (Sander and Joung 2014; Doudna and Charpentier 2014). Currently, the hottest point of gene editing for recombinant protein production is in improving the substrate utilization efficiency of microorganisms to prompt the industrial production of food protein (Salazar-Cerezo, de Vries, and Garrigues 2023; Kothari et al. 2023; Liu, Garrigues, and de Vries 2023). For example, microalgae and photosynthetic bacteria produce food protein by using carbon dioxide as carbon source through the CBB cycle. The utilization efficiency of carbon dioxide is improved through the catalytic efficiency increase of their key enzyme, RuBisCO, by genetic engineering. The methylotrophic yeast is edited to knock out the genes formaldehyde dehydrogenase gene (*FLD*), S-formyl glutathione hydrolase gene (*FGH*), and formate dehydrogenase gene (*FDH*) of the methanol dissimilation pathway while overexpressing the genes dihydroxyacetone synthetase genes (*DAS1*, *DAS2*) and dihydroxyacetone kinase gene (*DAK*) of the assimilation pathway, promoting central metabolism and protein production (Liu et al. 2021; Geier et al. 2015). Food protein production is significantly improved by the knockout of protein secretion–related genes (*GOS1*, *TDA3*, *VPS5*, *HDA2*) and overexpression of the genes *PDI1* and *COG5* in *S. cerevisiae* (Fotso et al. 2005; Jin, Seo, and Kim 2024). The gene editing of filamentous fungi mainly involves the screening of promoters (*GLA*, *EXL*, *GPD*) and the mining of transcriptional activators (*CIR1*, *ClR2*, *XINR*, *ACE2*) to improve the expression of food proteins (Wang et al. 2022b; Maumela et al. 2021; Aro et al. 2001; Andersen et al. 2008).

18.3.2 Metabolic Pathway Modification

Metabolic reactions are catalyzed by many enzymes continuously and sequentially. Therefore, the modification of metabolic pathways is expected to maximize the yield of target products and reduce costs. Due to the limited phylogenetic diversity and genetic transformation methods of microalgae, the progress of microalgae metabolic engineering is slow. The metabolic engineering of microalgae is mainly carried out through single gene insertion and knockout only. It is difficult to obtain commercial production of algae strains by single gene manipulation at the condition of a complex metabolic pathway. For example, the knockout of pyruvate dehydrogenase kinase in *Chlorophyta* increases lipid content by 82%, while the overexpression of phosphate starvation response 1 increases intracellular lipid content (Ma et al. 2014; Murakami et al. 2020). At present, researchers are reconstructing the primary metabolic network of *Chlorellococcus* and conducting applications by combining metabolomics and metabolic flow analysis. They have found that the absorption of ammonia nitrogen under autotrophic conditions is mainly through the glutamate dehydrogenase pathway, while nitrogen assimilation under heterotrophic conditions is almost completely dependent on the coupling reaction catalyzed by glutamine synthetase and glutamate synthetase (Zhu et al. 2024; Patel et al. 2015). Therefore, more analysis of the metabolic regulation mechanism of microalgae is expected to improve the expression of microalgae food protein.

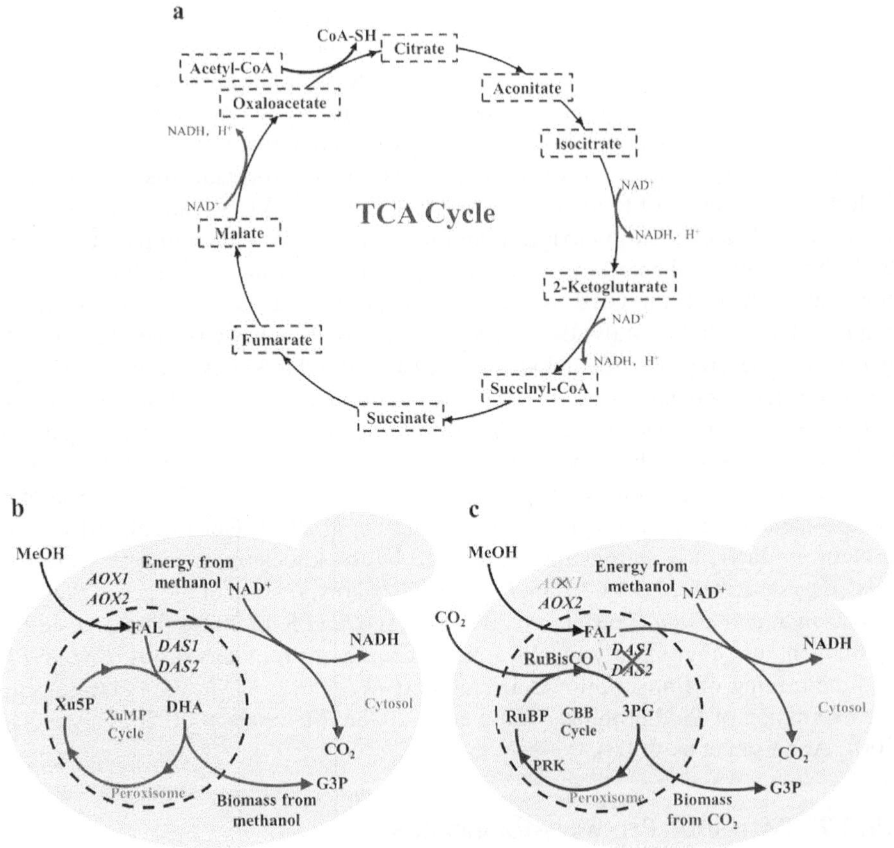

FIGURE 18.4 Schematic diagram of the modification of different microbial metabolic pathways: (a) TCA cycle; (b, c) CBB cycle inserted into *K. phaffii*.

The central metabolic pathways of yeast and filamentous fungi include the pentose phosphate pathway, tricarboxylic acid cycle, and glycolysis pathway (Figure 18.4a), which can be modified through gene knockout, gene expression, and new pathway construction. For example, the accumulation of microbial endogenous substances can be achieved through the knockout of key genes in other metabolic pathways. For example, the knockout of key genes of TCA and glycolysis pathways enhances the intracellular pentose phosphate pathway and enhances the biomass yield. In addition, the CBB cycle is introduced into *K. phaffii* to fix carbon dioxide by adding 8 heterologous genes and deleting 3 native genes (*AOX1*, *DAS1*, *DAS2*). Finally, *K. phaffii* is switched into autotrophic microorganisms (Figure 18.4b, c). Cofactor engineering increases the metabolic flux of related pathways and enhances biomass yield (Kawanishi and Matsunaga 2021; Baumschabl et al. 2022). At present, metabolic engineering is conducted via rational design, including computer design and simulation, and the predication of modification strategies and target genes, and is named as "precise metabolic engineering" for food protein production.

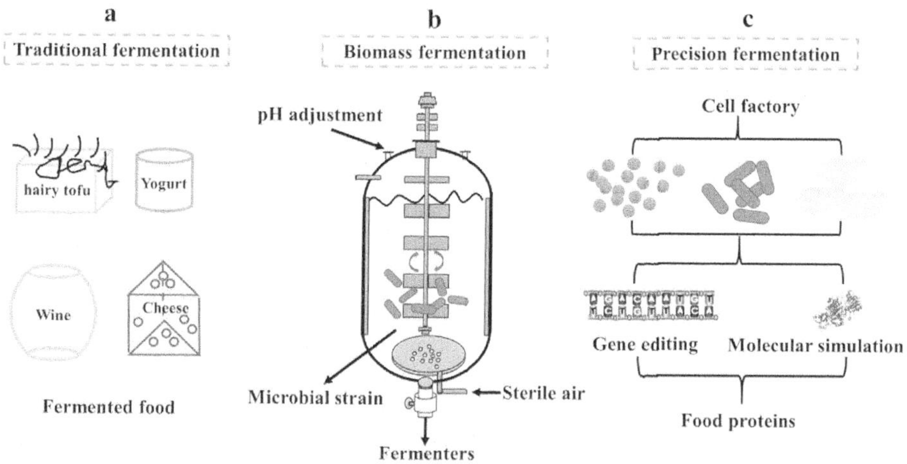

FIGURE 18.5 Diagram of different fermentation technologies: (a) Traditional fermentation; (b) biomass fermentation; (c) precision fermentation.

18.3.3 Precision Fermentation Technology

Fermentation is a necessary step to obtain large-scale production of food protein (Singh et al. 2022; Chai et al. 2022). Currently, the fermentation technology is generally divided into three types: traditional fermentation, biomass fermentation, and precision fermentation (Boukid et al. 2023; Knychala et al. 2024; Hilgendorf et al. 2024) (Figure 18.5). Traditional fermentation uses some natural microorganisms acting on the substrate to obtain products of unique flavor and nutritional characteristics, such as tempeh using *Rhizopus* and soybeans, hairy tofu using *Mucor* and tofu, cheese and yogurt using lactic acid and milk. Biomass fermentation uses microorganisms to produce a large amount of protein products which can be directly used as the food component. Precision fermentation uses the microbial host as a "cell factory" to produce specific food proteins, such as artificial milk protein, artificial soy protein, synthetic chicken protein, and heme protein (Nielsen, Meyer, and Arnau 2024; Hoppenreijs et al. 2024; Wang et al. 2022a). In addition, gene editing and metabolic engineering can also be used to modify the relevant enzymes in the microbial cell factory, so that their higher metabolic activity, new substrate specificity, higher stability, and the durability of enzymes trigger cost reduction. Precision fermentation is achieved through precise molecular transformation that benefits protein production.

18.4 CONCLUSION

Traditional food protein production methods not only cannot meet the human requirement for it but also make it difficult to achieve environmental friendliness and resource saving. With the development of technology, an alternative protein with microbial protein is an inevitable trend for the food protein industry. Microbes can achieve the large-scale, low-cost, and sustainable production of food protein. Further, recombinant microorganisms can produce high-quality single-cell proteins from carbon dioxide, methane, methanol, and other substrates. The recombinant microbial technology guarantees the protein supply and demand, improving the sensory food quality on the basis of protein nutrition and function.

For the future recombinant microbial technology to manufacture food protein, it warrants building cell factories, designing and modifying metabolic pathways, and improving production efficiency. An efficient cell factory of microalgae, bacillus, yeast, and filamentous fungi is expected based on the mining of new microbial species, key promoters, and transcription factors. Food protein synthesis and secretion by different microbial cells is another option through metabolic pathway design and regulation strategies. Finally, with the help of fermentation engineering platforms, large-scale and low-cost production of food protein and some specific proteins will be realized, and synthetic protein and functional protein products can be developed through synthetic pathways to establish a complete system for food protein production.

REFERENCES

Achor, I. M., T. Richardson, and N. R. Draper. 1981. "Effect of treating Candida utilis with acid or alkali, to remove nucleic acids, on the quality of the protein." *Journal of Agricultural and Food Chemistry* 29 (1):27–33. https://doi.org/10.1021/jf00103a008.

Alves, Samara Cardoso, Erick Diaz-Ruiz, Bruna Lisboa, Minaxi Sharma, Solange I. Mussatto, Vijay Kumar Thakur, Deepak M. Kalaskar, Vijai Kumar Gupta, and Anuj K. Chandel. 2023. "Microbial meat: A sustainable vegan protein source produced from agri-waste to feed the world." *Food Research International* 166:112596. https://doi.org/10.1016/j.foodres.2023.112596.

Amara, Amro A. A., and Nawal Abd El-Baky. 2023. "Fungi as a source of edible proteins and animal feed." *Journal of Fungi* 9 (1):73. https://doi.org/10.3390/jof9010073.

Andersen, Mikael R., Wanwipa Vongsangnak, Gianni Panagiotou, Margarita P. Salazar, Linda Lehmann, and Jens Nielsen. 2008. "A trispecies *Aspergillus* microarray: Comparative transcriptomics of three *Aspergillus* species." *Proceedings of the National Academy of Sciences of the United States of America* 105 (11):4387–4392. https://doi.org/10.1073/pnas.0709964105.

Areniello, Marica, Silvio Matassa, Giovanni Esposito, and Piet N. L. Lens. 2023a. "Biowaste upcycling into second-generation microbial protein through mixed-culture fermentation." *Trends in Biotechnology* 41 (2):197–213. https://doi.org/10.1016/j.tibtech.2022.07.008.

Areniello, Marica, Silvio Matassa, Giovanni Esposito, and Piet N. L. Lens. 2023b. "Microbial protein production from sulfide-rich biogas through an enrichment of methane- and sulfur-oxidizing bacteria." *Bioresource Technology* 383:129237. https://doi.org/10.1016/j.biortech.2023.129237.

Aro, N., A. Saloheimo, M. Ilmén, and M. Penttilä. 2001. "ACEII, a novel transcriptional activator involved in regulation of cellulase and xylanase genes of Trichoderma reesei." *Journal of Biological Chemistry* 276 (26):24309–24314. https://doi.org/10.1074/jbc.M003624200.

Awasthi, Mukesh Kumar, Vinay Kumar, Coralie Hellwig, Rachma Wikandari, Sharareh Harirchi, Taner Sar, Steven Wainaina, Raveendran Sindhu, Parameswaran Binod, Zengqiang Zhang, and Mohammad J. Taherzadeh. 2023. "Filamentous fungi for sustainable vegan food production systems within a circular economy: Present status and future prospects." *Food Research International* 164:112318. https://doi.org/10.1016/j.foodres.2022.112318.

Banks, Mason, Rob Johnson, Lori Giver, Geoff Bryant, and Miao Guo. 2022. "Industrial production of microbial protein products." *Current Opinion in Biotechnology* 75:102707. https://doi.org/10.1016/j.copbio.2022.102707.

Baumschabl, Michael, Oezge Ata, Bernd M. Mitic, Lisa Lutz, Thomas Gassler, Christina Troyer, Stephan Hann, and Diethard Mattanovich. 2022. "Conversion of CO2 into organic acids by engineered autotrophic yeast." *Proceedings of the National Academy of Sciences of the United States of America* 119 (47):e2211827119. https://doi.org/10.1073/pnas.2211827119.

Boukid, Fatma, Seedhabadee Ganeshan, Yingxin Wang, Mehmet Caglar Tulbek, and Michael T. Nickerson. 2023. "Bioengineered enzymes and precision fermentation in the food industry." *International Journal of Molecular Sciences* 24 (12):10156. https://doi.org/10.3390/ijms241210156.

Carroll, Dana. 2014. "Genome engineering with targetable nucleases." *Faseb Journal* 28 (1):409–439. https://doi.org/10.1146/annurev-biochem-060713-035418

Chai, Kong F., Kuan R. Ng, Malsha Samarasiri, and Wei N. Chen. 2022. "Precision fermentation to advance fungal food fermentations." *Current Opinion in Food Science* 47:100881. https://doi.org/10.1016/j.cofs.2022.100881.

Chaudhary, Kulbhushan, Dharmendra Pratap, and Pradeep K. Sharma. 2016. "Transcription activator-like effector nucleases (TALENs): An efficient tool for plant genome editing." *Engineering in Life Sciences* 16 (4):330–337. https://doi.org/10.1002/elsc.201500126.

Chen, Yanzhen, Miaomiao Li, Mingchen Yan, Yong Chen, Muhammad Saeed, Zhong Ni, Zhen Fang, and Huayou Chen. 2024. "*Bacillus subtilis*: Current and future modification strategies as a protein secreting factory." *World Journal of Microbiology & Biotechnology* 40 (6):195. https://doi.org/10.1007/s11274-024-03997-x.

Choi, Yoonsil, Geonjin Park, and Jae Jak Nam. 2022. "Effect of alternative protein food market growth on green house gas emissions in livestock sector." *Journal of Climate Change Research* 13 (3):299–304. https://doi.org/10.15531/ksccr.2022.13.3.299.

Demarco, Mariana, Jaqueline Oliveira de Moraes, Angelo Paggi Matos, Roberto Bianchini Derner, Fabio de Farias Neves, and Giustino Tribuzi. 2022. "Digestibility, bioaccessibility and bioactivity of compounds from algae." *Trends in Food Science & Technology* 121:114–128. https://doi.org/10.1016/j.tifs.2022.02.004.

Derbyshire, Emma. 2022. "Fungal-derived mycoprotein and health across the lifespan: A narrative review." *Journal of Fungi* 8 (7):653. https://doi.org/10.3390/jof8070653.

Doudna, Jennifer A., and Emmanuelle Charpentier. 2014. "The new frontier of genome engineering with CRISPR-Cas9." *Science* 346 (6213):1077. https://doi.org/10.1126/science.1258096.

Durkin, Alex, Tim Finnigan, Rob Johnson, John Kazer, Jianfeng Yu, David Stuckey, and Miao Guo. 2022. "Can closed-loop microbial protein provide sustainable protein security against the hunger pandemic?" *Current Research in Biotechnology* 4:365–376. https://doi.org/10.1016/j.crbiot.2022.09.001.

Eskandari, Azadeh, Nima Ghahremani Nezhad, Thean Chor Leow, Mohd Basyaruddin Abdul Rahman, and Siti Nurbaya Oslan. 2024. "Current achievements, strategies, obstacles, and overcoming the challenges of the protein engineering in *Pichia pastoris* expression system." *World Journal of Microbiology & Biotechnology* 40 (1):39. https://doi.org/10.1007/s11274-023-03851-6.

Falcon, Walter P., Rosamond L. Naylor, and Nikhil D. Shankar. 2022. "Rethinking global food demand for 2050." *Population and Development Review* 48 (4):921–957. https://doi.org/10.1111/padr.12508.

Fotso, P., Y. Koryakina, O. Pavliv, A. B. Tsiomenko, and V. V. Lupashin. 2005. "Cog1p plays a central role in the organization of the yeast conserved oligomeric golgi complex." *Journal of Biological Chemistry* 280 (30):27613–27623. https://doi.org/10.1074/jbc.M504597200.

Geada, Pedro, Catarina Moreira, Maria Silva, Rafaela Nunes, Leandro Madureira, Cristina M. R. Rocha, Ricardo N. Pereira, Antonio A. Vicente, and Jose A. Teixeira. 2021. "Algal proteins: Production strategies and nutritional and functional properties." *Bioresource Technology* 332:125125. https://doi.org/10.1016/j.biortech.2021.125125.

Geier, Martina, Christoph Brandner, Gernot A. Strohmeier, Melanie Hall, Franz S. Hartner, and Anton Glieder. 2015. "Engineering *Pichia pastoris* for improved NADH regeneration: A novel chassis strain for whole-cell catalysis." *Beilstein Journal of Organic Chemistry* 11:1741–1748. https://doi.org/10.3762/bjoc.11.190.

Gupta, Rajat M., and Kiran Musunuru. 2014. "Expanding the genetic editing tool kit: ZFNs, TALENs, and CRISPR-Cas9." *Journal of Clinical Investigation* 124 (10):4154–4161. https://doi.org/10.1172/jci72992.

He, JinTao, Min Tang, FeiFei Zhong, Jing Deng, Wen Li, Lin Zhang, QinLu Lin, Xu Xia, Juan Li, and Ting Guo. 2024. "Current trends and possibilities of typical microbial protein production approaches: A review." *Critical Reviews in Biotechnology*. https://doi.org/10.1080/07388551.2024.2332927.

Hilgendorf, Karson, Yirong Wang, Michael J. Miller, and Yong-Su Jin. 2024. "Precision fermentation for improving the quality, flavor, safety, and sustainability of foods."*Current Opinion in Biotechnology* 86:103084. https://doi.org/10.1016/j.copbio.2024.103084.

Hoppenreijs, L. J. G., A. Annibal, G. J. C. Vreeke, R. M. Boom, and J. K. Keppler. 2024. "Food proteins from yeast-based precision fermentation: Simple purification of recombinant β-lactoglobulin using polyphosphate." *Food Research International* 176:113801. https://doi.org/10.1016/j.foodres.2023.113801.

Ijaola, Ahmed Olanrewaju, Damilola O. Akamo, Toyosi Timilehin George, Ayse Sengul, Musefiu Yemi Adedeji, and Eylem Asmatulu. 2024. "Algae as a potential source of protein: A review on cultivation, harvesting, extraction, and applications." *Algal Research-Biomass Biofuels and Bioproducts* 77. https://doi.org/10.1016/j.algal.2023.103329.

Jach, Monika Elzbieta, Anna Serefko, Maria Ziaja, and Marek Kieliszek. 2022. "Yeast protein as an easily accessible food source." *Metabolites* 12 (1):63. https://doi.org/10.3390/metabo12010063.

Jain, Surbhi, James Heffernan, Jitendra Joshi, Thomas Watts, Esteban Marcellin, and Chris Greening. 2023. "Microbial conversion of waste gases into single-cell protein." *Microbiology Australia* 44 (1):27–30. https://doi.org/10.1071/ma23007.

Jin, Kyoung Chan, Seung-Oh Seo, and Sun-Ki Kim. 2024. "Animal-free production of hen egg ovalbumin in engineered *Saccharomyces cerevisiae* via precision fermentation." *International Journal of Biological Macromolecules* 271:132479. https://doi.org/10.1016/j.ijbiomac.2024.132479.

Karabulut, Gulsah, Aleksandra Purkiewicz, and Gulden Goksen. 2024. "Recent developments and challenges in algal protein and peptide extraction strategies, functional and technological properties, bioaccessibility, and commercial applications." *Comprehensive Reviews in Food Science and Food Safety* 23 (3):e13372. https://doi.org/10.1111/1541-4337.13372.

Kawanishi, Yuki, and Sachihiro Matsunaga. 2021. "Synthetic carbon fixation: Conversion of heterotrophs into autotrophs by Calvin-Benson-Bassham cycle induction." *Cytologia* 86 (4):277–281. https://doi.org/10.1508/cytologia.86.277.

Kerckhof, Frederiek-Maarten, Myrsini Sakarika, Marie Van Giel, Maarten Muys, Pieter Vermeir, Jo De Vrieze, Siegfried E. Vlaeminck, Korneel Rabaey, and Nico Boon. 2021. "From biogas and hydrogen to microbial protein through co-cultivation of methane and hydrogen oxidizing bacteria." *Frontiers in Bioengineering and Biotechnology* 9:733753. https://doi.org/10.3389/fbioe.2021.733753.

Khoshnevisan, Benyamin, Li He, Mingyi Xu, Borja Valverde-Perez, Jani Sillman, Georgia-Christina Mitraka, Panagiotis G. Kougias, Yifeng Zhang, Shuiping Yan, Long Ji,

Michael Carbajales-Dale, Seyedeh Nashmin Elyasi, Hadis Marami, Panagiotis Tsapekos, Hongbin Liu, and Irini Angelidaki. 2022. "From renewable energy to sustainable protein sources: Advancement, challenges, and future roadmaps." *Renewable & Sustainable Energy Reviews* 157:112041. https://doi.org/10.1016/j.rser.2021.112041.

Kim, Hyongbum, and Jin-Soo Kim. 2014. "A guide to genome engineering with programmable nucleases." *Nature Reviews Genetics* 15 (5):321–334. https://doi.org/10.1038/nrg3686.

Knychala, Marilia M., Larissa A. Boing, Jaciane L. Ienczak, Debora Trichez, and Boris U. Stambuk. 2024. "Precision fermentation as an alternative to animal protein, a review." *Fermentation-Basel* 10 (6):315. https://doi.org/10.3390/fermentation10060315.

Kothari, Richa, Har Mohan Singh, Rifat Azam, Kajol Goria, Anu Bharti, Anita Singh, Somvir Bajar, Ashish Pathak, A. K. Pandey, and V. V. Tyagi. 2023. "Potential avenue of genetic engineered algal derived bioactive compounds: Influencing parameters, challenges and future prospects." *Phytochemistry Reviews* 22 (4):935–968. https://doi.org/10.1007/s11101-023-09859-y.

Leger, Dorian, Silvio Matassa, Elad Noor, Alon Shepon, Ron Milo, and Arren Bar-Even. 2021. "Photovoltaic-driven microbial protein production can use land and sunlight more efficiently than conventional crops." *Proceedings of the National Academy of Sciences of the United States of America* 118 (26):e2015025118. https://doi.org/10.1073/pnas.2015025118.

Li, Ku, Kaina Qiao, Jian Xiong, Hui Guo, Yuyu Zhang, and Pedro Vicente Martinez-Culebras. 2023a. "Nutritional values and bio-functional properties of fungal proteins: Applications in foods as a sustainable source." *Foods* 12 (24):4388. https://doi.org/10.3390/foods12244388.

Li, Rui, Xiaolei Fan, Yufeng Jiang, Ruonan Wang, Rongbo Guo, Yifeng Zhang, and Shanfei Fu. 2023b. "From anaerobic digestion to single cell protein synthesis: A promising route beyond biogas utilization." *Water Research* 243:120417. https://doi.org/10.1016/j.watres.2023.120417.

Linder, Tomas. 2023. "Beyond agriculture how microorganisms can revolutionize global food production." *ACS Food Science & Technology* 3 (7):1144–1152. https://doi.org/10.1021/acsfoodscitech.3c00099.

Liu, Dujuan, Sandra Garrigues, and Ronald P. de Vries. 2023. "Heterologous protein production in filamentous fungi." *Applied Microbiology and Biotechnology* 107 (16):5019–5033. https://doi.org/10.1007/s00253-023-12660-8.

Liu, Taiyu, Baolin Liu, Hualan Zhou, and Jianguo Zhang. 2021. "Knockout of the DAS gene increases *S*-adenosylmethionine production in *Komagataella phaffii*." *Biotechnology & Biotechnological Equipment* 35 (1):29–36. https://doi.org/10.1080/13102818.2020.1837012.

Liu, Yaozheng, William R. Aimutis, and MaryAnne Drake. 2024. "Dairy, plant, and novel proteins: Scientific and technological aspects." *Foods* 13 (7):1010. https://doi.org/10.3390/foods13071010.

Lubeck, Mette, and Peter Stephensen Lubeck. 2022. "Fungal cell factories for efficient and sustainable production of proteins and peptides." *Microorganisms* 10 (4):753. https://doi.org/10.3390/microorganisms10040753.

Ma, Junrui, Yifei Sun, Demei Meng, Zhongkai Zhou, Yuyu Zhang, and Rui Yang. 2023. "Yeast proteins: The novel and sustainable alternative protein in food applications." *Trends in Food Science & Technology* 135:190–201. https://doi.org/10.1016/j.tifs.2023.04.003.

Ma, Yu-Han, Xiang Wang, Ying-Fang Niu, Zhi-Kai Yang, Meng-Han Zhang, Zhong-Ming Wang, Wei-Dong Yang, Jie-Sheng Liu, and Hong-Ye Li. 2014. "Antisense knockdown of pyruvate dehydrogenase kinase promotes the neutral lipid accumulation in the diatom *Phaeodactylum tricornutum*." *Microbial Cell Factories* 13 (1):100. https://doi.org/10.1186/s12934-014-0100-9.

Manuel, Benedetti, Valeria Vecchi, Simone Barera, and Luca Dall'Osto.2018. "Biomass from microalgae: The potential of domestication towards sustainable biofactories." *Microbial Cell Factories* 17:173. https://doi.org/10.1186/s12934-018-1019-3.

Maumela, Pfariso, Shaunita Rose, Eugene van Rensburg, Annie Fabian Abel Chimphango, and Johann Ferdinand Gorgens. 2021. "Bioprocess optimisation for high cell density endoinulinase production from recombinant *Aspergilhts niger*." *Applied Biochemistry and Biotechnology* 193 (10):3271–3286. https://doi.org/10.1007/s12010-021-03592-y.

Mosibo, Ornella Kongi, Giovanna Ferrentino, and Chibuike C. Udenigwe. 2024. "Microalgae proteins as sustainable ingredients in novel foods: Recent developments and challenges." *Foods* 13 (5):733. https://doi.org/10.3390/foods13050733.

Murakami, Hiroki, Natsue Kakutani, Yunato Kuroyanagi, Masako Iwai, Koichi Hori, Mie Shimojima, and Hiroyuki Ohta. 2020. "MYB-like transcription factor NoPSR1 is crucial for membrane lipid remodeling under phosphate starvation in the oleaginous microalga *Nannochloropsis oceanica*." *FEBS Letters* 594 (20):3384–3394. https://doi.org/10.1002/1873-3468.13902.

Nielsen, M. B., A. S. Meyer, and J. Arnau. 2024. "The next food revolution is here: Recombinant microbial production of milk and egg proteins by precision fermentation." *Annual Review of Food Science and Technology* 15 (1):173–187. https://doi.org/10.1146/annurev-food-072023-034256.

Patel, Anil K., Eric L. Huang, Etienne Low-Decarie, and Mark G. Lefsrud. 2015. "Comparative shotgun proteomic analysis of wastewater-cultured microalgae: Nitrogen sensing and carbon fixation for frowth and nutrient removal in *Chlamydomonas reinhardtii*." *Journal of Proteome Research* 14 (8):3051–3067. https://doi.org/10.1021/pr501316h.

Salazar-Cerezo, Sonia, Ronald P. de Vries, and Sandra Garrigues. 2023. "Strategies for the development of industrial fungal producing strains." *Journal of Fungi* 9 (8):835. https://doi.org/10.3390/jof9080834.

Sander, Jeffry D., and J. Keith Joung. 2014. "CRISPR-Cas systems for editing, regulating and targeting genomes." *Nature Biotechnology* 32 (4):347–355. https://doi.org/10.1038/nbt.2842.

Shahzad, Hafiz Muhammad Aamir, Fares Almomani, Asif Shahzad, Khaled A. Mahmoud, and Kashif Rasool. 2024. "Challenges and opportunities in biogas conversion to microbial protein: A pathway for sustainable resource recovery from organic waste." *Process Safety and Environmental Protection* 185:644–659. https://doi.org/10.1016/j.psep.2024.03.055.

Singh, Satnam, Wee Swan Yap, Xiao Yu Ge, Veronica Lee Xi Min, and Deepak Choudhury. 2022. "Cultured meat production fuelled by fermentation." *Trends in Food Science & Technology* 120:48–58. https://doi.org/10.1016/j.tifs.2021.12.028.

Song, Hanwu, Jingjing Li, Qihui Su, Hongwu Li, Xujie Guo, Shengxi Shao, Liangliang Fan, Peilun Xu, Wenguang Zhou, and Jun Qian. 2024. "Insight into the mechanism of nitrogen sufficiency conversion strategy for microalgae-based ammonium-rich wastewater treatment." *Chemosphere* 349:140904. https://doi.org/10.1016/j.chemosphere.2023.140904.

Teng, Maofang, Shujie Ma, Yuanyang Zou, Jingwen Zhou, Jianghua Li, Guocheng Du, and Guoqiang Zhang. 2024. "Molecular modification and food-grade system construction for protein-glutaminase production in *Bacillus subtilis*." *Food Bioscience* 59:103932. https://doi.org/10.1016/j.fbio.2024.103932.

Tsuda, Masashi, and Koichi Nonaka. 2024. "Recent progress on heterologous protein production in methylotrophic yeast systems." *World Journal of Microbiology & Biotechnology* 40 (7):200. https://doi.org/10.1007/s11274-024-04008-9.

Wang, Dong-Hui, Meng-Yuan Zhu, Shu-Juan Lian, Hua Zou, Shan-Fei Fu, and Rong-Bo Guo. 2022a. "Conversion of renewable biogas into single-cell protein using a combined microalga- and methane-oxidizing bacterial system." *ACS Es&T Engineering* 2 (12):2317–2325. https://doi.org/10.1021/acsestengg.2c00237.

Wang, Jian, Zhenyu Huang, Quanjin Jiang, Hynek Roubik, Qihao Xu, Adem Gharsallaoui, Ming Cai, Kai Yang, and Peilong Sun. 2023. "Fungal solid-state fermentation of crops and their by-products to obtain protein resources: The next frontier of food industry." *Trends in Food Science & Technology* 138:628–644. https://doi.org/10.1016/j.tifs.2023.06.020.

Wang, Shuo, Qixin Liu, Ji Li, and Zhiwu Wang. 2021. "Methane in wastewater treatment plants: Status, characteristics, and bioconversion feasibility by methane oxidizing bacteria for high value-added chemicals production and wastewater treatment." *Water Research* 198:117122. https://doi.org/10.1016/j.watres.2021.117122.

Wang, Xiuwen, Junhuan Yang, Hassan Mohamed, Aabid Manzoor Shah, Shaoqi Li, Shuxian Pang, Chen Wu, Futing Xue, Wenyue Shi, Beenish Sadaqat, and Yuanda Song. 2022b. "Simultaneous overexpression of increment 6-, increment 12-and increment 9-desaturases enhanced the production of γ-linolenic acid in *Mucor circinelloides* WJ11." *Frontiers in Microbiology* 13:1078157. https://doi.org/10.3389/fmicb.2022.1078157.

Woern, Carlos, and Lutz Grossmann. 2023. "Microbial gas fermentation technology for sustainable food protein production." *Biotechnology Advances* 69:108240. https://doi.org/10.1016/j.biotechadv.2023.108240.

Yang, Hui, Chaonan Song, Chengwei Liu, and Pengchao Wang. 2024. "Synthetic biology tools for engineering *Aspergillus oryzae*." *Journal of Fungi* 10 (1):34. https://doi.org/10.3390/jof10010034.

Zeng, Baohong, Kajsa Nilsson, Paulo Gonsalves Teixeira, and Bjorn Bergenstahl. 2023. "Study of mycoprotein extraction methods and its functional properties." *Colloids and Surfaces a-Physicochemical and Engineering Aspects* 659:130800. https://doi.org/10.1016/j.colsurfa.2022.130800.

Zhang, Quanwei, Yaokang Wu, Mengyue Gong, Hongzhi Zhang, Yanfeng Liu, Xueqin Lv, Jianghua Li, Guocheng Du, and Long Liu. 2021. "Production of proteins and commodity chemicals using engineered *Bacillus subtilis* platform strain." In *Microbial Cell Factories-Book*, edited by D. Mattanovich and P. I. Nikel, 173–185. https://doi.org/10.1042/EBC20210011

Zhao, Yan, Zhaowei Han, Xuchun Zhu, Bingyu Chen, Linyi Zhou, Xiaoyong Liu, and Hongzhi Liu. 2024. "Yeast proteins: Proteomics, extraction, modification, functional characterization, and structure: A review." *Journal of Agricultural and Food Chemistry*. https://doi.org/10.1021/acs.jafc.4c04821.

Zhou, Yingdong, Li Liu, Mingyu Li, and Changwei Hu. 2022. "Algal biomass valorisation to high-value chemicals and bioproducts: Recent advances, opportunities and challenges." *Bioresource Technology* 344. https://doi.org/10.1016/j.biortech.2021.126371.

Zhu, Yuqi, Yili Zhang, Hui Chen, Lisha Zhang, and Chensi Shen. 2024. "Stress responses and ammonia nitrogen removal efficiency of *Oocystis lacustris* in saline ammonium-contaminated wastewater treatment." *Toxics* 12 (5):353. https://doi.org/10.3390/toxics12050353.

19 Nanomaterials-Incorporated Intelligent Food Packing for the Detection and Prevention of Microbial Contamination and Spoilage

Seyma Dadi, Nimet Temur, Emine Ozdemir, Mustafa Nisari, Ugur Azizoglu, and Ismail Ocsoy

19.1 INTRODUCTION

Food packaging plays a vital role in preserving and maintaining the nutritional quality and safety of food until it reaches consumers. In addition to that, food packaging enables the distribution of food and also facilitates communication at the consumer level, providing information about food such as storage conditions and ingredients (Mihindukulasuriya & Lim, 2014). Therefore, the food packaging industry is in high demand, and it is one of the fastest-growing commercial sectors in the global market (Cheng et al., 2022). The global food packaging market size was valued at USD 362.9 billion in 2022, and its estimated annual growth rate of 5.7% from 2023 to 2030 highlights its significant position in the global market (Sezer et al., 2024).

The primary purpose of traditional food packaging is to provide containment and protect them from internal and external unfavorable conditions, such as oxygen, microorganisms, light, and heat (Shivakumar et al., 2018). Temperature fluctuations disrupt the humidity balance inside the food package, leading to food spoilage or pathogenic microbial growth during storage (Bibi et al., 2016; Aloui & Khwaldia, 2016). A variety of pathogenic microorganisms, including *Salmonella*, *Escherichia coli*, *Bacillus cereus*, and *Clostridium perfringens*, in association with the toxic substances they produce, can result in foodborne illnesses (Nummer et al., 2012). An inadequate or compromised packaging can facilitate the entry and growth of microorganisms, posing risks to food safety and quality. To address these problems,

 DOI: 10.1201/9781003540113-19

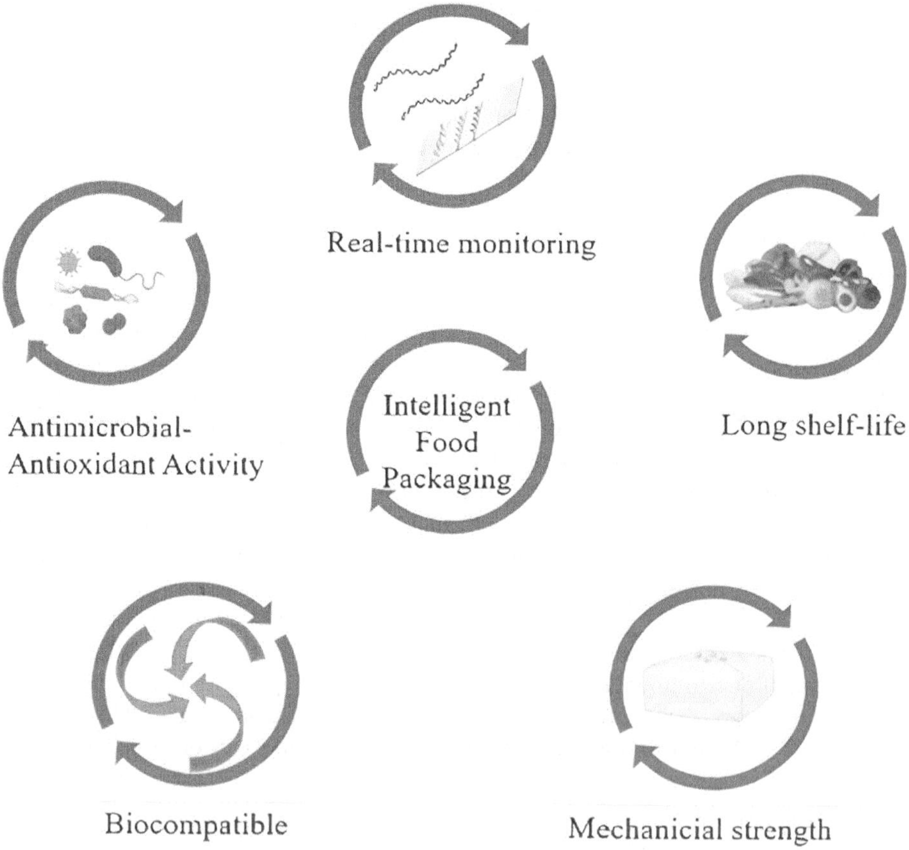

FIGURE 19.1 Benefits of intelligent food packaging.

innovative food packaging solutions have been developed to detect and report food quality (Brody et al., 2008). Active and intelligent packaging, as innovative food packaging, has represented significant progress in the field of packaging technology (Carina et al., 2021). These new techniques provide for extending shelf life, improving food quality and safety, and minimizing the need to use food preservatives (Perera et al., 2023; Dobrucka & Przekop, 2019). An *intelligent packaging* is a system that offers detailed information on internal and external environmental conditions, such as temperature, freshness, and more through a sensor and indicator technology (Perera et al., 2023; Firouz et al., 2021). The following diagram illustrates the benefits of intelligent food packaging (Figure 19.1).

In recent years, nanomaterials (NMs), such as silver (Ag), zinc oxide (ZnO), and carbon nanotubes (CNTs), have been involved in the food packaging technology. In addition to extending the shelf life of food by releasing antioxidants or flavor ingredients, NMs improve the barrier properties of food packaging against oxygen, moisture, and other gases. In this chapter, we provide an extensive overview on NMs-based intelligent food package systems.

19.2 DEVELOPMENT OF NANOMATERIALS FOR INTELLIGENT FOOD PACKAGE

Traditional food packaging results in a short shelf life and degradation due to microbial contamination (Chelliah et al., 2021; Mihindukulasuriya & Lim, 2014; Vanderroost et al., 2014). By using intelligent food packaging under proper storage conditions, food is preserved for longer periods of time and potential foodborne illnesses are prevented (Jafarzadeh et al., 2024; Moustafa et al., 2024). The use of various indicators in intelligent food packaging allows the consumer to monitor food quality in real time in terms of detection of food spoilage, microbial contamination, and environmental changes, including temperature, pH values, etc. (Babu, 2022; Cheng et al., 2022; Dodero et al., 2021).

NMs exhibit many different properties, such as electrical, optical, and catalytic effects, with high surface reactivity with increased surface area/volume ratio compared to bulk materials (Bumbudsanpharoke et al., 2015; Correa et al., 2017). They can be used in designing various biosensor applications (Mustafa & Andreescu, 2020; Siddiqui et al., 2022; Thirumalai et al., 2023). In particular, the durability of metallic NMs and their suitability for sensor applications, the use of polymers in the production of environmentally friendly and healthy packaging, and their widespread use with commercial potential are attracting attention (Chelliah et al., 2022; Chelliah et al., 2021). Organic-inorganic, metallic, polymeric, and carbon-based NMs are widely used in intelligent food packaging (Ahmad et al., 2024; Ghosh et al., 2022; Moustafa et al., 2024).

19.2.1 Metallic Nanomaterials in Intelligent Food Packaging

NMs have the ability to exert antimicrobial, antibiofilm, and antioxidant activity due to their unique physicochemical and biological properties (Hosseini et al., 2021; Nikolic et al., 2021; Siddiqui et al., 2022). With the use of metal oxide NMs in intelligent food packaging, oxidative stress occurs as a result of the accumulation of free hydroxyl ions and free oxygen radicals around microorganisms, disrupting their vital functions by causing DNA damage, cell wall damage, and protein denaturation (Omerović et al., 2021; Stuparu-Cretu et al., 2023). In addition, the use of metal oxide NMs in packaging contributes to long shelf life food technology by producing more robust, more durable, and less permeable packaging in contact with the outside (Nikolic et al., 2021). Ag NMs are attracting attention in the field of food packaging due to their strong antimicrobial and antioxidant properties (Rai et al., 2019). They are widely used for sensors that can obtain colorimetric responses with high reactivity, catalytic, and enzyme-like activities (Gao, 2023; Mustafa & Andreescu, 2020). In a previous study, rational Ag/TiO_2/CS nanocomposites (NCs) were synthesized using a practical method called photoreduction. The chitosan used in the synthesized NCs exhibited Ag^+ reduction under a TiO_2 catalyst, and its activity against microorganisms was observed by its long-term stability (Lin et al., 2015). TiO_2 NCs were developed to increase biocompatibility, avoid toxic substances, and prevent biofilm formation with chitosan, and antimicrobial effect against *Staphylococcus aureus* and *Escherichia coli* bacteria (Shafie et al., 2024).

19.2.2 Polymeric Nanomaterials in Intelligent Food Packaging

The use of biocompatible polymer materials in packaging that comes into direct contact with food is becoming increasingly important (Qin et al., 2014; Rydz et al., 2018; Seok, 2021). Non-biodegradable polymers have been widely used in packaging. However, this causes both food degradation and environmental pollution and poses a health threat. For this reason, interest in the production of biodegradable, environmentally friendly, and renewable food packaging is increasing day by day (Sultana et al., 2022). Biocompatible active polymers can have superior antimicrobial and antioxidant effects (Nasution et al., 2023). Thus, they prevent food contamination and ensure long shelf life of foods (Ludwicka et al., 2020). In addition, biodegradable polymers can gain new properties by being functionalized with other NMs, and intelligent food packaging makes it possible to produce multifunctional food packages (Makvandi et al., 2021; Vieira et al., 2022). Biocompatible, renewable biopolymer materials are derived from natural sources, such as cellulose, chitosan, starch, vegetable oils, and proteins (Liu et al., 2021; Mohamed et al., 2020; Shah et al., 2024). Non-toxic carboxymethyl cellulose (CMC) and double-layered hydroxide nanosheets (LDH-NSs) were developed against petroleum-based packaging (Zhao et al., 2024).

19.3 NANOMATERIAL-BASED MICROBIAL CONTAMINATION AND SPOILAGE SENSING TECHNOLOGIES IN INTELLIGENT FOOD PACKAGE

Food loss and waste is one of the global issues that significantly affect food security and sustainability. Food waste not only is an economic issue but also has environmental effect in terms of global warming potential (Amicarelli et al., 2021). The United Nations Environment Programme's (UNEP) Food Waste Index Report highlights the alarming scale of food waste. Whereas an estimated 931 million tons of food were wasted in 2019 alone, this value increased to 1.05 billion tons in 2022 (UNEP, 2021, 2024). In recent years, significant studies have been conducted on food contamination and the detection of food pathogens. This section provides a summary about some of these studies.

19.3.1 Biogenic Amine Monitoring

In general, the quantification of biogenic amines is carried out with analytical techniques, such as gas chromatography, high-pressure liquid chromatography, and capillary electrophoresis. Although quantification of biogenic amines can be achieved using these methods sensitively, they demand costly equipment, specialized personnel, and complex, multi-step experimental procedures (Khan et al., 2024).

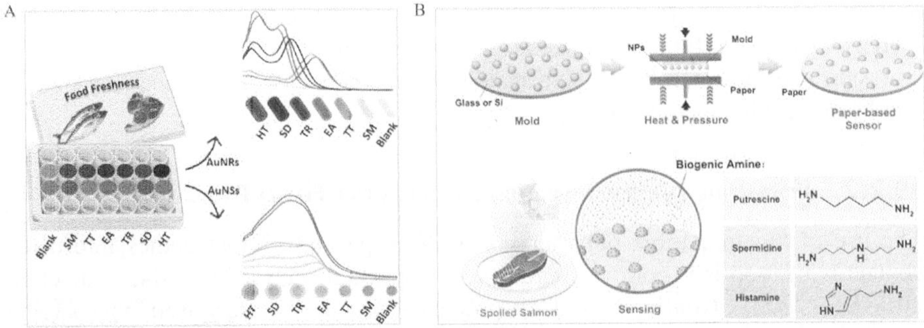

FIGURE 19.2 (a) UV-Vis spectra and images of the sensor elements AuNRs and AuNSs. Reproduced with permission from Reference (Orouji et al., 2021). (b) Schematic illustration of the imprinting of metal NPs onto an inkjet paper and NP-embedded paper providing gas sensor for the detection of biogenic amine released from spoiled salmon. Reproduced with permission from Reference (Tseng et al., 2017).

Gholampour et al. reported on a new colorimetric sensor based on a reaction of the hydrolysis-induced cadmium metallization by biogenic amines. This reaction led to the tuning of the LSPR peak of gold (Au) NPs. The color change from red to blue was correlated with the generated biogenic amine in meat during spoilage. The authors argued that this system is instrument-free, sensitive, cost-effective, and easy (Gholampour et al., 2021). The multicolor sensor array was generated by Orouji et al. for the discrimination and determination of biogenic amines, such as spermine, spermidine, ethylenediamine, tryptamine, histamine, and tyramine, during food spoilage. As shown in Figure 19.2a, the Ag metallization of Au NRs and Au NSs causes a blue shift of the LSPR peak, resulting in a rainbowlike color variation in the solution. The developed multicolorimetric sensor array was able to detect the biogenic amines during food storage, thus enabling in situ monitoring of food spoilage with the naked eye in a rapid manner (Orouji et al., 2021).

Tseng et al. (2017) generated refractometric sensors through the embedding of Ag NPs, Au NPs, and hollow Au-Ag NPs onto flexible paper using reversal nanoimprint lithography (Figure 19.2b). They found that hollow Au-Ag NPs containing paper had a critical wavelength dip shift upon the adsorption of putrescine and vapors, resulting in a LOD of 13.8 ppm for spiked putrescine (Tseng et al., 2017).

Recently, fluorescence sensing technology having high sensitivity and good stability represents a significant advancement in real-time monitoring of food spoilage with the naked eye. Song et al. developed a ratiometric fluorescence sensor with dual-emission carbon quantum dot (CQD) for monitoring food spoilage. Two different hydrophobic electrospun film CQD/polyvinylidene fluoride (PVDF) film and CQD@PVDF film represented fluorescence color change from yellow-green to blue in the presence of trimethylamine (TMA) with LODs of 1.04 μM and 2.1 μM, respectively (Song et al., 2024). In another study, a ratiometric fluorescence sensor strip was prepared by using D-penicillamine-capped Ag nanoclusters (DPA-Ag NCs) as indicator and red-emitting bimetallic AuAg NCs as internal reference for real-time monitoring of food spoilage (Zhang et al., 2023).

19.3.2 pH Monitoring

Aerobic and anaerobic microorganisms proliferating during the storage period in food packaging can affect the pH of the environment due to the accumulation of their metabolic products, such as lactic acid and acetic acid, produced as a result of glucose fermentation (Yousefi et al., 2019; Xia et al., 2024). pH monitoring is critical as it provides information to consumers about food quality before purchase or during consumption (Balbinot-Alfaro et al., 2019). Yan et al. developed pH-responsive dual-emission ratiometric fluorescent sensor for the detection of eight kinds of biogenic amines. The sensor was constructed from fluorescent filter paper containing pH-responsive carbon dots and Rhodamine B. Under 365 nm light irradiation, the fluorescent filter paper–based sensor exhibited a color change, indicating the presence of biogenic amine vapor released from spoiled mushroom and shrimp (Yan et al., 2023).

The anthocyanin group, known as pH indicator, has been actively used in colorimetric tests due to their easy and rapid pH-dependent protonation and deprotonation (Celik et al., 2021, 2020a, 2020b, 2023). Recently, pH-sensitive natural dyes like anthocyanin have been incorporated into polymeric films composed of natural polymers, including chitosan and cellulose, as well as synthetic polymers, such as polyvinyl alcohol (PVA), to develop intelligent package film. The color of the film changed from red to green during the process of fish spoilage, demonstrating its potential use as a packaging label (Ma et al., 2018). The anthocyanin/pectin was filled with MIL-100(Fe) to form the color of the films. They were exposed to ammonia vapor, then their purple color turned to blue-violet and to green (Huang et al., 2024).

19.3.3 Biosensor for Pathogen Detection

Microorganisms are among the leading reasons for food contamination during livestock handling, food processing, and food packaging (Yousefi et al., 2018). Microbial detection methods thus require advanced analytical laboratory techniques, including immunological analysis, bacterial culture, polymerase chain reaction (PCR), and enzyme-linked immunosorbent assays (ELISA) (Kim & Kim, 2021). Although these methods are highly sensitive, they are not practical for daily use and are very time-consuming and costly. However, there are two methods for microorganism detection that can be used in intelligent food packaging: (1) lateral flow immunoassay (LFIA) technique and (2) DNAzyme probe technology (Cheng et al., 2022).

The working principle of LFIA is based on the fact that immunochemical reactions are performed in a chromatographic membrane/pad system (Kasetsirikul et al., 2020; Nan et al., 2023). Wen et al. developed LFIA test systems combined with Au-Fe_3O_4 NPs for dual-mode detection of *Salmonella typhimurium.* Au provided colorimetric signal due to having LSPR property, and the Fe_3O_4 part was responsible for magnetic separation, enabling improved detection sensitivity (Wen et al., 2023).

Another study about *Salmonella* detection in food was conducted by Silva et al. An Au NP–based LFIA test system showed a LOD of 10^3 CFU·mL^{-1} and a LOQ of 1 CFU·mL^{-1}. Additionally, the test system accomplished the detection of *Salmonella* contamination in chicken, egg, milk, chocolate, and black pepper (Silva et al., 2024).

Tasbasi et al. generated LFIA by using aptamer-gated and 3, 3′, 5, 5′-Tetramethylbenzidine (TMB)–loaded mesoporous silica NPs for the detection of *Listeria monocytogenes* in minced chicken samples. Aptamer bacteria interaction led to the release of TMB molecule in the conjugation pad, and TMB molecules migrated to the test line immobilized horseradish peroxidase (HRP) and generated a blue color by oxidation reaction in the presence of hydrogen peroxide (Tasbasi et al., 2019).

The DNAzyme sensor can also be used for the in situ detection of bacteria in food (Ma et al., 2021). Qin et al. developed a fluorescent DNAzyme biosensor for the detection of *Pseudomonas aeruginosa* in food samples. DNAzyme was immobilized onto a polystyrene board to construct an indicator plate sensor. The resultant platform demonstrated a LOD of 1.2 CFU mL^{-1} after 10 minutes (Qin et al., 2021). A DNAzyme-based luminescent biosensor was constructed for the detection of *Salmonella*. The color change of the samples containing 1.5 copies/μL of *Salmonella* genome was observed with the naked eye in 10 minutes (Li et al., 2020). You et al. designed a phage-based fluorescent biosensor for the detection of *E. coli* O157:H7. The phage@DNAzyme probe displayed specificity toward *E. coli* O157:H7 and catalyzed the copper(II)-based azide-alkyne cycloaddition click reaction to generate a fluorescent signal. The signal was measured by a smartphone to quantify the *E. coli* with a LOD of 50 CFU mL^{-1} (You et al., 2023).

19.4 NANOMATERIAL-BASED PREVENTION OF MICROBIAL CONTAMINATION AND SPOILAGE IN INTELLIGENT FOOD PACKAGE

The advances in nanotechnology have made food packaging more effective and intelligent (Sethunga et al., 2024; Sultana et al., 2022). The incorporation of NMs in food packaging reduces environmental impact, enhances food safety, prevents quality degradation, and prolongs freshness (Sultana et al., 2022). Sethunga et al. modified cellulose acetate nanofibers (CANFs) with cinnamon and clove oils, reducing microbial spoilage in fruits and vegetables by 50% (Sethunga et al., 2024). MOFs effectively regulate ethylene concentrations to slow ripening, especially in ethylene-sensitive fruits (Sultana et al., 2022). Meat and dairy products, which are highly susceptible to microbial contamination, benefit significantly from antimicrobial and sensor packaging. For instance, AgNP coatings and pH indicators are critical in preserving the freshness of these products (Musaev et al., 2024). Liao et al. systematically investigated the synthesis of conductive materials, such as Au, Ag, and Cu, in conductive inks. They proposed the development of low-cost printed sensors and their use as indicators of food spoilage in packaging. Conductive electrodes are applied to the food package and convert the signal resulting from chemical reactions into an electrical signal, providing an effective sensing mechanism for food safety (Liao et al., 2019). The production and integration of NMs into packaging systems remain costly. However, advancements in production technologies are expected to reduce these costs (Dodero et al., 2021; Rathee et al., 2024).

Nano-sized components or antioxidant-loaded carrier systems have been developed to enhance the antioxidant activity in food packaging through several mechanisms (Alp-Erbay, 2022). Vera et al. demonstrated the antioxidant capacity of selenium nanoparticles (SeNPs) by producing an antioxidant-layered packaging and tested the packaging efficiency by coating hazelnut and walnut packages with antioxidant films (Vera et al., 2018). Naturally derived starch-based nanofilms are used in packaging materials to prevent or retard food oxidation. Hosseini et al. carried out an antioxidant evaluation of starch-based nanofilms in the presence of magnesium oxide NPs (MgO NPs) (Hosseini et al., 2021). Rizzotto et al. synthesized and evaluated Fe_2TiO_5 NPs to limit oxygen transfer and free radicals that cause food spoilage. DPPH (1,1-Diphenyl-2-picrylhydrazyl) antioxidant assay test showed strong radical scavenging. No cytotoxicity was observed in NPs. This packaging film can be used in food packaging due to both antimicrobial and antioxidant activity (Rizzotto et al., 2022).

19.5 CONCLUSION AND FUTURE PERSPECTIVES

Microbial contamination and food spoilage are a significant concern as these can influence food quality, food safety, and nutritional value. Active and intelligent packaging, as an innovative approach to food packaging, has marked a significant progress in the field of packaging technology. Intelligent food packaging technology aims not only to protect food and reduce food waste but also to provide information about food quality to consumers during transportation, distribution, and storage. Moreover, packaging materials with antimicrobial and antioxidant properties in intelligent packaging help prevent the proliferation of pathogenic microorganisms in food, thereby enhancing food safety. NMs have been integrated into intelligent food packaging, leading to the development of more effective packaging systems. NM-incorporated intelligent packaging extends the shelf life of food by inhibiting the growth of pathogenic microorganisms through the antimicrobial properties of NMs. In addition, nanosensors used for pathogen detection provide information to the consumer about food safety by real-time monitoring of factors, such as pH and gas level, of the food. In the future, the development of new, cost-reducing technologies will be necessary to enable its widespread use. Moreover, cost-effective intelligent packaging systems are expected to contribute to a sustainable food supply by ensuring efficient resource utilization and reducing food waste.

REFERENCES

Ahmad, K., Li, Y., Tu, C., Guo, Y., Yang, X., Xia, C., & Hou, H. (2024). Nanotechnology in food packaging with implications for sustainable outlook and safety concerns. *Food Bioscience*, 103625.

Aloui, H., & Khwaldia, K. (2016). Natural antimicrobial edible coatings for microbial safety and food quality enhancement. *Comprehensive Reviews in Food Science and Food Safety*, *15*(6), 1080–1103.

Alp-Erbay, E. (2022). Nanomaterials utilized in food packaging: State-of-the-art. *Food Engineering Reviews*, *14*(4), 629–654.

Amicarelli, V., Lagioia, G., & Bux, C. (2021). Global warming potential of food waste through the life cycle assessment: An analytical review. *Environmental Impact Assessment Review*, *91*, 106677.

Babu, P. J. (2022). Nanotechnology mediated intelligent and improved food packaging. *International Nano Letters*, *12*(1), 1–14.

Balbinot-Alfaro, E., Craveiro, D. V., Lima, K. O., Costa, H. L. G., Lopes, D. R., & Prentice, C. (2019). Intelligent packaging with pH indicator potential. *Food Engineering Reviews*, *11*, 235–244.

Bibi, F., Guillaume, C., Vena, A., Gontard, N., & Sorli, B. (2016). Wheat gluten, a bio-polymer layer to monitor relative humidity in food packaging: Electric and dielectric characterization. *Sensors and Actuators A: Physical*, *247*, 355–367.

Brody, A. L., Bugusu, B., Han, J. H., Sand, C. K., & McHugh, T. H. (2008). Innovative food packaging solutions. *Journal of Food Science*, *73*(8), 107–116.

Bumbudsanpharoke, N., Choi, J., & Ko, S. (2015). Applications of nanomaterials in food packaging. *Journal of Nanoscience and Nanotechnology*, *15*(9), 6357–6372.

Carina, D., Sharma, S., Jaiswal, A. K., & Jaiswal, S. (2021). Seaweeds polysaccharides in active food packaging: A review of recent progress. *Trends in Food Science & Technology*, *110*, 559–572.

Celik, C., Can Sezgin, G., Kocabas, U. G., Gursoy, S., Ildiz, N., Tan, W., & Ocsoy, I. (2021). Novel anthocyanin-based colorimetric assay for the rapid, sensitive, and quantitative detection of helicobacter pylori. *Analytical Chemistry*, *93*(15), 6246–6253.

Celik, C., Ildiz, N., Kaya, M. Z., Kilic, A. B., & Ocsoy, I. (2020a). Preparation of natural indicator incorporated media and its logical use as a colorimetric biosensor for rapid and sensitive detection of Methicillin-resistant Staphylococcus aureus. *Analytica Chimica Acta*, *1128*, 80–89.

Celik, C., Ildiz, N., Sagiroglu, P., Atalay, M. A., Yazici, C., & Ocsoy, I. (2020b). Preparation of nature inspired indicator based agar for detection and identification of MRSA and MRSE. *Talanta*, *219*, 121292.

Celik, C., Kalin, G., Cetinkaya, Z., Ildiz, N., & Ocsoy, I. (2023). Recent advances in colorimetric tests for the detection of infectious diseases and antimicrobial resistance. *Diagnostics*, *13*(14), 2427.

Chelliah, R., Khan, I., Wei, S., Madar, I. H., Sultan, G., Daliri, E. B.-M., . . . Oh, D. H. (2022). Intelligent packaging systems: Food quality and intelligent medicine box based on nano-sensors. In *Smart Nanomaterials in Biomedical Applications* (pp. 555–587): Springer.

Chelliah, R., Wei, S., Daliri, E. B.-M., Rubab, M., Elahi, F., Yeon, S.-J., . . . Oh, D. H. (2021). Development of nanosensors based intelligent packaging systems: Food quality and medicine. *Nanomaterials*, *11*(6), 1515.

Cheng, H., Xu, H., McClements, D. J., Chen, L., Jiao, A., Tian, Y., . . . Jin, Z. (2022). Recent advances in intelligent food packaging materials: Principles, preparation and applications. *Food Chemistry*, *375*, 131738.

Correa, D. S., Pavinatto, A., Mercante, L. A., Mattoso, L. H., Oliveira, J. E., & Riul Jr, A. (2017). Chemical sensors based on hybrid nanomaterials for food analysis. In *Nanobiosensors* (pp. 205–244): Elsevier.

Dobrucka, R., & Przekop, R. (2019). New perspectives in active and intelligent food packaging. *Journal of Food Processing and Preservation*, *43*(11), e14194.

Dodero, A., Escher, A., Bertucci, S., Castellano, M., & Lova, P. (2021). Intelligent packaging for real-time monitoring of food-quality: Current and future developments. *Applied Sciences*, *11*(8), 3532.

Firouz, M. S., Mohi-Alden, K., & Omid, M. (2021). A critical review on intelligent and active packaging in the food industry: Research and development. *Food Research International*, *141*, 110113.

Gao, Y. (2023). *Point-of-care colorimetric sensors for disease diagnosis and food monitoring*. UNSW Sydney.

Gholampour, S., Jalali, H., Zhiani, R., Rashidi, H., & Motavalizadehkakhky, A. (2021). Biogenic amines to tune the LSPR adsorption peak of gold NPs for intelligent packaging application. *Inorganic Chemistry Communications*, *123*, 108334.

Ghosh, T., Raj, G. B., & Dash, K. K. (2022). A comprehensive review on nanotechnology based sensors for monitoring quality and shelf life of food products. *Measurement: Food*, *7*, 100049.

Hosseini, S. N., Pirsa, S., & Farzi, J. (2021). Biodegradable nano composite film based on modified starch-albumin/MgO; antibacterial, antioxidant and structural properties. *Polymer Testing*, *97*, 107182.

Huang, K., Wang, L., Deng, Y., Zheng, H., Wu, S., Li, Z., . . . Guo, Z. (2024). Development of amine-sensitive intelligent film with MIL-100 (Fe) as function filler based on anthocyanins/pectin for monitoring chilled meat freshness. *International Journal of Biological Macromolecules*, *270*, 132463.

Jafarzadeh, S., Nooshkam, M., Qazanfarzadeh, Z., Oladzadabbasabadi, N., Strachowski, P., Rabiee, N., . . . Naebe, M. (2024). Unlocking the potential of 2D nanomaterials for sustainable intelligent packaging. *Chemical Engineering Journal*, *490*, 151711.

Kasetsirikul, S., Shiddiky, M. J., & Nguyen, N.-T. (2020). Challenges and perspectives in the development of paper-based lateral flow assays. *Microfluidics and Nanofluidics*, *24*, 1–18.

Khan, S., Monteiro, J. K., Prasad, A., Filipe, C. D., Li, Y., & Didar, T. F. (2024). Material breakthroughs in smart food monitoring: Intelligent packaging and on-site testing technologies for spoilage and contamination detection. *Advanced Materials*, *36*(1), 2300875.

Kim, S. O., & Kim, S. S. (2021). Bacterial pathogen detection by conventional culture-based and recent alternative (polymerase chain reaction, isothermal amplification, enzyme linked immunosorbent assay, bacteriophage amplification, and gold nanoparticle aggregation) methods in food samples: A review. *Journal of Food Safety*, *41*(1), e12870.

Li, S., Zhang, Y., Tian, J., & Xu, W. (2020). Luminescent DNAzyme and universal blocking linker Super Polymerase Chain Reaction visual biosensor for the detection of *Salmonella*. *Food Chemistry*, *324*, 126859.

Liao, Y., Zhang, R., & Qian, J. (2019). Printed electronics based on inorganic conductive nanomaterials and their applications in intelligent food packaging. *RSC Advances*, *9*(50), 29154–29172.

Lin, B., Luo, Y., Teng, Z., Zhang, B., Zhou, B., & Wang, Q. (2015). Development of silver/titanium dioxide/chitosan adipate nanocomposite as an antibacterial coating for fruit storage. *LWT-Food Science and Technology*, *63*(2), 1206–1213.

Liu, Y., Ahmed, S., Sameen, D. E., Wang, Y., Lu, R., Dai, J., . . . Qin, W. (2021). A review of cellulose and its derivatives in biopolymer-based for food packaging application. *Trends in Food Science & Technology*, *112*, 532–546.

Ludwicka, K., Kaczmarek, M., & Białkowska, A. (2020). Bacterial nanocellulose—A biobased polymer for active and intelligent food packaging applications: Recent advances and developments. *Polymers*, *12*(10), 2209.

Ma, Q., Liang, T., Cao, L., & Wang, L. (2018). Intelligent poly (vinyl alcohol)-chitosan nanoparticles-mulberry extracts films capable of monitoring pH variations. *International Journal of Biological Macromolecules*, *108*, 576–584.

Ma, X., Ding, W., Wang, C., Wu, H., Tian, X., Lyu, M., & Wang, S. (2021). DNAzyme biosensors for the detection of pathogenic bacteria. *Sensors and Actuators B: Chemical*, *331*, 129422.

Makvandi, P., Iftekhar, S., Pizzetti, F., Zarepour, A., Zare, E. N., Ashrafizadeh, M., . . . Sillanpaa, M. (2021). Functionalization of polymers and nanomaterials for water treatment, food packaging, textile and biomedical applications: A review. *Environmental Chemistry Letters*, *19*, 583–611.

Mihindukulasuriya, S., & Lim, L.-T. (2014). Nanotechnology development in food packaging: A review. *Trends in Food Science & Technology*, *40*(2), 149–167.

Mohamed, S. A., El-Sakhawy, M., & El-Sakhawy, M. A.-M. (2020). Polysaccharides, protein and lipid-based natural edible films in food packaging: A review. *Carbohydrate Polymers*, *238*, 116178.

Moustafa, H., Hemida, M. H., Nour, M. A., & Abou-Kandil, A. I. (2024). Intelligent packaging films based on two-dimensional nanomaterials for food safety and quality monitoring: Future insights and roadblocks. *Journal of Thermoplastic Composite Materials*, 08927057241264802.

Musaev, E., Cantù, E., Soprani, M., Serpelloni, M., Sardini, E., Ponzoni, A., . . . Baratto, C. (2024). Paper sensors for advanced smart packaging: Route to detection on the shelf and in real ambient. *IEEE Sensors Journal*, *20*, 31598–31605.

Mustafa, F., & Andreescu, S. (2020). Nanotechnology-based approaches for food sensing and packaging applications. *RSC Advances*, *10*(33), 19309–19336.

Nan, X., Yang, L., & Cui, Y. (2023). Lateral flow immunoassay for proteins. *Clinica Chimica Acta*, *544*, 117337.

Nasution, H., Harahap, H., Julianti, E., Safitri, A., & Jaafar, M. (2023). Smart packaging based on polylactic acid: The effects of antibacterial and antioxidant agents from natural extracts on physical–mechanical properties, colony reduction, perishable food shelf life, and future prospective. *Polymers*, *15*(20), 4103.

Nikolic, M. V., Vasiljevic, Z. Z., Auger, S., & Vidic, J. (2021). Metal oxide nanoparticles for safe active and intelligent food packaging. *Trends in Food Science & Technology*, *116*, 655–668.

Nummer, B., Shrestha, S., & Smith, J. (2012). Survival of Salmonella in a high sugar, low water-activity, peanut butter flavored candy fondant. *Food Control*, *27*(1), 184–187.

Omerović, N., Djisalov, M., Živojević, K., Mladenović, M., Vunduk, J., Milenković, I., . . . Vidić, J. (2021). Antimicrobial nanoparticles and biodegradable polymer composites for active food packaging applications. *Comprehensive Reviews in Food Science and Food Safety*, *20*(3), 2428–2454.

Orouji, A., Ghasemi, F., Bigdeli, A., & Hormozi-Nezhad, M. R. (2021). Providing multicolor plasmonic patterns with Au@ Ag core–shell nanostructures for visual discrimination of biogenic amines. *ACS Applied Materials & Interfaces*, *13*(17), 20865–20874.

Perera, K. Y., Pradhan, D., Rafferty, A., Jaiswal, A. K., & Jaiswal, S. (2023). A comprehensive review on metal oxide-nanocellulose composites in sustainable active and intelligent food packaging. *Food Chemistry Advances*, 100436.

Qin, M., Ma, X., Fan, S., Wu, H., Yan, W., Tian, X., . . . Wang, S. (2021). Rapid detection of *Pseudomonas aeruginosa* using a DNAzyme-based sensor. *Food Science & Nutrition*, *9*(7), 3873–3884.

Qin, Y., Howlader, M. M., Deen, M. J., Haddara, Y. M., & Selvaganapathy, P. R. (2014). Polymer integration for packaging of implantable sensors. *Sensors and Actuators B: Chemical*, *202*, 758–778.

Rai, M., Ingle, A. P., Gupta, I., Pandit, R., Paralikar, P., Gade, A., . . . dos Santos, C. A. (2019). Smart nanopackaging for the enhancement of food shelf life. *Environmental Chemistry Letters*, *17*, 277–290.

Rathee, S., Singh, K. R., Mallick, S., Singh, J., Pandey, S. S., Ojha, A., & Singh, R. P. (2024). Smart alginate nanomaterials: Revolutionizing food across delivery, preservation, packaging, safety, and waste upcycling. *Carbohydrate Polymer Technologies and Applications*, 100568.

Rizzotto, F., Vasiljevic, Z. Z., Stanojevic, G., Dojcinovic, M. P., Jankovic-Castvan, I., Vujancevic, J. D., . . . Vidic, J. (2022). Antioxidant and cell-friendly Fe2TiO5 nanoparticles for food packaging application. *Food Chemistry*, *390*, 133198.

Rydz, J., Musioł, M., Zawidlak-Węgrzyńska, B., & Sikorska, W. (2018). Present and future of biodegradable polymers for food packaging applications. *Biopolymers for Food Design*, 431–467.

Seok, S. (2021). Polymer-based biocompatible packaging for implantable devices: Packaging method, materials, and reliability simulation. *Micromachines*, *12*(9), 1020.

Sethunga, M., Gunathilake, K. D. P. P., Ranaweera, K. K. D. S., & Munaweera, I. (2024). Antimicrobial and antioxidative electrospun cellulose acetate-essential oils nanofibrous membranes for active food packaging to extend the shelf life of perishable fruits. *Innovative Food Science & Emerging Technologies*, *97*, 103802.

Sezer, M., Isgoren, M., Veli, S., Topkaya, E., & Arslan, A. (2024). Removal of microplastics in food packaging industry wastewaters with electrocoagulation process: Optimization by Box-Behnken design. *Chemosphere*, *352*, 141314.

Shafie, N. F. A., Zulkifli, N. S., Ali, R. R., Tarmizi, Z. I., & Faizo, S. M. (2024). Green synthesis of titanium dioxide nanoparticles using extraction of Psidium Guajava for smart packaging application. *Journal of Research in Nanoscience and Nanotechnology*, *11*(1), 16–23.

Shah, Y. A., Bhatia, S., Al-Harrasi, A., Oz, F., Khan, M. H., Roy, S., . . . Pratap-Singh, A. (2024). Thermal properties of biopolymer films: Insights for sustainable food packaging applications. *Food Engineering Reviews*, 1–16.

Shivakumar, N., Madhusudan, P., & Daniel, S. K. (2018). Nanomaterials for smart food packaging. In *Handbook of nanomaterials for industrial applications* (pp. 260–270): Elsevier.

Siddiqui, J., Taheri, M., Alam, A. U., & Deen, M. J. (2022). Nanomaterials in smart packaging applications: A review. *Small*, *18*(1), 2101171.

Silva, G. B., Alvarez, L. A., Campos, F. V., Guimarães, M. C., & Oliveira, J. P. (2024). A sensitive gold nanoparticle-based lateral flow immunoassay for quantitative on-site detection of Salmonella in foods. *Microchemical Journal*, *199*, 109952.

Song, W., Zhai, X., Shi, J., Zou, X., Xue, Y., Sun, Y., . . .& Povey, M. (2024). A ratiometric fluorescence amine sensor based on carbon quantum dot-loaded electrospun polyvinylidene fluoride film for visual monitoring of food freshness. *Food Chemistry* 434, 137423.

Stuparu-Cretu, M., Braniste, G., Necula, G.-A., Stanciu, S., Stoica, D., & Stoica, M. (2023). Metal oxide nanoparticles in food packaging and their influence on human health. *Foods*, *12*(9), 1882.

Sultana, A., Kathuria, A., & Gaikwad, K. K. (2022). Metal–organic frameworks for active food packaging. A review. *Environmental Chemistry Letters*, 1–17.

Tasbasi, B. B., Guner, B. C., Sudagidan, M., Ucak, S., Kavruk, M., & Ozalp, V. C. (2019). Label-free lateral flow assay for Listeria monocytogenes by aptamer-gated release of signal molecules. *Analytical biochemistry*, *587*, 113449.

Thirumalai, A., Harini, K., Pallavi, P., Gowtham, P., Girigoswami, K., & Girigoswami, A. (2023). Nanotechnology driven improvement of smart food packaging. *Materials Research Innovations*, *27*(4), 223–232.

Tseng, S.-Y., Li, S.-Y., Yi, S.-Y., Sun, A. Y., Gao, D.-Y., & Wan, D. (2017). Food quality monitor: Paper-based plasmonic sensors prepared through reversal nanoimprinting for rapid detection of biogenic amine odorants. *ACS Applied Materials & Interfaces*, *9*(20), 17306–17316.

United Nations Environment Program (UNEP). (2021). *Food waste index report, 2021*. https://catalogue.unccd.int/1679_FoodWaste.pdf

United Nations Environment Program (UNEP). (2024). *Food waste index report, 2024*. https://sdg2advocacyhub.org/wp-content/uploads/2024/03/food_waste_index_report_2024.pdf

Vanderroost, M., Ragaert, P., Devlieghere, F., & De Meulenaer, B. (2014). Intelligent food packaging: The next generation. *Trends in Food Science & Technology*, *39*(1), 47–62.

Vera, P., Canellas, E., & Nerín, C. (2018). New antioxidant multilayer packaging with nanoselenium to enhance the shelf-life of market food products. *Nanomaterials*, *8*(10), 837.

Vieira, I. R. S., de Carvalho, A. P. A. d., & Conte-Junior, C. A. (2022). Recent advances in biobased and biodegradable polymer nanocomposites, nanoparticles, and natural antioxidants for antibacterial and antioxidant food packaging applications. *Comprehensive Reviews in Food Science and Food Safety*, *21*(4), 3673–3716.

Wen, C. Y., Zhao, L. J., Wang, Y., Wang, K., Li, H. W., Li, X., . . .& Zeng, J. B. (2023). Colorimetric and photothermal dual-mode lateral flow immunoassay based on Au-Fe_3O_4 multifunctional nanoparticles for detection of *Salmonella typhimurium*. *Microchimica Acta* 190, 57.

Xia, F., Shi, W., Yu, Y., Jiang, C., Zhang, H., Zhang, P., & Sun, W. (2024). Intelligent pH-responsive ratiometric fluorescent nanofiber films for visual and real-time seafood freshness monitoring. *Food Chemistry*, 142240.

Yan, J., Fu, Q., Zhang, S., Shi, X., Zhang, Y., Hou, J., . . .& Ai, S. 2023. Fluorescent filter paper with pH-responsive carbon dots for the on-site detection of biogenic amines in food. *New Journal of Chemistry* 47(16), 7588–7594.

You, H., Wang, M., Wang, S., Xu, J., Hu, S., Li, T., . . . Gan, N. (2023). Ultrasensitive and specific phage@ DNAzyme probe-triggered fluorescent click chemistry for on-site detection of foodborne pathogens using a smartphone. *Analytical Chemistry*, *95*(30), 11211–11218.

Yousefi, H., Ali, M. M., Su, H.-M., Filipe, C. D., & Didar, T. F. (2018). Sentinel wraps: Real-time monitoring of food contamination by printing DNAzyme probes on food packaging. *ACS Nano*, *12*(4), 3287–3294.

Yousefi, H., Su, H.-M., Imani, S. M., Alkhaldi, K., M. Filipe, C. D., & Didar, T. F. (2019). Intelligent food packaging: A review of smart sensing technologies for monitoring food quality. *ACS Sensors*, *4*(4), 808–821.

Zhang, W., Ma, J., & Sun, D.-W. (2023). Dual-signal fluorescent test strips for spoilage sensing of packaged seafood: Visual monitoring of volatile basic nitrogens. *Food Chemistry*, *416*, 135725.

Zhao, M., Zhang, W., Ruan, C.-Q., & Zeng, K. (2024). Carboxymethyl cellulose/layered double hydroxide nanocomposite films with high barrier property and transparency. *Cellulose*, *31*(12), 7493–7505.

Index